理论力学

中英双语

Theoretical Mechanics
Chinese and English

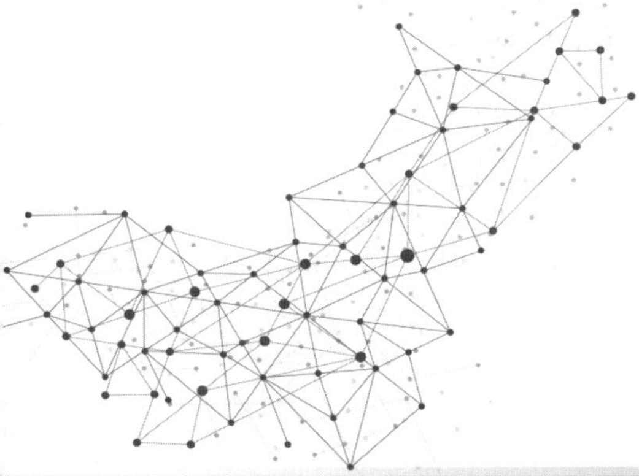

主 编 李顺才
副主编 刘 玉 董树亮 田 晶 王繁生 乔昂子

哈爾濱工業大學出版社
HARBIN INSTITUTE OF TECHNOLOGY PRESS

内容简介

本书内容分为三篇：静力学、运动学和动力学。静力学包括四章，分别是静力学公理和物体的受力分析、平面力系、空间力系、摩擦；运动学包括四章，分别是点的运动学、刚体的简单运动、点的合成运动、刚体的平面运动；动力学包括两章，分别是质点动力学的基本方程、动量定理・动量矩定理・动能定理。

本书对基本的概念、公理及主要知识点采用中英双语排版，即一段英文，一段对应的中文，用中文适当地扩充了英语内容不够丰富的知识点及例题，中文为主，英文为辅。

本书可作为高等学校机械、土建、水利、航空、航天等专业理论力学课程的教材，尤其适合作为中外合作办学高校工科专业理论力学教材及参考书，也可作为高职高专、成人高校相应专业的自学和函授教材，亦可供相关工程技术人员参考。

图书在版编目(CIP)数据

理论力学：汉、英/李顺才主编. —哈尔滨：哈尔滨工业大学出版社，2025.6

ISBN 978-7-5767-1926-0

Ⅰ. O31

中国国家版本馆 CIP 数据核字第 202511KN75 号

策划编辑　王桂芝
责任编辑　周一瞳
出版发行　哈尔滨工业大学出版社
社　　址　哈尔滨市南岗区复华四道街 10 号　邮编 150006
传　　真　0451-86414749
网　　址　http://hitpress.hit.edu.cn
印　　刷　哈尔滨博奇印刷有限公司
开　　本　787mm×1092mm　1/16　印张 22　字数 550 千字
版　　次　2025 年 6 月第 1 版　2025 年 6 月第 1 次印刷
书　　号　ISBN 978-7-5767-1926-0
定　　价　69.00 元

前　　言

为主动应对新一轮科技革命与产业变革，支撑服务创新驱动发展等一系列国家战略，2017年2月以来，教育部积极推进新工科建设，先后形成了“复旦共识”“天大行动”和“北京指南”，构成了新工科建设的“三部曲”，奏响了人才培养主旋律，指出了新工科建设需要借鉴国际经验，加强国际合作。国家“十四五”规划提出：坚持实施更大范围、更宽领域、更深层次对外开放，促进国际合作，实现互利共赢，推动共建“一带一路”行稳致远。而加强教育的国际交流与合作，要借鉴国际上先进的教育理念和教育经验，适应国家经济社会对外开放的要求，培养大批具有国际视野、通晓国际规则、能够参与国际事务和国际竞争的“专业+语言”国际化人才。

单纯语言的教学及教材难以满足“专业+语言”人才培养的需要。教材是教学的基本保证，由于国内针对机械、交通、土木类等工科专业理论力学课程的中英双语教材很少，因此建设“专业+语言”的双语教材是当务之急。

本书参考了多个版本的中文和外文理论力学方面的教材，组织专业老师和语言老师一起编写。在此过程中，语言老师也熟悉了相应专业词汇的中文及英文表达方法。本书对基本的概念、公理及主要知识点采用中英双语排版，即一段英文，一段对应的中文，用中文适当地扩充了英文内容不够丰富的知识点及例题，中文为主，英文为辅。

本书章节的编排参照哈尔滨工业大学理论力学教研室编写的《理论力学》教材，内容分为三篇：静力学、运动学和动力学。静力学包括四章，分别是静力学公理和物体的受力分析、平面力系、空间力系、摩擦；运动学包括四章，分别是点的运动学、刚体的简单运动、点的合成运动、刚体的平面运动；动力学包括两章，分别是质点动力学的基本方程、动量定理 · 动量矩定理 · 动能定理。

本书可作为高等学校机械、土建、水利、航空、航天等专业理论力学课程的教材，尤其适合作为中外合作办学高校工科专业理论力学教材及参考书，也可作为高职高专、成人高校相应专业的自学和函授教材，亦可供有关工程技术人员参考。

本书各章中文部分知识点由李顺才老师执笔；第1章、第6章、第10章翻译初稿由刘玉老师完成；目录、第2章、第3章、第4章、第5章、第7章、第8章、第9章翻译初稿由李顺才老师完成；王繁生老师审核了中文部分，乔昂子老师审核了翻译初稿；第1~8章的思考题、习题及其答案（包括制图）由华北理工大学董树亮博士完成；第9章的思考题、习题及其答案（包括制图）由徐州工程学院田晶博士完成。全书最终的审核及定稿由李顺才负责。研究生喻秋、梁丽、李大权参与了部分绘图工作，研究生李松原、李游、季晓丰、李卫、岳磊、刘保鹏、陈单敏协助完成了部分内容的翻译工作。

本书的写作和出版得到了国家自然科学基金项目（编号:52034007）和江苏师范大学教材出版基金（编号:JYJC202308）的资助。

本书的出版得到了江苏师范大学、华北理工大学、徐州工程学院、中国矿业大学等单位的大力支持和协作，在此深表感谢！

本书在编写过程中参阅了众多国内外公开出版的理论力学及工程力学方面的教材、习题指导书，以及网上发行的相关资料等，特此说明，并向原编者表示衷心感谢！

限于作者水平，书中疏漏、遗误之处在所难免，敬请同行专家和广大读者批评指正。

作　者

2025 年 4 月

Contents

(目　录)

Part 2　Kinematics (第 2 篇　运动学)

Part 3 Dynamics (第3篇 动力学)

Introduction(绪论)

The emergence and development of all sciences are closely related to the development of social production, the same is the development of mechanics. Mechanics is one of the oldest sciences. It develops with the accumulation of human knowledge and meets various practical needs in time. The ancient man created some simple tools by accumulating experience through labor, and improved them in practice. (一切科学的产生和发展都与社会生产的发展紧密联系,力学的发展也是如此.力学是最古老的科学之一,它随着人类知识的不断积累而发展,及时满足各种现实需求.远古人类通过劳动积累经验,创造了一些简单工具,并在实践中不断改进.)

When the pyramids were built in ancient Egypt, people used levers, slopes and sliders, and these empirical knowledge helped to discover the mechanics laws. Science was not divided into branches in ancient times, so mechanics, philosophy and natural sciences were all components of natural and social sciences. It was not until the 4th century BC that social sciences began to be distinguished from natural sciences. (在古埃及建造金字塔时,人们就已经开就始使用杠杆、斜面和滑块,这些经验知识有助于发现力学定律.古代没有对科学划分分支,所以力学、哲学和自然科学都是自然和社会科学的组成部分.直到公元前4世纪,社会科学与自然科学才开始区分.)

0.1 A Brief History of the Development of Theoretical Mechanics (理论力学的发展简史)

1. The period of establishing the foundation of theoretical mechanics (理论力学基础建立时期)

Archimedes(287 BC–212 BC) is considered to be the founder of the science of mechanics. He established the principle of lever balance, developed the method of measuring the center of gravity of an object, developed the theory of equilibrium of floating bodies in liquids, and his work *On Specific Gravity* laid the foundation of statics. The *Book of Mohist Canon*, written by Mo Di (468 BC–382 BC), is the earliest work involving the concept of force and the understanding of the balance principle of leverage. (阿基米德(公元前287—前212年)被认为是力学这门科学的奠基人,他确定了杠杆平衡原理,得到了测量物体重心的方法,提出了液体中浮体的平衡理论,他所著的《论比重》奠定了静力学基础.我国的墨翟(公元前468—前382年)所著的《墨经》是最早涉及力的概念及杠杆平衡原理认识的著作.)

During the Renaissance (14th – 16th century), the famous Italian artist, scientist and engineer Leonardo Da Vinci (1452 – 1519) made great contributions to the development of mechanics. He studied sliding friction, free-falling motion, and first proposed the concept of moment of force.(在文艺复兴时期(14—16 世纪),著名的意大利艺术家、科学家和工程师达芬奇(1452—1519)为力学的发展做出了巨大贡献.他研究了滑动摩擦和自由落体运动,并首次提出力矩的概念.)

The revolution in natural science was made possible by the great discovery of Copernicus of Poland (1473-1543), who proposed the heliocentric theory to replace the geocentric theory. According to Copernicus' heliocentric theory, German Kepler (1571-1630) proposed the three laws of planetary motion, and Italian Galileo (1564-1642) proposed the law of free fall, the law of inertia and the concept of acceleration. On this basis, Newton (1643-1727), a great British scientist, published the *Mathematical Principles of Natural Philosophy* in 1687, which was a great success. He put forward the three basic laws of dynamics and the law of gravitation, which attracted worldwide attention, and promoted the development of celestial mechanics. Galileo and Newton laid the foundations for the study of dynamics. (随着波兰的哥白尼(1473—1543)的伟大发现,自然科学的革命得以实现,他提出日心说来取代地心说.根据哥白尼的日心说,德国的开普勒(1571—1630)提出了行星运动三大定律,意大利的伽利略(1564—1642)提出了自由落体规律、惯性定律及加速度的概念.在此基础上,英国伟大科学家牛顿(1643—1727)在1687 年出版的《自然哲学的数学原理》一书中总其大成,提出了举世瞩目的动力学三个基本定律和万有引力定律,推动了天体力学的发展.伽利略和牛顿为动力学研究奠定了基础.)

2. Development period of theoretical mechanics(理论力学的发展期)

In 18th century, the general principles of classical mechanics were developed. At the same time, people had also began the research in the fields of solid mechanics, fluid dynamics, celestial mechanics and other fields. (18 世纪,经典力学的一般原理被提出.同时,人们也开始了对固体力学、流体动力学和天体力学等领域的研究.)

Bernoulli(1667-1748) of Switzerland defined the principle of virtual work. Euler (1707-1783) of Switzerland wrote *Mechanics*, which used differential equations to study mechanics. D'Alembert (1717-1785) of France wrote *Monographs on Dynamics*, which put forward the famous D'Alembert principle. Lagrange (1736-1813) in France proposed the second kind of Lagrange equation. British Hamilton (1805-1865) proposed the Hamiltonian principle. (瑞士的伯努利(1667—1748)定义了虚功原理.瑞士的欧拉(1707—1783)著有《力学》,用微分方程研究力学.法国的达朗贝尔(1717—1785)著有《动力学专论》,提出了著名的达朗贝尔原理.法国的拉格朗日(1736—1813)提出了第二类拉格朗日方程.英国的哈密顿(1805—1865)提出了哈密顿原理.)

In 1725, the Russian Academy of Sciences was founded on the initiative of Peter the Great of Russia. The development of mechanics was influenced by Academician Lomonosov (1711-1765) and the Swiss Euler (1707-1783), a famous mathematician, astronomer and physicist. Euler worked at the Russian Academy of Sciences for more than 30 years. His numerous works in the fields of mathematics, fluid mechanics and celestial mechanics refer to the mechanical study of rigid and elastic bodies. Euler's theory of resistance and elasticity of materials laid the foundation for the emerging discipline. (1725 年,在俄罗斯彼得大帝的倡议下,俄罗斯科学院成立.力学的发展受到罗蒙诺索夫院士(1711—1765)以及瑞士著名数学家、天文学家和物理学家欧拉

(1707—1783)的影响. 欧拉在俄罗斯科学院工作了 30 多年,他在数学、流体力学和天体力学领域的众多著作中都提到了对刚体和弹性体的力学研究. 欧拉提出了材料的阻力和弹性理论,为新兴学科奠定了基础.)

In the 18th and 19th centuries, the greatest foreign scientists in the field of mechanics were John Bernoulli, Daniel Bernoulli, D'Alembert and Lagrange. In the works of French scientists Varignon and Pansau, it was written that with the development of dynamics, statics was further developed. (18 世纪和 19 世纪,力学领域最伟大的外国科学家有约翰·伯努利、丹尼尔·伯努利、达朗贝尔和拉格朗日. 其中,法国科学家瓦里尼翁和潘索在著作中写道:随着动力学的发展,静力学也得到进一步发展.)

3. Modern development period of theoretical mechanics(理论力学的现代发展期)

Physicist Albert Einstein (1879 – 1955) established relativistic mechanics and made an epoch-making contribution to the development of mechanics. The further development of mechanics was made possible by Russian scientists of the 19th and 20th centuries, such as Ostrogradsky, Chebyshev, Kovalevskaya, Lyapnov, Meshelsky, Tsiolkovsky, Krylov, Zhukovsky and others, whose work was also very important. (物理学家爱因斯坦(1879—1955)创立了相对论力学,为力学学科的发展做出了划时代的贡献. 力学的进一步发展离不开 19 世纪和 20 世纪的俄罗斯科学家,如奥斯特罗格拉德斯基、切比雪夫、柯瓦列夫斯卡娅、李雅普诺夫、梅谢尔斯基、齐奥尔科夫斯基、克雷洛夫、茹科夫斯基等,他们的研究也非常重要.)

The discipline of mechanics intersects and penetrates with mathematics, physics, chemistry, astronomy, earth science, biology and almost all engineering disciplines, forming a large number of new interdisciplinary disciplines, which makes the discipline of mechanics maintain vigorous vitality. (力学学科极其广泛地与数学、物理、化学、天文、地学、生物等基础学科和几乎所有的工程学科相交叉、渗透,形成了大量的新兴交叉学科,使力学学科保持着旺盛的生命力.)

With the development of modern science and technology, the research content of mechanics has been infiltrated into other scientific fields. Such as solid mechanics and fluid mechanics theory has been used to study the strength of human bones, the rule of blood flow, and the transport of nutrients in plants, forming the biomechanics; the theory of fluid mechanics is used to study the motion of plasma in the magnetic field, forming the electromagnetic hydrodynamics; as well as explosion mechanics, physical mechanics, etc., are the marginal sciences formed by the combination of mechanics and other disciplines. The establishment of these new disciplines must be based on a solid knowledge of theoretical mechanics. (随着现代科学技术的发展,力学的研究内容已渗入到其他科学领域. 例如,固体力学和流体力学的理论被用来研究人体骨骼的强度、血液流动的规律及植物中营养的输送问题等,形成了生物力学;流体力学的理论被用来研究等离子体在磁场中的运动,形成了电磁流体力学;还有爆炸力学、物理力学等都是力学与其他学科结合而形成的边缘科学. 这些新兴学科的建立都必须以坚实的理论力学知识为基础.)

0.2 The Research Object and Content of Theoretical Mechanics（理论力学的研究对象和内容）

Theoretical mechanics is the science of studying the most general laws of mechanical motion of objects.（理论力学是研究物体机械运动最一般规律的科学.）

The change of the position of the object in space with time is called mechanical movement. Mechanical movement is the most needed movement in people's life and production practice. Equilibrium is a special case of mechanical motion.（物体在空间的位置随时间的改变称为机械运动.机械运动是人们生活和生产实践中最需见的一种运动.平衡是机械运动的特殊情况.）

In the objective world, there are all kinds of material movements, including the mechanical movement of macroscopic objects, the physical movement of molecules, the chemical movement of atoms, the life movement of protein bodies and the social movement of human beings. Of all the forms of motion of matter, mechanical motion is the simplest. Various forms of motion of matter can be transformed into each other under certain conditions, and in the advanced and complex motion, there are often simple mechanical motion.（在客观世界中，存在各种各样的物质运动，物质运动包括宏观物体的机械运动、分子的物理运动、原子的化学运动、蛋白体的生命运动及人的社会运动等基本形式.在物质的各种运动形式中，机械运动是最简单的一种.物质的各种运动形式在一定的条件下可以相互转化，而且在高级和复杂的运动中，往往存在着简单的机械运动.）

The research object of theoretical mechanics is the macroscopic object（particle system and rigid body）whose speed is much less than the speed of light. The research content is based on the basic laws summarized by Galileo and Newton, which belongs to the category of classical mechanics. The motion of objects and elementary particles approaching the speed of light can only be perfected and explained with the ideas of relativity and quantum mechanics. The motion of macroscopic object with velocity much less than the speed of light is the most common encounter in daily life and general engineering. Classical mechanics has the most extensive application. Theoretical mechanics deals with the most general and simple laws of this motion, and is the foundation of all branches of mechanics.（理论力学的研究对象是速度远小于光速的宏观物体（质点系及刚体），研究内容以伽利略和牛顿总结的基本定律为基础，属于古典力学的范畴.速度接近光速的物体和基本粒子的运动则必须用相对论和量子力学的观点才能完善并予以解释.宏观物体远小于光速的运动是日常生活及一般工程中最常遇到的.古典力学有着最广泛的应用，理论力学所研究的则是这种运动中最一般、最简单的规律，是各门力学分支的基础.）

Theoretical mechanics is divided into statics, kinematics and dynamics.（理论力学分为静力学、运动学和动力学.）

Statics mainly studies the conditions under which the force should be satisfied when a body under stress is in equilibrium. At the same time, it also studies the analysis method of the force of the object, and the method of simplifying the force system, etc.(静力学主要研究受力物体平衡时作用力所应满足的条件,同时也研究物体受力的分析方法,以及力系简化的方法等.)

Kinematics is the study of the motion of an object (such as trajectory, velocity, acceleration, etc.) from a geometrical point of view, rather than the physical causes of the object's motion.(运动学只从几何的角度来研究物体的运动(如轨迹、速度和加速度等),而不研究引起物体运动的物理原因.)

Dynamics is the study of the relationship between the motion of a body under force and the forces applied.(动力学研究受力物体的运动与作用力之间的关系.)

0.3 Research Methods of Theoretical Mechanics (理论力学的研究方法)

The research method of theoretical mechanics is to obtain axioms and laws through abstract synthesis of practical knowledge, and then obtain theorems and conclusions through mathematical deduction and logical reasoning, which in turn guide life and production practice.(理论力学的研究方法是从实践知识经过抽象综合得到公理、规律等,再经过数学演绎及逻辑推理等得到定理及结论,这些结论再反过来指导生活及生产实践.)

(1) Mechanical models such as particle and rigid body are obtained by abstraction.(由抽象化得到质点和刚体等力学模型.)

(2) The application of calculus, differential equations and other theories of higher numbers in mathematical reasoning.(数学推理中应用高数的微积分、微分方程等理论.)

0.4 The Purpose of Learning Theoretical Mechanics (学习理论力学的目的)

Theoretical mechanics is a professional basic course with strong theory. The purpose of learning theoretical mechanics is as follows.(理论力学是一门理论性较强的专业基础课.学习理论力学的目的如下.)

(1) Learn the general rules of mechanical motion of particle system and rigid body to lay a solid foundation for subsequent courses. These subsequent courses include mechanics of materials, mechanical principles, mechanical design, structural mechanics, elastoplastic mechanics, fluid mechanics, flight mechanics, vibration theory, fracture mechanics and many other professional courses.(学习质点系和刚体机械运动的一般规律,为后续课程打下坚实基础.这些后续课程包括材料力学、机械原理、机械设计、结构力学、弹塑性力学、流体力学、飞行力学、振动理论、断裂力学等许多专业课程.)

(2) Apply the theory to solve some simple practical problems.(能应用所学理论,解决一些较简单的实际问题.)

(3) Cultivate a dialectical materialist worldview and improve the ability to analyze and solve problems.(培养辩证唯物主义世界观,提高分析问题与解决问题的能力.)

0.5 Some Points for Attention in Learning Theoretical Mechanics (学习理论力学的几点注意事项)

(1) Combining theory with practice.(理论联系实际.)

(2) Pay attention to the cultivation of scientific logical thinking methods.(注意培养科学的逻辑思维方法.)

(3) Pay attention to the meaning and units of each physical quantity in the expression.(注意表达式中各物理量的意义及单位.)

(4) Taking homework seriously.(认真对待作业.)

(5) Learning methods: make notes of lectures; review in time, review old and learn new.(学习方法:做听课笔记;及时复习,温故而知新.)

(6) Learning attitude: serious and pragmatic.(学习态度:认真、务实.)

Part 1 Statics（第 1 篇　静力学）

Introduction（引言）

Statics is the branch of theoretical mechanics that deals with the equilibrium of objects under the action of forces.（静力学是理论力学中的一个分支,研究物体在力作用下的平衡规律.）

A system of forces is a collection of forces acting on an object.（力系是指作用在物体上的力的集合.）

Equilibrium is when an object remains stationary or moves in a straight line with respect to an inertial reference frame(such as the ground).（平衡是指物体相对于惯性参考系(如地面)保持静止或做匀速直线运动.）

Statics mainly studies the following three problems:(静力学主要研究以下三个问题:)

(1) The force analysis of the object;(物体的受力分析;)

(2) Equivalent replacement or simplification of force systems;（力系的等效替换或简化;）

(3) Establish the equilibrium conditions of the various systems of forces.（建立各种力系的平衡条件.）

The equilibrium condition of force system is very important in engineering, which is the basis of statics calculation when designing components, structures and mechanical parts. Therefore, statics has a widc rangc of applicationѕ in engineering.（力系的平衡条件在工程中有着十分重要的意义,是设计构件、结构和机械零件时静力学计算的基础. 因此,静力学在工程中有着广泛的应用.）

Chapter 1　Statics Axioms and Force Analysis of Objects (静力学公理和物体的受力分析)

The basic concepts and axioms of statics and the force analysis of objects are the basis of studying statics. This chapter firstly introduces the basic concepts and axioms of statics such as rigid body, free body, force, force system, resultant force, balance, etc. , then introduces the common constraint types and the expression of constraint reaction in engineering, and finally introduces the force analysis and force diagram.(静力学的基本概念、公理及物体的受力分析是研究静力学的基础. 本章先介绍刚体、自由体、力、力系、合力、平衡等基本概念及静力学公理,然后介绍工程中常见的约束类型及约束反力的表示,最后介绍受力分析及受力图.)

1.1　Basic Concepts and Axioms of Statics (静力学的基本概念与公理)

1.1.1　General definition (一般定义)

Statics is a part of theoretical mechanics that deals with the equilibrium of a system of particles under force. Objects in nature undergo a certain degree of deformation under external forces, that is, they change shape due to changes in the relative position of the particles of which they are composed. In general, however, the deformation of most solids is negligible. Only when the durability of the corresponding structure is considered, the calculation has practical significance, which is the subject of material mechanics. Most rigid body deformations can be ignored when only general equilibrium conditions are considered. Therefore, the concept of "absolute rigid body" was introduced in mechanics.(静力学是理论力学的一部分,研究质点系受力作用时的平衡规律. 自然界中物体在外力作用下会发生一定程度的变形,即因组成它们的颗粒的相对位置变化而改变形状. 然而,在一般情况下,大多数固体的变形是忽略不计的,只有在考虑相应结构的耐久性问题时,其计算才具有实际意义,这是材料力学学科研究的课题. 当只考虑一般平衡条件时,可以忽略大多数刚体的变形. 因此,在力学中引入了"绝对刚体"的概念.)

1. Absolute rigid body (绝对刚体)

An absolute rigid body is one in which the distance between two particles remains constant under the action of a force.(绝对刚体是指在力的作用下两个质点之间的距离始终保持不变的物体.)

As shown in Fig. 1.1, an object that is always $AB=\text{const}$ from any two internal points can be called a rigid body.(如图 1.1 所示,一个内部任意两点距离始终为 $AB = \text{const}$ 的物体称为刚体.)

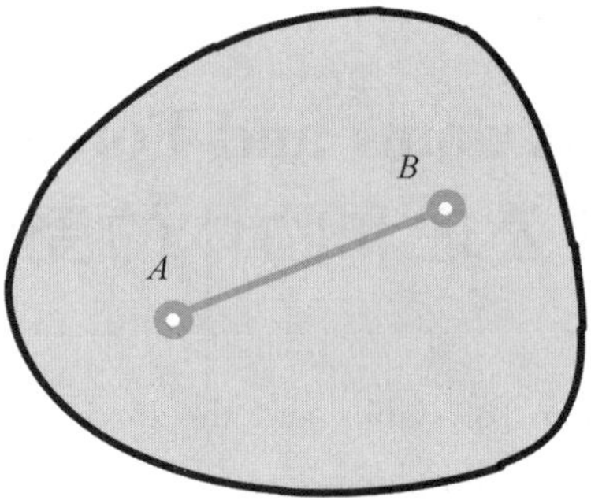

Fig. 1.1 Rigid body(图 1.1 刚体)

2. Free bodies and non-free bodies (自由体和非自由体)

An object that can move freely from a particular position in any direction in space is called a free body, while an object whose motion is restricted in a certain direction is called a non-free body.(可以从某个特定位置向空间任何方向自由运动的物体称为自由体,而某方向的运动受到限制的物体称为非自由体.)

3. Force (力)

Force is the main measure of the interaction between bodies. Force is a vector, it can be expressed as $\boldsymbol{F}, \boldsymbol{P}, \boldsymbol{Q}, \boldsymbol{S}, \boldsymbol{T}$, etc.(力是物体间相互作用的主要度量.力是矢量,可表示为 $\boldsymbol{F}$、$\boldsymbol{P}$、$\boldsymbol{Q}$、$\boldsymbol{S}$、$\boldsymbol{T}$ 等.)

This action of force can change the mechanical motion of the object. Mechanical interactions between objects can be roughly divided into two categories.(力对物体的作用可以使物体的机械运动状态发生变化.物体间的机械作用大致可分为两类.)

The first type is direct contact action; the second type is indirect action, such as "field" on the object is indirect action, including the earth on the object of gravity, and electric field on the charge of gravity or repulsive force. The effect of force on the object is generally divided into two aspects: ① the motion state of the object is changed, also known as the motion effect or external effect; ② make the object deform, also known as deformation effect or internal effect. Theoretical mechanics treats objects as rigid bodies and studies only the external effects of forces. The mechanics of materials deals with the internal effects of objects.(第一类是直接接触作用;第二类是间接作用,如"场"对物体的作用即间接作用,包括地球对物体的引力、电场对电荷的引力或斥力等.力对物体产生的效应一般分为两个方面:①使物体的运动状态发生改变,又称运动效应或外效应;②使物体发生变形,又称变形效应或内效应.理论力学把物体视为刚体,只研究力的外效应.而材料力学研究物体的内效应.)

The action of force on an object consists of the following three elements (Fig. 1.2).(力对物体的作用包括以下三个要素(图 1.2).)

(1) Numerical value (magnitude of force).(数值(力的大小).)

(2) Direction of force.(力的方向.)

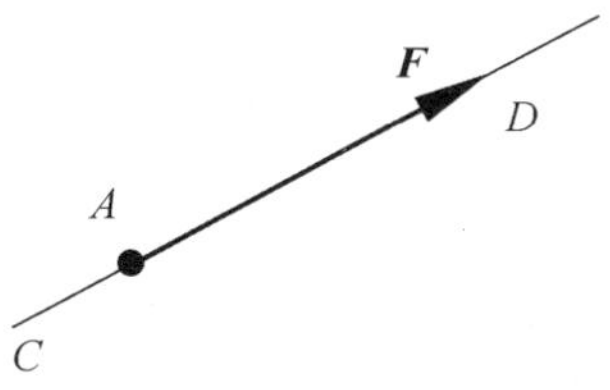

Fig. 1.2 Three elements of the force (图 1.2 力的三要素)

(3) Application point (point A as shown in the Fig. 1.2). (作用点(如图 1.2 中点 A 所示).)

(4) Line of action. (作用线.)

The line along the direction of force action is called the line of force action. (过力的作用点沿力作用方向的直线称为力的作用线.)

4. Unit of force(力的单位)

The basic unit of measurement for force in the International System of Units (SI) is 1 Newton (N). (国际单位制(SI)中力的基本测量单位是 1 牛顿(N).)

5. External and internal forces (外力和内力)

(1) External force. (外力.)

External force refers to the force acting on the system (objects in the system) by objects outside the system (not involving the interaction of objects in the system). (外力是指系统外的物体作用于该系统(系统内物体)上的力(不涉及该系统内物体的相互作用力).)

(2) Internal force. (内力.)

Internal forces are the forces between parts of a body (objects in a system). (内力是指物体内各部分(系统内各物体)之间的作用力.)

6. Concentrating force and distributing force (集中力与分布力)

(1) Concentrating force. (集中力.)

The force applied to a point on a body is called the concentration force. If the force has a small area of action (compared to the size of the entire member), it can be considered to act centrally on a single point of the member and is called a concentrated force. (施加在物体上某一点的力称为集中力. 如果力的作用面积(与整个构件的尺寸相比)不大,则可以认为它集中地作用在构件的一点,称为集中力.)

(2) Distributing force. (分布力.)

A force (load) distributed over a given length, area or volume of a given body is called a distributed force. (分布在给定物体一定长度、面积或体积上的力(载荷)称为分布力.)

The set degree of the distributed load is expressed by q, the value $[q]=\text{N/m}$ (N/m^2, N/m^3). (分布载荷的集度用 q 表示,值$[q]=\text{N/m}$(N/m^2、N/m^3).)

Types of distributed loads is as follows. (分布载荷的种类如下.)

①Uniform load ($q=\text{const}$). It can be replaced by a concentrated force, acting right in the middle of the distribution (Fig. 1.3). (均布载荷($q=\text{const}$). 其可用集中力 $Q=q\cdot l$ 代替,作用在分布区段的正中间(图 1.3).)

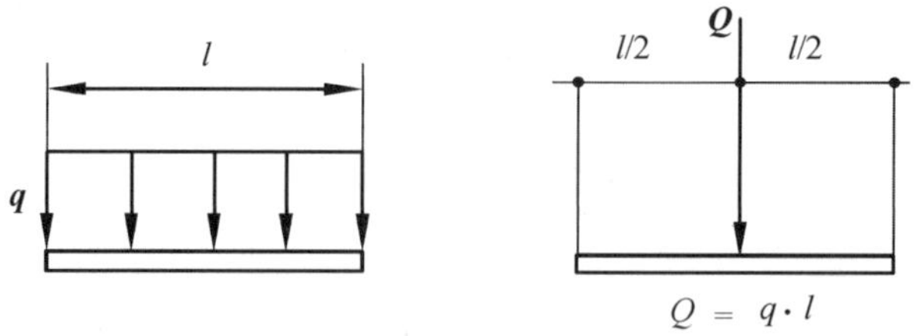

(a) Before replace(替换前)　　(b) After replace(替换后)

Fig. 1.3　Uniform load (图 1.3　均布载荷)

②Linear distributed load $q(x)$. It can be replaced by the concentrated force $Q = q \cdot l/2$, and the force is applied at the end of the distance distribution section $l/3$, where $q = q_{max}$ (Fig. 1.4). (线性分布载荷 $q(x)$. 可由集中力 $Q = q \cdot l/2$ 代替,力施加在距离分布区段末端 $l/3$ 处. 其中,$q = q_{max}$(图 1.4).)

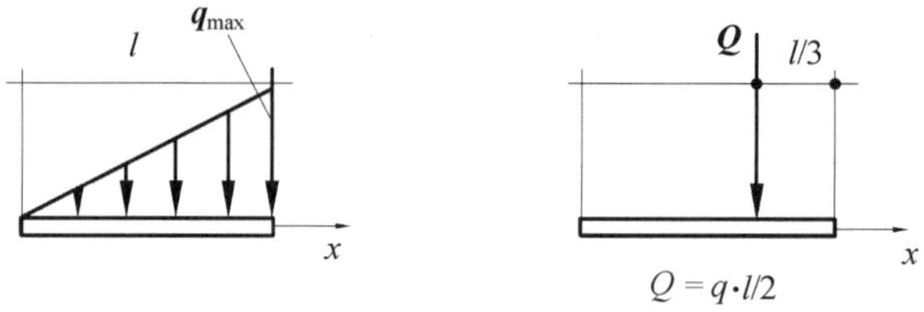

Fig. 1.4　Linearly distributed load (图 1.4　线性分布载荷)

③ Arbitrarily distributed load $q(x)$. It can be replaced by concentrated force equivalent $Q = \int_0^l q(x)\mathrm{d}x$. The Force Q act on the centre of gravity C of the figure enclosed by axial Ox and $q(x)$ (Fig. 1.5). (任意分布载荷 $q(x)$. 可用集中力 $Q = \int_0^l q(x)\mathrm{d}x$ 等效替换. 力 $\boldsymbol{Q}$ 作用在轴 Ox 和 $q(x)$ 所围成图形的重心 C 上(图 1.5).)

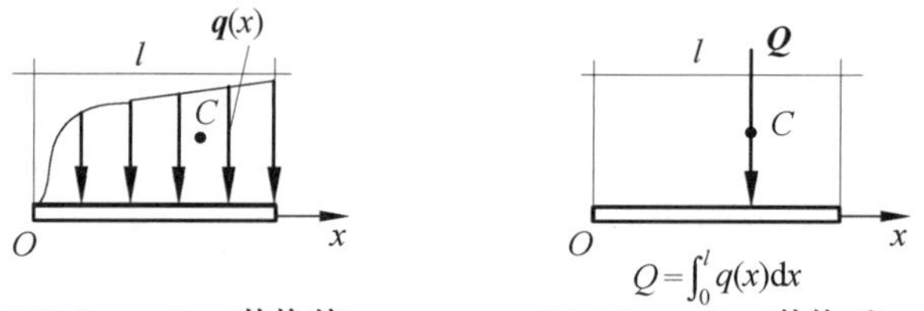

(a) Before replace(替换前)　　(b) After replace(替换后)

Fig. 1.5　Arbitrarily distributed load (图 1.5　任意分布载荷)

7. System of forces(力系)

The set of forces acting on a body is called a system of forces. (作用在物体上所有力的集合称为力系.)

The types of forces is as follows. (力系的种类如下.)

(1) If the lines of action of the forces are in the same plane, the system of forces is called the plane system of forces; and if it's not in the same plane, it's called a space system of forces. (如果所有力的作用线在同一平面上,则该力系称为平面力系;而如果不在同一个平面上,则称为空间力系.)

(2) If the lines of action of a force intersect at a point, the force system is called a confluence system.(如果力的作用线相交于一点,则该力系称为汇交力系.)

(3) If the lines of action of forces are parallel to each other, they are called parallel systems of forces.(如果力的作用线相互平行,则称为平行力系.)

(4) A force system acting on a free rigid body is called an equivalent force system if it can be replaced by another force system without changing the state of rest or motion in which it is located.(如果作用在自由刚体上的力系可以被另一个力系代替而不改变其所处的静止或运动状态,那么这两个力系称为等效力系.)

(5) A force system that acts on a free rigid body and keeps it in mechanical equilibrium is called an equilibrium force system.(作用于自由刚体并使它保持力学平衡状态的力系称为平衡力系.)

8. Resultant force and equilibrium force(合力与平衡力)

(1) If a force is equivalent to the force system, it is the resultant force of the force system.(如果某力与该力系等效,则此力为该力系的合力.)

(2) A force acting on the same body, of equal magnitude, in opposite directions, and with collinear lines of action is called an equilibrium force.(作用在同一个物体上,大小相等、方向相反、作用线共线的力称为平衡力.)

1.1.2 Research tasks in statics (静力学的研究任务)

Task 1: Study the force analysis of the object and the equivalent simplification of the force system.(任务1:研究物体的受力分析与力系的等效简化.)

Task 2: Study the equilibrium conditions of a system of forces acting on a free rigid body.(任务2:研究作用于自由刚体上力系的平衡条件.)

1.1.3 Basic principles of statics (axioms and theorems) (静力学的基本原理(公理与定理))

All theorems and static equations are based on several fundamental principles called axioms that are not proved mathematically. Statics axioms are the result of human knowledge accumulation and reflect the objective process. Many experiments and observations have confirmed the correctness of these axioms.(所有定理和静力学方程都基于不采用数学证明的且被称为公理的几条基本原理.静力学公理是人类知识积累的成果,反映了客观过程.许多实验和观测都证实了这些公理的正确性.)

1. Axiom of two-force balance (二力平衡公理)

If two forces are acting on a free absolute rigid body, then the rigid body can be in equilibrium if and only if the forces are equal in magnitude ($F_1 = F_2$), opposite in direction, and on the same straight line (Fig. 1.6), we have(如果两个力作用在一个自由的绝对刚体上,那么当且仅当两个力的大小相等($F_1 = F_2$)、方向相反并且在同一条直线上时,刚体才能处于平衡状态(图1.6),有)

$$\boldsymbol{F}_1 = -\boldsymbol{F}_2$$

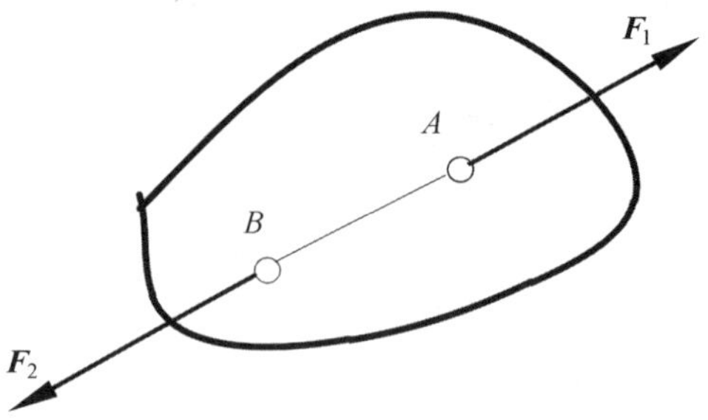

Fig. 1 6　Two-force equilibrium condition(图 1.6　二力平衡条件)

2. Axiom of equilibrium system of forces (加减平衡力系公理)

Adding or subtracting any equilibrium force system to any known force system acting on a rigid body does not change the effect of the original force system on the rigid body. (在作用于刚体的任意已知力系中,加上或减去任意的一个平衡力系,不改变原力系对刚体的作用效果.)

According to the principle of additive and subtractive equilibrium force system, the pair of equilibrium forces acting at point *B* in Fig. 1.7 can be removed. (根据加减平衡力系原理,图 1.7中作用在 *B* 点的一对平衡力可以去掉.)

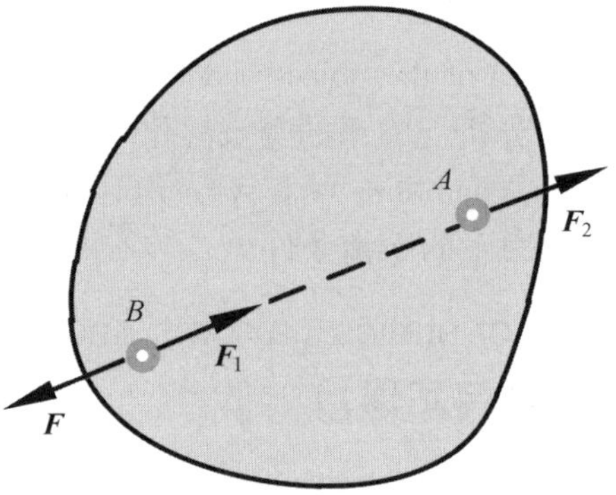

Fig. 1.7　Addition and subtraction equilibrium forces system(图 1.7　加减平衡力系)

From these two axioms, follows the following inference. (由上述两个公理可导出以下推论.)

A force acting on a rigid body moves to any point along its line of action in the rigid body without changing the effect of the force on the rigid body. That is, force is a sliding vector (Fig. 1.8). (作用于刚体上的力在刚体内沿着它的作用线移到任意一点,并不改变该力对刚体的作用效果,即力是滑动矢量(图 1.8).)

Note: The principle of force transfer applies only to rigid bodies, not to deformable bodies. (注意:力的可传性原理仅适用于刚体,对变形体不成立.)

3. The parallelogram rule(平行四边形法则)

If two forces are intersected by lines at a point, the resultant force also acts on the same point, and the magnitude of the resultant force can be expressed by the diagonal of a parallelogram

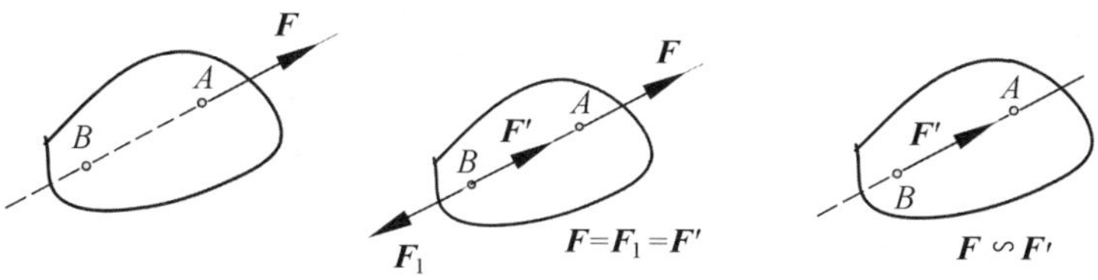

Fig. 1.8 Proof of principle of force transmittability (图 1.8 力的可传性原理证明)

with sides of the two forces (Fig. 1.9), that is(如果两个力作用线汇交于一点,则其合力也作用于同一点,且合力的大小可用以这两个力为边的平行四边形的对角线来表示(图 1.9),即)

$$\boldsymbol{F}_R = \boldsymbol{F}_1 + \boldsymbol{F}_2$$

Fig. 1.9 Synthesis of two co-point forces (图 1.9 两个共点力的合成)

Corollary: Three force equilibrium confluence theorem. (推论:三力平衡汇交定理.)

Three balanced forces acting on a rigid body must be in the same plane if the lines of action of two of them converge at a point, and the lines of action of the third force pass through the point of intersection (Fig. 1.10). (作用于刚体上三个相互平衡的力,若其中两个力的作用线汇交于一点,则此三力必在同一平面内,且第三个力的作用线通过汇交点(图 1.10).)

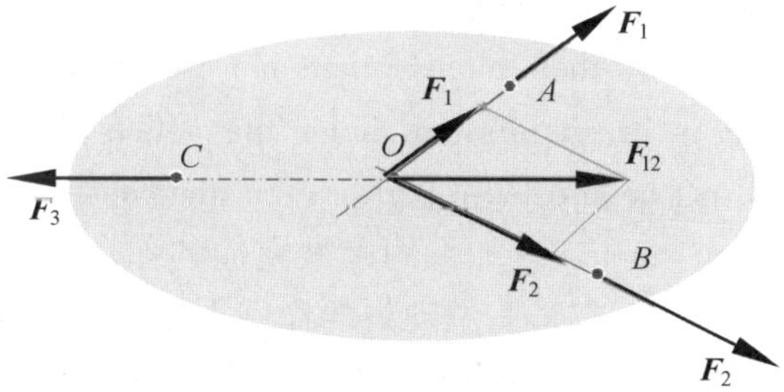

Fig. 1.10 Three force equilibrium confluence theorem(图 1.10 三力平衡汇交定理)

4. Law of action and reaction (作用与反作用定律)

The action and reaction forces have equal and opposite directions and act along the same straight line, but the two forces act on different bodies respectively (Fig. 1.11), so the two forces cannot form an equilibrium system of forces, we have(作用力与反作用力大小相等、方向相反,沿着同一直线作用,但两个力分别作用在不同的物体上(图 1.11),因此这两个力不能构成平衡力系,有)

$$\boldsymbol{F}' = -\boldsymbol{F}$$

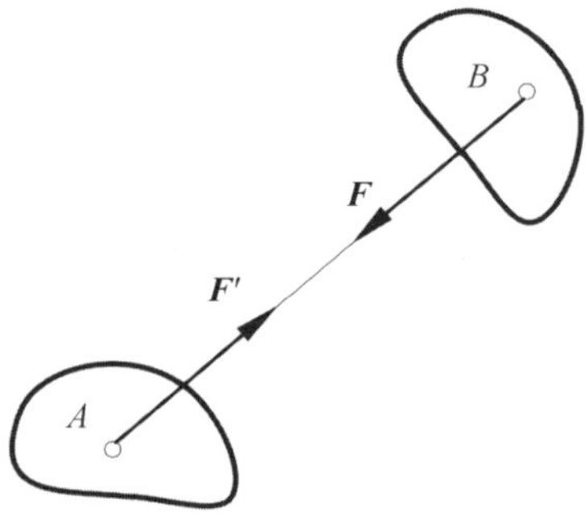

Fig. 1.11 Action and reaction force (图 1.11 作用力与反作用力)

Characteristics of internal forces: From the axiom, it can be ignored when the sum of internal forces acting on an absolutely rigid body constitutes a balanced system of forces. In other words, when we look at the equilibrium conditions of an object, we only need to consider the external forces. (内力的特性:由公理得出,作用于绝对刚体的内力总和构成了一个平衡力系时,可以将其忽略. 也就是说,在研究物体的平衡条件时,只需考虑外力.)

5. Rigid axiom(刚化公理)

When a deformable body in equilibrium is under the action of a known system of forces, if the deformable body is regarded as a rigid body (rigid), the equilibrium state remains unchanged. (当变形体在已知力系作用下处于平衡时,若将此变形体视为刚体(刚化),则平衡状态保持不变.)

Or, in equilibrium, the forces acting on any deformable or deformed structure satisfy the same conditions as for an absolutely rigid body. (或者说,在平衡状态下,作用于任何变形体或变形结构的力满足与绝对刚体相同的条件.)

6. Axiom for unbinding rigid bodies (unbinding axiom)
(解除刚体约束的公理(解除约束公理))

If constraints are replaced by their forces, then non-free bodies can be considered free. When the constrained object is in equilibrium under the action of some principal force, the equilibrium of the object will not be affected if part or all of the constraint is removed and replaced by the corresponding constraint reaction. (如果约束被它们的作用力取代,那么非自由体可以被认为是自由的. 当受约束的物体在某些主动力的作用下处于平衡时,若将其部分或全部约束解除,代之以相应的约束反力,则物体的平衡不受影响.)

1.2 Constraints and Constraint Reactions (约束和约束反力)

1.2.1 General definition (一般定义)

1. Definition of constraint (约束的定义)

All conditions that restrict the movement of objects in space are called constraints. (所有限

制空间内物体移动的条件称为约束.)

2. Constraint reaction（约束反力）

The restraining action of a constraint on the motion of a constrained object is the action of a force called the constraint reaction.（约束对被约束物体运动的阻碍作用是一种力的作用，这种力称为约束反力.）

3. Constraint reaction direction（general rule）（约束反力方向（一般规则））

The direction of the constraint reaction is opposite to the direction of motion that the constraint can prevent.（约束反力的方向与约束所能阻止的运动方向相反.）

The forces acting on an object are divided into applying forces（the main force）and restraining forces（being forces）.（作用在物体上的力可分为施加力（主动力）和约束反力（被动力）.）

The principal force is characterized by the fact that the magnitude and direction of each force is known in advance and does not depend on the action of other forces applied to the body. For example：human muscle strength，gravity and given elasticity.（主动力的特点是每个力的大小和方向是预先已知的，并且不依赖于施加在物体上的其他力的作用，如人的肌力、重力和给定的弹力.）

The constraint body obstructs the free movement of the restricted object and changes the motion state of the object. Therefore，the constraint body must bear the force of the object and give the object equal and opposite reaction force at the same time. This force is called binding force or reaction force and belongs to the forced force. The binding force prevents the movement of the object through the contact between the restraining body and the object，and its direction is opposite to the direction of the motion that the restraining body can prevent.（约束体阻碍物体的自由运动，改变了物体的运动状态，因此约束体必然承受物体的作用力，同时给予物体以相等、相反的反作用力，这种力称为约束力或反力，属于被动力. 约束力是通过约束体与物体间相互接触来阻止物体运动的，它的方向与约束体所能阻止的运动方向相反.）

1.2.2 Some basic constraint types（一些基本的约束类型）

1. Smooth contact surface constraint or support（光滑接触面约束或支承）

The friction of the smooth contact surface is negligible.（光滑接触面的摩擦力可忽略不计.）

Smooth contact surface constraint or support constraint reaction through the contact point，the direction along the common normal of the contact surface and toward the object（Fig. 1. 12，Fig. 1. 13）.（光滑接触面约束或支承的约束反力通过接触点，方向沿接触面的公法线并指向物体（图 1.12，图 1.13）.）

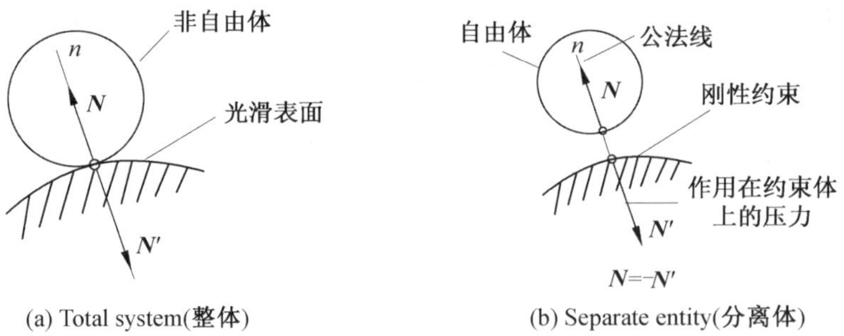

Fig. 1. 12　Contact surface is a curved surface（图 1. 12　接触面为曲面）

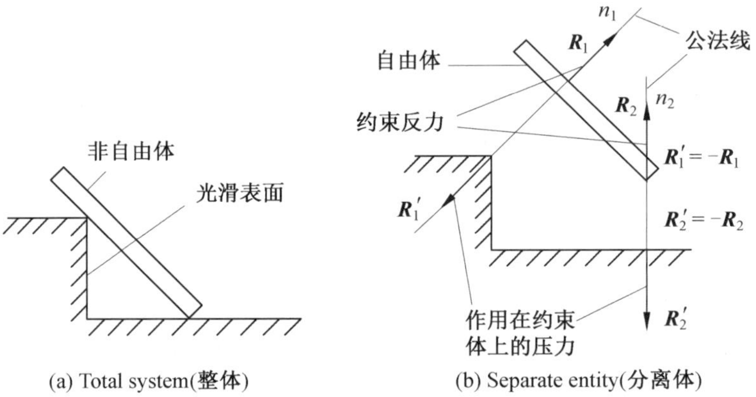

Fig. 1. 13　Contact surface is a plane（图 1. 13　接触面为平面）

Constraint characteristics：Only the movement of the object along the common normal tends to the direction of the supporting plane.（约束特点：只限制物体沿公法线趋向于支承面方向的运动.）

Constraint reaction characteristic：Along the common normal at the contact point，pointing towards the object（object compression）.（约束反力特征：沿接触处的公法线，指向物体（物体受压）.）

The restrained reaction forces of two meshing gears are F_R and F'_R，as shown in Fig. 1. 14.（两个相互啮合的齿轮，其约束反力为 $\boldsymbol{F}_R$ 和 $\boldsymbol{F}'_R$，如图 1. 14 所示.）

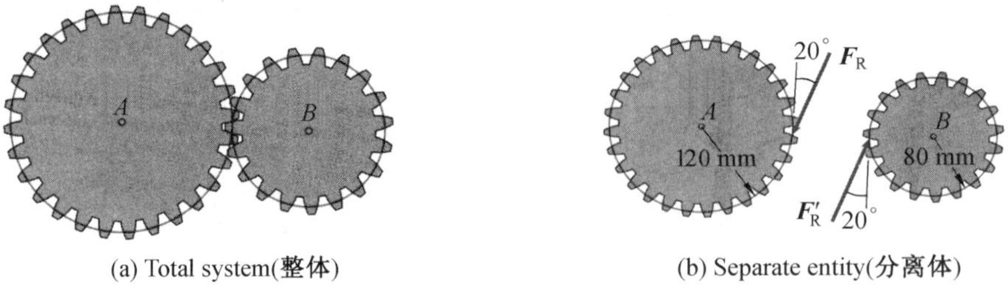

Fig. 1. 14　Constrained reaction of gear meshing（图 1. 14　齿轮啮合时的约束反力）

Fig. 1. 15 – 1. 17 are some examples of smooth surface constrained reaction drawing.（图 1. 15 ~ 1. 17是一些光滑面约束反力画法的实例.）

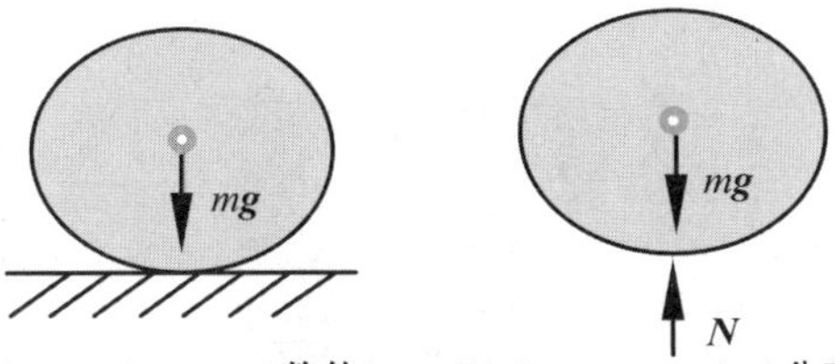

(a) Total system(整体)　(b) Separate entity(分离体)

Fig. 1.15　Constraint reaction drawing method 1 of smooth surface (图 1.15　光滑面约束反力画法 1)

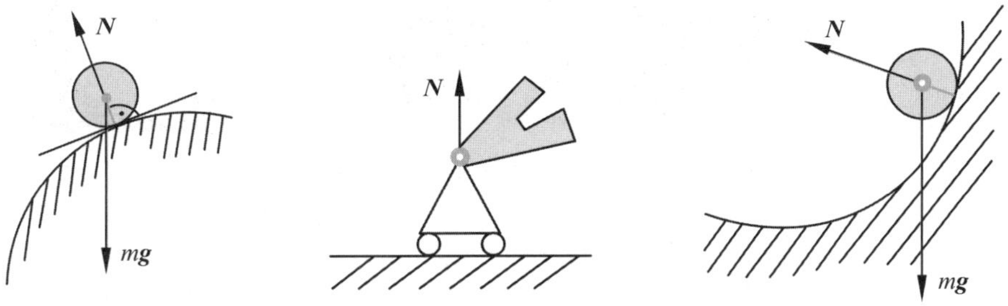

(a) Convex surface(凸面)　(b) Sliding support(滑动支座)　(c) Camber concave(凹面)

Fig. 1.16　Constraint reaction drawing method 2 of smooth surface(图 1.16　光滑面约束反力画法 2)

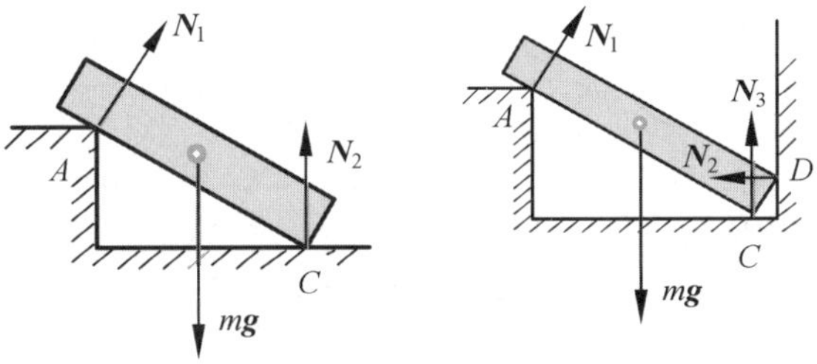

(a) Contact surface 1(接触面1)　(b) Contact surface 2(接触面2)

Fig. 1.17　Constraint reaction drawing method 3 of smooth surface(图 1.17　光滑面约束反力画法 3)

2. Flexible body constraint(柔性体约束)

A flexible body constraint is a constraint that does not allow the object to move away from the suspension point of the rope (Fig. 1.18). (柔性体约束是一种不允许物体从绳索悬挂点移开的约束(图 1.18).)

In other words, constraints formed by flexible, non-extendable ropes, belts, chains and other flexible bodies on rigid bodies are called flexible body constraints. (也就是说,由柔性的、不可伸长的绳索、皮带及链条等柔性体对刚体所构成的约束称为柔性体约束.)

Constraint feature: Only the movement of the object along the extension direction of the flexible body is restricted. (约束特征:只限制物体沿柔性体伸长方向的运动.)

Constraint reaction characteristics are as follows. (约束反力特征如下.)

①Azimuth. Along the axis of the flexible body. (方位. 沿柔性体轴线.)

②Toward. Away from the bound object (tension). (指向. 背离被约束物体(受拉).)

The restraining reaction T of the non-extendable rope being pulled points along the rope towards the suspension point (Fig. 1.19). (受拉的不可伸长的绳索的约束反力 T 沿着绳索指

向悬挂点(图 1.19).)

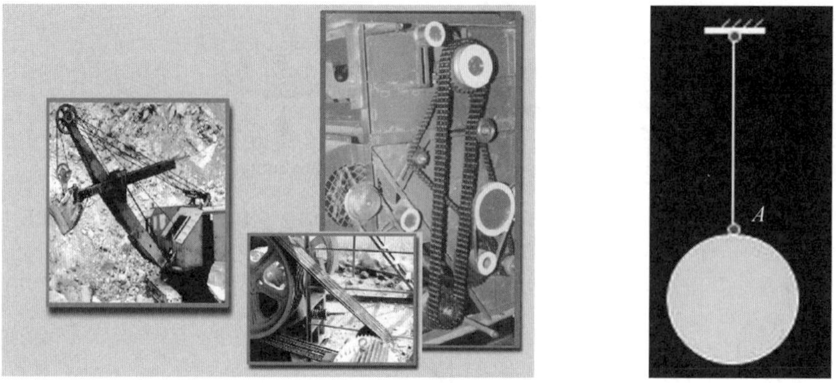

Fig. 1.18　Flexible body constraints (图 1.18　柔性体约束)

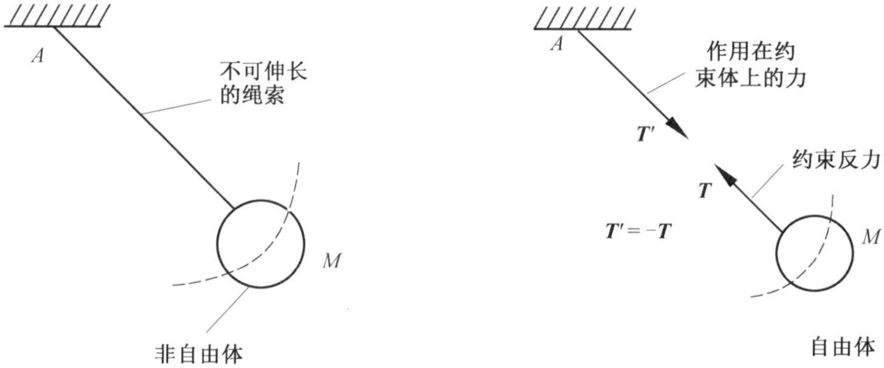

(a) Rope(绳索)

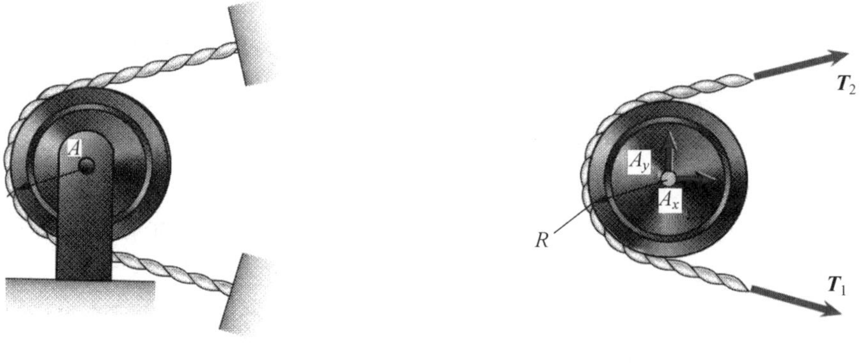

(b) Chain(链条)

Fig. 1.19　Constraint reaction of flexible body(图 1.19　柔性体约束反力)

3. Smooth cylindrical hinge restraint(光滑圆柱形铰链约束)

Two objects are respectively drilled with cylindrical holes of the same diameter and connected with a cylindrical pin. When friction is not considered, smooth cylindrical hinge constraints are formed, referred to as hinge constraints(Fig. 1.20). (两物体分别钻有直径相同的圆柱形孔,用一个圆柱形销钉连接起来,在不计摩擦时,即构成光滑圆柱形铰链约束,简称铰链约束(图 1.20).)

Cylindrical hinge constraints are generally divided into fixed hinge supports, intermediate hinges and movable hinge supports(Fig 1.21).(圆柱形铰链约束一般分为固定铰链支座、中间铰链和可动铰链支座(图 1.21).)

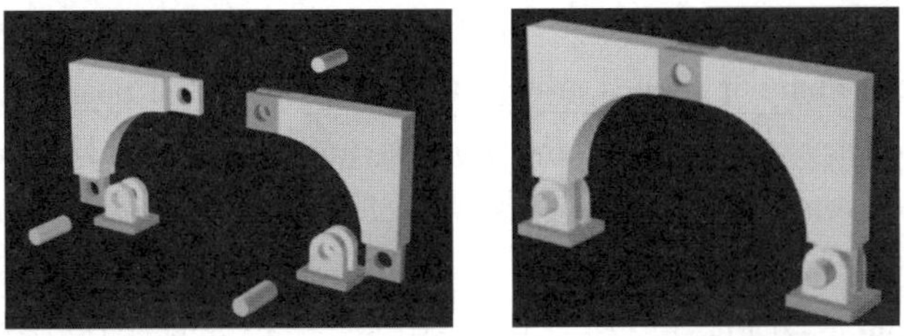

Fig 1.20 Hinged constraint structures(图 1.20 铰链约束结构)

(a) Fixed hinge support (固定铰链支座) (b) Intermediate hinge (中间铰链) (c) Movable hinge support (可动铰链支座)

Fig. 1.21 Three common hinge restraints(图 1.21 常见的三种铰链约束)

The two members in the intermediate hinge restraint are connected by cylindrical pins and neither is fixed. A fixed hinge support is formed if one of the two hinged members is connected to a foundation or frame. A row of rollers is installed at the bottom of the hinge support, which can make the support move along the fixed supporting surface and constitute the constraint of the movable hinge support. The constraint of movable hinge support is very common in bridge, building and other engineering structures. The property of this support is similar to that of the restrained reaction on a smooth surface.(中间铰链约束中两构件用圆柱形销钉连接且均不固定.如果用铰链连接的两个构件中有一个构件与地基或机架相连,便构成固定铰链支座.在铰链支座的底部安装一排滚轮,可使支座沿固定支承面移动,就构成可动铰链支座约束.可动铰链支座约束在桥梁、房屋等工程结构中很常见,这种支座的性质与光滑面约束反力相似.)

(1) Fixed cylindrical hinge (bearing) constraints.(固定圆柱铰链(轴承)约束.)

A cylindrical fixed hinge (or simply fixed hinge) is a constraint that causes an object to rotate around a common axis called the hinge axis (Fig. 1.22).(一个圆柱形固定铰链(或简称固定铰链)是一种约束,该约束可使一个物体绕公共轴线转动,公共轴称为铰链轴(图 1.22).)

Constraint feature: Only the radial movement of the object along the cylindrical hinge is restricted. Its movement along the hinge axis and rotation around the axis are not restricted.(约束特征:只限制物体沿圆柱形铰链径向的移动,不限制其沿着铰链轴向移动和绕轴线的

转动.)

Because the contact surface is smooth, the cylindrical hinge still belongs to the smooth surface constraint. Smooth surface constraint reaction can be represented by a resultant force along the common normal direction (radial) of the contact surface through the contact point (Fig. 1.22). When the member rotates around the hinge axis, the position of the contact point K changes, so the action line of the resultant force is undetermined. However, regardless of the direction of the line of action of the force, its line of action must pass through the center of the cross section of the pin, so the resultant force can be represented by two orthogonal components ($\boldsymbol{X}_0$ and $\boldsymbol{Y}_0$) of the center O (Fig. 1.22). (由于接触面是光滑的,因此圆柱形铰链仍然属于光滑面约束. 光滑面约束反力可以用通过接触点、沿着接触面的公法线方向(径向)的一个合力 $\boldsymbol{F}_R$ 表示(图 1.22). 构件在绕铰链轴线的转动过程中,接触点 K 的位置在变化,故合力 $\boldsymbol{F}_R$ 的作用线待定. 但无论该力的作用线方位如何,其作用线都要通过销钉横截面的圆心,故合力 $\boldsymbol{F}_R$ 可用通过圆心 O 的两个正交分力 $\boldsymbol{X}_0$ 和 $\boldsymbol{Y}_0$ 表示(图 1.22).)

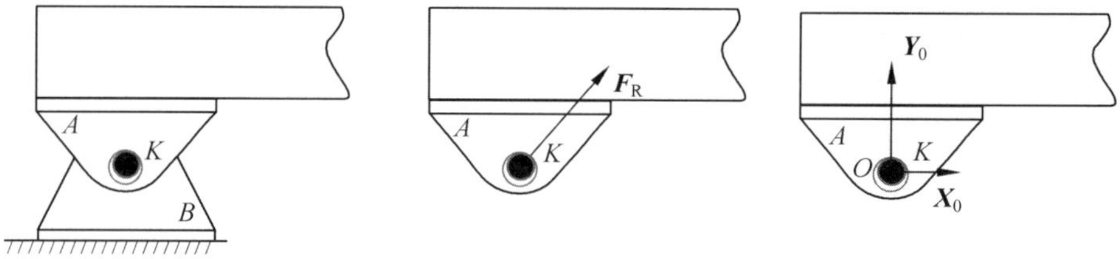

Fig. 1.22 Representation of constraint reaction on fixed hinges(图 1.22 固定铰链约束反力的表示)

The constraint reaction of a fixed cylindrical hinge can be decomposed into two mutually perpendicular forces along the coordinate axis. Typically, they are represented by $\boldsymbol{X}_A$ and $\boldsymbol{Y}_A$ (Fig. 1.23). (固定圆柱铰链的约束反力可以分解为沿着坐标轴的两个相互垂直的力,通常用 $\boldsymbol{X}_A$ 和 $\boldsymbol{Y}_A$ 表示(图 1.23).)

The constraint reaction of a fixed hinge (Fig. 1.23) can be expressed as the reaction $\boldsymbol{X}_A$ and $\boldsymbol{Y}_A$ resultant force $\boldsymbol{R}_A$. It is difficult to express the binding reaction $\boldsymbol{R}_A$ of a fixed hinge as a net force, because its direction is undetermined. (固定铰链的约束反力(图 1.23)可以表示为由反力 $\boldsymbol{X}_A$ 和 $\boldsymbol{Y}_A$ 产生的合力 $\boldsymbol{R}_A$. 以合力 $\boldsymbol{R}_A$ 表示固定铰链的约束反力是困难的,因为它的方向待定.)

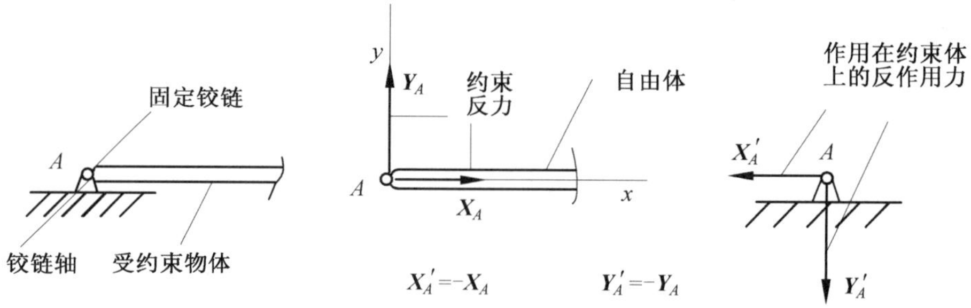

Fig. 1.23 Binding and constraint reaction forces for fixed cylindrical hinges
(图 1.23 固定圆柱铰链的约束力与约束反作用力)

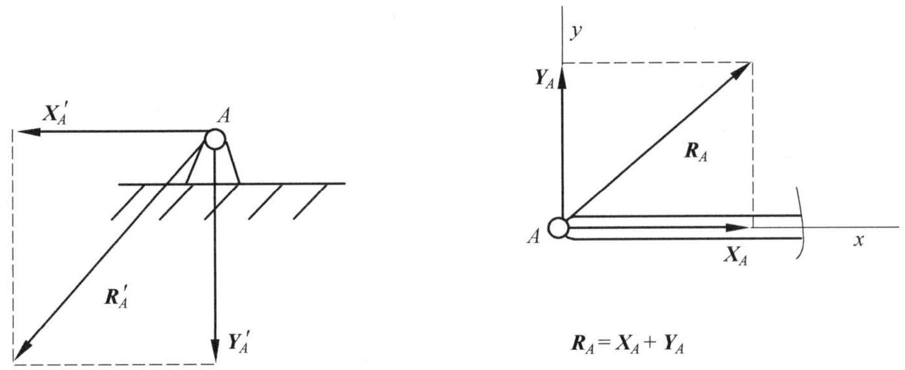

Fig. 1.23(continued)(图 1.23(续))

(2) Middle hinge restraint (pin).(中间铰链约束(销钉).)

As shown in Fig 1.24, both members connected by intermediate hinge constraints can rotate about the hinge axis, but the radial movement of members along the hinge is restricted. The constraint reaction of the hinge on either member at the opening can be expressed either by a radial net force through the center of the circle (direction to be determined) or by two orthogonal components.(如图 1.24 所示,中间铰链约束连接的两个构件都可以绕铰链轴转动,但构件沿铰链径向的移动都被限制了.铰链对其中任何一个构件在开孔处的约束反力既可以用一个通过圆心的径向合力表示(指向待定),也可以用两个正交分力表示.)

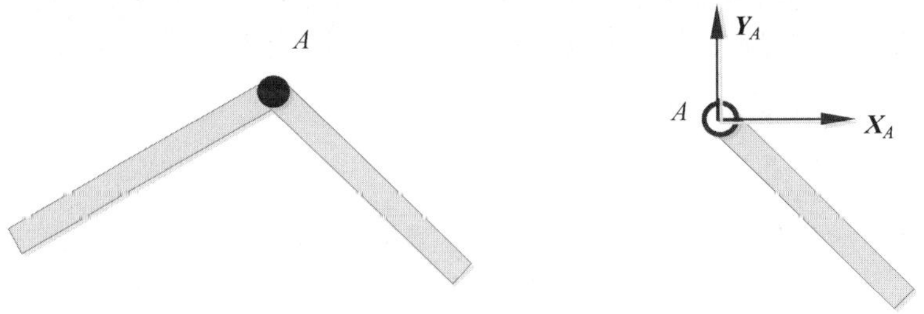

Fig. 1.24 Constraint reaction of middle hinge(图 1.24 中间铰链约束反力表示)

(3) Movable cylindrical hinge.(可动圆柱形铰链.)

A movable cylindrical hinge (or simply a movable hinge) is a cylindrical hinge that allows the object to move along the reference plane (Fig. 1.25).(可动圆柱形铰链(或者简单地说是可移动的铰链)是指允许物体沿着参考平面移动的圆柱形铰链(图 1.25).)

The hinge constraint reaction acts as a wire through the center of the pin, perpendicular to the supporting surface.(铰链约束反力作用线 $\boldsymbol{R}_A$ 通过销钉中心,垂直于支承面.)

4. Ball hinge constraint(球铰链约束)

The constraints that hold the two members together by means of a ball and socket are called ball hinges(Fig. 1.26). The restraint effect of the ball and socket on the ball is to restrict its translation in any direction, but to allow it to rotate around the center of the ball. The essence of this action is smooth surface constraint. The constraint force is applied to the contact point in a radial direction towards the center of the sphere.(通过圆球和球窝将两个构件连接在一起的约

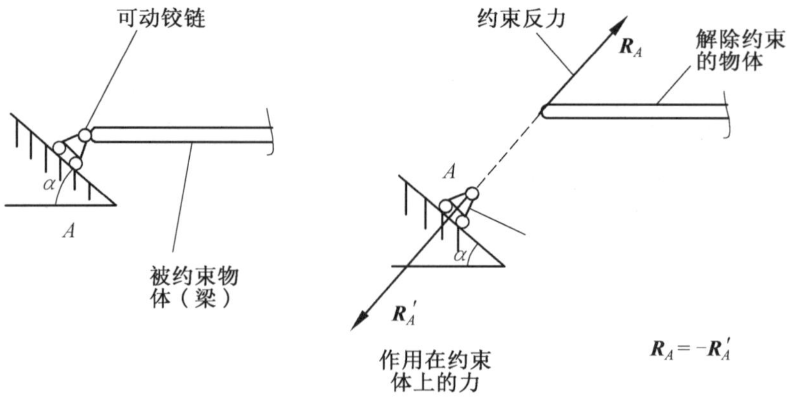

Fig. 1.25　Movable cylindrical hinge(图 1.25　可动圆柱形铰链)

束称为球铰链(图 1.26).球窝对球的约束作用是限制其沿任意方向的平移,而只允许其绕球心转动.这种作用的实质是光滑面约束,约束力作用于接触点,方向沿径向指向球心.)

A ball-hinge constraint is a constraint that allows a member to rotate around the center of a spherical chain. The direction of the resultant constraint reaction $\boldsymbol{R}_A$ in space on the ball hinge is arbitrary. The resultant constraint reaction force on the ball hinge can be decomposed into three component forces $\boldsymbol{X}_A$, $\boldsymbol{Y}_A$, $\boldsymbol{Z}_A$ along the coordinate axis. The ball-hinge constraint can be replaced by three constraint reactions along the coordinate axis. (球铰链约束是指允许构件围绕球形链中心旋转的约束.球铰链约束反力合力 $\boldsymbol{R}_A$ 在空间中的方向是任意的.球铰链约束反力合力可以分解成沿坐标轴方向的三个分力 $\boldsymbol{X}_A$、$\boldsymbol{Y}_A$、$\boldsymbol{Z}_A$.球铰链约束可被沿坐标轴方向的三个约束反力替代.)

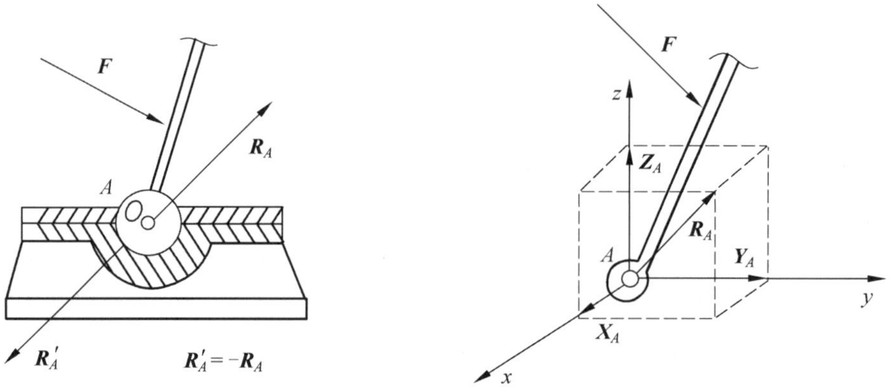

Fig. 1.26　Hinge constraint(图 1.26　铰链约束)

5. Bearing constraint(轴承约束)

Bearing constraint is a common constraint in machinery. Commonly used radial bearings and thrust bearings. Fig. 1.27 shows the radial bearing device. The central shaft of the constraint is the constrained body, and the bearing seat is the constraint body. The shaft can rotate arbitrarily in the bearing seat hole, but also can move along the center line of the hole, that is, the bearing does not limit its axial movement and rotation around the shaft. But the bearing limits the radial outward translation of the rotating shaft, which is equivalent to the hinge support. The schematic diagram and constraint reaction force are shown in Fig. 1.27. (轴承约束是机械中常见的约束，常用的有径向轴承及止推轴承. 图 1.27 所示为径向轴承装置，该约束中轴是被约束体，轴承座是约束体. 轴可在轴承座孔内任意转动，也可以沿孔的中心线移动，即轴承并不限制它的轴向运动和绕轴转动. 但是轴承限制了转轴径向向外的平移，相当于铰链支座. 其简图及约束反力如图 1.27 所示.)

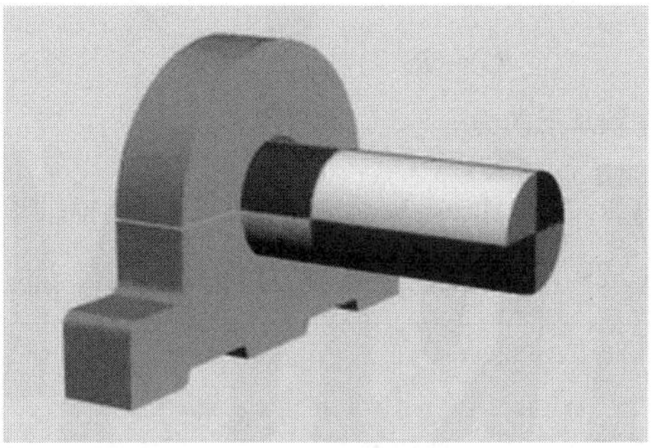

(a) Bearing constraint(轴承约束)

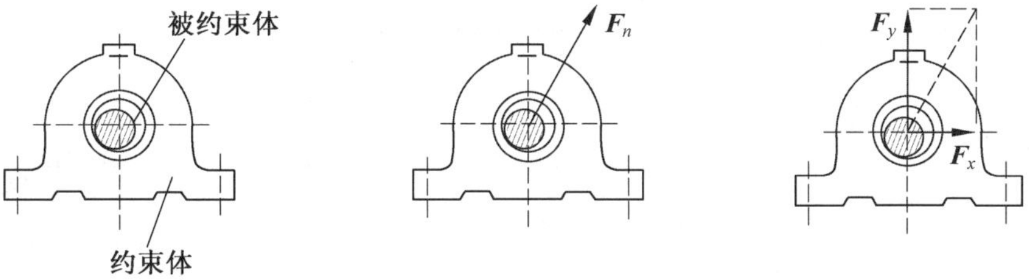

(b) Diagram(简图) (c) Resultant(合力) (d) Decomposition of the resultant (合力分解)

Fig. 1.27 Radial bearing (radial bearing) (图 1.27 向心轴承(径向轴承))

The other is the thrust bearing (Fig. 1.28), which limits the radial translation of the shaft and its axial movement, allowing only rotation about the shaft. Compared with the radial bearing constraint, the thrust bearing has one more constraint reaction along the axial direction. (另一种是止推轴承(图 1.28)，它限制转轴沿径向的平移，又限制它的轴向移动，只允许绕轴转动. 与径向轴承约束相比，止推轴承多了一个沿轴向的约束反力.)

In Fig. 1.28, the direction of the resultant bearing constraint reaction force in space is arbitrary. The constrained reaction force of the bearing (same as the constrained reaction force of the ball hinge) can be decomposed into three component forces $\boldsymbol{X}_A$, $\boldsymbol{Y}_A$, $\boldsymbol{Z}_A$ along the axis, and the bearing is replaced by three constrained reactions along the axis. (图 1.28 中，轴承约束反力合力 $\boldsymbol{R}_A$ 在空间中的方向是任意的. 轴承的约束反力合力(与球铰链约束反力合力一样)可以分解成沿坐标轴方向的三个分力 $\boldsymbol{X}_A$、$\boldsymbol{Y}_A$、$\boldsymbol{Z}_A$，轴承用沿坐标轴方向的三个约束反力替代.)

6. Variant of movable cylindrical hinge—slider constraint (可动圆柱形铰链的变体——滑块约束)

The slider constraint reaction line $\boldsymbol{R}_A$ passes through the center of the pin, perpendicular to

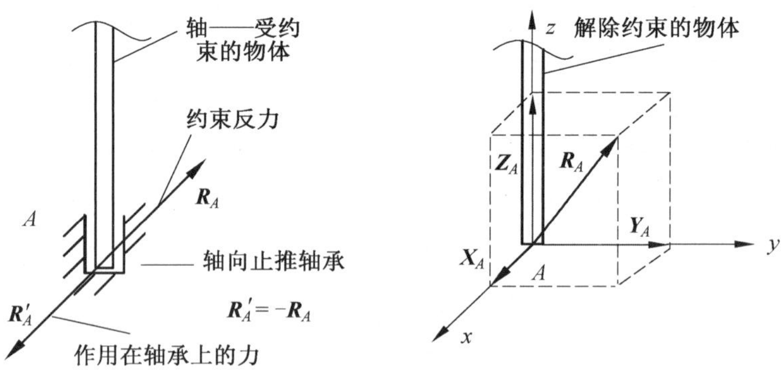

Fig. 1.28 Thrust bearing (图 1.28 止推轴承)

the supporting surface(Fig. 1.29).(滑块约束反力作用线 R_A 通过销钉中心,垂直于支承面(图 1.29).)

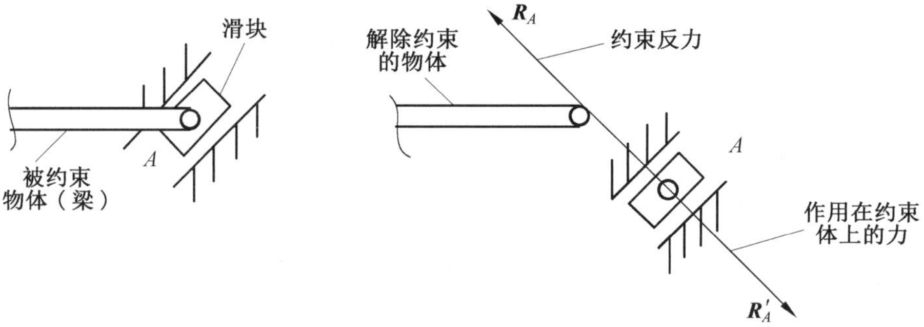

Fig. 1.29 Slider (图 1.29 滑块)

7. Rigid fixed end (or fixed support) constraints(刚性固定端(或固定支座)约束)

The rigid fixed end is the constraint that restricts any movement of the object(Fig. 1.30).(刚性固定端是限制物体任何运动的约束(图 1.30).)

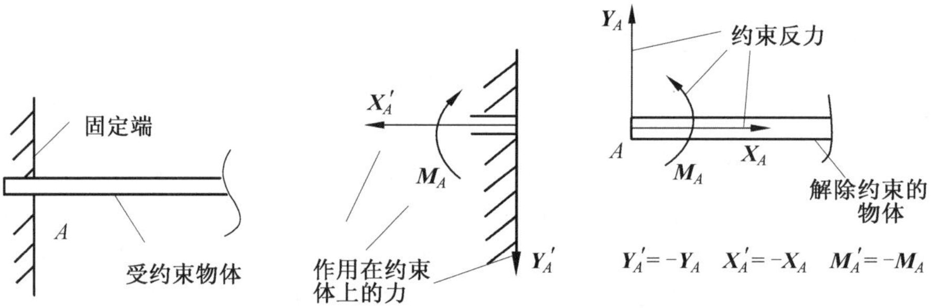

Fig. 1.30 Constraint on fixed end of plane(图 1.30 平面固定端约束)

Note: The fixed support is different from the fixed hinge support. The fixed support restricts the translation and rotation of the object at the joint; fixed hinges do not limit rotation and can be rotated around the hinge axis.(注意:固定支座与固定铰链支座不同,固定支座限制物体在连接处的平移和转动;而固定铰链不限制转动,可以绕铰链轴转动.)

The rigid fixed end mainly includes plane, space and sliding bearings.（刚性固定端主要包括平面、空间和滑动的定向支座.）

（1）Plane fixed end and space fixed end.（平面固定端与空间固定端.）

The action of a planar fixed support（Fig. 1.30）can be replaced by an unknown constraint reaction $\boldsymbol{R}_A$ in any direction in the plane and an unknown torque $\boldsymbol{M}_A$（constraint reaction）, where the constraint reaction $\boldsymbol{R}_A$ restricts movement and the constraint moment $\boldsymbol{M}_A$ restricts rotation（平面固定支座（图 1.30）的作用可以用一个在该平面内任意方向的未知约束反力 $\boldsymbol{R}_A$ 及一个未知力矩 $\boldsymbol{M}_A$（约束反力矩）代替. 其中，约束反力 $\boldsymbol{R}_A$ 限制移动，约束力矩 $\boldsymbol{M}_A$ 限制转动.）

The constrained reaction $\boldsymbol{R}_A$ is usually represented by the sum of two unknown forces $\boldsymbol{X}_A$ and $\boldsymbol{Y}_A$ along the axis, so the fixed end of the plane is replaced by the sum of two constrained reactions $\boldsymbol{X}_A$ and $\boldsymbol{Y}_A$ and the moment of constraint reaction $\boldsymbol{M}_A$.（约束反力 $\boldsymbol{R}_A$ 通常用两个沿着坐标轴方向的未知力 $\boldsymbol{X}_A$ 和 $\boldsymbol{Y}_A$ 表示，因此平面固定端被两种约束反力 $\boldsymbol{X}_A$ 和 $\boldsymbol{Y}_A$ 及约束反力矩 $\boldsymbol{M}_A$ 代替.）

The action of a spatially fixed support（Fig. 1.31）can be replaced by an unknown reaction force $\boldsymbol{R}_A$ in any direction in space and an unknown moment $\boldsymbol{M}_A$（constraint reaction moment）.（空间固定支座（图 1.31）的作用可以用一个在空间有任意方向的未知反力 $\boldsymbol{R}_A$ 及一个未知力矩 $\boldsymbol{M}_A$（约束反力矩）代替.）

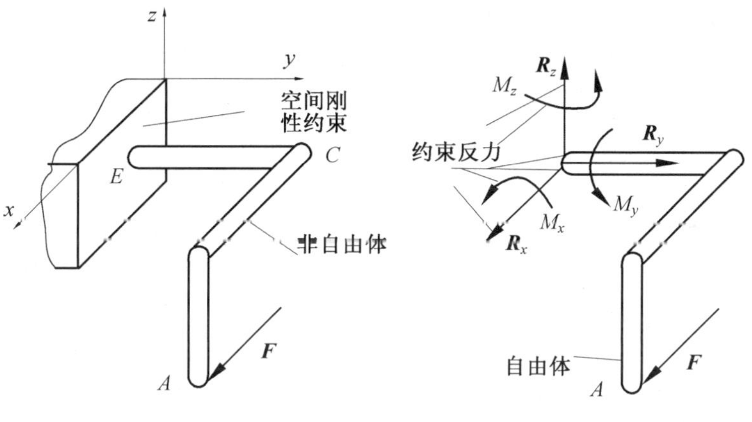

Fig. 1.31 Fixed end of space（图 1.31 空间固定端）

Note: Fig. 1.31 does not reflect the force of the object on the restrained body.

（注：图 1.31 没体现出物体对约束体的作用力.）

The constraint reaction of a spatial fixed force $\boldsymbol{R}_A$ can be expressed in terms of the constraint reaction $\boldsymbol{R}_x$, $\boldsymbol{R}_y$, $\boldsymbol{R}_z$ and moment M_A in the direction of three coordinate axes and the projected moments M_x, M_y, M_z in the direction of the corresponding coordinate axes. Therefore, the space fixed support is replaced by three forces $\boldsymbol{R}_x$, $\boldsymbol{R}_y$, $\boldsymbol{R}_z$ and three constraint moments M_x, M_y, M_z.（空间固定支座的约束反力可以用 $\boldsymbol{R}_A$ 在三个坐标轴方向的约束反力 $\boldsymbol{R}_x$、$\boldsymbol{R}_y$、$\boldsymbol{R}_z$ 和力矩 M_A 在三个相应坐标轴方向投影的力矩 M_x、M_y、M_z 表示. 因此，空间固定支座被三个作用力 $\boldsymbol{R}_x$、$\boldsymbol{R}_y$、$\boldsymbol{R}_z$ 和三个约束力矩 M_x、M_y、M_z 取代.）

(2) Rigid sliding support (plane). (刚性滑动支座(平面).)

The sliding support is a kind of support that "restricts" the rotation, but it does not restrict the translation movement along the guide rail. The support can slide on the guide rail (Fig. 1.32). Sliding support constraint reaction: $\boldsymbol{R}_A$ is the constraint reaction perpendicular to the supporting surface; M_A is the constraint moment. (滑动支座是一种"限制"转动的支座,但其不限制沿导轨的平移运动,支座可以在导轨上滑动(图1.32). 滑动支座约束反力: $\boldsymbol{R}_A$ 是垂直于支承面的约束反力; M_A 是约束力矩.)

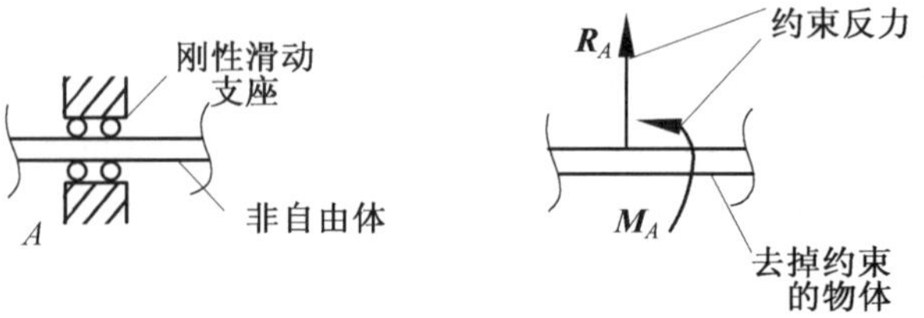

Fig. 1.32 Rigid sliding support (directional support) (图1.32 刚性滑动支座(定向支座))

Note: The difference between sliding support (directional support) and movable hinge support, the former limits rotation, but the latter does not.

(注:滑动支座(定向支座)与可动铰链支座的区别,前者限制转动,后者不限制转动.)

8. Linear massless bar (two-force bar) constraint (直线型无质量杆(二力杆)约束)

Linear massless bar: A bar whose axis is straight and its weight is negligible compared to its perceived load (Fig. 1.33). (直线型无质量杆:其轴线为直线且其质量与其感知的载荷相比可忽略不计(图1.33).)

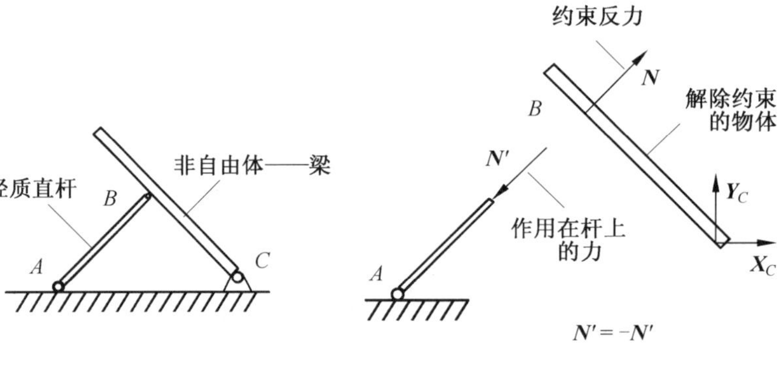

Fig. 1.33 Two-force bar(图1.33 二力杆)

The constraining force N of an articulated linear bar acts along the axis of the bar. (铰接式直线杆的约束反力 N 沿杆的轴线作用.)

The action of an articulated straight bar at the joint can be replaced by a force S in the bar. In this case, the force is assumed to act and be strained along the axis of the bar (Fig. 1.34), i.e. the force is assumed to be positive (if the rod is extended, then $S>0$; if it is compressed, then $S<0$). (铰接式直杆在连接处的作用可以用杆中的力 S 来代替. 在这种情况下,力被假定沿着杆的轴线作用受拉(图1.34),即力被假定为正(如果杆伸长,则 $S>0$;如果杆压缩,则 $S<0$).)

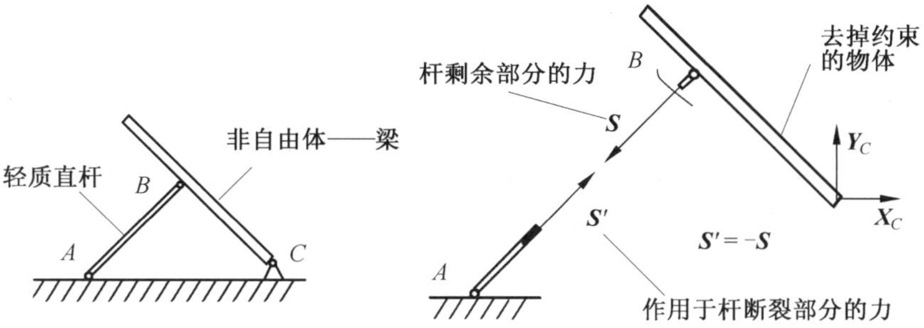

Fig. 1.34 Articulated straight pole(图 1.34 铰接式直杆)

1.3 Force Analysis and Force Diagrams of Objects (物体的受力分析和受力图)

In engineering practice, it is often necessary to apply equilibrium condition to solve unknown force according to known force. To this end, the first need to carry on the force analysis of the object, determine the object by several forces, each force position and direction of action.(在工程实际中,常常需要根据已知力,应用平衡条件求解未知力.为此,首先需要对物体进行受力分析,确定物体受了几个力,以及每个力的作用位置和作用方向.)

The forces acting on a body include the main force and the forced force.(作用在物体上的力包括主动力和被动力.)

①Principal force. The force that moves or tends to move a body, the magnitude and direction of which are known, such as gravity, water pressure, etc.(主动力.使物体运动或有运动趋势的力,其大小和方向都已知,如重力、水压力等.)

②Forced force. A force caused by and varying with a dominant force, often of unknown magnitude and direction, like constraining reaction.(被动力.由主动力引起并随其变化的力,其大小和方向往往未知,如约束反力.)

In order to clearly represent the force of an object, we separate the object to be studied (the loaded body) from the surrounding object (the loading body). This step is called the selection of the research object or the selection of the separation. Then draw all the forces (including the main force and the constraint reaction) of the force body on the research object, and draw the force diagram of the force body separately.(为清晰地表示物体的受力情况,把需要研究的物体(受力体)从周围的物体(施力体)中分离出来,这个步骤称为选取研究对象或选取分离体.然后把施力体对研究对象的作用力(包括主动力及约束反力)全部画出来,单独画出受力体的受力简图.)

A force diagram is a schematic diagram of all the forces on the object of study.(受力图是研究对象所受全部作用力的简图.)

When drawing the force diagram of the separated body, we should pay attention to remove the constraint acting on it and replace it with the corresponding constraint reaction. When drawing a force diagram, we should follow the principle of easy before difficult. We should draw the main force before the subject force, and only draw the force on the object.(画分离体的受力图时,要注意把作用在其上的约束去掉,用相应的约束反力替换.画受力图时应遵循先易后难的原则,先画主动力,后画被动力,只画物体所受的力.)

1.3.1 Steps and precautions for drawing a receiver diagram (画受力图的步骤和注意事项)

1. Steps(步骤)

(1) Determine the research object.(确定研究对象.)

(2) Draw the main driving force.(画主动力.)

(3) Draw the corresponding constraint reaction correctly according to the constraint type.(根据约束类型,正确地画出相应的约束反力.)

It is not necessary to infer the direction of constraint reaction by subjective conjecture or simply by the direction of the main force.(切不可单凭主观臆想或简单地根据主动力的方向去推测约束反力的方向.)

2. Precautions(注意事项)

(1) Do not omit the constraint reaction.(不要漏画约束反力.)

(2) Do not draw more force, for each force drawn should be clear its force body.(不要多画力,对于画出的每一个力,应明确其施力体.)

(3) When analyzing the interaction force between two bodies, pay attention to check whether the arrows of these forces conform to the relationship between the force and the reaction force.(在分析两物体间的相互作用力时,要注意检查这些力的箭头是否符合作用力与反作用力的关系.)

(4) When studying the balance of the system, only the external force (external force) on the study object is drawn on the force diagram, not the internal force.(当研究系统平衡时,在受力图上只画出外部物体对研究对象的作用力(外力),不画成对出现的内力.)

1.3.2 Sample questions(例题)

Example 1: An object of weight $\boldsymbol{P}$ is held against a smooth wall by a rope, as shown in Fig. 1.35(a). Draw the force diagram of the weight.(**例题 1**:一个重力为 $\boldsymbol{P}$ 的物体在绳子作用下靠在光滑的墙上,如图 1.35(a)所示,请画出重物的受力图.)

Solution:(**解**:)

(1) Take the weight as the object of study, and draw its schematic diagram separately, that is, take the separated body.(取重物为研究对象,并单独画出其简图,即取分离体.)

(2) Draw the main driving force. It has its own weight $\boldsymbol{P}$.(画主动力,其有物体本身自重 $\boldsymbol{P}$.)

(3) Draw constrained reaction force. At point A, the object is constrained by the flexibility of the rope to the object, so at point A, it is subjected to a pull $\boldsymbol{F}_A$ along the direction of the rope away from the object. At point B, the object is constrained by the smooth contact between the wall and the object, so it is affected by the normal reaction $\boldsymbol{F}_B$ of the wall at point B and points to the

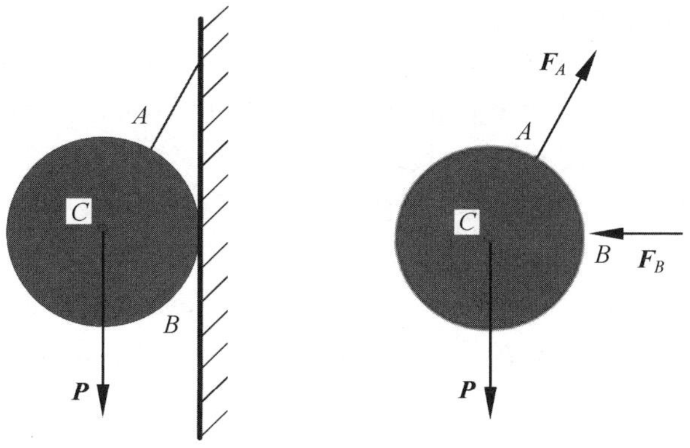

(a) Original diagram(原图)　　(b) Force diagram(受力图)

Fig. 1.35　The ball hangs on the smooth wall(图 1.35　小球挂在光滑的墙上)

center of the circle along the common normal direction of the contact point between the object and the wall.(画约束反力. 物体在 A 点受到绳子对物体的柔性体约束,故在 A 点受到沿着绳索的方向背离物体的拉力 $\boldsymbol{F}_A$. 物体在 B 点受到墙面对物体的光滑接触约束,故在 B 点受到墙面法向反力 $\boldsymbol{F}_B$ 的作用,沿着物体和墙面的接触点的公法线方向指向圆心.)

The force diagram of the object is shown in Fig 1.35 (b). (物体受力图如图 1.35(b)所示.)

Example 2: Draw the force diagram of the beam in Fig. 1.36(a), where beam AB does not count its own weight. (**例题 2**: 画出图 1.36(a)中梁的受力图,其中梁 AB 不计自重.)

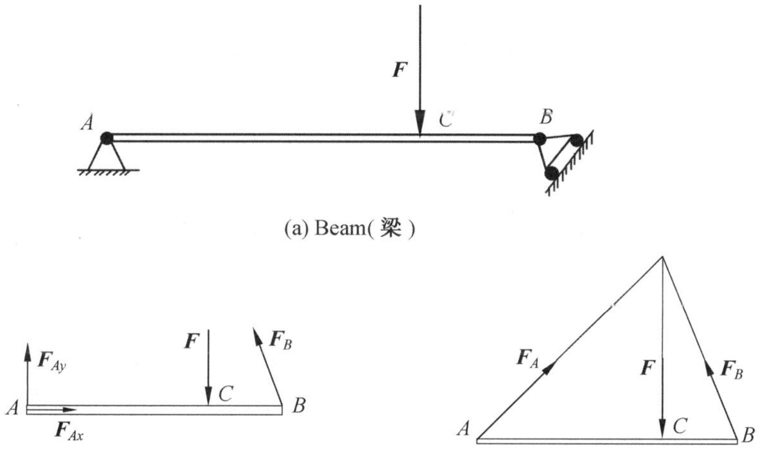

Fig. 1.36　simply supported beams(图 1.36　简支梁)

Solution: (**解**:)

(1) Take beam AB as the research object and draw its schematic diagram separately. (取梁 AB 为研究对象,并单独画出其简图.)

(2) Draw the main driving force. Because AB doesn't care about its own weight, it only has an external force $\boldsymbol{F}$ at point C. (画主动力. 因 AB 不计自重,故仅在 C 点受到外力 $\boldsymbol{F}$.)

(3) Draw constrained reaction force. A is a fixed hinge constraint whose binding force can be expressed by two orthogonal components $\boldsymbol{F}_{Ax}$ and $\boldsymbol{F}_{Ay}$ of unknown magnitude. B is the movable hinge constraint, whose binding force is perpendicular to the contact surface, and is denoted by $\boldsymbol{F}_B$. (画约束力. A 处为固定铰链约束,其约束力可用两个大小未知的正交分力 $\boldsymbol{F}_{Ax}$ 和 $\boldsymbol{F}_{Ay}$ 表示. B 处为可动铰链约束,其约束力垂直于接触面,用 $\boldsymbol{F}_B$ 表示.)

The force diagram of the beam is shown in Fig. 1.36 (b). (梁的受力如图 1.36(b)所示).

Further analysis shows that since only A, B, C are stressed on the beam AB and maintain balance, the direction of the stress $\boldsymbol{F}_A$ at point A can be determined according to the confluence theorem of the equilibrium of the three forces. The action line of $\boldsymbol{F}$ and $\boldsymbol{F}_B$ intersect at one point. When the beam AB is equilibrium, the action line of $\boldsymbol{F}_A$ must pass through the intersection point between $\boldsymbol{F}$ and $\boldsymbol{F}_B$, as shown in Fig. 1.36 (c). As for the direction of $\boldsymbol{F}_A$, it is tempory as shown in Fig. 1.36, later determined by the equilibrium conditions. (进一步分析可知,由于梁 AB 上仅有 A、B、C 三点受力,并保持平衡,因此根据三力平衡汇交定理,可确定 A 点的受力 $\boldsymbol{F}_A$ 的方向. $\boldsymbol{F}$ 和 $\boldsymbol{F}_B$ 的作用线相交于一点,梁 AB 平衡时, $\boldsymbol{F}_A$ 的作用线必过 $\boldsymbol{F}$ 与 $\boldsymbol{F}_B$ 的交点,如图 1.36(c)所示. $\boldsymbol{F}_A$ 的指向暂定如图 1.36 所示,以后由平衡条件确定.)

Example 3: A three-hinged arch bridge is shown in Fig. 1.37, which is hinged by the left and right arches (Fig. 1.37(a)). Assuming that the weight of each arch is not considered, a load $\boldsymbol{P}$ is applied on the left arch. Try to draw the force diagram of the left and right arch and the whole system (**例题 3**: 图 1.37 所示为三铰拱桥,由左、右两拱铰接而成(图 1.37(a)). 设各拱自重不计,在左拱上作用有载荷 P. 试分别画出左、右拱及整体的受力图.)

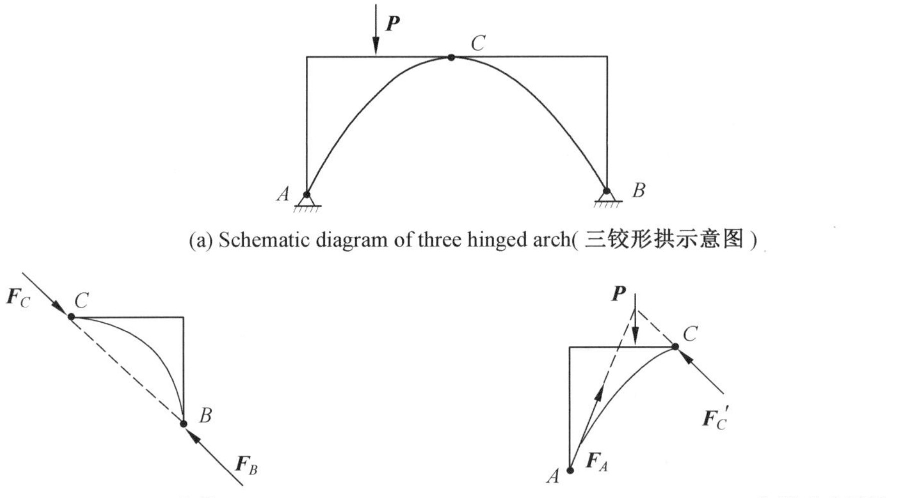

(a) Schematic diagram of three hinged arch(三铰形拱示意图)

(b) Right arch(右拱)

(c) Drawing method 1 for left arch force(左拱受力画法 1)

Fig. 1.37 Force diagram of three-hinged arch bridge(图 1.37 三铰拱桥受力图)

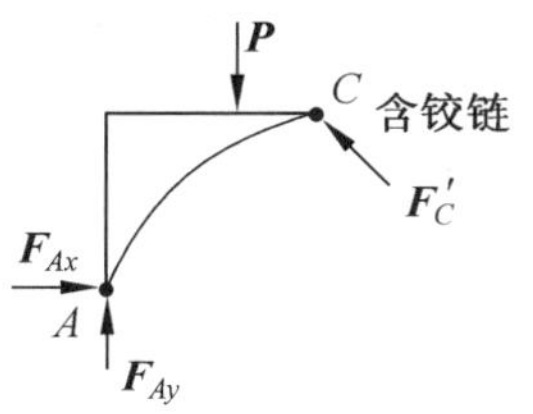

(d) Drawing method 2 for left arch force
(左拱受力画法 2)

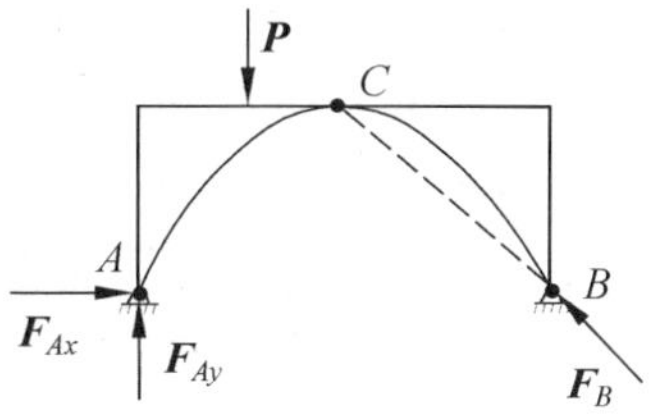

(e) Total force diagram
(整体受力图)

Fig. 1.37(continued)(图 1.37(续))

Solution:(解:)

(1)The force of arch BC is analyzed. Since arch BC has no weight and is only constrained by hinges at B and C, arch BC is a two-force member. At B and C, they are acted on by $\boldsymbol{F}_B$ and $\boldsymbol{F}_C$ respectively, and $\boldsymbol{F}_B = -\boldsymbol{F}_C$, the two forces are in equal size, reverse direction and collinear, as shown in Fig. 1.37(b). (分析拱 BC 的受力. 由于拱 BC 自重不计,且只在 B、C 两处受到铰链约束,因此拱 BC 为二力构件. 在 B、C 处分别受 $\boldsymbol{F}_B$ 和 $\boldsymbol{F}_C$ 的作用,且 $\boldsymbol{F}_B = -\boldsymbol{F}_C$,这两个力等值、反向、共线,如图 1.37(b)所示.)

(2) Arch AC was selected as the research object. The active force is only the load $\boldsymbol{P}$, regardless of weight. Arch AC at hinge C is subjected a constrained reaction force $\boldsymbol{F}_C$ given by arch BC, according to the laws of action and reaction, $\boldsymbol{F}'_C = -\boldsymbol{F}_C$. The arch AC has the force $\boldsymbol{F}_A$ of the fixed hinge constraint at point A. Since arch AC is in equilibrium under the action of three forces $\boldsymbol{F}_A$, $\boldsymbol{P}$, $\boldsymbol{F}_C$, the direction of force $\boldsymbol{F}_A$ at point A can be determined according to the confluence theorem of three forces equilibrium. Action lines of $\boldsymbol{P}$ and $\boldsymbol{F}_C$ intersect at one point, when arch AC is in equilibrium, action lines of $\boldsymbol{F}_A$ must cross the intersection point of $\boldsymbol{P}$ and $\boldsymbol{F}_C$, as shown in Fig. 1.37(c). As for the direction of $\boldsymbol{F}_A$, it is temporarily shown in Fig. 1.37 (c), later determined by the equilibrium conditions. (取拱 AC 为研究对象. 由于不计自重,因此主动力仅有载荷 $\boldsymbol{P}$. 拱 AC 在铰链 C 处,有拱 BC 给它的约束力 $\boldsymbol{F}_C$,根据作用和反作用定律有 $\boldsymbol{F}'_C = -\boldsymbol{F}_C$. 拱 AC 在 A 处有固定铰链约束给它的约束力 $\boldsymbol{F}_A$. 由于拱 AC 在 $\boldsymbol{F}_A$、$\boldsymbol{P}$、$\boldsymbol{F}_C$ 三个力作用下平衡,因此根据三力平衡汇交定理,可确定 A 点的受力 $\boldsymbol{F}_A$ 的方向. $\boldsymbol{P}$ 和 $\boldsymbol{F}_C$ 的作用线相交于一点,在拱 AC 平衡时,$\boldsymbol{F}_A$ 的作用线必过 $\boldsymbol{P}$ 与 $\boldsymbol{F}_C$ 的交点,如图 1.37(c)所示. $\boldsymbol{F}_A$ 的指向暂定如图 1.37(c)所示,以后由平衡条件确定.)

Since point A is a fixed hinge constraint, the force $\boldsymbol{F}_A$ at point A can also be decomposed into two orthogonal component forces $\boldsymbol{F}_{Ax}$ and $\boldsymbol{F}_{Ay}$ of unknown magnitude, as shown in Fig 1.37(d). (由于 A 点为固定铰链约束,因此也可以将 A 点的受力 $\boldsymbol{F}_A$ 分解为两个大小未知的正交分力 $\boldsymbol{F}_{Ax}$ 和 $\boldsymbol{F}_{Ay}$,如图 1.37(d)所示.)

(3) Force analysis of the whole system. When the whole system is selected as the object of study, the balanced whole structure can be hardened into a rigid body. Since at hinge C, it is satisfied $\boldsymbol{F}'_C = -\boldsymbol{F}_C$, these two forces appear in pairs throughout the system, called internal forces. The effect of internal forces on the system cancels out each other, so it can be eliminated

without affecting the balance of the whole system. Therefore, the internal force does not need to be drawn when the overall system is analyzed. In a global force diagram, we only need to draw the forces on the system from objects outside the system. These forces are called external forces, such as $\boldsymbol{P}$, $\boldsymbol{F}_{Ax}$, $\boldsymbol{F}_{Ay}$, $\boldsymbol{F}_B$. (整个系统受力分析.当选整个系统为研究对象时,可把平衡的整体结构刚化为刚体.由于铰链 C 处满足 $\boldsymbol{F}'_C=-\boldsymbol{F}_C$,因此这两个力成对地出现在整个系统内,称为内力.内力对系统的作用效应相互抵消,因此可以消除而不影响整个系统的平衡.故内力在整体系统受力分析时不必画出.在整体受力图中只需画出系统以外的物体对系统的作用力,这种力称为外力,如 $\boldsymbol{P}$ 、$\boldsymbol{F}_{Ax}$ 、$\boldsymbol{F}_{Ay}$ 、$\boldsymbol{F}_B$.)

The force of the whole system is shown in Fig 1.37(e). (整个系统受力如图 1.37(e)所示.)

Example 4: Try to draw the force diagram of each object and the whole system in Fig. 1.38(a) respectively. The weight of each component is not considered. (**例题 4**: 试分别画出图 1.38(a)中每个物体及整体的受力图,各构件自重不计.)

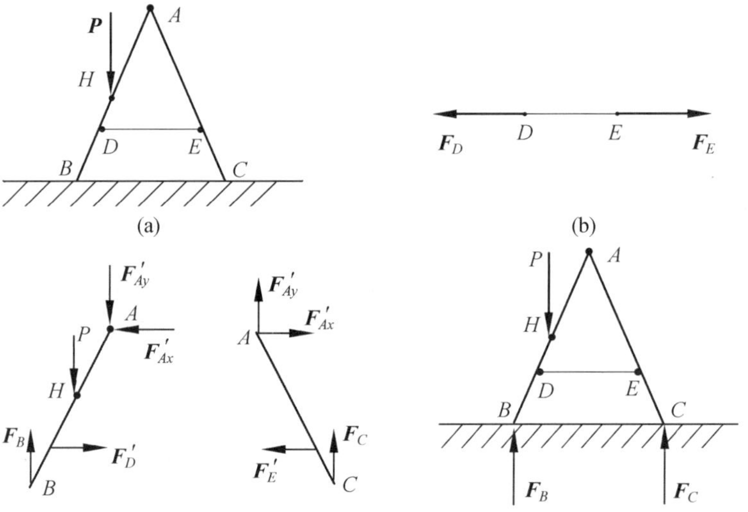

(a) (b)

(c) Force diagram of AB (d) Force diagram of AC (e) Force diagram of whole system

Fig. 1.38 Ladder(图 1.38 梯子)

Solution: (**解**:)

(1) Stress analysis of rope DE. The two ends of the rope D and E are respectively subjected to the pull of the ladder $\boldsymbol{F}_D$ and $\boldsymbol{F}_E$ on them, and the two forces are equivalent, reverse and collinear. The line of action is the line DE, and the two forces deviate from the rope DE along the direction of the line of action(Fig 1.38(b)). (绳子 DE 的受力分析.绳子两端 D 和 E 分别受到梯子对它的拉力 $\boldsymbol{F}_D$ 和 $\boldsymbol{F}_E$ 的作用,两个力等值、反向、共线,作用线为 DE 连线,且两个力沿着作用线的方向背离绳子 DE(图 1.38(b).)

(2) Force analysis of AB part of ladder. AB is subjected to the load P at H, and is subjected to the constrained reaction force $\boldsymbol{F}_{Ax}$, $\boldsymbol{F}_{Ay}$ by the AC part at hinge A. There is the tensional force $\boldsymbol{F}'_D$ of the rope DE on it at D, and $\boldsymbol{F}'_D$ is the reaction force of $\boldsymbol{F}_D$. At point B, the smooth surface is exerting a normal reaction $\boldsymbol{F}_B$ on it. (梯子 AB 部分的受力分析. AB 在 H 处受到载荷 P 的作

用,在铰链 A 处受到 AC 部分给它的约束力 $\boldsymbol{F}_{Ax}$ 和 $\boldsymbol{F}_{Ay}$. 在 D 点受绳子 DE 对它的拉力 $\boldsymbol{F}'_D$, $\boldsymbol{F}'_D$ 是 $\boldsymbol{F}_D$ 的反作用力. 在 B 点受光滑体面对它的法向反力 $\boldsymbol{F}_B$.)

The force on part AB of ladder is shown in Fig. 1. 38(c).(梯子 AB 部分的受力如图 1.38(c)所示.)

(3) Force analysis of AC part of ladder. It is subjected to the constrained reaction force by the AB part at hinge A, $\boldsymbol{F}_{Ax}$ and $\boldsymbol{F}_{Ay}$, $\boldsymbol{F}'_{Ax}$ and $\boldsymbol{F}'_{Ay}$ are the reaction forces of $\boldsymbol{F}_{Ax}$ and $\boldsymbol{F}_{Ay}$, respectively. There is the tension $\boldsymbol{F}'_E$ of the rope DE on it at point E, and $\boldsymbol{F}'_E$ is the reaction force of $\boldsymbol{F}_E$. There is the normal reaction force $\boldsymbol{F}_C$ at C by the smooth ground.(梯子 AC 部分的受力分析. 在铰链 A 处受 AB 部分对它的约束力 $\boldsymbol{F}_{Ax}$ 和 $\boldsymbol{F}_{Ay}$, $\boldsymbol{F}'_{Ax}$ 和 $\boldsymbol{F}'_{Ay}$ 分别是 $\boldsymbol{F}_{Ax}$ 和 $\boldsymbol{F}_{Ay}$ 的反作用力. 在 E 点受绳子 DE 对它的拉力 $\boldsymbol{F}'_E$, $\boldsymbol{F}'_E$ 是 $\boldsymbol{F}_E$ 的反作用力. 在 C 处受光滑地面对它的法向反力 $\boldsymbol{F}_C$.)

The force of ladder AC is shown in Fig. 1.38(d).(梯子 AC 的受力如图 1.38(d)所示.)

(4) Force analysis of the whole system. When the whole system is selected as the object of study, the whole structure of equilibrium can be hardened into a rigid body. Because the force on hinge A is satisfied with $\boldsymbol{F}_{Ax} = -\boldsymbol{F}'_{Ax}$, $\boldsymbol{F}_{Ay} = -\boldsymbol{F}'_{Ay}$, and the forces received by the connecting points D and E of the rope and ladder are also satisfied $\boldsymbol{F}_D = -\boldsymbol{F}'_D$, $\boldsymbol{F}_E = -\boldsymbol{F}'_E$, respectively. These forces are internal forces of the system and can be eliminated out. Therefore, the forces of the whole system is shown in Fig. 1.38(e).(整个系统的受力分析. 当选整个系统为研究对象时,可把平衡的整个结构刚化为刚体. 由于铰链 A 处所受的力满足 $\boldsymbol{F}_{Ax} = -\boldsymbol{F}'_{Ax}$, $\boldsymbol{F}_{Ay} = -\boldsymbol{F}'_{Ay}$, 绳子与梯子连接点 D 和 E 所受的力也分别满足 $\boldsymbol{F}_D = -\boldsymbol{F}'_D$, $\boldsymbol{F}_E = -\boldsymbol{F}'_E$, 这些力均为系统的内力,可以消除,因此整个系统受力如图 1.38(e)所示.)

It should be pointed out that the difference between internal and external forces is not absolute. For example, when we take ladder AC as the research object, $\boldsymbol{F}_{Ax}$, $\boldsymbol{F}_{Ay}$ and $\boldsymbol{F}'_E$ all belong to external forces, but when taking the whole system as the research object, $\boldsymbol{F}_{Ax}$, $\boldsymbol{F}_{Ay}$ and $\boldsymbol{F}'_E$ also become internal forces. It can be seen that the difference between internal and external forces is only meaningful relative to a specific research object.(应该指出,内力与外力的区别不是绝对的. 例如,当把梯子 AC 作为研究对象时, $\boldsymbol{F}_{Ax}$、$\boldsymbol{F}_{Ay}$、$\boldsymbol{F}'_E$ 均属于外力,但取整体为研究对象时, $\boldsymbol{F}_{Ax}$、$\boldsymbol{F}_{Ay}$、$\boldsymbol{F}'_E$ 又成为内力. 可见,内力与外力的区别只有相对于某一确定的研究对象时才有意义.)

Example 5: As shown in Fig. 1.39(a), when pin C is attached to rod BC, try to draw the force diagram of each rod and the whole system. The weight of each component is not considered.(**例题 5**: 如图 1.39(a)所示,当销钉 C 附于 BC 杆时,试分别画出各杆及整体的受力图,各构件自重不计.)

Solution:(**解**:)

(1) Study the force analysis of AC rod first. Both ends of the rod A and C are constrained by hinges respectively, so rod AC is a two-force member. At point A and C, $\boldsymbol{F}_A$ and $\boldsymbol{F}_C$ is acted respectively and $\boldsymbol{F}_A = -\boldsymbol{F}_C$, and these two forces are equivalent, reverse and collinear, as shown in Fig. 1.39(b).(先研究杆 AC 的受力分析. 杆两端 A、C 分别受到铰链约束的作用,因此杆 AC 为二力构件. 在 A、C 处分别受 $\boldsymbol{F}_A$ 和 $\boldsymbol{F}_C$ 的作用,且 $\boldsymbol{F}_A = -\boldsymbol{F}_C$, 这两个力等值、反向、共线,

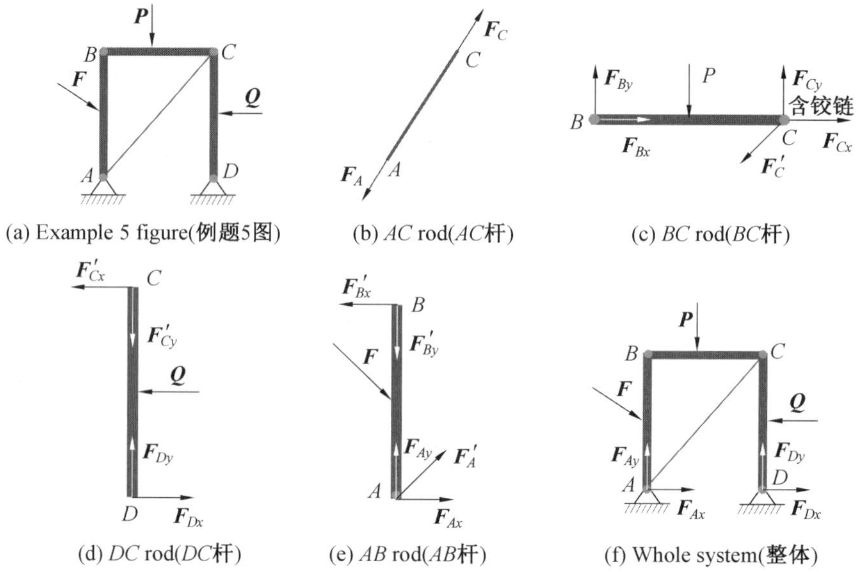

Fig. 1.39　Force analysis of Example 5(图 1.39　例题 5 受力分析)

如图 1.39(b)所示.)

(2) Force analysis of bar *BC*. *BC* is subject to load $\boldsymbol{P}$, and hinge *C* is subject to the constrained reaction force $\boldsymbol{F}'_C$ from rod *AC*. Since pin *C* is attached to *BC*, pin is subject to reaction force $\boldsymbol{F}_{Cx}$ and $\boldsymbol{F}_{Cy}$ from rod *CD*. At *B*, it is the hinge constraint, and $\boldsymbol{F}_{Bx}$ and $\boldsymbol{F}_{By}$ at *B* is subject to the constrained reaction force from bar *AB*. (杆 *BC* 的受力分析. *BC* 受到载荷 $\boldsymbol{P}$ 的作用,在铰链 *C* 处受到杆 *AC* 给它的约束力 $\boldsymbol{F}'_C$. 由于销钉 *C* 附于 *BC* 上,因此销钉受杆 *CD* 给它的约束力 $\boldsymbol{F}_{Cx}$ 和 $\boldsymbol{F}_{Cy}$. *B* 处为铰链约束,在 *B* 点受杆 *AB* 对它的约束力 $\boldsymbol{F}_{Bx}$ 和 $\boldsymbol{F}_{By}$.)

The force of rod *BC* is shown in Fig 1.39(c).(杆 *BC* 的受力如图 1.39(c)所示.)

(3) Force analysis of rod *CD*. *CD* is subjected to a load $\boldsymbol{Q}$, at hinge *C* rod *CD* is subjected to the constrained reaction force $\boldsymbol{F}'_{Cx}$ and $\boldsymbol{F}'_{Cy}$ from pin *C*, $\boldsymbol{F}'_{Cx}$ and $\boldsymbol{F}'_{Cy}$ are the reaction forces of $\boldsymbol{F}_{Cx}$ and $\boldsymbol{F}_{Cy}$, respectively. *D* is the fixed hinge constraint, whose binding force can be expressed by two orthogonal components of unknown magnitude $\boldsymbol{F}_{Dx}$ and $\boldsymbol{F}_{Dy}$. (杆 *CD* 的受力分析. *CD* 受载荷 $\boldsymbol{Q}$ 的作用,在铰链 *C* 处受销钉 *C* 对它的约束力 $\boldsymbol{F}'_{Cx}$ 和 $\boldsymbol{F}'_{Cy}$, $\boldsymbol{F}'_{Cx}$ 和 $\boldsymbol{F}'_{Cy}$ 分别是 $\boldsymbol{F}_{Cx}$ 和 $\boldsymbol{F}_{Cy}$ 的反作用力. *D* 处为固定铰链约束,其约束力可用两个大小未知的正交分力 $\boldsymbol{F}_{Dx}$ 和 $\boldsymbol{F}_{Dy}$ 表示.)

The force of rod *CD* is shown in Fig. 1.39(d).(杆 *CD* 的受力如图 1.39(d)所示.)

(4) Stress analysis of bar *AB*. *AB* is subjected to the load $\boldsymbol{F}$, and hinge *B* is subjected to the constrained reaction force $\boldsymbol{F}'_{Bx}$ and $\boldsymbol{F}'_{By}$ from *BC* bar, $\boldsymbol{F}'_{Bx}$ and $\boldsymbol{F}'_{By}$ are the reaction forces of $\boldsymbol{F}_{Bx}$ and $\boldsymbol{F}_{By}$, respectively. *A* is a fixed hinge constraint whose binding force can be expressed by two orthogonal components of unknown magnitude. *A* is also affected by the binding force $\boldsymbol{F}'_A$ from *AC* rod, which is the reaction force of $\boldsymbol{F}_A$. (杆 *AB* 的受力分析. *AB* 受载荷 *F* 的作用,在铰链 *B* 处受杆 *BC* 对它的约束力 $\boldsymbol{F}'_{Bx}$ 和 $\boldsymbol{F}'_{By}$, $\boldsymbol{F}'_{Bx}$ 和 $\boldsymbol{F}'_{By}$ 分别为 $\boldsymbol{F}_{Bx}$ 和 $\boldsymbol{F}_{By}$ 的反作用力. *A* 处为固定铰链约束,其约束力可用两个大小未知的正交分力 $\boldsymbol{F}_{Ax}$ 和 $\boldsymbol{F}_{Ay}$ 表示. *A* 处还受杆 *AC* 给它的约束力 $\boldsymbol{F}'_A$ 的作用, $\boldsymbol{F}'_A$ 是 $\boldsymbol{F}_A$ 的反作用力.)

The force of bar *AB* is shown in Fig. 1.39(e).(杆 *AB* 的受力如图 1.39(e)所示.)

(5) Force analysis of the whole system. $\boldsymbol{F}_A$, $\boldsymbol{F}'_A$, $\boldsymbol{F}_C$, $\boldsymbol{F}'_C$, $\boldsymbol{F}_{Bx}$, $\boldsymbol{F}'_{Bx}$, $\boldsymbol{F}_{By}$, $\boldsymbol{F}'_{By}$, $\boldsymbol{F}_{Cx}$, $\boldsymbol{F}'_{Cx}$, $\boldsymbol{F}_{Cy}$, $\boldsymbol{F}'_{Cy}$ are pairs of internal forces in the system that can cancel out each other. Therefore, the force of the whole system is shown in Fig. 1.39(f).(整个系统的受力分析. $\boldsymbol{F}_A$、$\boldsymbol{F}'_A$、$\boldsymbol{F}_C$、$\boldsymbol{F}'_C$、$\boldsymbol{F}_{Bx}$、$\boldsymbol{F}'_{Bx}$、$\boldsymbol{F}_{By}$、$\boldsymbol{F}'_{By}$、$\boldsymbol{F}_{Cx}$、$\boldsymbol{F}'_{Cx}$、$\boldsymbol{F}_{Cy}$、$\boldsymbol{F}'_{Cy}$ 均为成对出现的系统内力,可以互相抵消. 故整个系统受力如图 1.39(f)所示.)

1.3.3 Conclusion and discussion(结论与讨论)

(1) Statics studies the equilibrium conditions and applications of force systems acting on objects, which include the following points.(静力学研究作用于物体上力系的平衡条件及应用,包括以下几点.)

① Force analysis of the object.(物体的受力分析.)

② Equivalent replacement (or simplification) of force systems.(力系的等效替换(或简化).)

③ To establish the equilibrium conditions of various systems of forces.(建立各种力系的平衡条件.)

(2) Force is the mutual mechanical action between objects, which changes the state of mechanical motion of objects (including deformation).(力是物体间相互的机械作用,这种作用使物体的机械运动状态发生变化(包括变形).)

(3) Statics axioms are the most basic and universal objective laws of mechanics.(静力学公理是力学的最基本、最普遍的客观规律.)

Axiom 1: Parallelogram rule of force.(**公理** 1:力的平行四边形规则.)

Axiom 2: Two-force equilibrium condition.(**公理** 2:二力平衡条件.)

The above two axioms clarify the rules of composition and equilibrium conditions for the simplest system of forces acting on a body.(以上两个公理阐明了作用在一个物体上最简单的力系的合成规则及平衡条件.)

Axiom 3: Principle of adding or subtracting equilibrium systems of forces.(**公理** 3:加减平衡力系原理.)

This axiom is the basis for studying equivalent transformation of force system.(这个公理是研究力系等效变换的依据.)

Axiom 4: Law of action and reaction.(**公理** 4:作用与反作用定律.)

This axiom states the relationship between the interactions of two objects.(这个公理阐明了两个物体相互作用的关系.)

Axiom 5: Rigid principle.(**公理** 5:刚化原理.)

(4) Constraint and constraint reaction are as follows.(约束与约束反力如下.)

①Constraint. Objects around a non-free body that restrict certain displacements.(约束. 限制非自由体某些位移的周围物体.)

②Constraint reaction. The force exerted by a constraint on a non-free body.(约束反力. 约束对非自由体施加的力.)

(5) The force analysis and force diagram of the body are the premise of studying the balance and motion of the body.(物体的受力分析和受力图是研究物体平衡和运动的前提.)

1.4 Questions(思考题)

1. Explain the meaning and difference between the following expressions and words:(说明下列式子与文字的意义和区别:)

(1) $\boldsymbol{F}_1 = \boldsymbol{F}_2$;

(2) $F_1 = F_2$;

(3) Force F_1 is equivalent to force F_2.(力 F_1 等效于力 F_2.)

2. Try to distinguish between the two equations $\boldsymbol{F}_R = \boldsymbol{F}_1 + \boldsymbol{F}_2$ and $F_R = F_1 + F_2$.(试区别 $\boldsymbol{F}_R = \boldsymbol{F}_1 + \boldsymbol{F}_2$ 和 $F_R = F_1 + F_2$ 两个等式代表的意义.)

3. What is a two-force member? Is the two-force member related to the shape of the member? Is a bar that is hinged at both ends a two-force bar? Is a rigid bar regardless of its own weight a two-force bar? (什么是二力构件?二力构件与构件的形状有关吗?凡两端用铰链连接的杆都是二力杆吗?凡不计自重的刚性杆都是二力杆吗?)

4. Are there any errors in the force diagram of the objects in the Fig. 1.40–1.44? How to correct it? (图 1.40 ~ 1.44 中各物体的受力图是否有错误,如何改正?)

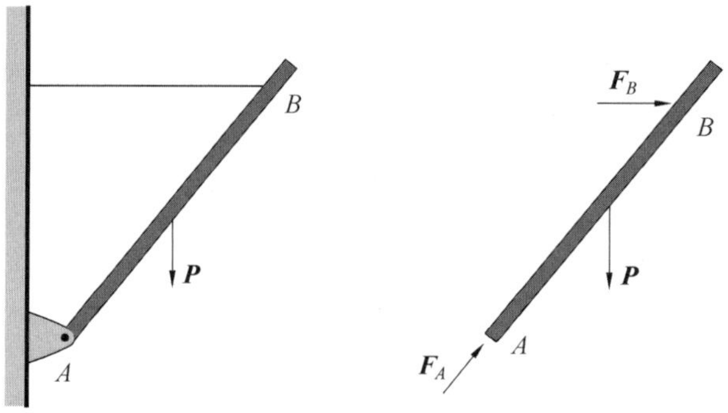

(a) Original Figure(原图)　　(b) Free-body diagram(受力图)

Fig. 1.40　Diagram of Question 4 (1)(图 1.40　思考题 4 图(1))

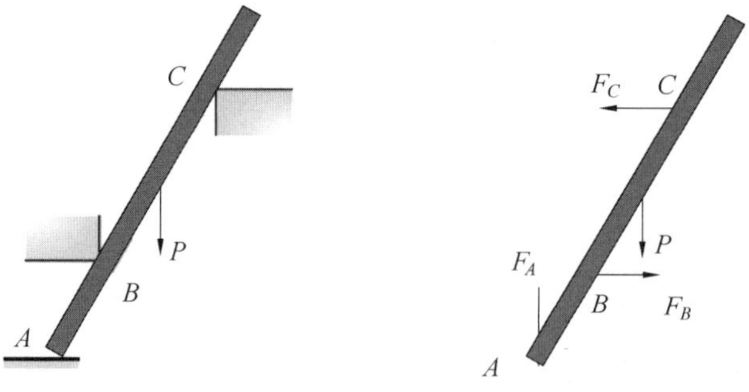

(a) Original Figure(原图)　　(b) Free-body diagram(受力图)

Fig. 1.41　Diagram of Question 4 (2)(图 1.41　思考题 4 图(2))

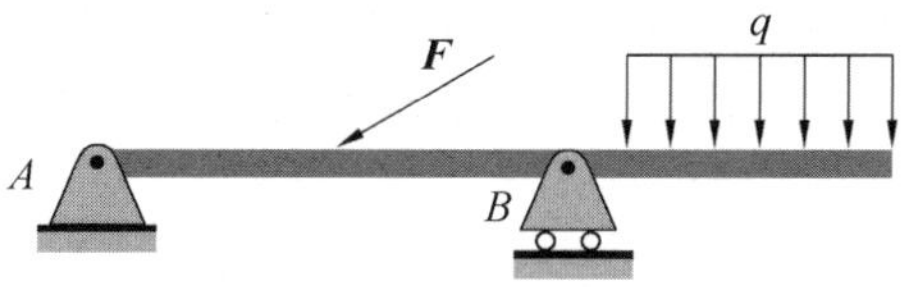

(a) Original Figure(原图)

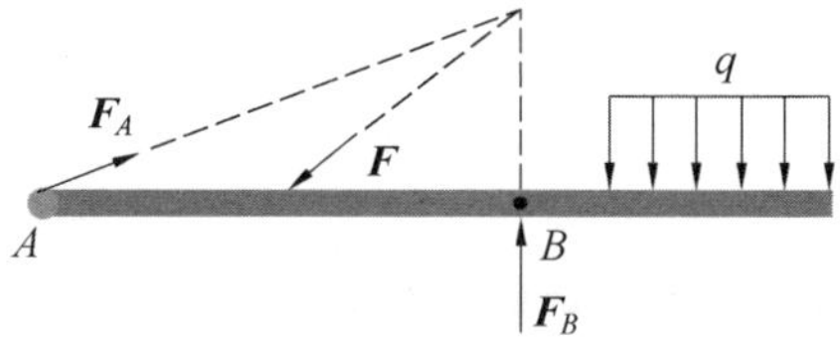

(b) Free-body diagram(受力图)

Fig. 1.42 Diagram of Question 4 (3)(图 1.42 思考题 4 图(3))

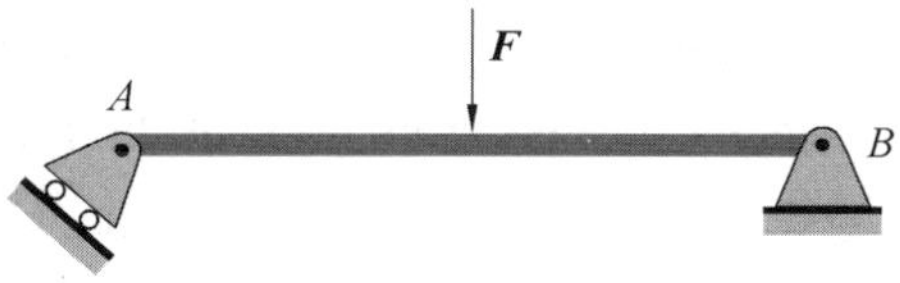

(a) Original Figure(原图)

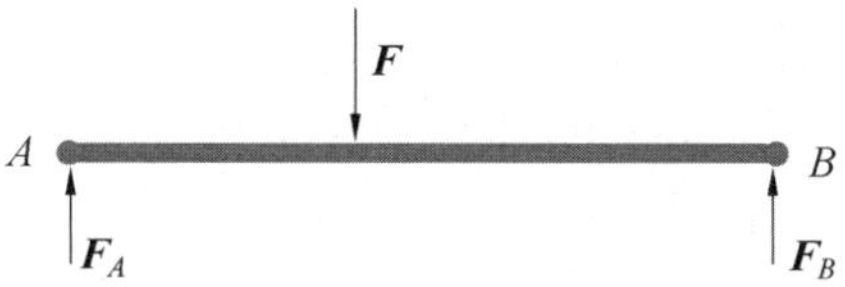

(b) Free-body diagram(受力图)

Fig. 1.43 Diagram of Question 4 (4)(图 1.43 思考题 4 图(4))

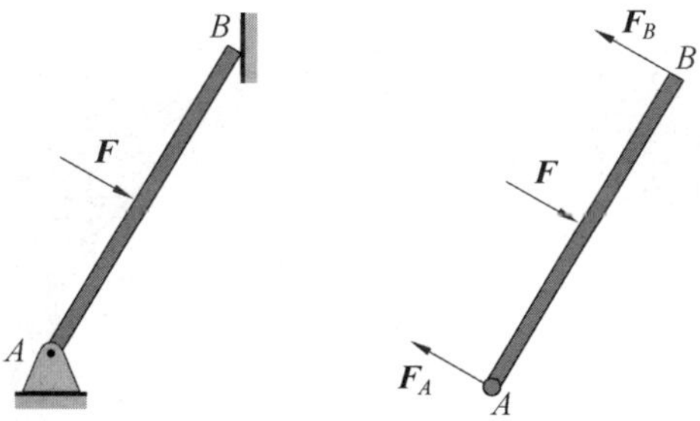

(a) Original Figure(原图) (b) Free-body diagram(受力图)

Fig. 1.44 Diagram of Question 4 (5)(图 1.44 思考题 4 图(5))

5. The force $\boldsymbol{F}$ is exerted on point A of the rigid body, as shown in Fig. 1.45. Can a force applied to point B make the the rigid body balance? Why? (刚体上 A 点受力 $\boldsymbol{F}$ 作用,如图 1.45 所示,能否在 B 点加一个力使刚体平衡,为什么?)

6. As shown in Fig. 1.46, if the force $\boldsymbol{F}$ is applied to point B, can the system be balanced? If the force $\boldsymbol{F}$ is still acting at point B, but you can change the direction of the force $\boldsymbol{F}$ at will, in

what direction can the structure be in equilibrium? (如图 1.46 所示,若力 $\boldsymbol{F}$ 作用在 B 点,系统能否平衡? 若力 $\boldsymbol{F}$ 仍作用在 B 点,但可以任意改变力 $\boldsymbol{F}$ 的方向,$\boldsymbol{F}$ 作用在什么方向上结构能平衡?)

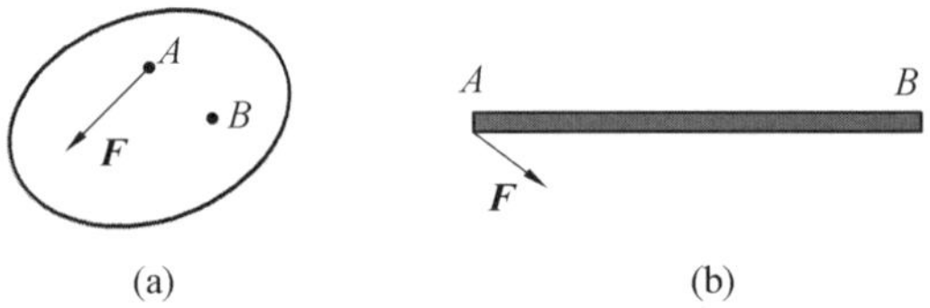

Fig. 1.45　Diagram of Question 5
(图 1.45　思考题 5 图)

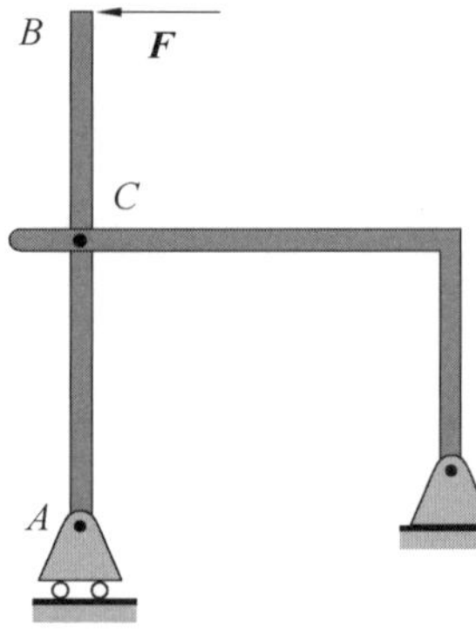

Fig. 1.46　Diagram of Question 6
(图 1.46　思考题 6 图)

7. In Fig. 1.47, the force acts on the pin at hinge C of the three-hinged arch, regardless of the weight of all objects. (图 1.47 中力 $\boldsymbol{F}$ 作用于三铰拱的铰链 C 处的销钉上,所有物体质量不计.)

(1) Try to draw the force diagram of the left and right arch and pin C respectively; (试分别画出左、右两拱及销钉 C 的受力图;)

(2) If pin C belongs to AC, draw the force diagram of the left and right arches respectively; (若销钉 C 属于 AC,分别画出左、右两拱的受力图;)

(3) If pin C belongs to BC, draw the force diagram of the left and right arches respectively. (若销钉 C 属于 BC,分别画出左、右两拱的受力图.)

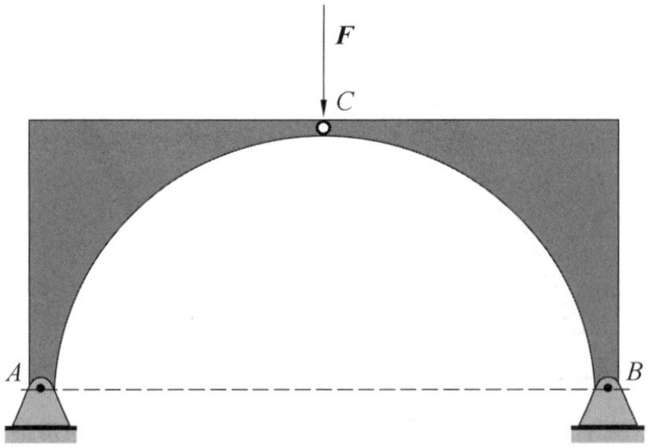

Fig. 1.47　Diagram of Question 7(图 1.47　思考题 7 图)

1.5　Exercises(习题)

1. Draw the force diagram of each stressed member in Fig. 1.48. The body weight of components without gravity is not considered, and all contacts are smooth. (画出图 1.48 中各受

力构件的受力图.未画重力的构件自重不计,所有接触均为光滑接触.)

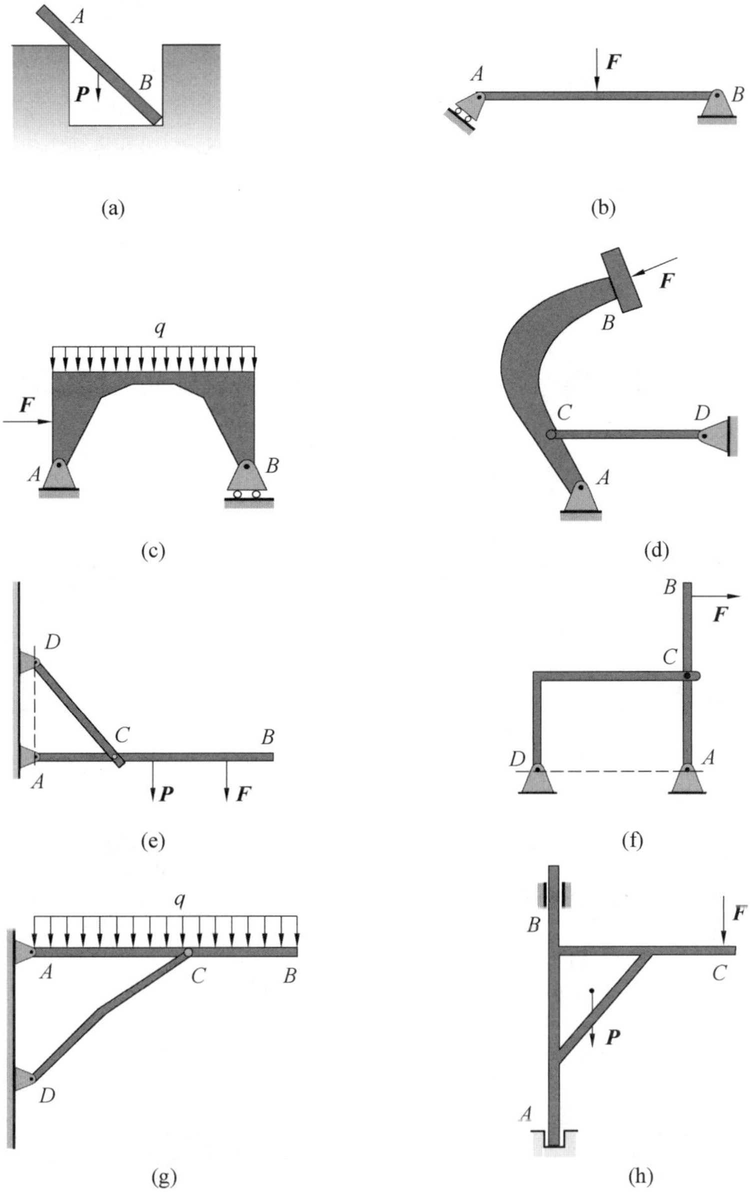

Fig. 1.48 Diagram of Exercise 1(图 1.48 习题 1 图)

2. Draw the force diagram of each object marked with the following characters (excluding pins, supports and foundations) and the overall force diagram of the system in Fig. 1.49. The body weight of components without gravity is not considered, and all contacts are smooth.(画出图 1.49 中每个标注字符的物体(不包含销钉、支座和基础)的受力图和系统整体受力图.未画重力的构件自重不计,所有接触均为光滑接触.)

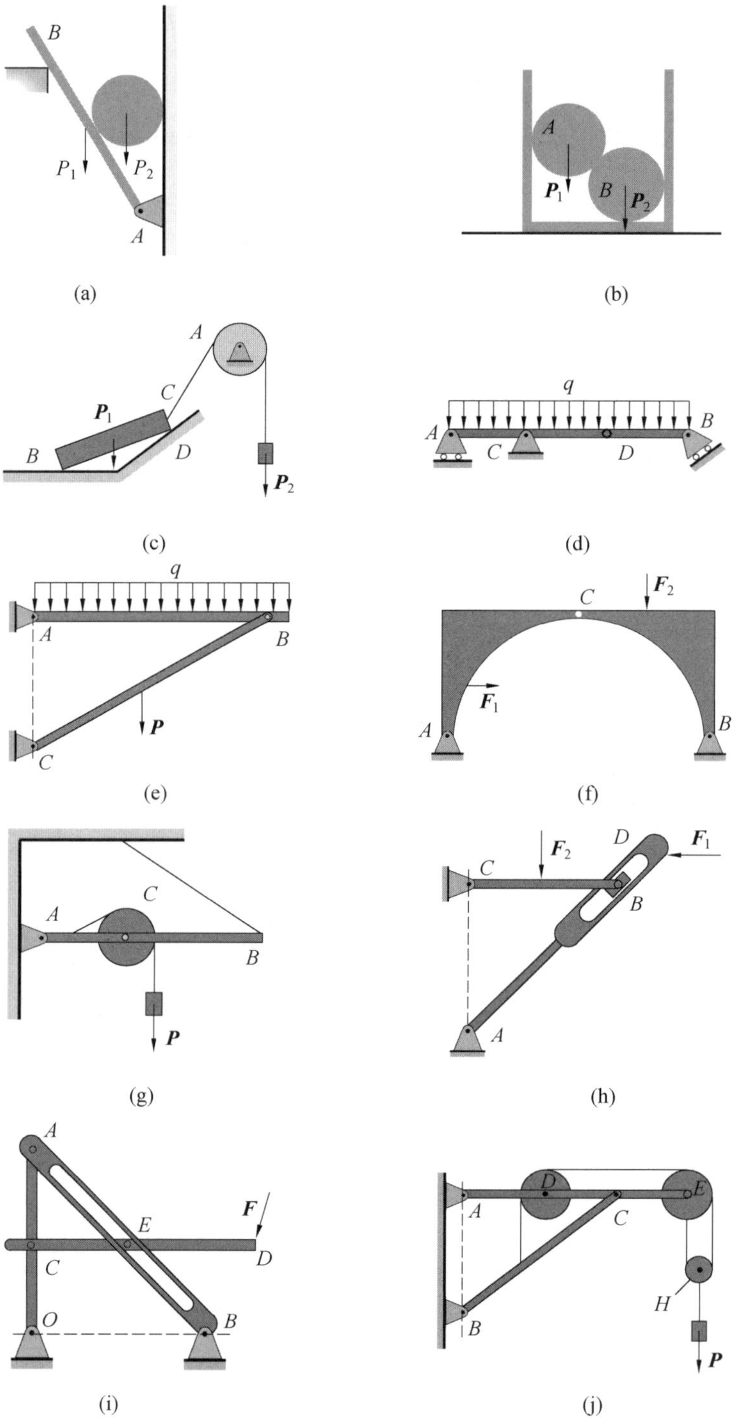

Fig. 1.49 Diagram of Exercise 2(图 1.49 习题 2 图)

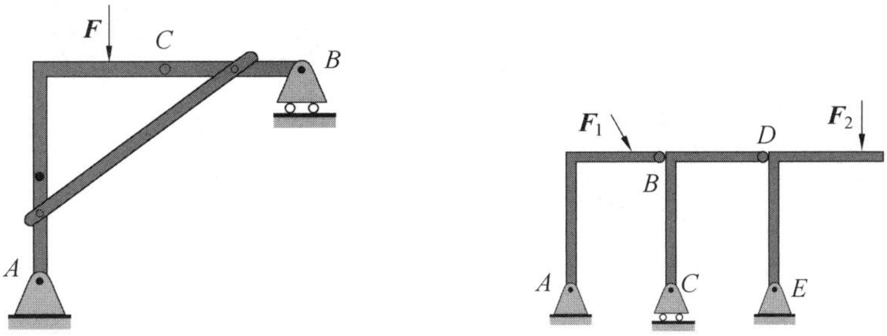

(k) (l)

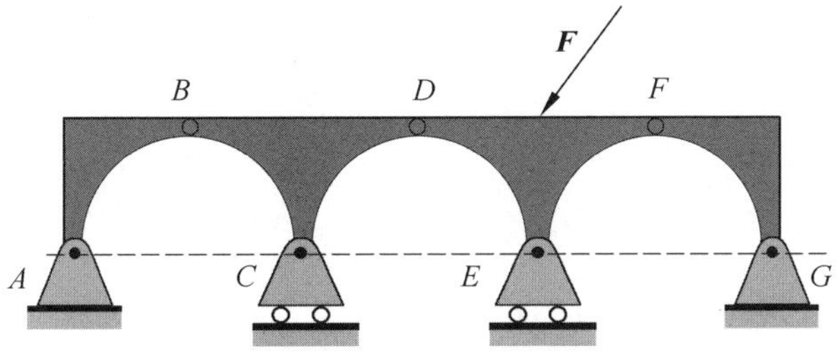

(m)

Fig. 1.49(continued)(图 1.49(续))

3. Draw the force diagram of each of the following marked characters (excluding pins, supports and foundations), the overall stress diagram of each problem and the force diagram of pin A (pin A penetrating each component) in Fig. 1.50. The body weight of components without gravity is not considered, and all contacts are smooth. (画出图 1.50 中每个标注字符的物体(不包含销钉、支座和基础)的受力图,以及各题的整体受力图和销钉 A(销钉 A 穿透各构件)的受力图. 未画重力的构件自重不计,所有接触均为光滑接触.)

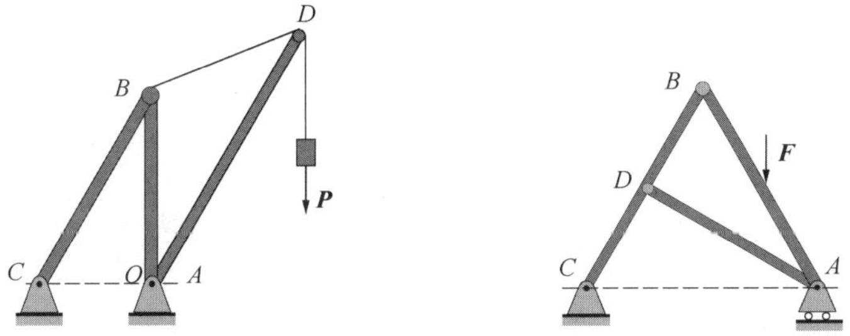

(a) (b)

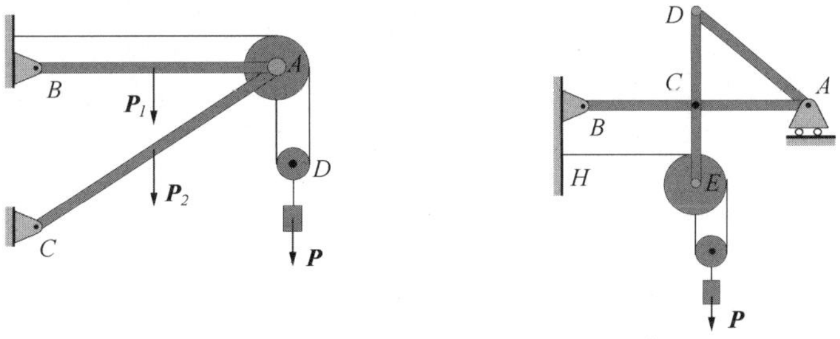

(c) (d)

Fig. 1.50 Diagram of Exercise 3(图 1.50 习题 3 图)

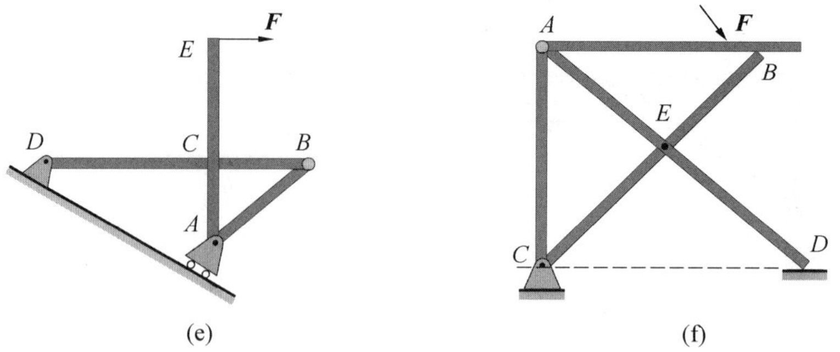

Fig. 1.50(continued)(图 1.50(续))

Chapter 2　Planar Force Systems
(平面力系)

A force system in which the action lines of the forces are in the same plane is called a plane force system. Planar force systems include coplanar concurrent forces system, planar force couple systems and planar arbitrary force systems.（各力作用线在同一平面内的力系称为平面力系. 平面力系包括平面汇交力系、平面力偶系和平面任意力系.）

2.1　Synthesis and Decomposition of Forces(力的合成与分解)

2.1.1　Synthesis of forces(力的合成)

1. Two-force synthesis(二力合成)

According to the parallelogram rule for forces (Fig. 2.1) and the triangle rule (Fig. 2.2), the two forces $\boldsymbol{F}_1$ and $\boldsymbol{F}_2$ are combined to obtain the combined force $\boldsymbol{F}_R$.（根据力的平行四边形法则(图 2.1)和三角形法则(图 2.2),两个力 $\boldsymbol{F}_1$ 和 $\boldsymbol{F}_2$ 合成得到合力 $\boldsymbol{F}_R$.）

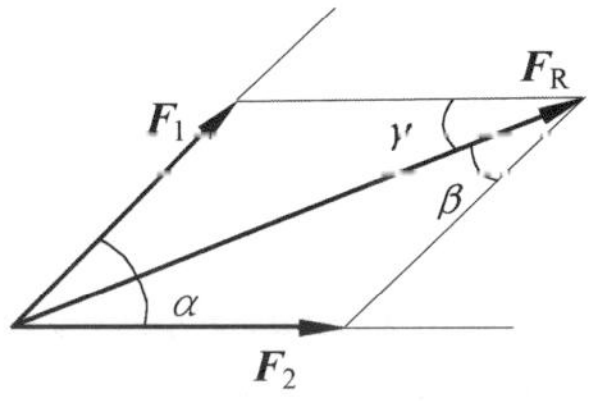

Fig. 2.1　The parallelogram rule of force
（图 2.1　力的平行四边形法则）

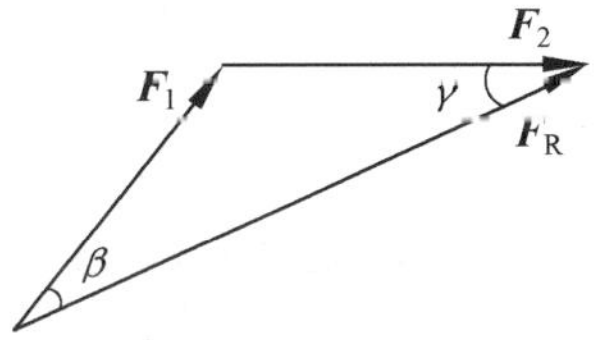

Fig. 2.2　The triangle rule of force
（图 2.2　力的三角形法则）

The formula for the magnitude of the force is expressed by(力的大小用公式表示为)

$$|\boldsymbol{F}_R| = \sqrt{F_1^2 + F_2^2 + 2F_1F_2\cos\alpha}$$

The angle of clamping is expressed as(夹角表示为)

$$F_1/\sin\gamma = F_2/\sin\beta = F_R/\sin\alpha$$

2. Synthesis of three forces that are not in the same plane
(不在同一平面内的三力合成)

The three forces $\boldsymbol{F}_1$, $\boldsymbol{F}_2$, $\boldsymbol{F}_3$ do not lie in the same plane, and the combined force $\boldsymbol{F}_R$ is determined by successive application of the parallelepiped rule (Fig. 2.3) and is represented by the diagonal of the parallelepiped constructed by these forces (according to the parallelogram rule).（三个作用力 $\boldsymbol{F}_1$、$\boldsymbol{F}_2$、$\boldsymbol{F}_3$ 不在同一平面上,合力 $\boldsymbol{F}_R$ 通过连续运用平行四边形法则来

确定(图 2.3),并由这些力(按平行六面体规则)构造的平行六面体的对角线表示.)

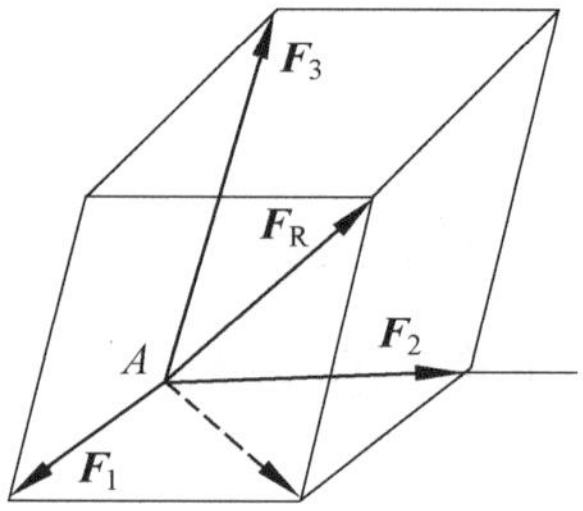

Fig. 2.3 Parallel hexahedron rule(图 2.3 平行六面体法则)

2.1.2 Decomposition of forces(力的分解)

1. Decomposition of a force in two specified directions(力在两个指定方向上的分解)

To decompose the force $\boldsymbol{F}_R$ in the two given directions AB and AD, construct a parallelogram (Fig. 2.4) with the force $\boldsymbol{F}_R$ as a diagonal and the direction of its division parallel to the given direction. The forces $\boldsymbol{F}_1$ and $\boldsymbol{F}_2$ are obtained by decomposing along the two sides of the parallelogram. (若要在两个给定方向 AB 和 AD 上分解力 $\boldsymbol{F}_R$，需构造平行四边形(图 2.4),力 $\boldsymbol{F}_R$ 为对角线,其分力方向与给定方向平行. 沿着平行四边形的两边分解得到力 $\boldsymbol{F}_1$ 和 $\boldsymbol{F}_2$.)

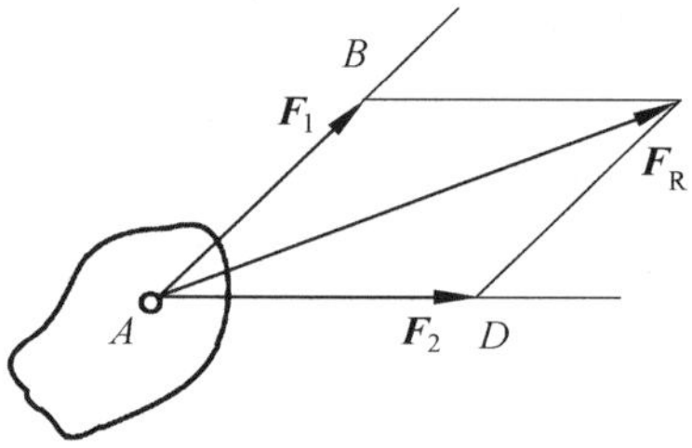

Fig. 2.4 Decomposition of forces along two directions(图 2.4 力沿两个方向的分解)

2. Decomposition of forces in three specified directions(力在三个指定方向上的分解)

To decompose the force $\boldsymbol{F}_R$ along three directions that are not in the same plane, a parallel hexahedron is constructed, where the diagonals represent the given force. The component forces $\boldsymbol{F}_1$, $\boldsymbol{F}_2$, $\boldsymbol{F}_3$ will then be represented along the edges of this parallel hexahedron (Fig. 2.5). (若要将力 $\boldsymbol{F}_R$ 沿不在同一平面上的三个方向进行分解,需要构造一个平行六面体,其中对角线表示给定的力. 然后,分力 $\boldsymbol{F}_1$、$\boldsymbol{F}_2$、$\boldsymbol{F}_3$ 将沿着这个平行六面体的棱边表示(图 2.5).)

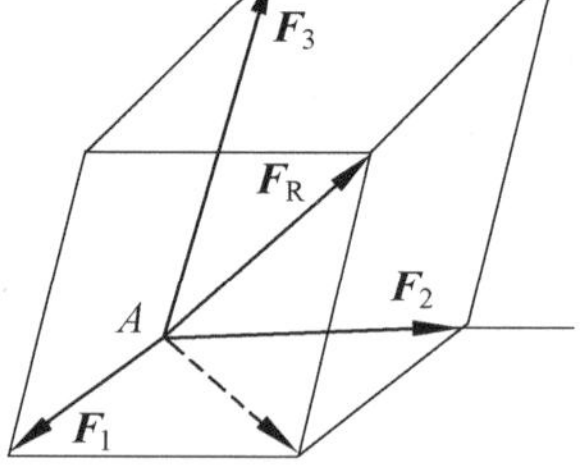

Fig. 2.5 Decomposition of forces in three directions in space(图 2.5 力在空间三个方向的分解)

2.1.3 Projection of forces on axes and planes(力在轴和平面上的投影)

1. Projection of forces on an axis(力在轴上的投影)

The projection of a force (or any other vector quantity) onto an axis is an algebraic quantity equal to the product of the cosine of the angle between the force and the positive direction of the axis and the force (Fig. 2.6), we have(力(或任何其他矢量)在轴上的投影是一个代数量,它等于力与轴的正方向之间夹角的余弦与力的乘积(图 2.6),有)

$$\begin{cases} F_x = F\cos\alpha = ab \\ Q_x = Q\cos\alpha_1 = -Q\cos\beta = -dc \\ P_x = 0 \end{cases}$$

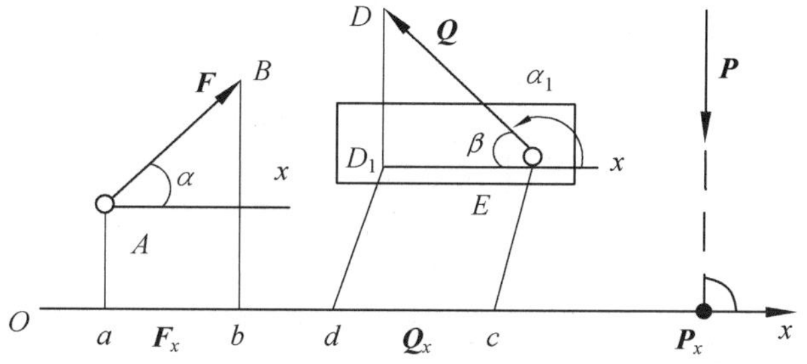

Fig. 2.6 Direct projection(图 2.6 直接投影)

2. Projection of forces on a plane (secondary projection)(力在平面上的投影(二次投影))

The projection of the force $\boldsymbol{F}$ on the Oxy plane $\boldsymbol{F}_{xy}$ is a vector, $\boldsymbol{F}_{xy} = \overrightarrow{OB_1}$, i. e. the length of the directed line segment between the starting projection and the end projection of the force $\boldsymbol{F}$ on that plane (Fig. 2.7), we have(力 $\boldsymbol{F}$ 在 Oxy 平面上的投影 $\boldsymbol{F}_{xy}$ 为矢量, $\boldsymbol{F}_{xy} = \overrightarrow{OB_1}$,即力 $\boldsymbol{F}$ 在该平面上起点投影到终点投影之间的有向线段长度(图 2.7),有)

$$|\boldsymbol{F}_{xy}| = F\cos\theta$$

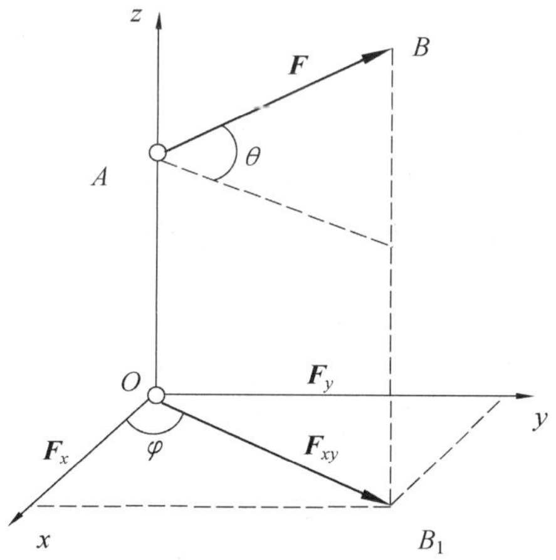

Fig. 2.7 Secondary projection(图 2.7 二次投影)

In some cases, in order to facilitate finding the projection of a force on an axis, it is possible to first find its projection on a plane and then project the projection of the force on the plane back onto that axis (Fig. 2.7), we have(在某些情况下,为便于找到力在轴上的投影,可以首先找到它在平面上的投影,然后将力在平面上的投影再投影到该轴上(图 2.7),有)

$$\begin{cases} F_x = F_{xy}\cos\varphi = F\cos\theta\cos\varphi \\ F_y = F_{xy}\sin\varphi = F\cos\theta\sin\varphi \end{cases}$$

Sometimes, it is more convenient to use the direct projection method in order to find the projection of the force on the axis (Fig. 2.8), we have (有时,为找到力在轴上的投影,使用直接投影法更方便(图 2.8),有)

$$\begin{cases} F_x = F\cos\alpha \\ F_y = F\cos\beta \\ F_z = F\cos\gamma \end{cases}$$

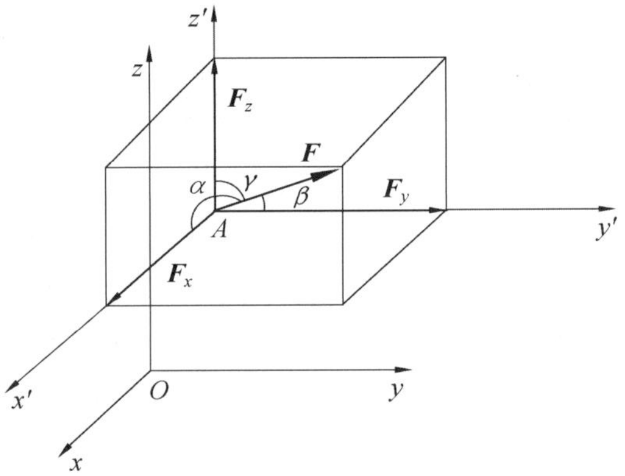

Fig. 2.8 Direct projection method(图 2.8 直接投影法)

2.1.4 Analytical method(解析法)

In order to determine the projection of the force by the analytical method, it is sufficient to project it onto the axes of the coordinate system. (为用解析法确定力的投影,只需将其投影到坐标系轴上.)

1. Spatial coordinate system(空间坐标系)

If the projection F_x, F_y, F_z is given, then the magnitude of the force is determined by (Fig. 2.8)(如果给出投影 F_x、F_y、F_z,那么力的大小为(图 2.8))

$$|\boldsymbol{F}| = \sqrt{F_x^{\ 2} + F_y^{\ 2} + F_z^{\ 2}}$$

The direction of the force is determined with the help of a directional cosine angle calculation (the angle between the force and the coordinate axis), that is(力的方向是借助方向余弦角度计算(力与坐标轴的夹角)来确定的,即)

$$\cos\alpha = \boldsymbol{F}_x/\boldsymbol{F},\ \cos\beta = \boldsymbol{F}_y/\boldsymbol{F},\ \cos\gamma = \boldsymbol{F}_z/\boldsymbol{F}$$

$$\alpha = \arccos(\boldsymbol{F}_x/\boldsymbol{F}),\ \beta = \arccos(\boldsymbol{F}_y/\boldsymbol{F}),\ \gamma = \arccos(\boldsymbol{F}_z/\boldsymbol{F})$$

2. Plane coordinate system(平面坐标系)

If the projections $\boldsymbol{F}_x$ and $\boldsymbol{F}_y$ are given, the magnitude and direction of the force is determined by(Fig. 2.9)(如果给出投影 F_x 和 F_y,则力的大小和方向为(图2.9))

$$|\boldsymbol{F}| = \sqrt{F_x^2 + F_y^2},\ \cos\alpha = \boldsymbol{F}_x/\boldsymbol{F},\ \cos\beta = \boldsymbol{F}_y/\boldsymbol{F}$$

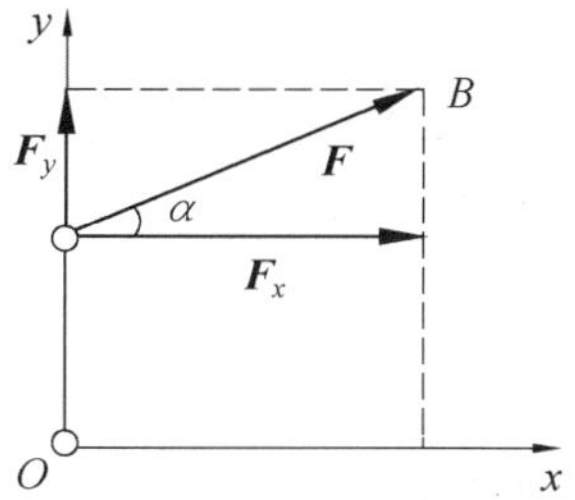

Fig. 2.9 Analytical method in a plane coordinate system(图 2.9 平面坐标系下的解析法)

2.1.5 Analytical method of force synthesis(力合成的解析法)

The analytical method of force synthesis is based on the theorem of the projection of the unity vector: the projection of the unity vector on any axis is equal to the algebraic sum of the projections of the component vectors on the same axis. (力合成的解析法基于合矢量投影定理:合矢量在任意轴上的投影等于各分矢量在同一轴上投影的代数和.)

The projection theorem is often applied to the calculation of combined force vectors for the synthesis of spatial or planar force systems. (投影定理常应用于空间或平面力系合成的合力矢量的计算.)

1. Synthesis of space force systems(空间力系的合成)

The resolution of the force $\boldsymbol{F}_1, \boldsymbol{F}_2, \ldots, \boldsymbol{F}_n$ gives the projection of each force onto the known axes as(对力 $\boldsymbol{F}_1, \boldsymbol{F}_2, \cdots, \boldsymbol{F}_n$ 进行解析,则可得到各力在已知坐标轴上的投影,即)

$$\boldsymbol{F}_{1x}, \boldsymbol{F}_{2x}, \cdots, \boldsymbol{F}_{nx}; \boldsymbol{F}_{1y}, \boldsymbol{F}_{2y}, \cdots, \boldsymbol{F}_{ny}; \boldsymbol{F}_{1z}, \boldsymbol{F}_{2z}, \cdots, \boldsymbol{F}_{nz}$$

By the projection theorem, if $\boldsymbol{F}_R$ is the vector sum of $\boldsymbol{F}_1, \boldsymbol{F}_2, \ldots, \boldsymbol{F}_n$, which is $\boldsymbol{F}_R = \sum \boldsymbol{F}_k$, then(根据投影定理,如果 $\boldsymbol{F}_R$ 是 $\boldsymbol{F}_1, \boldsymbol{F}_2, \cdots, \boldsymbol{F}_n$ 的矢量和,也就是 $\boldsymbol{F}_R = \sum \boldsymbol{F}_k$,则)

$$F_{Rx} = \sum F_{kx}, F_{Ry} = \sum F_{ky}, F_{Rz} = \sum F_{kz}$$

Calculating $\boldsymbol{F}_{Rx}, \boldsymbol{F}_{Ry}, \boldsymbol{F}_{Rz}$, we can obtain the magnitude of the combined force $\boldsymbol{F}_R$ (the combined vector), that is (计算 F_{Rx}、F_{Ry}、F_{Rz},可以得到合力 $\boldsymbol{F}_R$ (合矢量)的大小,即)

$$|\boldsymbol{F}_R| = \sqrt{F_{Rx}^2 + F_{Ry}^2 + F_{Rz}^2}$$

The direction of the combined force $\boldsymbol{F}_R$ can be calculated from the cosine of the angle α, β, γ (the angle between the combined force and the coordinate axis), that is(合力 $\boldsymbol{F}_R$ 的方向可以通过角 α、β、γ (合力与坐标轴的夹角)的余弦计算得出,即)

$$\cos\alpha = \boldsymbol{F}_{Rx}/\boldsymbol{F}_R,\ \cos\beta = \boldsymbol{F}_{Ry}/\boldsymbol{F}_R,\ \cos\gamma = \boldsymbol{F}_{Rz}/\boldsymbol{F}_R$$

$$\alpha = \arccos(\boldsymbol{F}_{Rx}/\boldsymbol{F}_R),\ \beta = \arccos(\boldsymbol{F}_{Ry}/\boldsymbol{F}_R),\ \gamma = \arccos(\boldsymbol{F}_{Rz}/\boldsymbol{F}_R)$$

2. Synthesis of planar force systems(平面力系的合成)

For forces in the same plane, $\boldsymbol{F}_1$, $\boldsymbol{F}_2$, ..., $\boldsymbol{F}_n$ are analyzed, i. e. the projection of the force on the given coordinate axis is $\boldsymbol{F}_{1x}$, $\boldsymbol{F}_{2x}$, ..., $\boldsymbol{F}_{nx}$ and $\boldsymbol{F}_{1y}$, $\boldsymbol{F}_{2y}$, ..., $\boldsymbol{F}_{ny}$. (对在同一平面上

的力 $\boldsymbol{F}_1$, $\boldsymbol{F}_2$,⋯, $\boldsymbol{F}_n$ 进行解析,即力在给定坐标轴上的投影为 $\boldsymbol{F}_{1x}$,$\boldsymbol{F}_{2x}$,⋯,$\boldsymbol{F}_{nx}$ 和 $\boldsymbol{F}_{1y}$, $\boldsymbol{F}_{2y}$,⋯,$\boldsymbol{F}_{ny}$.)

According to the projection theorem, we have(根据投影定理,有)

$$\boldsymbol{F}_{Rx} = \sum \boldsymbol{F}_{kx},\ \boldsymbol{F}_{Ry} = \sum \boldsymbol{F}_{ky}$$

The magnitude and direction of the force $\boldsymbol{F}_R$ are(力 $\boldsymbol{F}_R$ 的大小和方向为)

$$|\boldsymbol{F}_R| = \sqrt{F_{Rx}^2 + F_{Ry}^2},\ \cos\alpha = \boldsymbol{F}_{Rx}/\boldsymbol{F}_R,\ \alpha = \arccos(\boldsymbol{F}_{Rx}/\boldsymbol{F}_R)$$

2.2 Coplanar Concurrent Forces System(平面汇交力系)

A force system in which the lines of action of forces intersect at a single point is called a coplanar concurrent forces system (Fig. 2. 10). There are two methods of studying coplanar concurrent force system: the geometric method (Fig. 2. 10(b)) and the analytical method (Fig. 2. 10(c)).(力的作用线相交于一点的力系称为平面汇交力系(图 2. 10). 研究平面力系有两种方法:几何法(图 2. 10(b))和解析法(图 2. 10(c)).)

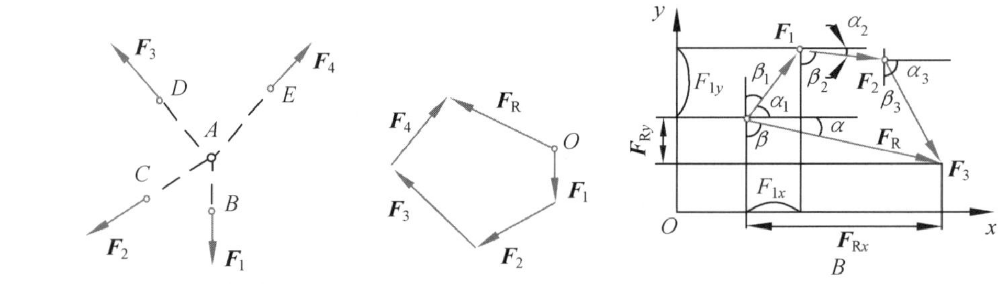

(a) Coplanar concurrent forces system (平面汇交力系) (b) Geometric methods (几何法) (c) Analytical methods (解析法)

Fig. 2. 10 Coplanar concurrent forces system and its methods(图 2. 10 平面汇交力系及其研究方法)

2.2.1 Geometric method for the synthesis of concurrent forces (汇交力合成的几何法)

The essence of force system synthesis is to obtain its principal vector. The principal vector of any force system is the vector sum of all the forces acting in the force system, we have (力系合成的实质在于获得它的主矢量. 任何力系的主矢量是力系中所有作用力的矢量和,有)

$$\boldsymbol{F}_R = \boldsymbol{F}_1 + \boldsymbol{F}_2 + \cdots + \boldsymbol{F}_n = \sum \boldsymbol{F}_k \tag{2.1}$$

There are two ways of finding the master vector.(求主矢量有两种方法.)

(1) The parallelogram rule or the triangle rule is used continuously(Fig. 2. 11).(连续利用平行四边形法则或三角形法则(图 2. 11).)

(2) Constructing a force polygon by translating the force (Fig. 2. 12). In this case, each subsequent force is added at the end of the previous force. The closed edge of the polygon is the principal vector $\boldsymbol{F}_R$.(通过力的平移构造力的多边形(图 2. 12). 在这种情况下,每个后续的

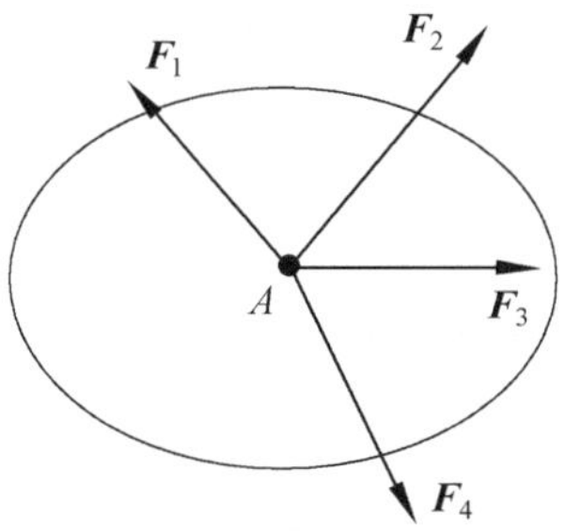

(a) Four planar converging forces
(四个平面汇交力)

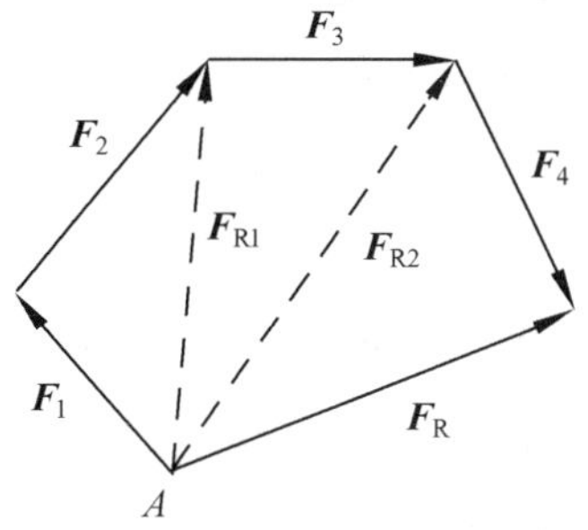

(b) Multiple successive times using the triangle rule
(连续多次利用三角形法则)

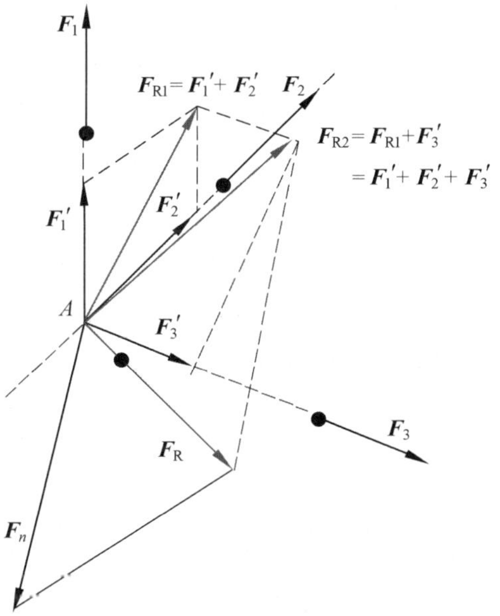

(c) Using the quadrilateral rule several times in succession
(连续多次利用四边形法则)

Fig. 2. 11 Geometric method (图 2. 11 几何法)

力都在前一个力的末尾处添加. 多边形的封闭边是主矢量 $\boldsymbol{F}_R$.)

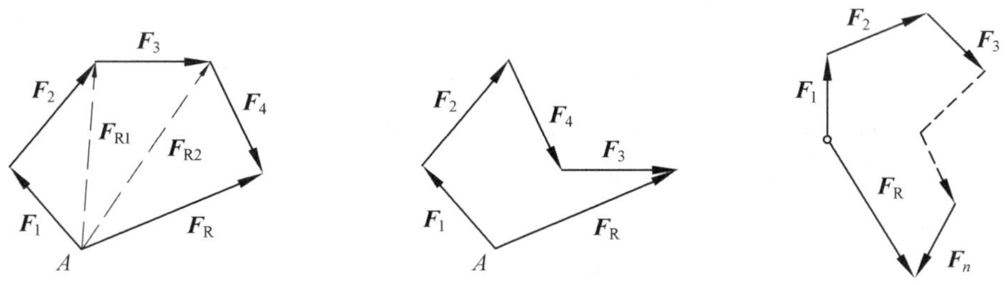

Fig. 2. 12 Polygon law of forces(图 2. 12 力的多边形法则)

Conclusion: The system of coplanar concurrent forces can be reduced to a combined force, the magnitude and direction of which are equal to the vector sum of the component forces, the line of action of which passes through the point of confluence and whose magnitude is expressed by the closed edge of the force polygon, that is (**结论**:平面汇交力系可简化为一合力,其合力的大小与方向等于各分力的矢量和,合力的作用线通过汇交点,其大小用力多边形的封闭边表示,即)

$$\boldsymbol{F}_{\mathrm{R}}=\boldsymbol{F}_1+\boldsymbol{F}_2+\cdots+\boldsymbol{F}_n=\sum \boldsymbol{F}_i$$

2.2.2 Geometric method for the equilibrium of concurrent forces systems (汇交力系平衡的几何法)

The concurrent forces system $\boldsymbol{F}_1, \boldsymbol{F}_2, \ldots, \boldsymbol{F}_n$ is equivalent to a combined force $\boldsymbol{F}_{\mathrm{R}}$ and is called equivalent force $\boldsymbol{F}_{\mathrm{R}}$. (汇交力系 $\boldsymbol{F}_1, \boldsymbol{F}_2, \cdots, \boldsymbol{F}_n$ 相当于一个合力 $\boldsymbol{F}_{\mathrm{R}}$,称为等效力 $\boldsymbol{F}_{\mathrm{R}}$.)

The equivalent force $\boldsymbol{F}_{\mathrm{R}}$ (and the principal vector) is equal to the geometric sum of all forces, that is $\boldsymbol{F}_{\mathrm{R}}=\sum \boldsymbol{F}_k$, which means for a system of confluent forces (only for this system), the resultant is equivalent to the principal vector. Thus, the equilibrium condition is(等效力 $\boldsymbol{F}_{\mathrm{R}}$(及主矢量)等于所有力的几何求和,即 $\boldsymbol{F}_{\mathrm{R}}=\sum \boldsymbol{F}_k$. 也就是说,对于一个汇交力系(仅限于此力系),其合力等同于主矢量. 因此,平衡条件为)

$$\boldsymbol{F}_{\mathrm{R}}=\sum \boldsymbol{F}_k=0 \tag{2.2}$$

Since $\boldsymbol{F}_{\mathrm{R}}=0$, the polygon $\boldsymbol{F}_1, \boldsymbol{F}_2, \ldots, \boldsymbol{F}_n$ is closed (Fig. 2. 13), i. e. the end point of the last force $\boldsymbol{F}_n$ coincides with the start point of the first force $\boldsymbol{F}_1$. These conditions (the polygon of forces is closed means that the combined forces of the force systems are equal to zero) are the geometrical conditions for the equilibrium of the concurrent forces systems. (因为 $\boldsymbol{F}_{\mathrm{R}}=0$,所以多边形 $\boldsymbol{F}_1, \boldsymbol{F}_2, \cdots, \boldsymbol{F}_n$ 封闭(图 2. 13),即最后一个力 $\boldsymbol{F}_n$ 的终点与第一个力 $\boldsymbol{F}_1$ 的起点重合. 这些条件(力的多边形封闭,即力系的合力等于零)是汇交力系平衡的几何条件.)

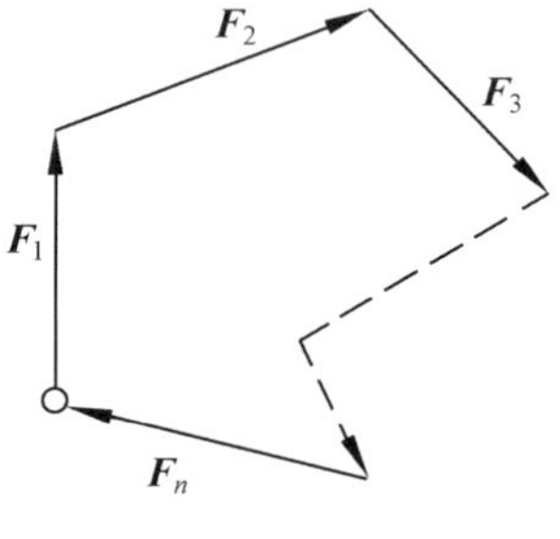

Fig. 2. 13 Force polygon closed by itself(图 2. 13 力多边形自行封闭)

Therefore, a necessary and sufficient condition for the equilibrium of a coplanar concurrent forces system is that the combined force of the system is equal to zero, we have (因此,平面汇交力系平衡的必要和充分条件是该力系的合力等于零,有)

$$\boldsymbol{F}_{\mathrm{R}}=\sum \boldsymbol{F}_k=0$$

Conclusion: A necessary and sufficient condition for the geometric method of equilibrium of a coplanar concurrent forces system is that the force polygon of that force system closes itself. (**结论**:平面汇交力系几何法平衡的必要和充分条件是该力系的力多边形自行封闭.)

Note the difference between Fig. 2. 14(a) and Fig. 2. 14(b). (请注意图 2. 14(a)与

图 2.14(b)的区别.)

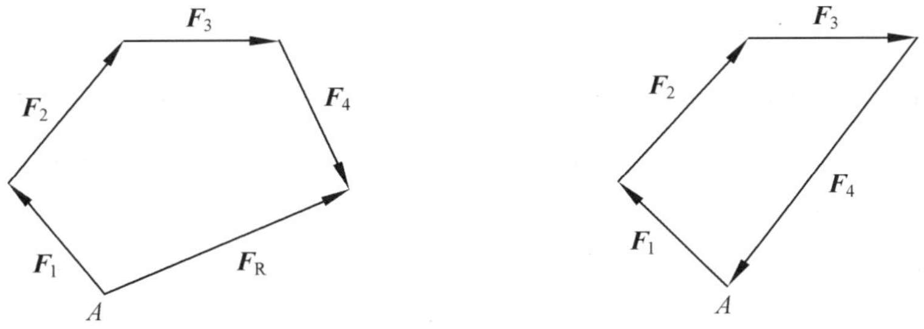

(a) Resultant force F_R(合力F_R) (b) Equilibrium(平衡)

Fig. 2.14 Combined forces and equilibrium for the geometric method of coplanar concurrent forces systems in the plane (图 2.14 平面汇交力系几何法的合力与平衡)

Example 1: As shown in Fig. 2.15, $\boldsymbol{P}$ and a are known. Find the restraint reaction force at A and B. (**例题** 1: 如图 2.15 所示,已知 $\boldsymbol{P}$ 和 a,求 A、B 处约束反力.)

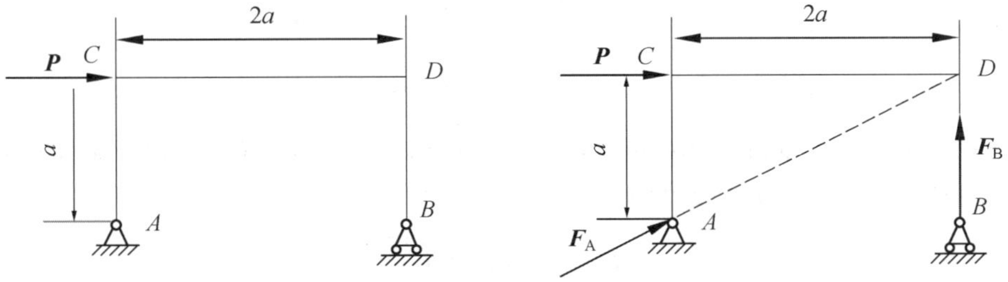

(a) Rigid frame(刚架) (b) Force diagram(受力图)

Fig. 2.15 Rigid frame force diagram (图 2.15 刚架受力图)

Solution: (**解**:)

(1) Rigid frame as the object of study. (取刚架为研究对象.)

(2) Draw a force diagram, as shown in Fig. 2.15(b). (画受力图,如图 2.15(b)所示.)

(3) Scaled graphing solutions. (按比例作图求解.)

From the geometric relationship in Fig. 2.16, we have(由图 2.16 中的几何关系得)

$$F_B = P\tan\alpha = 0.5P$$

$$F_A = \sqrt{P^2 + F_B^2} = \frac{\sqrt{5}}{2}P$$

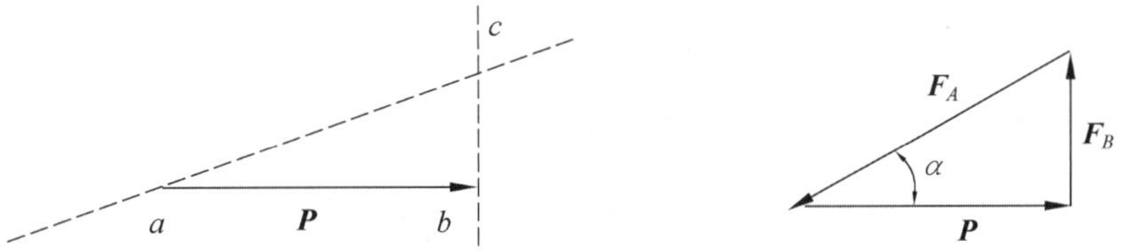

Fig. 2.16 Schematic diagram of geometric method solution(图 2.16 几何法求解示意图)

2.2.3 Analytical method for the synthesis of confluent force systems (汇交力系合成的解析法)

Deriving the analytical method for the concurrent forces system from $\boldsymbol{F}_R = \boldsymbol{F}_1 + \boldsymbol{F}_2 + ... + \boldsymbol{F}_n$: the projection of the combined vector on the coordinate axes is equal to the algebraic sum of the projections of the component forces on the same projection axis (Fig. 2.10(c)). We obtain two algebraic equations, those are (根据 $\boldsymbol{F}_R = \boldsymbol{F}_1 + \boldsymbol{F}_2 + \cdots + \boldsymbol{F}_n$ 推导汇交力系的解析方法:合矢量在坐标轴上的投影等于各分力在同一投影轴上投影的代数和(图 2.10(c).得到两个代数式,即)

$$\begin{cases} F_{Rx} = F_{1x} + F_{2x} + \cdots + F_{nx} = \sum_{i=1}^{n} F_{ix} \\ F_{Ry} = F_{1y} + F_{2y} + \cdots + F_{ny} = \sum_{i=1}^{n} F_{iy} \end{cases} \tag{2.3}$$

$$\begin{cases} F_R = \sqrt{F_{Rx}^2 + F_{Ry}^2} = \sqrt{(\sum F_{ix})^2 + (\sum F_{iy})^2} \\ \cos(\angle \boldsymbol{F}_R, \mathrm{i}) = \dfrac{F_{Rx}}{F_R}, \cos(\angle \boldsymbol{F}_R, \mathrm{j}) = \dfrac{F_{Ry}}{F_R} \end{cases} \tag{2.4}$$

Therefore, the magnitude of the combined force of all confluent forces can be determined as (因此,可以确定所有汇交力合力的大小,即)

$$|\boldsymbol{F}_R| = \sqrt{F_{Rx}^2 + F_{Ry}^2}$$

The direction of the combined vector is(合矢量的方向为)

$$\cos\alpha = F_x / F_R,\ \cos\beta = F_y / F_R$$

Example 2: If $F_1 = 200$ N, $F_2 = 300$ N, $F_3 = 100$ N and $F_4 = 250$ N are known, find the combined forces of the plane confluent force systems shown in Fig. 2. 17. (**例题 2**:已知 $F_1 = 200$ N,$F_2 = 300$ N,$F_3 = 100$ N,$F_4 = 250$ N,求图 2.17 所示平面汇交力系的合力.)

Solution: (**解**:)

$$\begin{aligned} F_{Rx} &= \sum_{i=1}^{4} F_{ix} = F_{1x} + F_{2x} + F_{3x} + F_{4x} \\ &= F_1\cos 30° - F_2\cos 60° - F_3\cos 45° + F_4\cos 45° = 129.3\ (\mathrm{N}) \end{aligned}$$

$$\begin{aligned} F_{Ry} &= \sum_{i=1}^{4} F_{iy} = F_{1y} + F_{2y} + F_{3y} + F_{4y} \\ &= F_1\cos 60° + F_2\cos 30° - F_3\cos 45° - F_4\cos 45° = 112.3\ (\mathrm{N}) \end{aligned}$$

$$F_R = \sqrt{F_{Rx}{}^2 + F_{Ry}{}^2} = \sqrt{129.3^2 + 112.3^2} = 171.3\ (\mathrm{N})$$

$$\cos(\angle \boldsymbol{F}_R, \mathrm{i}) = \frac{F_{Rx}}{F_R} = \frac{129.3}{171.3} = 0.754\ 8$$

$$\cos(\angle \boldsymbol{F}_R, \mathrm{j}) = \frac{F_{Ry}}{F_R} = \frac{112.3}{171.3} = 0.655\ 6$$

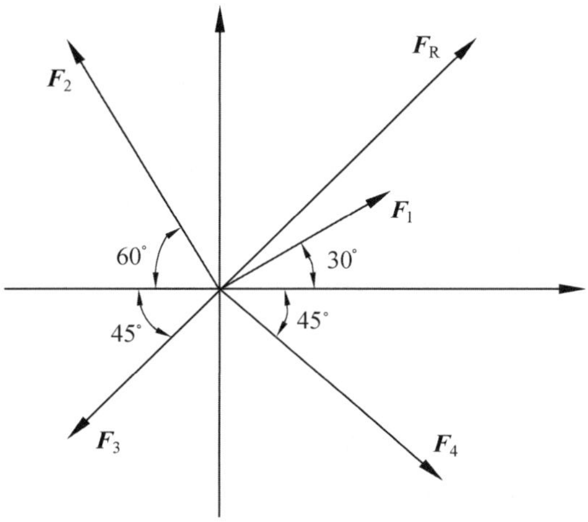

Fig. 2. 17 Coplanar concurrent forces system (图 2. 17 平面汇交力系)

2.2.4 Analytical method for the equilibrium of confluent force system (汇交力系平衡的解析法)

1. Coplanar concurrent forces system(平面汇交力系)

The condition for the equilibrium of a coplanar concurrent forces system is that the resulting $\boldsymbol{F}_{\mathrm{R}}$ must have a magnitude equal to zero, that is to say, the force polygon must be closed (using the geometric method) or the projection of the combined forces on the coordinate axes must be equal to zero ($F_{\mathrm{R}x} = F_{\mathrm{R}y} = 0$) (using the analytical method).(平面汇交力系平衡的条件是所得 $\boldsymbol{F}_{\mathrm{R}}$ 的大小等于零,即力多边形必须封闭(使用几何法)或在坐标轴上合力的投影必须等于零($F_{\mathrm{R}x} = F_{\mathrm{R}y} = 0$)(使用解析方法).)

Thus, for a system of coplanar concurrent forces, we obtain two balancing equations, those are(因此,对于平面汇交力系,得到两个平衡方程,即)

$$\sum F_{ix} = 0, \quad \sum F_{iy} = 0 \tag{2.5}$$

A sufficient condition for the equilibrium of a coplanar concurrent forces system is that the sum of the projections of the forces in the system on the two axes is equal to zero, those are(平面汇交力系平衡的充要条件为该力系中各力在这两个坐标轴上的投影之和等于零,即)

$$\sum F_{ix} = 0, \quad \sum F_{iy} = 0$$

2. Three-force equilibrium convergence theorem(三力平衡汇交定理)

Theorem: If the three forces acting on a rigid body are mutually balanced forces, then the combined forces of any two of them must satisfy the two-force equilibrium condition with the third force, and the three forces must be in the same plane, and the lines of action of the three forces must converge at a point (Fig. 2. 18).(**定理**:如果作用于刚体上的三个力是相互平衡的力,那么其中任意两个力的合力一定与第三个力满足二力平衡条件,同时此三力必在同一平面内,且三个力的作用线一定汇交于一点(图 2. 18).)

Example 3: Try the analytical method to find the restrained reaction forces at A and B in

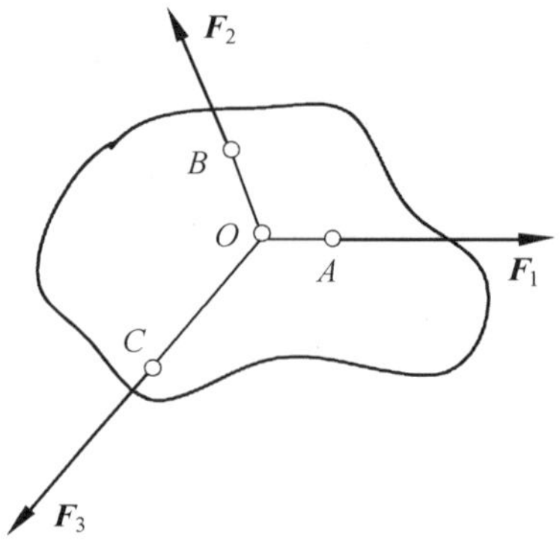

Fig. 2. 18 Three-force equilibrium confluence(图 2. 18 三力平衡汇交)

Example 1, $\boldsymbol{P}$ and a are konwn, as shown in Fig. 2. 19. (**例题** 3:试用解析法求例题 1 中 A、B 处约束反力,$\boldsymbol{P}$ 和 a 已知,如图 2. 19 所示.)

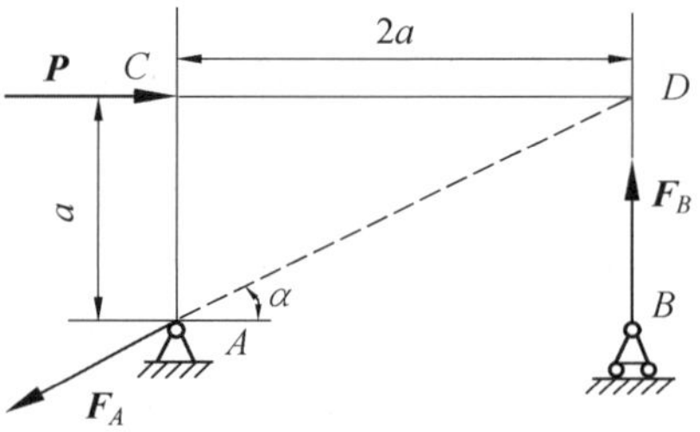

Fig. 2. 19 Analytical method of finding restrained reaction forces (图 2. 19 解析法求约束反力)

Solution:(解:)

(1) Take the rigid frame as the object of study. (取刚架为研究对象.)

(2) Draw the force diagram. (画受力图.)

(3) Set up a coordinate system and solve the equation, we have(建立坐标系,列方程求解,有)

$$\sum F_{ix} = 0, P - F_A\cos\alpha = 0$$

$$\sum F_{iy} = 0, F_B - F_A\sin\alpha = 0$$

Solve the above equation, we have(解上述方程,得到)

$$F_A = \frac{\sqrt{5}}{2}P,\ F_B = \frac{1}{2}P$$

Example 4: A simple diagram of a press is shown in Fig. 2. 20, $\boldsymbol{F}$ and α are known. Find the pressure on the object M. (**例题** 4:图 2. 20 所示为一压榨机压榨的简单示意图,F 和 α 已知. 求物块 M 的压力.)

Solution:(解:)

(1) It is known that both AB and BC rods are two-force rods and can be first drawn as shown in Fig. 2. 21 (a). Take the pin B as the object of study, its force diagram is shown in Fig. 2. 21(b), we have(由已知可知,AB 及 BC 杆均为二力杆,可先画受力图,如图 2. 21(a)所示. 取销钉 B 为研究对象,其受力图如图 2. 21(b)所示,有)

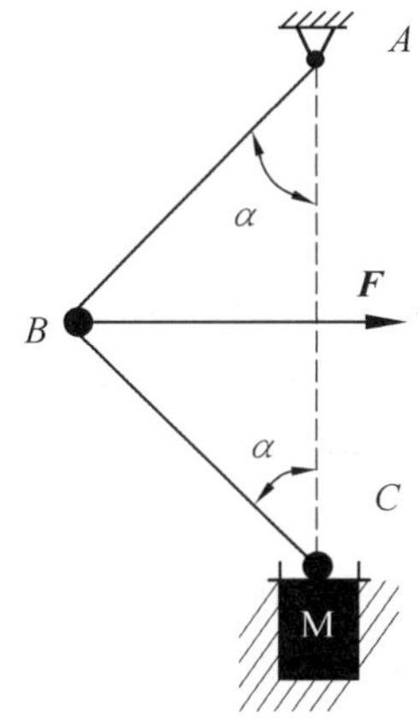

Fig. 2. 20 Simple diagram of the press (图 2. 20 压榨机受力图)

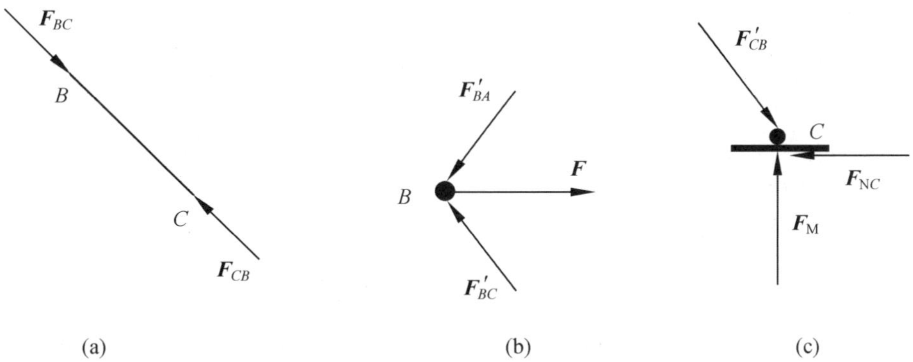

Fig. 2. 21 Force diagram (图 2. 21 受力图)

$$\sum F_{ix} = 0, F - (F_{BA} + F_{BC})\sin\alpha = 0$$

$$\sum F_{iy} = 0, F_{BC}\cos\alpha - F_{BA}\cos\alpha = 0$$

We have(解得)

$$F_{BC} = F_{BA} = \frac{F}{2\sin\alpha}$$

(2) Take the baffle C as the study object and its force diagram is shown in Fig. 2. 21(c), we have(取挡板 C 为研究对象,其受力图如图 2. 21(c)所示,有)

$$\sum F_{iy} = 0, F_M - F_{CB}\cos\alpha = 0$$

We have(解得)

$$F_M = F_{CB}\cos\alpha = \frac{F}{2}\cot\alpha$$

Example 5: Determine the tension in the rope when an object weighing 5 N is kept in equilibrium (Fig. 2. 22(a)). (**例题** 5:确定重力为 5 N 的物体保持平衡时绳索的张力(图 2. 22(a)).)

Solution:(解:)

There is a certain sequence that should be followed when solving hydrostatic problems. This example shows how to solve this type of problem in detail.(在解决静力学问题时，应遵循一定的顺序. 本示例详细介绍如何解决此类型的问题.)

To draw a schematic diagram of a structure, choose the object (node, rod or rigid body) whose equilibrium should be taken into account and which must be related to the required solution and known conditions. In this case, the source data (mass, angles α and β) and the required value (tension) are connected to an object weighing 5 N, which is the object of equilibrium.(绘制结构示意图,选择应考虑其平衡的对象(节点、杆或刚体)，并且必须与所需求解的和已知的条件相关联. 在此题中,源数据(质量、角度 α 和 β) 和所求的值(张力) 连接到重 5 N 的物体，该物体即为平衡的对象.)

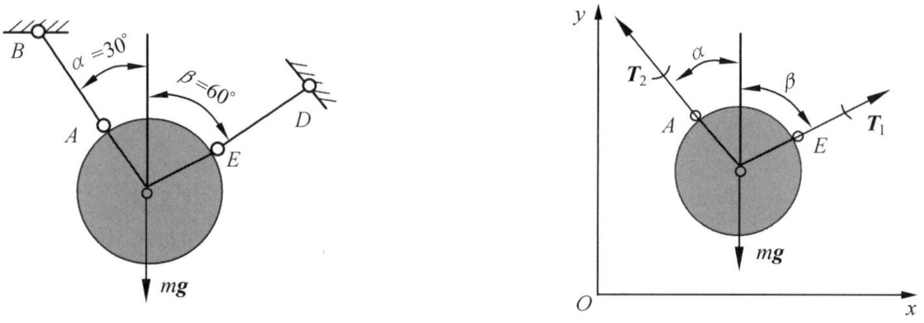

Fig. 2. 22 Figure of Example 5(图 2. 22 例题 5 图)

Remove the constraints and apply all forces and reactions to the equilibrium object under consideration. Particular care should be taken when solving for this. The balancing equations studied in statics only apply to free bodies, so you should carefully consider which restraining forces should be reflected in the force diagram during the removal of the restraints. In this problem, the objects are attached to ropes *AB* and *ED*. When the restraints are removed, we replace them with the tensions of the ropes $\boldsymbol{T}_2$ and $\boldsymbol{T}_1$ respectively (Fig. 2. 22(b)).(去掉约束，将所有作用力和反作用力应用于所考虑的平衡对象. 在求解时应当特别细心. 静力学中研究的平衡方程仅适用于自由体,因此应该仔细考虑在解除约束过程中哪些约束力应反映在受力图中. 在本题中,物体连接的是绳索 *AB* 和 *ED*. 当去掉约束时,分别用绳索的张力 $\boldsymbol{T}_2$ 和 $\boldsymbol{T}_1$ 替换它们(图 2. 22(b)).)

Analysis the force system obtained. The object is in equilibrium under the action of a system of coplanar concurrent forces (their lines of action intersect at the center of the ball). For this system of forces, two balancing equations can be formulated. The unknowns in these equations are also two, so that the problem is determined to be stationary.(分析所得到的力系. 物体在平面汇交力系的作用下处于平衡状态 (它们的作用线在球的中心相交). 对于这种力系，可以列出两个平衡方程. 这些等式中的未知数也是两个，因此确定是静定问题.)

List the equilibrium conditions for the geometric or analytical method. Find the unknown values and use the analytical method to solve the problem. We list the balancing equations for a system of plane confluent forces, those are(列出几何法或解析法的平衡条件,找到未知值使用

解析法求解该问题. 列出平面汇交力系的平衡方程,有)

$$\sum F_{ix} = 0,\ \sum F_{iy} = 0$$

$$-T_2\cos 60° + T_1\cos 30° = 0$$

$$T_2\cos 30° + T_1\cos 60° - mg = 0$$

After solving the resulting system of equations, we calculate the tension in the rope, and we have(求解所得的方程组后,计算出绳索的张力,有)

$$T_1 = 2.5\ \text{N},\ T_2 = 4.34\ \text{N}$$

2.3 Moment of a Plane Force to a Point, Plane Force Couple (平面力对点之矩,平面力偶)

2.3.1 Concept of moments(力矩的概念)

(1) Types of torque. (力矩的种类.)

The following types of torque are available. (有以下类型的力矩.)

① algebraic moments relative to the center. (相对于中心的代数力矩.)

② Vector moment relative to the center. (相对于中心的矢量力矩.)

③ Moment relative to the shaft. (相对于轴的力矩.)

(2) Algebraic moments. (代数力矩.)

Force acting on an object can not only move the object, but can also rotate around a point. Let the force $\boldsymbol{F}$ act at the point A and the object rotates around the point O (Fig. 2.23). Since a force can move along its line of action, the effect of the rotation of this force does not depend on the point at which it is applied, but is determined by the vertical distance h from the point O to the line of action of the force. (作用在物体上的力不仅可以使物体移动,而且还能围绕某一点转动. 设力 $\boldsymbol{F}$ 作用在 A 点,物体围绕 O 点转动(图 2.23). 由于力可以沿其作用线移动, 因此此力的转动效果不取决于施加此力的作用点, 而是由从 O 点到力的作用线的垂直距离 h 确定.)

(a) Counterclockwise rotation (逆时针转)　　(b) Clockwise rotation (顺时针转)

Fig. 2.23 Object rotating around a point O(图 2.23 绕 O 点转动的物体)

The algebraic moment of the force with respect to the center O is a scalar and is equal to the force multiplied by the scalar value of the force arm (Fig. 2.23). The equation is(力相对于中心 O 的代数力矩是标量,其等于力乘力臂的标量值(图 2.23),其公式为)

$$M_O(\boldsymbol{F}) = \pm F \cdot h \tag{2.6}$$

The central point O at which the algebraic moment is calculated is called the center of the moment. (计算代数力矩的中心点 O 称为力矩中心.)

The vertical distance from the center of the moment to the line of action of the force is called the force arm h (Fig. 2.23). (从力矩中心到力的作用线的垂直距离称为力臂 h (图 2.23).)

When the force turns the object around the center of the moment or the moment is counterclockwise, the positive sign "+" is taken (Fig. 2.23(a)); when it is clockwise, the negative sign "−" is taken (Fig. 2.23(b)) (当力使物体绕矩心的转向或力矩的转向为逆时针时,取正号"+"(图 2.23(a));当为顺时针时,则取负号"−"(图 2.23(b)).)

The algebraic moment of a force reflects its rotational effect (physical meaning of the algebraic moment of a force). (力的代数矩体现了它的转动效应(力的代数矩的物理意义.)

(3) The case where the algebraic moment of the force is equal to zero. (力的代数矩等于零的情况.)

The algebraic moment of the force with respect to the center is zero (Fig. 2.24) in the following cases. (在下列情况下,力相对于中心的代数矩为零(图 2.24).)

① The magnitude of the force is equal to zero. (力的大小等于零.)

② The line of action of the force passes through the center of the moment (force arm $h=0$). (力的作用线通过矩心(力臂 $h=0$).)

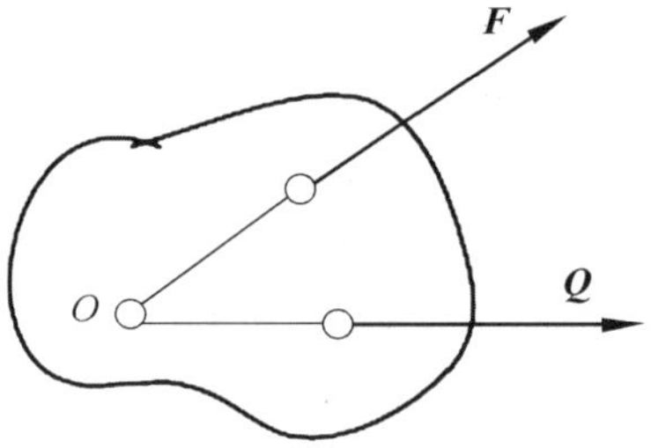

Fig. 2.24 Algebraic moments to zero(图 2.24 代数力矩为零)

2.3.2 Varignon's combined moment theorem (plane force system) (伐里农合力矩定理(平面力系))

There are several types of combined moment theorems. (合力矩定理有以下几种类型.)

① Planar force system. (平面力系.)

② Space force system. (空间力系.)

1. Combined moment theorem in plane force systems(平面力系中合力矩定理)

If a given system of forces has a combined force, its moment for any point OO is equal to the algebraic sum of the moments of the component forces relative to the same point. The formula is (如果给定的力系有一个合力,则它对于任一点 O 的力矩等于各分力相对于同一点的力矩的代数和,其公式为)

$$M_O(\boldsymbol{F}_R) = \sum M_O(\boldsymbol{F}) \tag{2.7}$$

2. Combined moment theorem in plane force systems (the case of two forces)
(平面力系中合力矩定理(两个力的情况))

If the combined force of two forces $\boldsymbol{F}_R = \boldsymbol{F}' + \boldsymbol{F}''$, then the moment $M_O(\boldsymbol{F}_R)$ of the combined force $\boldsymbol{F}_R$ for any point O is equal to the algebraic sum of the moments of the component forces $\boldsymbol{F}'$ and $\boldsymbol{F}''$ for the same point, i. e. $M_O(\boldsymbol{F}_R) = M_O(\boldsymbol{F}') + M_O(\boldsymbol{F}'')$. Typically, $\boldsymbol{F}'$ is parallel to $\boldsymbol{F}''$ in the Cartesian coordinate system (Fig. 2.25). (如果二力的合力 $\boldsymbol{F}_R = \boldsymbol{F}' + \boldsymbol{F}''$,那么合力 $\boldsymbol{F}_R$ 对于任一点 O 之矩 $M_O(\boldsymbol{F}_R)$ 等于分力 $\boldsymbol{F}'$ 与 $\boldsymbol{F}''$ 对于同一点之矩的代数和,即 $M_O(\boldsymbol{F}_R) = M_O(\boldsymbol{F}') + M_O(\boldsymbol{F}'')$. 通常情况下,$\boldsymbol{F}'$ 与 $\boldsymbol{F}''$ 平行于笛卡儿坐标系(图 2.25).)

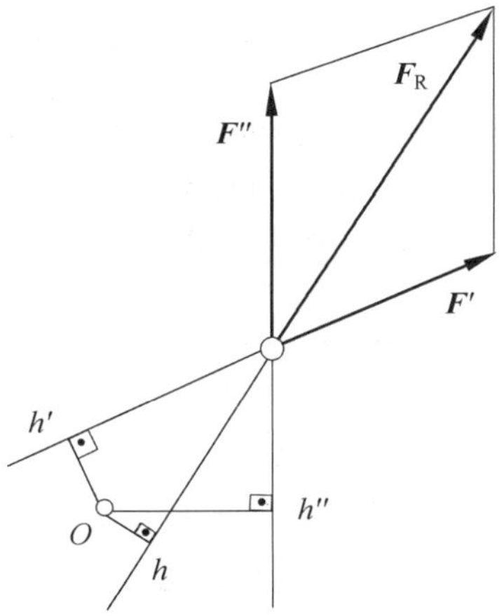

Fig. 2.25 Combined moment of a plane force system (图 2.25 平面力系合力矩)

Example 6: As shown in Fig. 2.26, $\boldsymbol{F}_N$, α, r are known. Find the moment of the force $\boldsymbol{F}_N$ on the wheel center O. (**例题 6**: 如图 2.26 所示,$\boldsymbol{F}_N$、α、r 已知. 求力 $\boldsymbol{F}_N$ 对轮心 O 的力矩.)

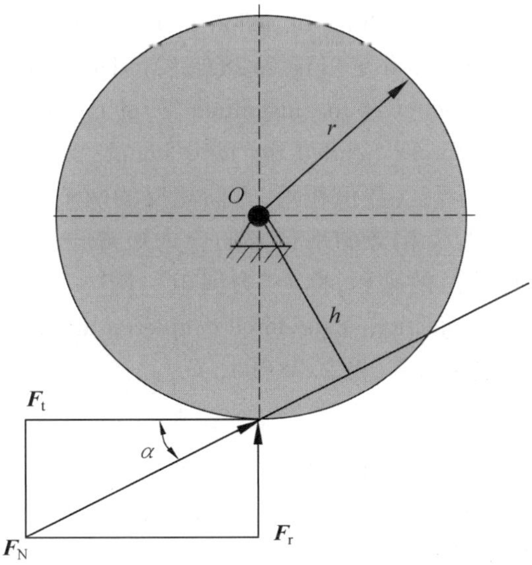

Fig. 2.26 Example 6 diagram (图 2.26 例题 6 图)

Solution：（解：）

（1） Direct calculation, we have（直接计算，有）

$$M_O(\boldsymbol{F}_N) = F_N h = F_N r\cos\alpha$$

（2） Calculation using theorem of moments of resultant force, we have（利用合力之矩定理计算，有）

$$\begin{aligned} M_O(\boldsymbol{F}_N) &= M_O(\boldsymbol{F}_r) + M_O(\boldsymbol{F}_t) \\ &= M_O(\boldsymbol{F}_t) \\ &= F_N r\cos\alpha \end{aligned}$$

2.3.3 Force couple, planar force couple system（力偶，平面力偶系）

Two parallel forces（ $\boldsymbol{F}, \boldsymbol{F}'$ ） of equal magnitude, opposite direction and not co-linear, acting on the same rigid body are called force couples（Fig. 2.27）.（作用在同一刚体上的两个大小相等、方向相反且不共线的平行力（ $\boldsymbol{F}, \boldsymbol{F}'$ ）称为力偶（图 2.27）.）

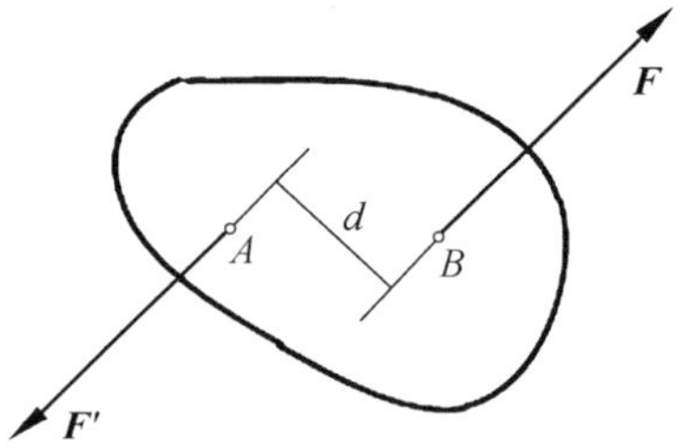

Fig. 2.27 Force couple（图 2.27 力偶）

The plane determined by the lines of action of the two forces is called the plane of action of the force couple.（两力作用线所决定的平面称为力偶的作用面.）

The vertical distance between the lines of action of two forces of equal value and opposite direction is called the force couple arm h（Fig. 2.28（a））. The combined moment of two equal parallel forces is equal to the product of the magnitude F of one of the parallel forces and the distance d between the parallel forces（called the force couple arm）, called the "force couple moment", denoted by "+" or "−".（数值相等、方向相反的两个力的作用线之间的垂直距离称为力偶臂 h（图 2.28（a））. 两个相等的平行力的合力矩等于平行力中的一个力的大小 F 与平行力之间距离 d（称力偶臂）的乘积，称为"力偶矩"，用"+" 或"−"表示.）

The algebraic sum of the projection of the force couple on any axis is zero（Fig. 2.28（b））.（力偶在任何轴上投影的代数和为零（图 2.28（b）.）

The proof is（证明为）

$$\sum F_x = F'\cos\alpha - F\cos\alpha = 0$$

According to Fig. 2.28（c）, the moment of the force couple to any point is（根据图 2.28（c），力偶对任一点取矩为）

$$M_O(\boldsymbol{F}, \boldsymbol{F}') = M_O(\boldsymbol{F}) + M_O(\boldsymbol{F}') = -F(x+d) + F'x = -Fd \text{ (clockwise)（顺时针）}$$

1. Types of couple moments（力偶矩的类型）

There are two types of couple moments.（力偶矩有两种类型.）

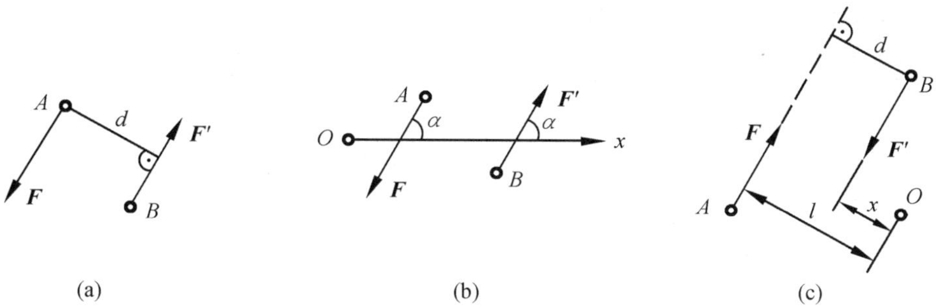

Fig. 2.28 Couple moment (图 2.28 力偶矩)

① Vector moments. (矢量矩.)

② Algebraic moments. (代数矩.)

2. Algebraic couple moments(代数力偶矩)

The moment of couple, as a measure of the effect of a couple on the rotation of an object, is equal to the product of the magnitude of either force in the couple and the length of the force couple's arm, with the corresponding plus or minus sign, i. e. $m = \pm \boldsymbol{F} \cdot d$. (力偶矩用来度量力偶使物体转动的效应,等于力偶中任一力的大小与力偶臂长度的乘积再冠以相应的正负号,即 $m = \pm \boldsymbol{F} \cdot d$.)

The positive and negative sign of the couple moment is the same as that of the force moment, i. e. the positive sign is taken when the couple makes the object turn counterclockwise, and the negative sign is taken when the opposite is true. (力偶矩的正负号规定与力矩相同,即力偶使物体逆时针旋转时取正号,反之取负号.)

In Fig. 2. 29, $m(\boldsymbol{F},\boldsymbol{F}') > 0$, $m(\boldsymbol{P},\boldsymbol{P}') < 0$. (图 2. 29 中, $m(\boldsymbol{F},\boldsymbol{F}') > 0$, $m(\boldsymbol{P},\boldsymbol{P}') < 0$.)

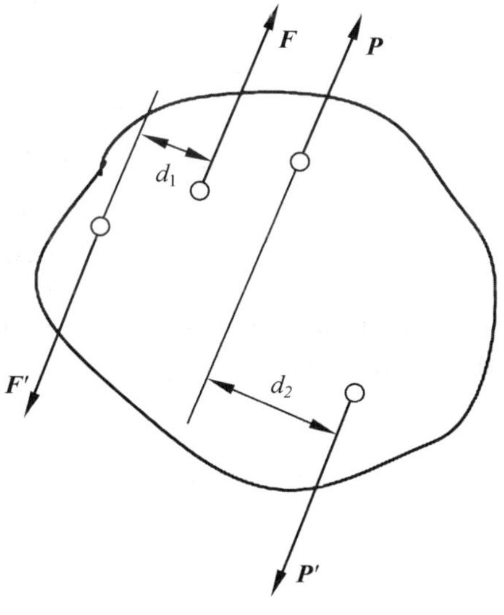

Fig. 2.29 Symbols for couple moments(图 2.29 力偶矩的符号)

3. Properties of couples(力偶的性质)

(1) The couple can move and rotate at will within its plane of action without affecting its effect on the rigid body. (力偶可以在其作用面内任意移动和转动,而不影响它对刚体的效应.)

(2) The couple may be transferred from that plane to any other plane in the body without affecting its effect on the rigid body.(在不影响其对刚体效应的情况下,可将力偶从该平面转至物体内任何其他平面.)

(3) The action of a force couple on a rigid body is entirely characterised by its moment, which is $m=F\cdot d$, where d is the force couple arm (the shortest distance between the lines of action of the force). (力偶对刚体的作用完全以其力矩为特征,其力偶矩为 $m=F\cdot d$,其中 d 是力偶臂(力的作用线之间的最短距离).)

(4)There is no combined force for couple. (力偶没有合力.)

4. Couples represented by arcs with arrows(用带箭头的弧线表示力偶)

The action of a couple is entirely characterized by its moment, so it is represented by an arc with an arrow indicating the steering of the couple (Fig. 2.30). In this case, the nature of the couple equivalent transformation can be applied to any part of the absolute rigid body.(力偶的作用完全以其力矩为特征,因此它以带箭头的弧线来表示,箭头表示力偶的转向(图 2.30).在这种情况下,力偶等效变换的性质可以应用于绝对刚体的任何部分.)

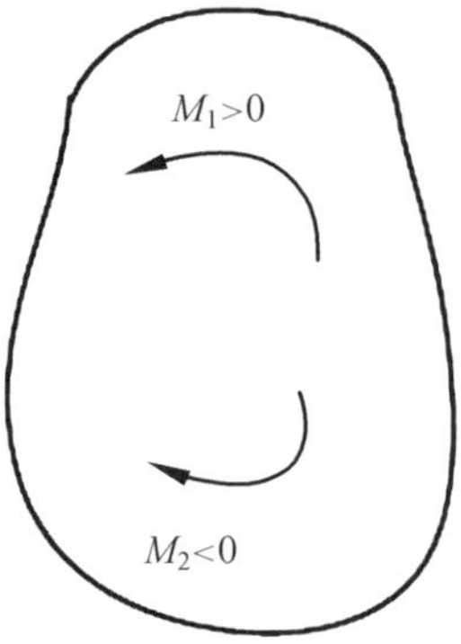

Fig. 2.30 Representation of couples as arcs with arrows(图 2.30 以带箭头的弧线表示力偶)

5. Synthesis theorem for planar couple systems(平面力偶系合成定理)

The synthesis of a couple system in the same plane results in a couple with a couple moment equal to the algebraic sum of the component couple moments.(在同一平面上的力偶系合成结果为一个力偶,其力偶矩等于各分力偶矩的代数和.)

If there are n couples acting in the same plane, then(若作用在同一平面内有 n 个力偶,则有)

$$M = m_1 + m_2 + \cdots + m_n = \sum m_K \tag{2.8}$$

6. The necessary and sufficient conditions for the equilibrium of a system of couples in the same plane (同一平面内力偶系平衡的必要与充分条件)

If the algebraic sum of the force couple moments in the force couple system is zero

(Fig. 2.31), then the force couple system acting on the absolute rigid body in one plane is equal to zero (in equilibrium balance), that is(如果力偶系中各力偶矩的代数和为零(图 2.31),则在一个平面上,对绝对刚体起作用的力偶系等于零(处于平衡状态),即)

$$\sum m_{\kappa} = 0 \tag{2.9}$$

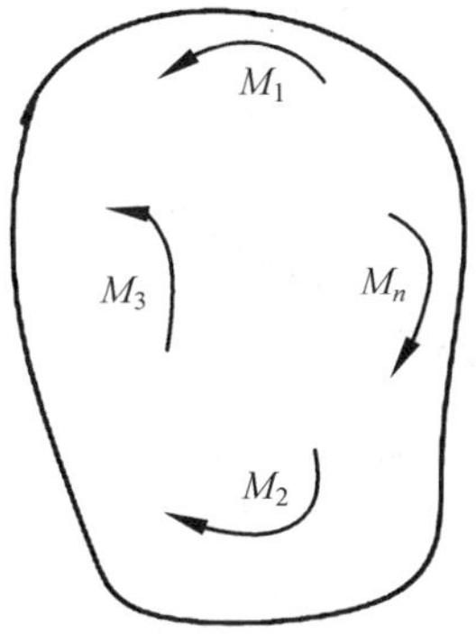

Fig. 2.31 Planar force couple system (图 2.31 平面力偶系)

Example 7: In Fig. 2.32, a and M are known. Find the restraint reaction forces at A and C (**例题** 7:图 2.32 中,a 和 M 已知. 求 A、C 处约束反力.)

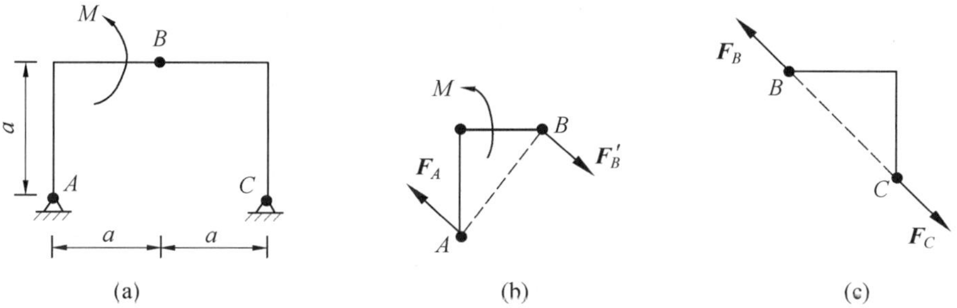

Fig. 2.32 Example 7 diagram (图 2.32 例题 7 图)

Solution: (**解**:)

(1) Take AB as the study object and its force diagram is shown in Fig. 2.32(b), we have (取 AB 为研究对象,其受力图如图 2.32(b)所示,有)

$$\sum M = 0,\ M - F_A\sqrt{2}a = 0$$

$$F_A = F'_B = \frac{\sqrt{2}}{2a}M$$

(2) Take BC as the study object and its force diagram is shown in Fig. 2.32(c), we have (取 BC 为研究对象,其受力图如图 2.32(c)所示,有)

$$F_C = F_B = F'_B = \frac{\sqrt{2}}{2a}M$$

If this couple is moved to the BC member, then find the restraint reaction forces at A and C. In this case, can the couple move within its plane of action? Is the moment of the couple to any point still equal to the moment of the couple? (若将此力偶移至 BC 构件上,再求 A、C 处约束

反力. 在此种情况下,力偶能否在其作用面内移动? 力偶对任意点之矩是否还等于力偶矩?)

Example 8: As shown in Fig. 2.33, it is known that $AB = CD = a$, $\angle BCD = 30°$. Find the relationship between M_1, M_2 at equilibrium. (**例题 8**: 如图 2.33 所示,已知 $AB = CD = a$, $\angle BCD = 30°$. 求平衡时 M_1、M_2之间的关系.)

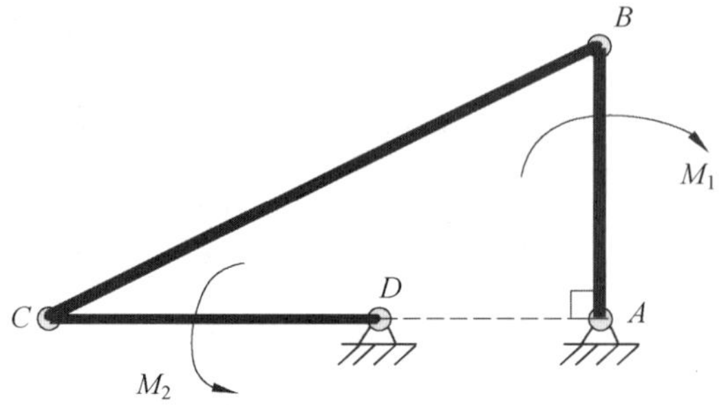

Fig. 2.33 Example 8 diagram (例题 8 图)

Solution:(**解**:)

(1) Take AB as the study object and its force diagram is shown in Fig. 2.34, we have (取 AB 为研究对象,其受力图如图 2.34 所示,有)

$$\sum M = 0,\ F_B a\cos 30° - M_1 = 0$$

So we have(解得)

$$M_1 = \frac{\sqrt{3}}{2} F_B A$$

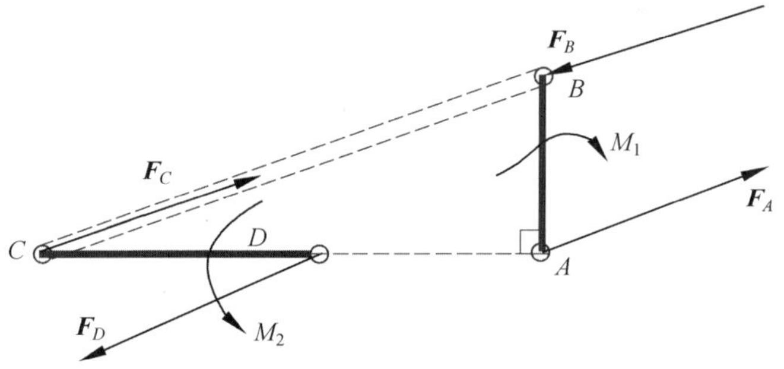

Fig. 2.34 Force diagram of AB and CD (AB 与 CD 受力图)

(2) Take CD as the study object and its force diagram is shown in Fig. 2.34, we have (取 CD 为研究对象,其受力图如图 2.34 所示,有)

$$\sum M = 0,\ M_2 - F_C a \sin 30° = 0$$

So we have(解得)

$$M_2 = \frac{1}{2} F_C a$$

Since(因为)

$$F_B = F_C$$

there is(所以)

$$\frac{M_1}{M_2} = \sqrt{3}$$

2.4 Simplification of Plane Arbitrary Force Systems (平面任意力系的简化)

The simplification of the force system is divided into the following two main cases. (力系简化主要分为以下两种情况.)

(1) Simplification of arbitrary force systems in space. (空间任意力系简化.)

(2) Simplification of plane arbitrary force systems. (平面任意力系简化.)

In order to simplify the system of arbitrary forces in the plane, a theorem needs to be introduced: the translation theorem of forces. (为对平面任意力系简化,需引入一个定理:力的平移定理.)

1. Translation theorem for forces(力的平移定理)

The force $\boldsymbol{F}$ acting at any point A on the rigid body can be moved parallel to any point B, but only if a couple is attached at the same time. The moment of this additional couple is equal to the moment of the original force $\boldsymbol{F}$ on the new point B, as shown in Fig. 2.35, of which(可以把作用于刚体上任意点 A 的力 $\boldsymbol{F}$ 平行移到任一点 B,但必须同时附加一个力偶,这个附加力偶的矩等于原来的力 $\boldsymbol{F}$ 对新作用点 B 的矩,如图 2.35 所示,其中)

$$M = F \cdot d = M_B(F)$$

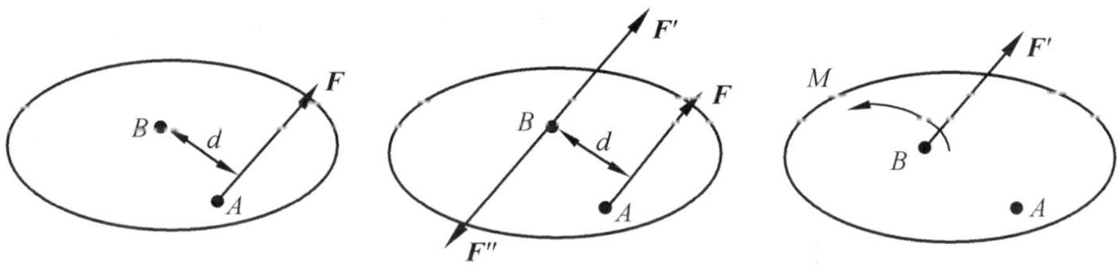

Fig. 2.35 Parallel movement of force (图 2.35 力的平行移动)

Think: As shown in Fig. 2.36, why do nails sometimes bend? (思考:如图 2.36 所示,为什么钉子有时会折弯?)

2. Simplification theorem for plane force systems(平面力系的简化定理)

The planar arbitrary force system in Fig. 2.37(a) is simplified towards point O to obtain a coplanar concurrent forces system and an additional planar couples system according to the force translation theorem, shown in Fig. 2.37(b), where(图 2.37(a)中平面任意力系向 O 点简化,根据力的平移定理得到一个平面汇交力系和一个附加的平面力偶系,如图 2.37(b)所示,其中)

$$F_1 = F'_1, F_2 = F'_2, F_3 = F'_3,\quad M_1 = M_O(\boldsymbol{F}_1), M_2 = M_O(\boldsymbol{F}_2), M_3 = M_O(\boldsymbol{F}_3)$$

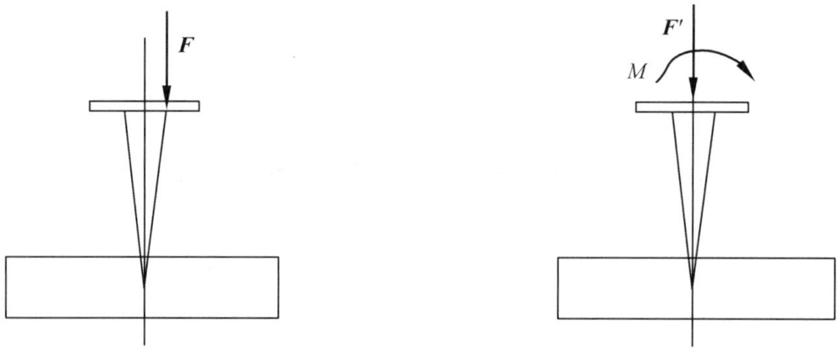

(a) Eccentric force(偏心力) (b) Additional bending effect(平移后附加弯曲效应)

Fig. 2.36 Bending of a nail (钉子钉弯)

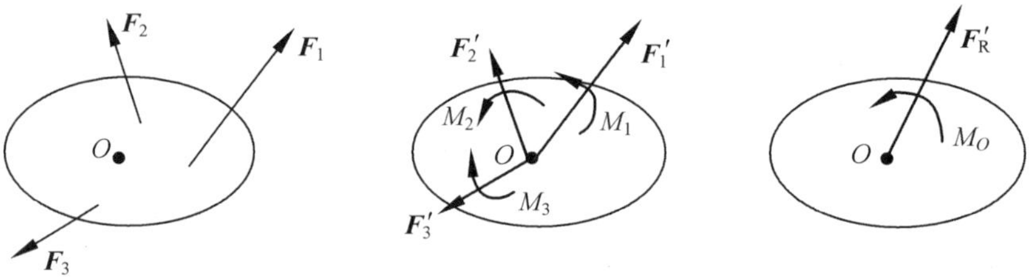

(a) Original force system (原力系) (b) Force system after translation (平移后力系) (c) Simplification result (简化结果)

Fig. 2.37 Simplification of an arbitrary force system in the plane(图 2.37 平面任意力系的简化)

$$\boldsymbol{F}'_{\mathrm{R}} = \boldsymbol{F}'_1 + \boldsymbol{F}'_2 + \boldsymbol{F}'_3 = \boldsymbol{F}_1 + \boldsymbol{F}_2 + \boldsymbol{F}_3 = \sum \boldsymbol{F}_i$$

$$M_O = M_1 + M_2 + M_3 = M_O(\boldsymbol{F}_1) + M_O(\boldsymbol{F}_2) + M_O(\boldsymbol{F}_3) = \sum M_O(\boldsymbol{F}_i)$$

That is(即)

$$\begin{aligned} \boldsymbol{F}'_{\mathrm{R}} &= \sum \boldsymbol{F}_i \\ M_O &= \sum M_O(\boldsymbol{F}_i) \end{aligned} \tag{2.10}$$

The coplanar concurrent forces system can be synthesized into a combined force $\boldsymbol{F}'_{\mathrm{R}}$, which is equal to the vector sum of the forces in the original force system, acting at the simplified center O, and $\boldsymbol{F}'_{\mathrm{R}}$ is called as the principal vector of the original force system. The planar couple system can be synthesized into a combined force couple, the magnitude of which is related to the position of the O point, so it is noted as M_O, whose magnitude is equal to the algebraic sum of the moments of the forces in the original planar arbitrary force system to the simplified center O. (平面汇交力系可合成为一个合力 $\boldsymbol{F}'_{\mathrm{R}}$,该合力等于原来力系中各力的矢量和,作用在简化中心 O 点,$\boldsymbol{F}'_{\mathrm{R}}$ 称为原力系的主矢量. 平面力偶系可合成为一个合力偶,该合力偶矩的大小与 O 点的位置有关,因此记为 M_O,其大小等于原来的平面任意力系中的各力对简化中心 O 点力矩的代数和.)

The vector $\boldsymbol{F}'_{\mathrm{R}}$, which is equal to the geometric sum of the forces, is the principal vector. The magnitude of the principal vector is independent of the direction and the simplified center at point O, i. e. $\boldsymbol{F}'_{\mathrm{R}}$ is invariant. (主矢量 $\boldsymbol{F}'_{\mathrm{R}}$ 等于所有力的几何和. 主矢量的大小与方向和简化

中心 O 点无关,也就是说 $\boldsymbol{F}'_R$ 是不变量.)

The calculation of the principal vector $\boldsymbol{F}'_R$ is more easily done by the analytical method. As shown in Fig. 2.38, a right-angle coordinate system is established to obtain the principal vector, that is(主矢量 $\boldsymbol{F}'_R$ 的计算用解析法比较方便. 如图 2.38 所示,建立直角坐标系,可得到主矢量,即)

$$F'_{Rx} = \sum F_x,\ F'_{Ry} = \sum F_y$$
$$F'_R = \sqrt{(\sum F_x)^2 + (\sum F_y)^2} \tag{2.11}$$
$$\cos < \boldsymbol{F}'_R, \mathrm{i} > = \frac{\sum F_x}{F'_R}, \cos < \boldsymbol{F}'_R, \mathrm{j} > = \frac{\sum F_y}{F'_R}$$

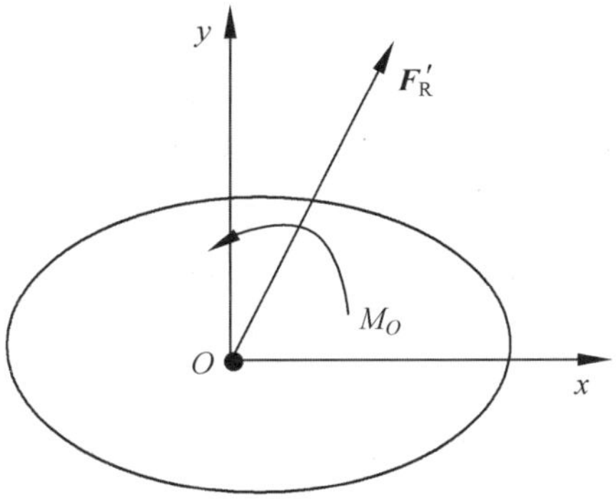

Fig. 2.38 Analytical method for calculating the master vector (图 2.38 解析法计算主矢量)

3. The concept of principal moments for arbitrary force systems in the plane (平面任意力系主矩的概念)

M_O is equal in magnitude to the algebraic sum of the moments of the forces at the simplified center O and is called the principal moment of the plane arbitrary force system at the simplified center O, we have(M_O大小等于各力对于简化中心 O 点力矩的代数和,称为平面任意力系对于简化中心 O 的主矩,有)

$$M_O = \sum m_O(\boldsymbol{F}_i)$$

Conclusion: A planar arbitrary force system acting on a rigid body, when simplified for an arbitrary centre O, can be replaced by a force $\boldsymbol{F}'_R$ and a force couple with a moment $\boldsymbol{M}_O$, where $\boldsymbol{F}'_R$ is the principal vector of the force system and act at point O; and $\boldsymbol{M}_O$ is the moment of the planar force system respect to the simplified centre O(Fig. 2.38). (**结论**:作用在刚体上的平面任意力系,当对于任意中心 O 点进行简化时,可用一个力 $\boldsymbol{F}'_R$ 及一个力矩为 $\boldsymbol{M}_0$ 的力偶进行替换, $\boldsymbol{F}'_R$ 为力系的主矢量且作用于 O 点. $\boldsymbol{M}_O$ 为平面力系对简化中心 O 点的矩(图 2.38).)

The principal vector and principal moment are calculated as(主矢量和主矩的计算公式为)

$$\begin{cases} F'_{Rx} = \sum F_x,\ F'_{Ry} = \sum F_y \\ F'_R = \sqrt{(\sum F_x)^2 + (\sum F_y)^2} \\ \cos(F'_R, i) = \dfrac{\sum F_x}{F'_R}, \cos(F'_R, j) = \dfrac{\sum F_y}{F'_R} \\ M_O = \sum m_O(\boldsymbol{F}_i) \end{cases} \tag{2.12}$$

4. Special circumstances(特殊情况)

(1) After the original force system has been simplified towards point O, the additional force couple system is equivalent to the original force system. The original force system is simplified to a combined force couple whose moment is equal to the algebraic sum of the moments of all force couples. Therefore, the principal moments are independent of the position of the center of simplification.(原力系向 O 点简化后,附加力偶系与原力系等效.原力系简化为一合力偶,该力偶的矩等于所有力偶矩的代数和.因此,主矩与简化中心的位置无关.)

(2) The original force system is reduced to a combined force acting at point O, in this case, the principal vector is the combined force, since the principal vector can replace the sum of all forces.(原力系简化为一个作用在 O 点的合力,此时主矢是合力,因为主矢量可代替所有力之和.)

(3) As shown in Fig. 2.39, this force system can be replaced by a combined force $\boldsymbol{F}_R$ acting at the new simplified center, which is at a distance $d = M_O/R$ from the former.(如图 2.39 所示,该力系可用一个作用于新的简化中心的合力 $\boldsymbol{F}_R$ 来代替,该简化中心与前者相距 $d = M_O/R$.)

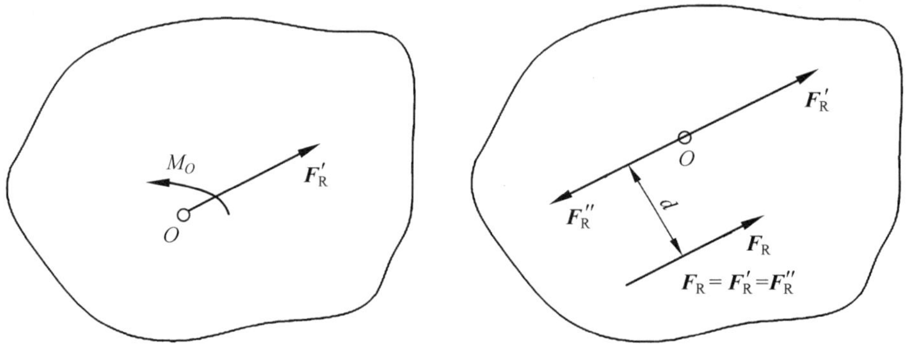

Fig. 2.39 Simplification of $F'_R \neq 0, M_O \neq 0$(图 2.39 $F'_R \neq 0, M_O \neq 0$ 的简化)

(4) The plane force system is in equilibrium.(平面力系处于平衡状态.)

Example 9: As shown in Fig. 2.40, a plane arbitrary force system with $F_1 = 150$ N, $F_2 = 200$ N, $F_3 = 300$ N and $F = F' = 200$ N. Find the resultant of simplifying the force system towards point O and the magnitude of the combined force and its distance d from the origin O.(**例题 9**:如图 2.40 所示,平面任意力系中 $F_1 = 150$ N,$F_2 = 200$ N,$F_3 = 300$ N,$F = F' = 200$ N.求力系向 O 点简化的结果,并求力系合力的大小及其与原点 O 的距离 d.)

Solution:(解:)

Calculation of the main vector, we have(主矢量的计算,有)

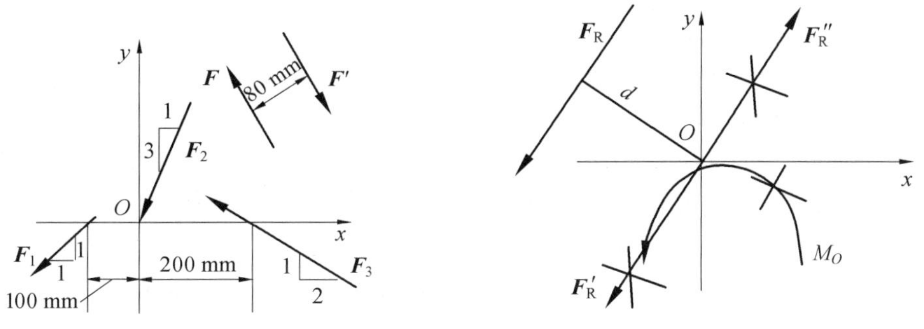

(a) Original system(原力系)　　(b) Simplification analysis(简化分析)

Fig. 2.40　Example 9 diagram (图 2.40　例题 9 图)

$$\boldsymbol{F}'_{Rx} = \sum F_x = -437.6\ \text{N}, \boldsymbol{F}'_{Ry} = \sum F_y = -161.639\ \text{N}$$

$$\boldsymbol{F}'_R = \sqrt{(\sum F_x)^2 + (\sum F_y)^2} = 466.5\ \text{N}$$

$$\cos < \boldsymbol{F}'_R, i > = \frac{\sum \boldsymbol{F}'_x}{\boldsymbol{F}'_R} = \frac{-437.64}{466.5} = -0.938$$

Calculation of principal moment, we have (主矩的计算,有)

$$M_O = \sum_{i=1}^{n} M_O(\boldsymbol{F}_i) = 21.44\ \text{N} \cdot \text{m}$$

$$d = \frac{M_O}{F_R} = 45.96\ \text{mm}$$

2.5　Equilibrium Conditions and Balancing Equations for Planar Arbitrary Force Systems (平面任意力系的平衡条件和平衡方程)

1. Geometric equilibrium conditions(几何平衡条件)

A necessary and sufficient condition for the equilibrium of an arbitrary force system in the plane is that both the principal vector of the force system and the principal moment of the force system for any point are equal to zero, that is(平面任意力系平衡的必要和充分条件是力系的主矢和力系对于任意点的主矩都等于零,即)

$$\boldsymbol{F}'_R = 0,\ M_O(\boldsymbol{F}_R) = \sum M_O(\boldsymbol{F}_i) = 0$$

2. The basic form of the analytical condition for the balancing equation of an arbitrary force system in the plane(平面任意力系平衡方程的基本形式)

The necessary and sufficient conditions for a system of arbitrary forces in the plane to be in equilibrium are that the algebraic sum of the projections of all the forces in the system on each of the two axes in the plane of action is equal to zero, and that the sum of the moments of the forces for any point in the plane is equal to zero, we have(平面任意力系平衡的必要与充分条件是力系中所有力在作用面内两个坐标轴中每一个轴上投影的代数和等于零,以及各力对于平面内任意点之矩的和等于零,有)

$$\sum F_x = 0, \ \sum F_y = 0, \ \sum M_O(\boldsymbol{F}_i) = 0 \tag{2.13}$$

These equations are often referred to as the basic balancing equations. Depending on the position of the forces, it is sometimes recommended to establish the equilibrium conditions in the form of two moment equations and a projection equation. (这些方程通常称为基本平衡方程.根据力的位置,有时建议以两个力矩方程和一个投影方程的形式建立平衡条件.)

A few notes are as follows. (以下是几点说明.)

(1) Three equations can be solved for only three unknown quantities. (三个方程只能求解三个未知量.)

(2) The two projected axes need not be perpendicular to each other, as long as they are not parallel. (两个投影坐标轴不一定互相垂直,只要不平行即可.)

(3) The projection axes are parallel or perpendicular to as many unknown forces as possible. (投影坐标轴尽可能与多个未知力平行或垂直.)

(4) In the moment equation, the center of moment is chosen to be the intersection of as many unknown forces as possible. (力矩方程中,矩心尽可能选为多个未知力的交点.)

In addition to the above basic forms, there are two other auxiliary forms of balancing equations for planar arbitrary force systems, the two-moment and three-moment balancing equations. (平面任意力系平衡方程除上述基本形式外,还有其他两种辅助形式,二力矩式及三力矩式的平衡方程.)

(1) The first auxiliary form of the analytical condition for the equilibrium of a planar arbitrary force system. (平面任意力系平衡解析条件的第一种辅助形式.)

Equilibrium condition of the two-moment equations is as follows. (两个力矩方程的平衡条件如下.)

A necessary and sufficient condition for the equilibrium of an arbitrary force system in the plane is that the sum of the moments of all these forces with respect to some centers A and B and the sum of their projections on the Ox axis not perpendicular to the line AB is zero, that is (平面任意力系平衡的必要与充分条件是所有这些力相对于某些中心 A 和 B 的力矩之和以及它们在不垂直于线段 AB 的 Ox 轴上的投影之和为零,即)

$$\sum \boldsymbol{F}_x = 0, \ \sum M_A(\boldsymbol{F}_k) = 0, \ \sum M_B(\boldsymbol{F}_k) = 0 \tag{2.14}$$

When using the two-moment equation, note that the chosen projection axis x-axis must not be perpendicular to the line of the two selected moment centers A and B, as shown in Fig. 2.41. (运用二矩式时注意所选择的投影轴 x 轴不得与选定的两个矩心 A、B 的连线垂直,如图 2.41 所示.)

(2) The second auxiliary form of the analytical condition for the equilibrium of a planar arbitrary force system. (平面任意力系平衡解析条件的第二种辅助形式.)

Equilibrium condition of the three-moment equations is as follows. (三个力矩方程的平衡条件如下.)

A necessary and sufficient condition for the equilibrium of an arbitrary force system in the plane is that the sum of the moments of all these forces with respect to any three centers A, B, C

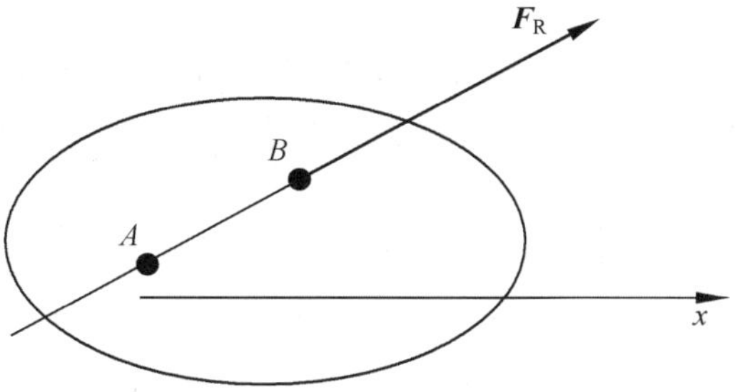

Fig. 2.41 Choice of projection axis when using the two-moment formula
(图 2.41 运用二矩式时投影轴的选择)

which are not in the same line is zero, that is (平面任意力系平衡的必要与充分条件是所有这些力相对于不在同一直线上的任何三个中心 A、B、C 的力矩之和为零,即)

$$\sum M_A(\boldsymbol{F}_k) = 0,\ \sum M_B(\boldsymbol{F}_k) = 0,\ \sum M_C(\boldsymbol{F}_k) = 0 \qquad (2.15)$$

Note: When applying the three-moment formula, the three points of the moment centers A, B, C must not be co-linear. (**注意**:应用三矩式时,矩心 A、B、C 三点不能共线.)

3. Analytical conditions for the equilibrium of planar parallel force systems (平面平行力系平衡的解析条件)

If all the forces are parallel to the axes (i. e. the Oy axis in Fig. 2.42), the analytical condition for the equilibrium of the force system is(如果所有力都平行于轴(如图 2.42 中 Oy 轴),则力系的平衡解析条件为)

$$\sum F_{ky} = 0,\ \sum M_O(\boldsymbol{F}_k) = 0 \qquad (2.16)$$

or(或)

$$\sum M_A(\boldsymbol{F}_k) = 0,\ \sum M_B(\boldsymbol{F}_k) = 0 \qquad (2.17)$$

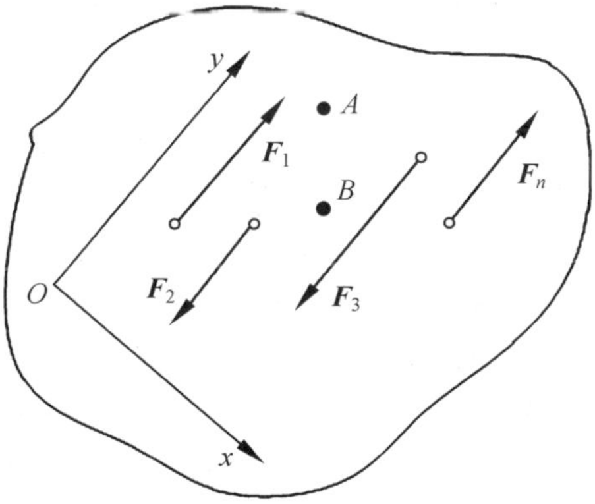

Fig. 2.42 Planar parallel force system (图 2.42 平面平行力系)

2.6 Equilibrium, Stationary and Super-stationary Problems of Body System(物体系的平衡、静定和超静定的问题)

1. External and internal constraints(外部和内部约束)

The constraints that connect the interactions between objects within a system are called internal constraints. The constraints acting on the system by external objects are called the external constraints of the system (Fig. 2.43). (连接系统内部各物体间相互作用的约束称为内部约束. 而外界物体作用于系统的约束称为该系统的外部约束(图 2.43).)

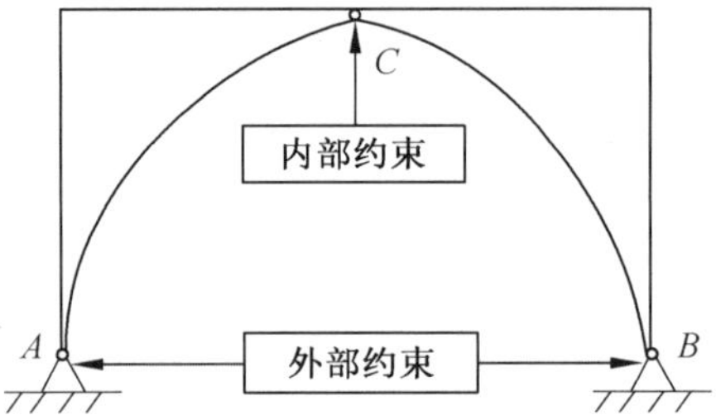

(a) External and internal constraints(外部和内部约束)

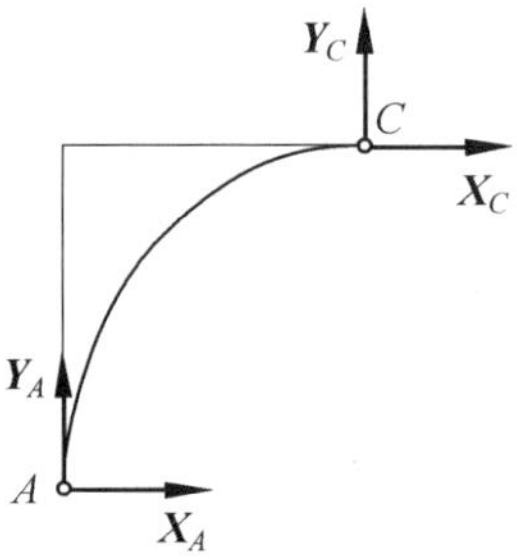

(b) Force analysis(受力分析)

Fig. 2.43 External and internal constraints and the force analysis(图 2.43 外部和内部约束及其受力分析)

2. Holistic approach(整体法)

One of the equilibrium conditions for the overall *ABC* structure was obtained when determining the restrained reaction forces, that is (在确定约束反力时,得到了 *ABC* 整体结构的平衡条件之一,即)

$$\sum F_{kx} = 0,\ \sum F_{ky} = 0,\ \sum M_A(\boldsymbol{F}_k) = 0$$

However, these equilibrium conditions are not sufficient to determine all the bound counter forces (four bound counter forces cannot be fully determined by three equations). Such systems are known as statically indeterminate systems. (然而,这些平衡条件不足以确定所有的约束反力(四个约束反力不能由三个方程完全确定). 这种系统称为静不定系统.)

By dividing the structure into parts through internal constraints, a supplementary balancing equation can be obtained by picking one of the parts (*AC* or *BC*) to solve this statically indeterminate problem (Fig. 2.43(b)). In this case, it is generally necessary to take one point, where the hinges are linked (here point *C*) to establish the moment equations, we have (通过内部约束将结构分成几部分,选取其中一个部件(*AC* 或 *BC*)得到一个补充的平衡方程,可以求解出这种静不定问题(图 2.43(b)). 在这种情况下,一般来说,有必要取一个点即铰链连接的地方(这里是 *C* 点)来建立力矩方程,有)

$$\sum M_C(\boldsymbol{F}_k) = 0$$

3. Splitting method(拆分法)

In determining the case of external and internal forces, the internal hinge joints are split at the interior and for each part (Fig. 2.44), the balancing equations are established (in one of three balancing equation forms).(在确定外力和内力的情况时,在内部铰链连接处被拆分,并且对于每个部分(图 2.44),建立平衡方程(以三种平衡方程形式之一).)

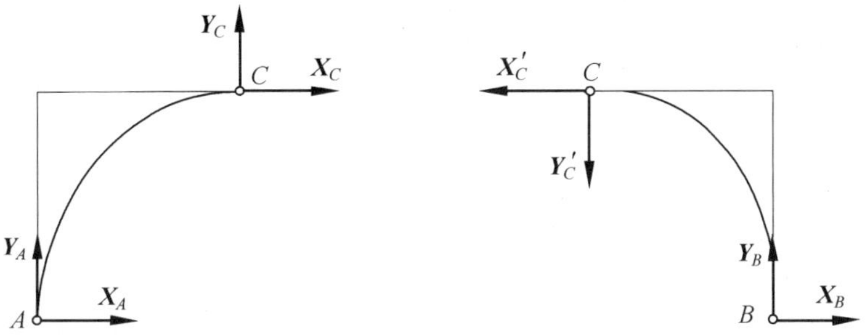

Fig. 2.44 Splitting method (图 2.44 拆分法)

For the AC part, we have(对 AC 部分,有)

$$\sum F_{kx}=0,\ \sum F_{ky}=0,\ \sum M_A(\boldsymbol{F}_k)=0$$

For the BC part, we have(对 BC 部分,有)

$$\sum F_{kx}=0,\ \sum M_B(\boldsymbol{F}_k)=0,\ \sum M_C(\boldsymbol{F}_k)=0$$

Considering $\boldsymbol{X}_C=-\boldsymbol{X}'_C$, $\boldsymbol{Y}_C=-\boldsymbol{Y}'_C$, and the size $X_C=X'_C$, $Y_C=Y'_C$, there will be six unknown counter forces in the system of six equations $\boldsymbol{X}_A$, $\boldsymbol{Y}_A$, $\boldsymbol{X}_B$, $\boldsymbol{Y}_B$, $\boldsymbol{X}_C$, $\boldsymbol{Y}_C$. In other words, the system is a static problem. (考虑到 $\boldsymbol{X}_C=-\boldsymbol{X}'_C$, $\boldsymbol{Y}_C=-\boldsymbol{Y}'_C$,而大小 $X_C=X'_C$, $Y_C=Y'_C$,在六个方程组中,将有六个未知反力 $\boldsymbol{X}_A$、$\boldsymbol{Y}_A$、$\boldsymbol{X}_B$、$\boldsymbol{Y}_B$、$\boldsymbol{X}_C$、$\boldsymbol{Y}_C$. 也就是说,系统是静定问题.)

To verify the correctness of the solution, the balancing equation can be written using the holistic method to test it. (为验证解的正确性,可以用整体法编写平衡方程来检验.)

Summary: The form of the balancing equation for an arbitrary system of forces in the plane is as follows. (**总结**:平面任意力系的平衡方程的形式如下.)

(1) Basic form. (基本形式.)

$$\sum F_x=0,\ \sum F_y=0,\ \sum M_A(\boldsymbol{F})=0$$

(2) Two moment equations type. (二力矩式.)

$$\sum F_x=0,\ \sum M_A(\boldsymbol{F})=0,\ \sum M_B(\boldsymbol{F})=0$$

The x-axis must not be perpendicular to the line between points A and B. (x 轴不得垂直于 A、B 两点的连线.)

(3) Three moment equations type. (三力矩式.)

$$\sum M_A(\boldsymbol{F})=0,\ \sum M_B(\boldsymbol{F})=0,\ \sum M_C(\boldsymbol{F})=0$$

Points A, B, C must not be co-linear. (A、B、C 三点不得共线.)

Consider: Is there a triple projection? (思考:是否存在三投影式?)

$$\begin{cases}\sum F_{x1}=0\\ \sum F_{x2}=0\\ \sum F_{x3}=0\end{cases}$$

Example 10: As in Fig. 2.45, it is known that $M=Pa$, find the restraint reaction force at A and B. (**例题** 10:如图 2.45 所示,已知 $M=Pa$, 求 A、B 处约束反力.)

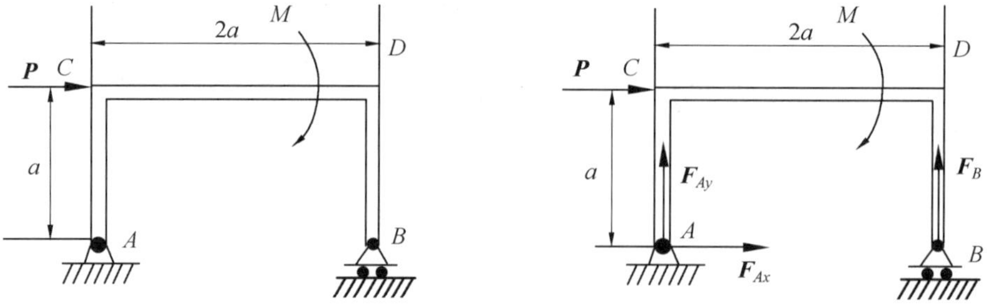

Fig. 2.45 Rigid frame and force diagram (图 2.45 刚架及受力图)

Solution 1:(**解法** 1:)

Choose the basic form of the balancing equation. (选取平衡方程的基本形式.)

(1) A rigid frame is taken as the study object. (取刚架为研究对象.)

(2) Draw the force diagram as shown in Fig. 2.45(b). (画受力图,如图 2.45(b)所示.)

(3) Set up a coordinate system and solve the equation, we have(建立坐标系,列方程求解,有)

$$\sum F_x = 0, F_{Ax} + P = 0$$

$$\sum F_y = 0, F_{Ay} + F_B = 0$$

$$\sum M_A(\boldsymbol{F}) = 0, F_B 2a - M - Pa = 0$$

Solving the above equation, we have(解上述方程,得到)

$$F_{Ax} = -P,\ F_{Ay} = -P,\ F_B = P$$

Solution 2:(**解法** 2:)

Two-moment balancing equations are(二矩式平衡方程为)

$$\sum F_x = 0, F_{Ax} + P = 0$$

$$\sum M_A(\boldsymbol{F}) = 0, F_B 2a - M - Pa = 0$$

$$\sum M_B(\boldsymbol{F}) = 0, -F_{Ay} 2a - Pa - M = 0$$

Solving the above equations, we have(解上述方程,得到)

$$F_{Ax} = -P, F_{Ay} = -P, F_B = P$$

Solution3:(**解法** 3:)

The three-moment form balancing equations are(三矩式平衡方程为)

$$\begin{cases} \sum M_A(\boldsymbol{F}) = 0, F_B 2a - M - Pa = 0 \\ \sum M_B(\boldsymbol{F}) = 0, -F_{Ay} 2a - Pa - M = 0 \\ \sum M_C(\boldsymbol{F}) = 0, F_{Ax} a + F_B 2a - M = 0 \end{cases}$$

Solving the above equations, we have(解上述方程,得到)

$$F_{Ax} = -P, F_{Ay} = -P, F_B = P$$

Example 11: Find the restraint reaction force at A in Fig. 2. 46 (a). (**例题 11**: 求图 2.46(a)中 A 处的约束反力.)

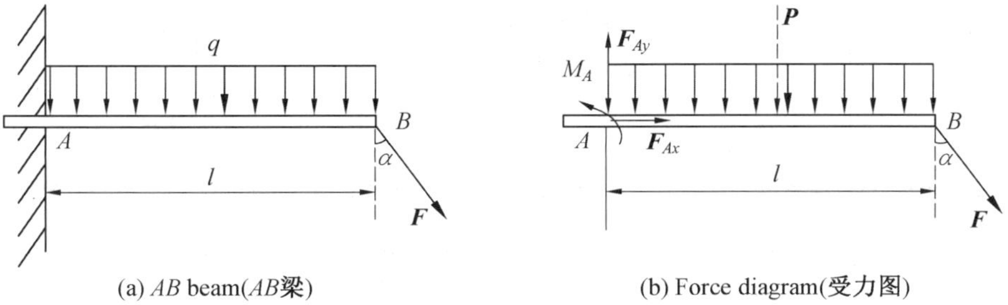

Fig. 2.46 Example 11 diagram (图 2.46 例题 11 图)

Solution: (解:)

The diagram shows a planar fixed end restraint or fixed end support at A, with distributed loads and concentrated forces acting on the beam. The representation of the reaction forces of the fixed end bearing has been described in Chapter 1, and the following is a brief explanation of the representation of the reaction forces and the simplification of the distributed loads. (图中 A 处为一个平面固定端约束或固定端支座,梁上作用有分布载荷及集中力. 在第 1 章中已经介绍过固定端支座反力的表示,下面再对其反力表示及分布载荷的简化做一些补充说明.)

(1) Fixed end support. (固定端支座.)

A constraint that can neither be moved nor rotated is also called a fixed end (insertion end) constraint. A sketch of this is shown in Fig. 2. 47 (a). The forces on the member in the plane fixed end are more complex and can be seen as the action of a plane arbitrary force system, as shown in Fig. 2. 47 (b). Simplifying the plane arbitrary force system towards end point A gives a force $\boldsymbol{F}_A$ and a force couple M_A where the force $\boldsymbol{F}_A$ limits the movement of the member and the force couple M_A limits the rotation of the member. Since the line of action of the force $\boldsymbol{F}_A$ cannot be determined, it is generally decomposed into two orthogonal components $\boldsymbol{F}_{Ax}$ and $\boldsymbol{F}_{Ay}$, as shown in Fig. 2. 47 (c). (既不能移动,又不能转动的约束又称固定端(插入端)约束,其简图如图 2.47(a)所示. 在平面固定端里构件的受力比较复杂,可看成一个平面任意力系的作用,如图 2.47(b)所示. 把平面任意力系向端点 A 简化,得到一个力 $\boldsymbol{F}_A$ 和一个力偶 M_A,其中力 $\boldsymbol{F}_A$ 限制构件的移动,力偶 M_A 限制构件的转动. 由于力 $\boldsymbol{F}_A$ 的作用线不能确定,因此一般分解为两个正交分力 $\boldsymbol{F}_{Ax}$ 和 $\boldsymbol{F}_{Ay}$,如图 2.47(c)所示.)

(2) Distributed loads. (分布载荷.)

In Fig. 2. 48, $q(x)$ is the set of distributed loads, and the concentrated force on a micro-segment $\mathrm{d}x$ is $\mathrm{d}P=q(x)\mathrm{d}x$, the combined force on the whole member is (图 2.48 中,$q(x)$ 为分布载荷的集度,在一个微段 $\mathrm{d}x$ 上的集中力为 $\mathrm{d}P=q(x)\mathrm{d}x$,则整个构件上的合力为)

$$P=\sum \mathrm{d}P=\int_0^l q(x)\,\mathrm{d}x$$

The line of action of the combined force is parallel to the direction of the distributed load, and the position of the line of action of the combined force can be determined by the theorem of the combined moment. Let the distance of the combined force P to the left end A be h. Then,

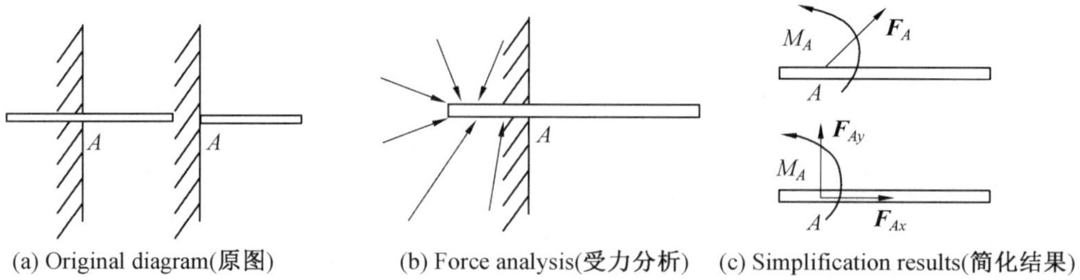

Fig. 2.47 Sketch of the fixed end constraint and decomposition of the constraint reaction force for a plane view (图 2.47 平面固定端约束简图及约束反力分解)

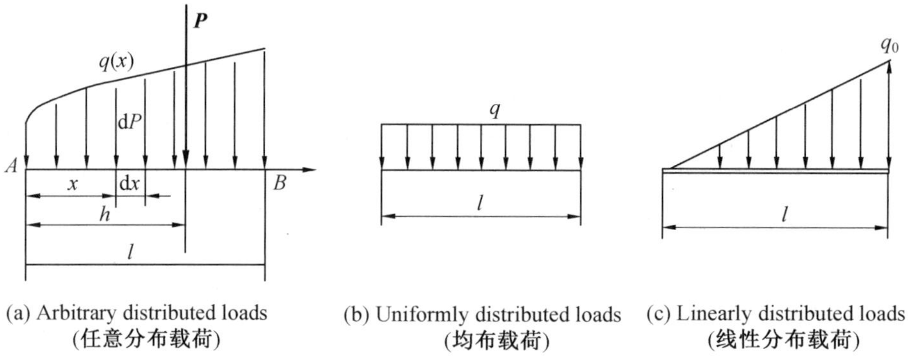

Fig. 2.48 Distributed loads (图 2.48 分布载荷)

from the combined moment theorem, we get(合力作用线与分布载荷方向平行,合力作用线位置可由合力矩定理来确定. 设合力 P 到左端 A 的距离为 h,则由合力矩定理得到)

$$Ph = \sum \mathrm{d}P \cdot x = \int_0^l q(x) x \mathrm{d}x$$

Location of the line of action of the combined forces is (合力作用线位置为)

$$h = \frac{\int_0^l q(x) x \mathrm{d}x}{\int_0^l q(x) \mathrm{d}x}$$

For the two common exceptions: (对于常见的两个特例:)

① Uniformly distributed load is(Fig. 2.48(b))(均布载荷为(图 2.48(b)))

$$P = \int_0^l q(x) \mathrm{d}x = ql \text{ , } h = \frac{\int_0^l q(x) x \mathrm{d}x}{\int_0^l q(x) \mathrm{d}x} = \frac{l}{2}$$

② Triangular distribution of loads is(Fig. 2.48(c))(三角形分布载荷为(图 2.48(c)))

$$q(x) = \frac{q_0}{l} x$$

$$P = \int_0^l q(x) \mathrm{d}x = \int_0^l \frac{q_0}{l} x \mathrm{d}x = \frac{1}{2} q_0 l \text{ , } h = \frac{\int_0^l q(x) x \mathrm{d}x}{\int_0^l q(x) \mathrm{d}x} = \frac{2l}{3}$$

The AB beam is taken as the study object and its force analysis is shown in Fig. 2.46(b). The balancing equation is given by the basic form of the equilibrium condition, we have(取 AB 梁为研究对象,其受力分析如图 2.46(b)所示. 通过平衡条件基本形式列出平衡方程,有)

$$\sum F_x = 0, F_{Ax} + F\sin\alpha = 0$$

$$\sum F_y = 0, F_{Ay} - ql - F\cos\alpha = 0$$

$$\sum M_A(\boldsymbol{F}) = 0, M_A - ql\frac{l}{2} - Fl\cos\alpha = 0$$

Solve the above equation, we have(解上述方程,得到)

$$F_{Ax} = -F\sin\alpha, F_{Ay} = ql + F\cos\alpha, M_A = Pl\cos\alpha + \frac{1}{2}ql^2$$

Example 12: A wind force $F = 20$ kN is applied to a truss weighing 100 kN as shown in Fig. 2.49. Find the support reaction force. (**例题** 12:如图 2.49 所示,风力 $F=20$ kN,作用于重为 100 kN 的桁架,求支反力.)

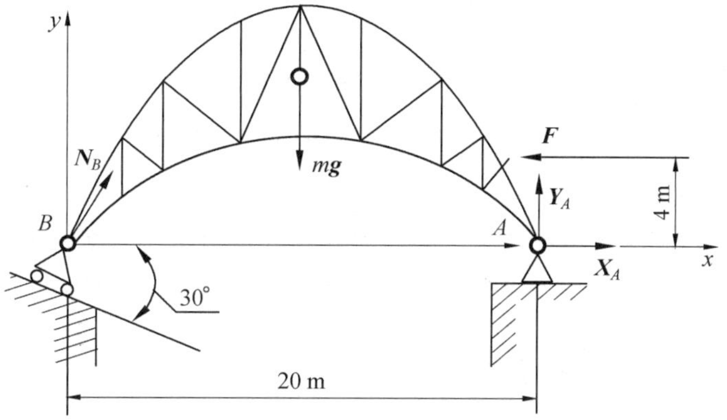

Fig. 2.49 Truss(图 2.49 桁架)

Solution: (**解**:)

(1) Selection of trusses as balancing objects. (选择桁架作为平衡对象.)

(2) Release constraints and replacing the restraint with a counterforce (Fig. 2.49). (解除约束,用反力代替约束(图 2.49).)

(3) Three balancing equations are listed by force analysis. Comparing the quantities of the unknown quantities N_B, X_A, Y_A with those of the equations, we conclude that the system is stationary. (通过受力分析,列出三个平衡方程. 将未知量 N_B、X_A、Y_A 的数量与方程的数量进行比较后,得出结论:该系统是静定的.)

(4) Write the balancing equation for the specific problem, we have(针对具体问题写出平衡方程,有)

$$\sum F_x = 0, N_B\cos 60° + X_A - 20 = 0$$

$$\sum F_y = 0, N_B\cos 30° + Y_A - 100 = 0$$

$$\sum M_A(\boldsymbol{F}) = 0, -100 \times 10 + Y_A \times 20 + 20 \times 4 = 0$$

(5) Solve the system of balancing equations, we have(求解平衡方程组,得到)

$$Y_A = 46\ \text{kN},\ N_B = 62.4\ \text{kN},\ X_A = -11.2\ \text{kN}$$

The negative result of the reaction force X_A indicates that its true direction is the opposite of the positive direction assumed in the diagram. (反力 X_A 结果为负,说明它的真实方向与图中所假设的正向相反.)

Example 13: As shown in Fig. 2.50, it is known that $P=0.4$ kN, $Q=1.5$ kN, $\sin\alpha=4/5$, $AB=BC=l$. Find the restrained reaction forces of the supports A and C. (**例题** 13:如图 2.50 所示,已知 $P=0.4$ kN,$Q=1.5$ kN,$\sin\alpha=4/5$,$AB=BC=l$,求支座 A、C 的约束反力.)

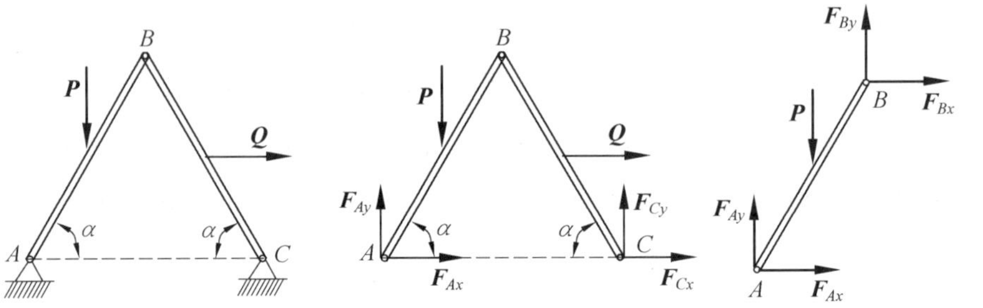

(a) Support(支座)　(b) Force diagram of system (整体受力图)　(c) Force diagram of AB (AB 受力图)

Fig. 2.50　Example 13 diagram (图 2.50　例题 13 图)

Solution: (**解**:)

(1) Taking the whole system as the study object, the force diagram is shown in Fig. 2.50(b) and the balancing equation is listed as(取整体为研究对象,受力图如图 2.50(b)所示,列出平衡方程,有)

$$\sum M_A(\boldsymbol{F}) = 0, F_{Cy}2l\cos\alpha - P\frac{l}{2}\cos\alpha - Q\frac{l}{2}\sin\alpha = 0 \tag{2.18}$$

$$\sum F_y = 0, F_{Ay} + F_{Cy} - P = 0 \tag{2.19}$$

$$\sum F_x = 0, F_{Ax} + F_{Cx} + Q = 0 \tag{2.20}$$

Solve the above equation, we have(解上述方程,得到)

$$F_{Ay} = -0.2\ \text{kN},\ F_{Cy} = 0.6\ \text{kN}$$

(2) Taking AB as the object of study, the force diagram is shown in Fig. 2.50(c) and the balancing equation is listed as(取 AB 为研究对象,受力图如图 2.50(c)所示,列出平衡方程,有)

$$\sum M_B(\boldsymbol{F}) = 0, F_{Ax}l\sin\alpha + P\frac{l}{2}\cos\alpha - F_{Ay}l\cos\alpha = 0 \tag{2.21}$$

We have(解得)

$$F_{Ax} = -0.3\ \text{kN}$$

Substituting F_{Ax} into Eq. (2.20), we get(把 F_{Ax} 代入式(2.20)中,得到)

$$F_{Cx} = -1.2\ \text{kN}$$

Example 14: As shown in Fig. 2.51(a), it is known that $M=10$ kN·m, $q=2$ kN/m, find the restrained reaction forces of supports A and C. (**例题** 14:如图 2.51(a)所示,已知 $M=$

10 kN · m, q=2 kN/m,求支座 A、C 的约束反力.)

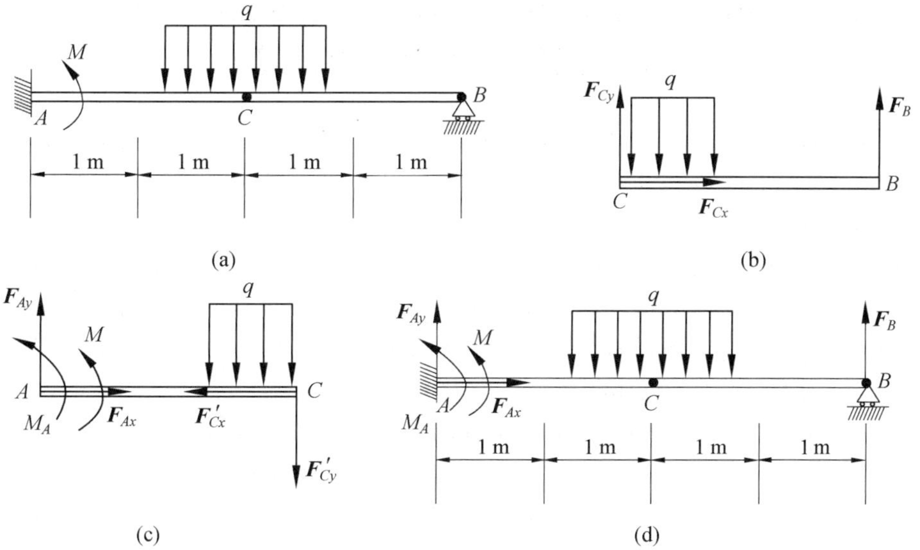

Fig. 2.51　Example 14 diagram(图 2.51　例题 14 图)

Solution:(解:)

(1) Take BC as the study object and the force diagram is shown in Fig. 2.51(b), then we have (取 BC 为研究对象,受力图如图 2.51(b)所示,则有)

$$\sum F_x = 0,\ F_{Cx} = 0$$

$$\sum M_C(\boldsymbol{F}) = 0, F_B \cdot 2 - q \cdot 1 \cdot 0.5 = 0$$

$$\sum F_y = 0, F_{Cy} + F_B - q \cdot 1 = 0$$

We have(解得)

$$F_B = 0.5\ \text{kN}, F_{Cx} = 0, F_{Cy} = 1.5\ \text{kN}$$

(2) The other unknown forces can be solved for by taking AC or the whole as the object of study, and the force diagrams are shown in Fig. 2.51(c) and Fig. 2.51(d). The following solution takes AC as an example and lists the balancing equations, those are(取 AC 或整体为研究对象可以求解其他的未知力,受力图如图 2.51(c)和图 2.51(d)所示. 下面的求解以 AC 为例,列出平衡方程,有)

$$\sum F_x = 0, F_{Ax} - F'_{Cx} = 0$$

$$\sum F_y = 0, F_{Ay} - F'_{Cy} - q \cdot 1 = 0$$

$$\sum M_A(\boldsymbol{F}) = 0, M_A + M - q \cdot 1 \cdot 1.5 - F'_{Cy} \cdot 2 = 0$$

We have(解得)

$$F_{Ax} = 0, F_{Ay} = 3.5\ \text{kN}, M_A = -4\ \text{kN} \cdot \text{m}$$

2.7 Calculation of Internal Forces of a Simple Planar Truss (平面简单桁架的内力计算)

In the engineering field, truss structures are commonly used in structures such as house buildings, bridges, cranes and TV towers. (在工程领域,房屋建筑、桥梁、起重机、电视塔等结构中常用桁架结构.)

A truss is a structure in which the bars are hinged to each other at both ends, and its geometry remains unchanged after the forces are applied. The hinged joints of the bars in a joist are called nodes. (桁架是一种将杆件彼此在两端用铰链相互连接而成的结构,它在受力后几何形状不变. 桁架中杆件的铰链接头称为节点.)

The advantages of trusses: the rods are mainly subjected to tension or pressure, which allows the full use of materials, saves materials and reduces the weight of the structure. In order to simplify the calculation of joists, the following assumptions are often used in engineering practice. (桁架的优点:杆件主要承受拉力或压力,可以充分发挥材料的作用,节约材料,减轻结构的重力. 为简化桁架的计算,工程实际中常采用以下几条假设.)

(1) The bars of the trusses are straight. (桁架的杆件都是直的.)

(2) The bars are connected by smooth hinges. (杆件用光滑的铰链连接.)

(3) The loads all act on the nodes. (载荷均作用在节点上.)

(4) The weight is distributed equally over the nodes. (质量平均分配在节点上.)

Such a truss is called an ideal truss. Of course, the actual joist differs somewhat from the above assumptions. However, these assumptions simplify the calculations and the results obtained meet the actual engineering needs. According to these assumptions, the rods of the truss can be regarded as diametrical rods. (这样的桁架称为理想桁架. 当然,实际的桁架与上述假设有一定差别. 但上述假设能够简化计算,而且所得结果符合工程实际需要. 根据这些假设,桁架的杆件都可看成二力杆.)

In this section, only the static joist in the truss is studied. If any rod is removed from the truss, the truss will be deformed movably, then this truss is called redundant rod truss, and it can be proved that only redundant rod truss is static truss (Fig. 2.52). If the removal of a few rods does not deform the joist, the joist is called a redundant rod joist (Fig. 2.53). The truss without residual rod shown in Fig. 2.52 is based on a triangular frame and requires two additional rods for each additional node, thus forming a truss also known as a planar simple truss, which can be proved to be a static truss. (本节只研究桁架中的静定桁架. 如果从桁架中任意去掉一根杆件,桁架就会活动变形,则这种桁架称为无余杆桁架,可以证明只有无余杆桁架才是静定桁架(图 2.52). 如果去掉几根杆件仍然不会使桁架活动变形,则这种桁架称为有余杆桁架(图 2.53). 图 2.52 所示的无余杆桁架以三角形框架为基础,每增加一个节点,需要增加两根杆件,这样构成的桁架又称平面简单桁架,可以证明这种桁架是静定桁架.)

Two methods of finding the internal forces of truss bars: the nodal method and the section method. (求桁架杆件内力的两种方法:节点法和截面法.)

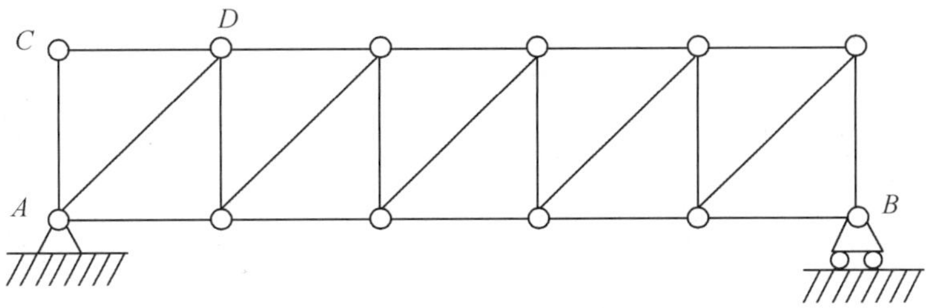

Fig. 2.52　Truss without spare rods（图 2.52　无余杆桁架）

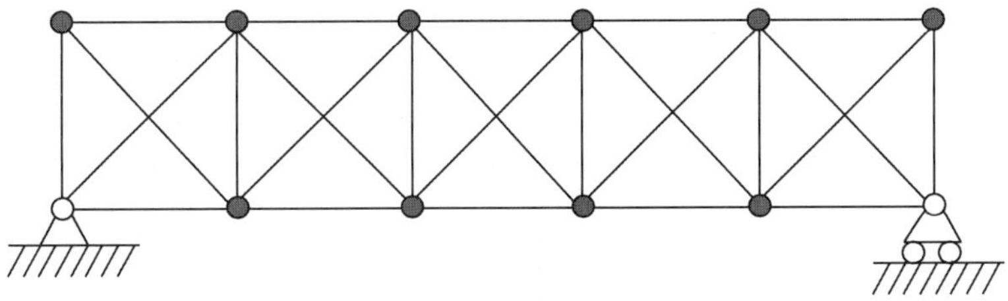

Fig. 2.53　Truss with surplus rods（图 2.53　有余杆桁架）

Example 15: Find the forces on each rod of the truss in Fig. 2.54.(**例题** 15:求图 2.54 中桁架各杆的力.)

Solution:(**解**:)

(1) Take the whole as the object of study, and its force diagram is shown in Fig. 2.54(b). List the balancing equations of the plane arbitrary force system, we have(取整体为研究对象,其受力图如图 2.54(b)所示.列出平面任意力系平衡方程,有)

$$\sum F_x = 0, F_{Ax} + 20 = 0$$

$$\sum F_y = 0, F_{Ay} + F_{By} - 40 = 0$$

$$\sum M_A(\boldsymbol{F}) = 0, F_{By} \cdot 5 - 120 = 0$$

We have(得到)

$$F_{Ax} = -20 \text{ kN},\ F_{Ay} = 16 \text{ kN},\ F_B = 24 \text{ kN}$$

(2)Take node C as the object of study, the force diagram is shown in Fig. 2.54(c), and list the balance equations of the plane arbitrary force system, we have (取节点 C 为研究对象,受力图如图 2.54(c)所示,列出平面任意力系平衡方程,有)

$$\sum F_x = 0, F_2 + 20 = 0$$

$$\sum F_y = 0, F_1 = 0$$

We have(得到)

$$F_1 = 0 \text{ N}, F_2 = -20 \text{ kN}$$

(3)Take node A as the object of study, its balance equations are(取节点 A 为研究对象,其平衡方程为)

$$\begin{cases} \sum F_x = 0, F_{Ax} + F_4 + F_3 \cos 45° = 0 \\ \sum F_y = 0, F_{Ay} + F_1 + F_3 \sin 45° = 0 \end{cases}$$

We have(得到)

$$F_3 = -16\sqrt{2}\ \text{kN}, F_4 = 36\ \text{kN}$$

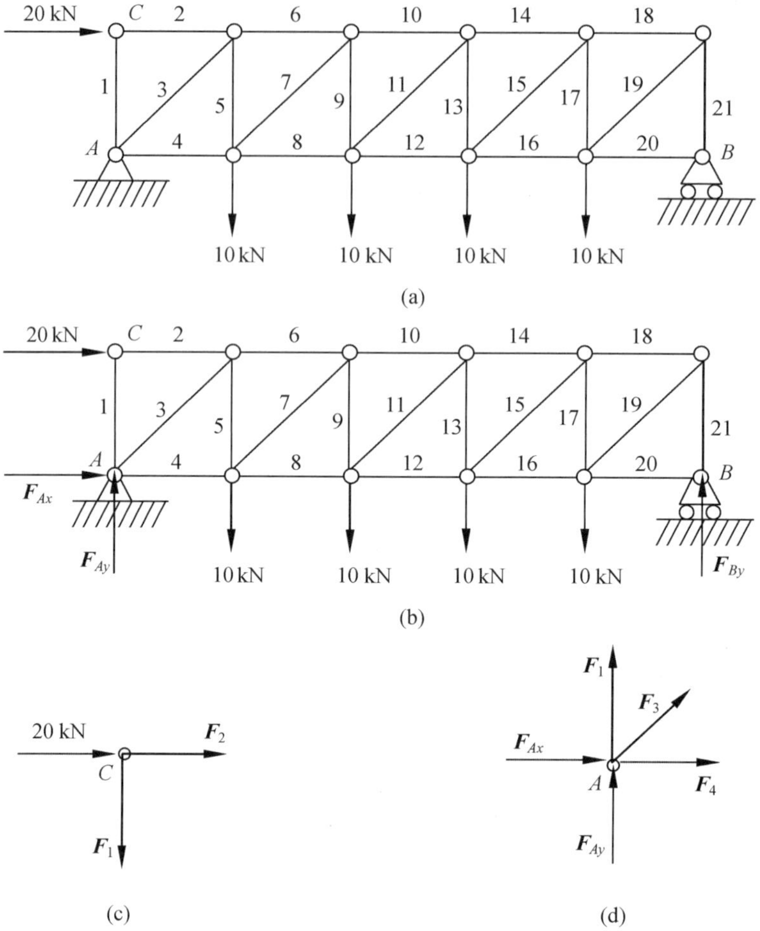

Fig. 2.54 Example 15 diagram (例题 15 图)

Example 16: Use the section method to find the forces in each of the rods of trusses 6, 7, 8 in Fig. 2.54(a). (**例题** 16:用截面法求图 2.54(a)中桁架 6、7、8 各杆的力.)

Solution: (**解**:)

(1) With reference to the above example, the external restraint reaction force is found for the whole body, $F_{Ax}=-20$ kN, $F_{Ay}=16$ kN, $F_B=24$ kN. (参照上例,以整体为研究对象,求得外部约束反力,$F_{Ax}=-20$ kN, $F_{Ay}=16$ kN, $F_B=24$ kN.)

(2) Assume a truncated system with one section, as required for the solution of the problem. Take a section as the object of study and calculate the internal forces of the requested bar. (根据解题的需要,假想用一截面截断体系. 取某一部分为研究对象,计算所求杆件内力.)

As shown in Fig. 2.55, suppose that the three rods 6, 7, 8 are truncated along a section *mm*, retaining the simple left part of the force as the object of study, draw a force diagram and list the balancing equations, we have(如图 2.55 所示,假想沿一截面 *mm* 截断 6、7、8 三杆,保留受力简单的左边部分为研究对象,画受力图,列平衡方程,有)

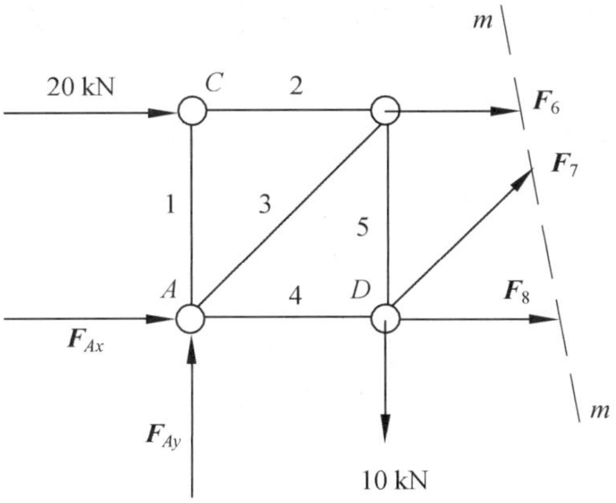

Fig. 2.55 Example 16 diagram（图 2.55 例题 16 图）

$$\sum F_y = 0,\ F_{Ay} + F_7 \sin 45° - 10 = 0$$

$$\sum M_D(\boldsymbol{F}) = 0,\ F_6 \cdot 1 + F_{Ay} \cdot 1 + 20 \cdot 1 = 0$$

$$\sum M_C(\boldsymbol{F}) = 0,\ F_8 \cdot 1 + F_7 \cdot \sqrt{2} + F_{Ax} \cdot 1 - 10 \cdot 1 = 0$$

Solving which we have(求解得到)

$$F_6 = -36\ \text{kN}, F_7 = -6\sqrt{2}\ \text{kN}, F_8 = -42\ \text{kN}$$

2.8 Questions(思考题)

1. Is a combined force always greater than a divided force? Are the three forces converging at a point, but on different sides, a balanced force system?（合力是否一定比分力大？三力汇交于一点，但不同面，是平衡力系吗？）

2. When using the analytical method to find the combined forces of a plane confluence system, are the combined forces obtained the same if different right-angle axes are taken?（用解析法求平面汇交力系的合力时，若取不同的直角坐标轴，所求得的合力是否相同？）

3. Do the x-axis and y-axis have to be perpendicular to each other when using the analytical method to find the equilibrium of a coplanar concurrent forces system? When the two axes are not perpendicular, can the balancing equations $\sum F_x = 0$ and $\sum F_y = 0$ satisfy the equilibrium conditions of the force system?（用解析法求平面汇交力系的平衡问题时，x 轴与 y 轴是否一定要相互垂直？当两轴不垂直时，建立的平衡方程 $\sum F_x = 0$ 和 $\sum F_y = 0$ 能否满足力系的平衡条件？）

4. When transmission lines have the same span, is it easier for the wire to break with less sag or is it easier for the wire to break with more sag and why?（输电线跨度相同时，是电线下垂量小，电线易于拉断，还是下垂量大，电线易于拉断，为什么？）

5. For the three structures shown in Fig. 2.56, the members have no self-weight and ignore friction, $\theta = 60°$. The same horizontal force $\boldsymbol{F}$ acts at B. Is the restraining force at hinge A the same? Draw a diagram to show their magnitude and direction.（图 2.56 所示的三种结构，构件

自重不计,忽略摩擦,$\theta = 60°$.B 处都作用相同的水平力 $\boldsymbol{F}$,铰链 A 处的约束力是否相同?画图表示其大小和方向.)

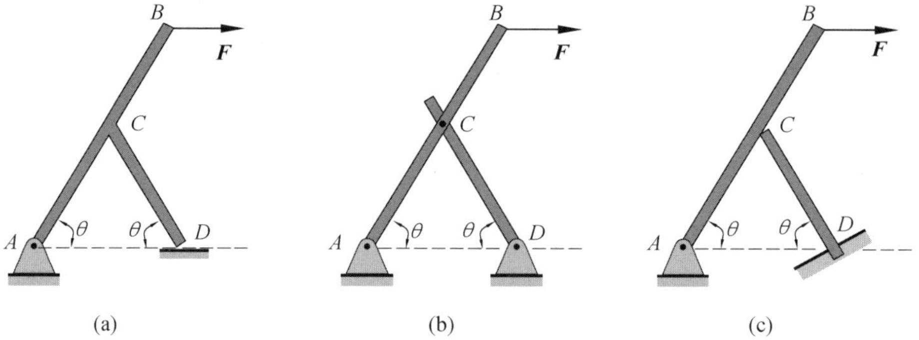

Fig. 2.56　Diagram of Question 5(图 2.56　思考题 5 图)

6. What are the differences and similarities between moment ofthe force and the moment of couple?(力矩与力偶矩有何异同?)

7. In each of the diagrams in Fig. 2.57, the moments of the forces or couples on point A are equal. Do they cause the same support restraint force?(在图 2.57 各图中,力或力偶对点 A 的矩都相等,它们引起的支座约束力是否相同?)

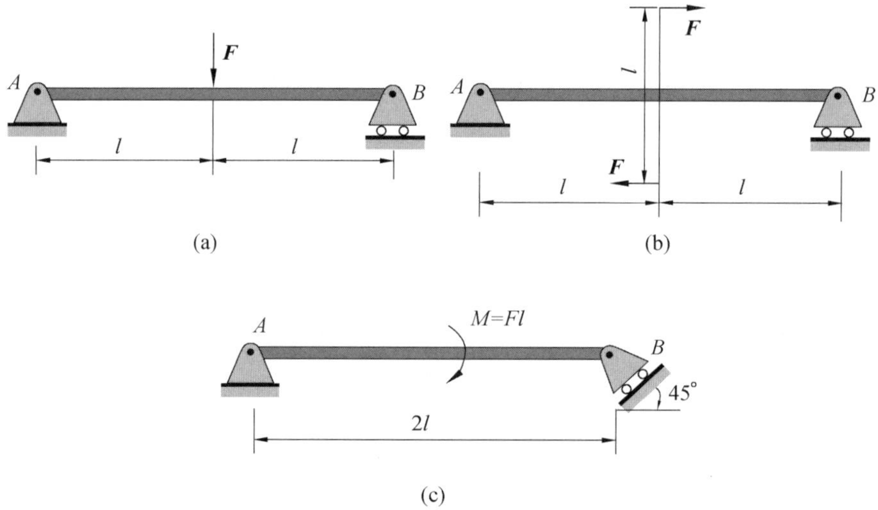

Fig. 2.57　Diagram of Question 7(图 2.57　思考题 7 图)

8. When constructing various bridges in engineering, there is a distinction between a two-column pier design and a single-column pier design, as shown in Fig. 2.58. If only the vehicle load $\boldsymbol{P}_1$ is considered, without considering the weight of the bridge, what will happen if the load exceeds the design limit? With a bridge of the same width, which design option is more reasonable if only the bridge is considered for overturning? Which is the more reasonable design solution if the bridge is considered to be of the same width and only whether the bridge overturns, taking into account the weight $\boldsymbol{P}_2$ of the bridge?(工程中修各种桥梁时,有双柱墩设计和单柱墩设计之分,如图 2.58 所示.若不考虑桥梁自重,只考虑车辆载荷 $\boldsymbol{P}_1$,载荷超过设计极限时,将产生

什么情况? 在桥同宽的情况下,若只考虑桥是否侧翻,哪种设计方案更合理? 而若考虑桥梁自重 $\boldsymbol{P}_2$,在桥同宽的情况下,只考虑桥是否侧翻,哪种设计方案更合理?)

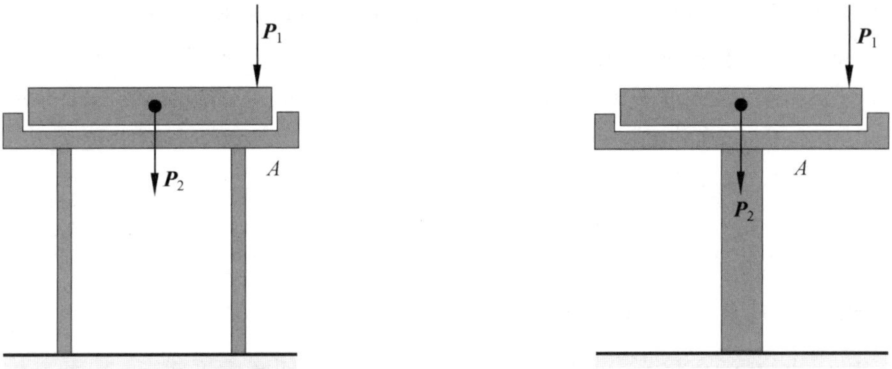

Fig. 2.58 Diagram of Question 8(图 2.58 思考题 8 图)

9. The principal moments of a planar force system simplified to points A and B are both zero. Could it be a force couple? Is it possible to balance? (某平面力系向 A、B 两点简化的主矩皆为零,此力系简化的最终结果可能是一个力吗,可能是一个力偶吗,可能平衡吗?)

10. Is it possible for a system of coplanar concurrent forces system to be simplified to a point beyond the point of confluence and the resultant force to be one force? Could it be a couple? Could it be a force and a couple? (平面汇交力系向汇交点以外一点简化,其结果可能是一个力吗,可能是一个力偶吗,可能是一个力和一个力偶吗?)

11. What might be the final result of simplifying a system of forces in a plane to any point in the same plane? (某平面力系向同平面内任一点简化的结果都相同,此力系简化的最终结果可能是什么?)

12. An arbitrary force system in the plane is simplified to a point A to obtain a force F'_{RA} ($F'_{RA} \neq 0$) and a force couple with moment $M_A \neq 0$ and B is another point in the plane. (某平面任意力系向点 A 简化得一个力 F'_{RA} ($F'_{RA} \neq 0$)与一个力矩为 $M_A \neq 0$ 的力偶,B 为平面内另一点.)

(1) Is it possible to simplify towards point B to obtain only one force couple? (向点 B 简化仅得一力偶是否可能?)

(2) Is it possible to get only one force by simplifying towards point B? (向点 B 简化仅得一力是否可能?)

(3) Is it possible to simplify to point B to get $F'_{RA} = F'_{RB}$, $M_A \neq M_B$? (向点 B 简化得 $F'_{RA} = F'_{RB}$, $M_A \neq M_B$ 是否可能?)

(4) Is it possible to simplify to point B to get $F'_{RA} = F'_{RB}$, $M_A = M_B$? (向点 B 简化得 $F'_{RA} = F'_{RB}$, $M_A = M_B$ 是否可能?)

(5) Is it possible to simplify to point B to get $F'_{RA} \neq F'_{RB}$, $M_A = M_B$? (向点 B 简化得 $F'_{RA} \neq F'_{RB}$, $M_A = M_B$ 是否可能?)

(6) Is it possible to simplify to point B to get , $F'_{RA} \neq F'_{RB}$ $M_A \neq M_B$? (向点 B 简化得 $F'_{RA} \neq F'_{RB}$, $M_A \neq M_B$ 是否可能?)

13. $OABC$ in Fig. 2. 59 is a square with side length a. A planar force system is known to have a principal vector of size F'_{RA} and a principal moment of unknown magnitude and direction when simplified to point A. The system is also known to have a combined force pointing to point O. It is also known that the force system simplifies to point B to give a combined force pointing to point O. Give the force system to point C to simplify the principal vector (magnitude and direction) and the principal moment (magnitude and direction). (图 2. 59 中 $OABC$ 为一边长为 a 的正方形，已知某平面力系向点 A 简化得一大小为 F'_{RA} 的主矢与一主矩，主矩大小、方向未知. 又已知该力系向点 B 简化得一合力，合力指向点 O. 请给出该力系向点 C 简化的主矢（大小和方向）和主矩（大小与转向）.）

14. In Fig. 2. 59, if an arbitrary force system in a plane satisfies $\sum F_y = 0$, $\sum M_B = 0$, then (determine the true or false):（在图 2. 59 中，若某平面任意力系满足 $\sum F_y = 0$，$\sum M_B = 0$，则（判断正误）：）

(1) Must have $\sum M_A = 0$;（必有 $\sum M_A = 0$；）

(2) Must have $\sum M_C = 0$;（必有 $\sum M_C = 0$；）

(3) There may be $\sum F_x = 0$, $\sum M_O \neq 0$;（可能有 $\sum F_x = 0$，$\sum M_O \neq 0$；）

(4) There may be $\sum F_x \neq 0$, $\sum M_O = 0$.（可能有 $\sum F_x \neq 0$，$\sum M_O = 0$.）

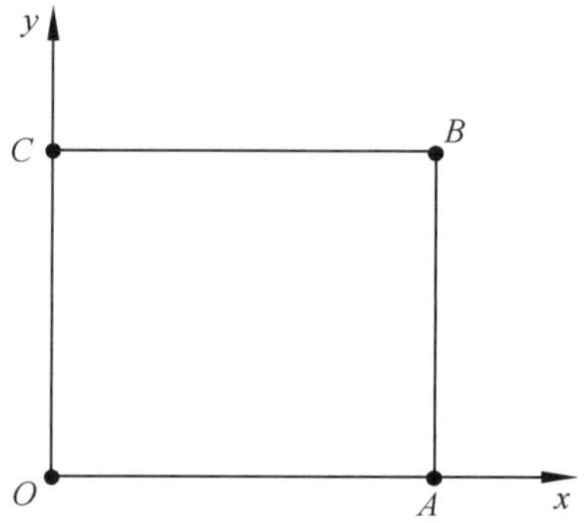

Fig. 2. 59　Diagram of Question 13 and 14（图 2. 59　思考题 13 题与 14 题图）

15. How do you determine between stationary and stationary indeterminate problems? Which of the six situations shown in Fig. 2. 60 are stationary and indeterminate problems?（怎么判断静定和静不定问题？图 2. 60 所示的六种情形中哪些是静不定问题？）

16. Fig. 2. 61 represents several cases of rod hinging in a truss, where no load acts on the nodes of Fig. 2. 61(a) and Fig. 2. 61(c) and an external force $\boldsymbol{F}$ acts on node B of Fig. 2. 61(b), the line of action of which is along the horizontal rod. Which of the seven bars in the diagram must have an internal force equal to zero (known as a zero bar)? Why?（图 2. 61 表示一桁架中杆件铰链的几种情况，图 2. 61(a) 和图 2. 61(c) 的节点上没有载荷作用，图 2. 61(b) 的节点 B 上受到外力 $\boldsymbol{F}$ 作用，该力作用线沿水平杆. 求图中七根杆件中哪些杆件的内力一定等于零（称为零杆），为什么？）

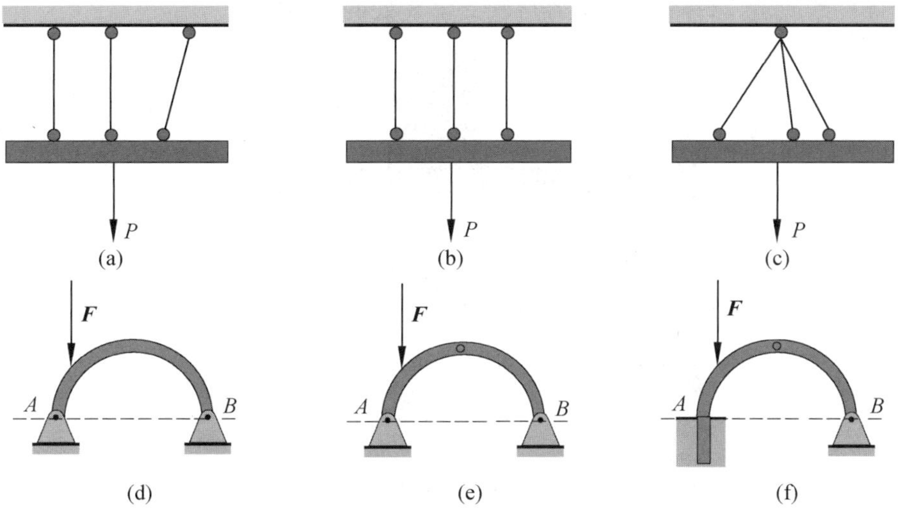

Fig. 2.60 Diagram of Question 15(图 2.60 思考题 15 题图)

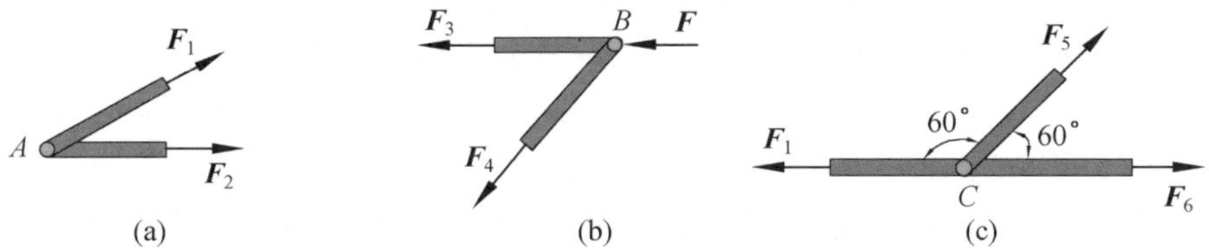

Fig. 2.61 Diagram of Question 16(图 2.61 思考题 16 题图)

2.9 Exercises(习题)

1. A horizontal force is applied at point B of the steel frame $\boldsymbol{F}$ in Fig. 2.62, and the weight of the frame can be omitted. Find the binding forces $\boldsymbol{F}_A$ and$\boldsymbol{F}_D$ at support A and D. (图 2.62 所示钢架的点 B 作用一水平力 $\boldsymbol{F}$,钢架的质量略去不计. 求支座 A 和 D 处的约束力 $\boldsymbol{F}_A$ 和 $\boldsymbol{F}_D$.)

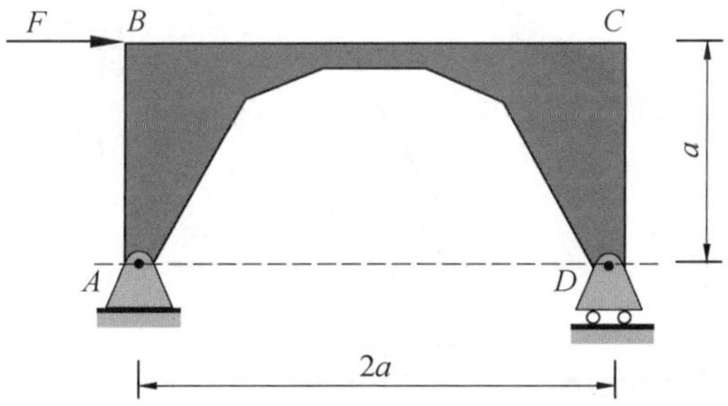

Fig. 2.62 Diagram of Exercise 1(图 2.62 习题 1 图)

2. The motor weighs $P=5$ kN and is placed in the center of the horizontal beam AC in Fig. 2.63, with support BC at an angle of 30° to the horizontal beam, neglecting the weight of the beam and the spar. Find the force on the brace BC with the restraining force at the hinge support A. (图 2.63 中电动机重 $P=5$ kN,放在水平梁 AC 的中央,支撑 BC 与水平梁的夹角为 30°,

忽略梁和撑杆的质量,求撑杆 BC 受力与铰支座 A 处的约束力.)

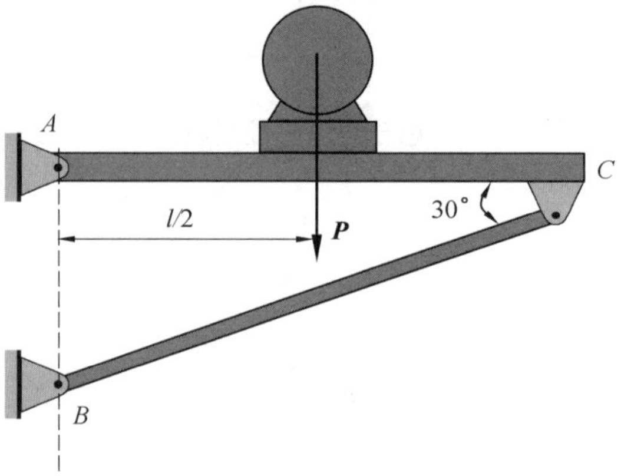

Fig. 2.63 Diagram of Exercise 2(图 2.63 习题 2 图)

3. The weight $P=20$ kN is hung by a rope from the pulley B of the stand and the other end of the rope is attached to the winch D as shown in Fig. 2.64. By turning the winch, the object is lifted. The size of the pulley, the self-weight of the rods AB and BC, and the friction at the pulley bearings are omitted, and A, B, C are hinged. When the object is in equilibrium, find the forces on the rods AB and BC. (重物 $P=20$ kN,用绳子挂在支架的滑轮 B 上,绳子的另一端接在绞车 D 上,如图 2.64 所示. 转动绞车,物体便能升起. 该滑轮的大小、杆 AB 与 BC 自重及滑轮轴承处摩擦略去不计,A、B、C 三处均为铰链连接. 当物体处于平衡状态时,求杆 AB 与 BC 所受的力.)

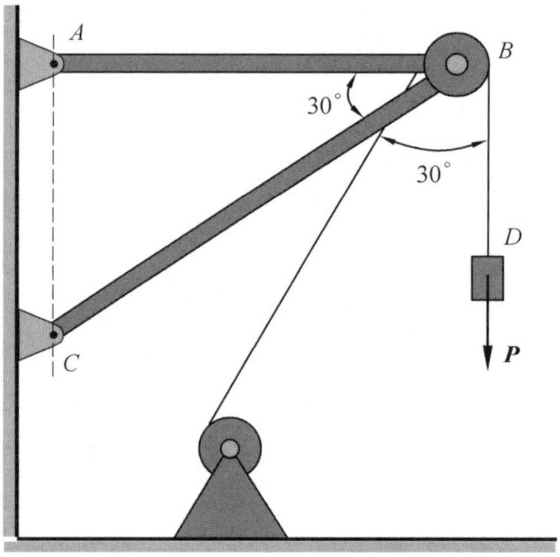

Fig. 2.64 Diagram of Exercise 3 (图 2.64 习题 3 图)

4. In Fig. 2.65, the ends of a rod AB are connected to the centers of two wheels A, B by smooth strands and they are placed on two smooth inclined surfaces. The weight of both wheels is $\boldsymbol{P}$ and the weight of the rod AB is not taken into account. Find the value of the angle θ at equilibrium. If the weight of wheel A is $P_A=300$ N, what should be the weight P_B of wheel B if

you want the rod AB to be in a horizontal position ($\theta=0°$) at equilibrium? (图2.65中,在杆AB的两端用光滑铰与两轮中心A、B连接,并将它们置于两光滑斜面上.两轮重力均为P,杆AB的重力不计,求平衡时角θ之值.若轮A重力$P_A=300$ N,欲使平衡时杆AB在水平位置($\theta=0°$),轮B重力P_B应为多少?)

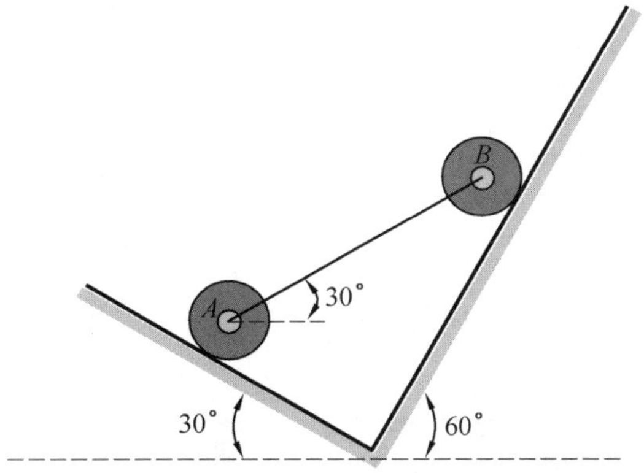

Fig. 2.65 Diagram of Exercise 4(图2.65 习题4图)

5. As shown in Fig. 2.66, only the main force $\boldsymbol{F}$ acts on each bar. Calculate the moment of the force $\boldsymbol{F}$ on point O in each of the following diagrams. (图2.66所示各杆件上只有主动力$\boldsymbol{F}$作用,计算下列各图中力$\boldsymbol{F}$对点O的矩.)

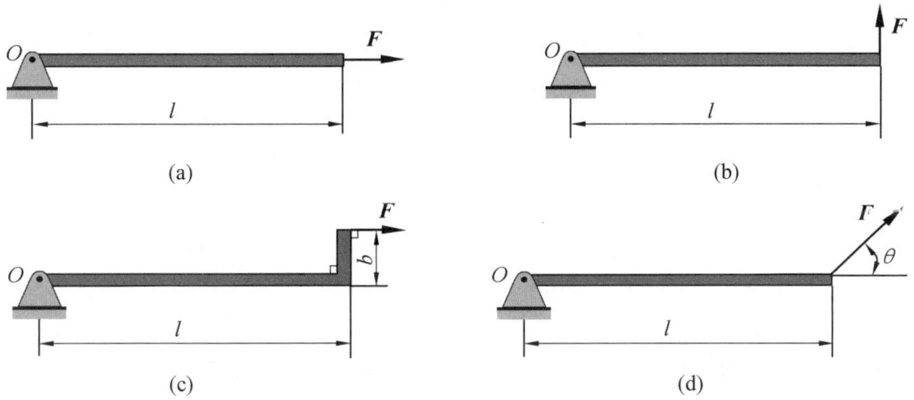

Fig. 2.66 Diagram of Exercise 5(图2.66 习题5图)

6. As shown in Fig. 2.67, there is a positive force $\boldsymbol{F}$ acting on the steel frame and find the moment of the force $\boldsymbol{F}$ on point A and point B. (如图2.67所示,钢架上作用有主动力$\boldsymbol{F}$,求力$\boldsymbol{F}$对点A和点B的力矩.)

7. In the structure of Fig. 2.68, the self-weight of the members is neglected. A force couple with moment M is applied to member AB to find the restraining force on supports A and C. (在图2.68所示的结构中,各构件的自重忽略不计.在构件AB上作用一力偶矩为M的力偶,求支座A和C的约束力.)

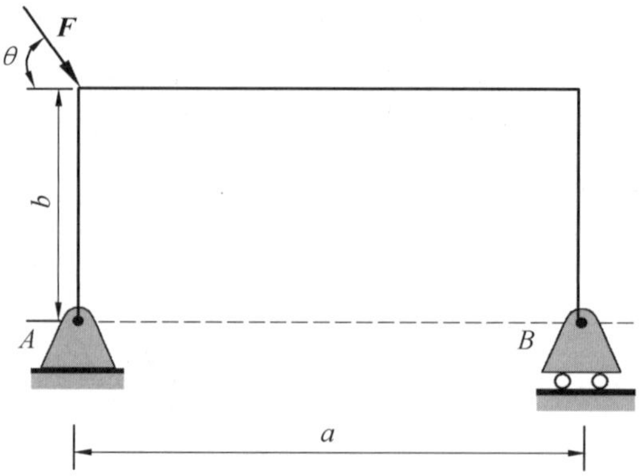

Fig. 2.67　Diagram of Exercise 6(图 2.67　习题 6 图)

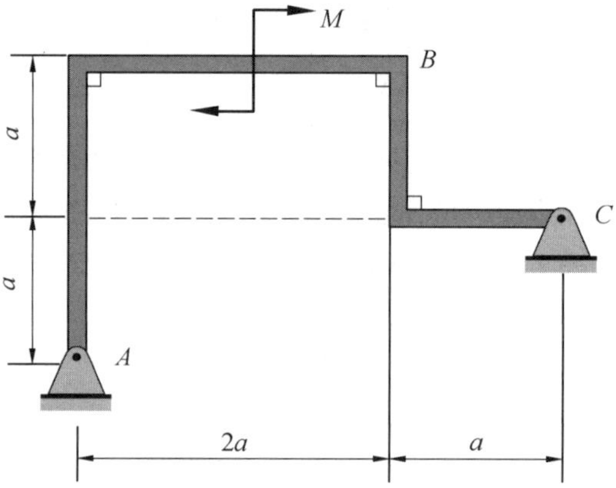

Fig. 2.68　Diagram of Exercise 7(图 2.68　习题 7 图)

8. In Fig. 2.69, the radii of the pitch circle of the two gears are r_1, r_2, the moment of the active force couple acting on wheel Ⅰ is M_1, the gear pressure angle is θ and the weight of the two gears is not taken into account. Determine the magnitude and direction of the binding force of bearings O_1 and $O_2$and the drag coupling moment M_2 on wheel Ⅱ to maintain uniform rotation of the two gears.(图 2.69 中,两齿轮的节圆半径分别为 r_1、r_2,作用于轮Ⅰ上的主动力偶的力偶矩为 M_1,齿轮压力角为 θ,不计两齿轮的质量. 求使二齿轮维持匀速转动时齿轮Ⅱ的阻力偶之矩 M_2与轴承 O_1、O_2的约束力大小和方向.)

9. The four-link mechanism O_1ABO_2 is balanced in the position of Fig. 2.70, $O_1A=0.4$ m, $O_2B=0.6$ m. The force dip moment $M_1=100$ N · m acts on rod O_1A. The weight of each rod is not accounted for. Find the magnitude of the moment M_2 and the force on rod AB.(四连杆机构 O_1ABO_2在图 2.70 位置平衡,$O_1A=0.4$ m,$O_2B=0.6$ m,作用在杆 O_1A 上的力偶矩 $M_1=100$ N · m,各杆的质量不计. 求力偶矩 M_2的大小和杆 AB 所受的力.)

10. In the steel frame of Fig. 2.71, it is known that $q=3$ kN/m, $F=6\sqrt{2}$ kN, $M=10$ kN · m and the self-weight of the frame is not taken into account. Find the restraining force at the fixed end A.(在图 2.71 所示的钢架中,已知 $q=3$ kN/m, $F=6\sqrt{2}$ kN,$M=10$ kN · m,不计钢架自重,求固定端 A 处的约束力.)

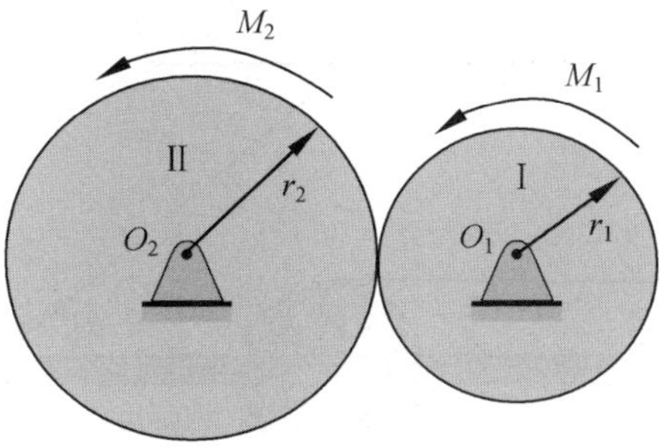

Fig. 2.69 Diagram of Exercise 8(图 2.69 习题 8 图)

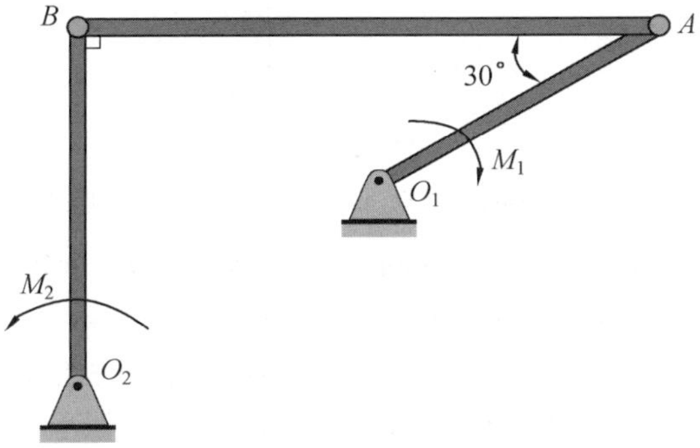

Fig. 2.70 Diagram of Exercise 9(图 2.70 习题 9 图)

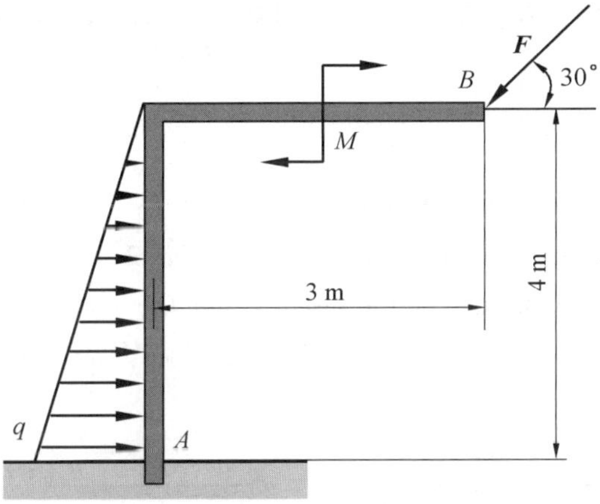

Fig. 2.71 Diagram of Exercise 10 (图 2.71 习题 10 图)

11. A horizontal beam AB is shown in Fig. 2.72 and a fixed pulley of radius $r=0.1$ m is fitted with a pin at D on the beam. There is a rope across the fixed pulley with one end tied horizontally to the wall and a weight weighing $P=1\ 800$ N suspended from the other end. $AD=0.2$ m, $BD=0.2$ m, $\theta=45°$, and the weight of the beam, rod, pulley and rope are not taken into account. Find the restraining force on the beam at hinge A and of rod BC. (水平梁 AB 如图 2.72 所示,在梁上 D 处用销子安装半径 $r=0.1$ m 的定滑轮. 有一跨过定滑轮的绳子,其一端水平地系于墙上,另一端悬挂有重 $P=1\ 800$ N 的重物. $AD=0.2$ m, $BD=0.2$ m, $\theta=45°$,不计梁、杆、滑轮和绳的质量,求铰链 A 处和杆 BC 对梁的约束力.)

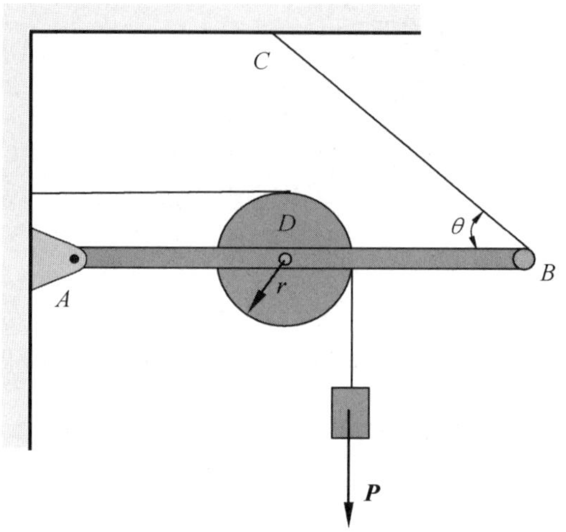

Fig. 2.72 Diagram of Exercise 11(图 2.72 习题 11 图)

12. A weightless horizontal beam is supported and loaded as shown in Fig. 2.73. A force $\boldsymbol{F}$, a force couple with moment M and a uniform load of strength q are known. Find the restraining forces at supports A and B.(无重水平梁的支撑和载荷如图 2.73 所示. 已知力 $\boldsymbol{F}$、力偶矩为 M 的力偶和强度为 q 的均布载荷. 求支座 A 和 B 处的约束力.)

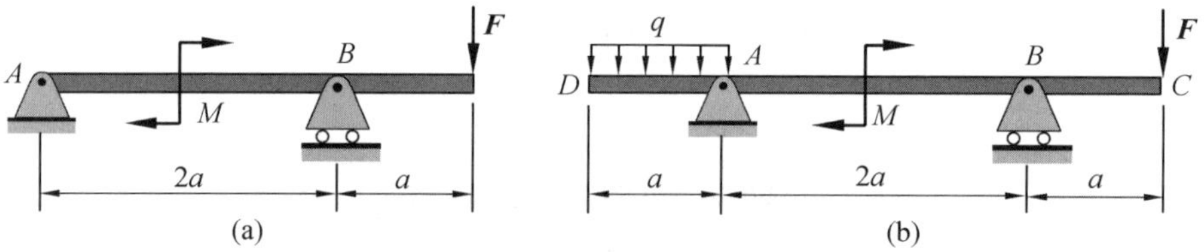

Fig. 2.73 Diagram of Exercise 12(图 2.73 习题 12 图)

13. In the two continuous beams a, b as shown in Fig. 2.74, with known q, M, a and θ, and without regard to the self-weight of the beam, find the restraining force of the continuous beam at A, C.(在图 2.74 所示的 a、b 两连续梁中,已知 q、M、a 和 θ,不计梁的自重,求连续梁在 A、C 处的约束力.)

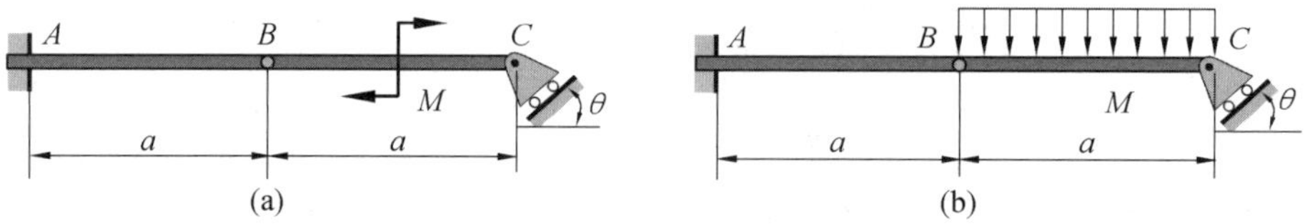

Fig. 2.74 Diagram of exercise 13(图 2.74 习题 13 图)

14. Fig. 2.75 is a transmission system for a gate opening and closing facility. The radii of each gear are known to be r_1, r_2, r_3, r_4, the radius of the drum wheel is r, the gate weight is $\boldsymbol{P}$, the pressure angle of the gear is θ, and the weight of each gear is not accounted for. Find the minimum opening dipole moment M and the restraining force at the shaft O_3.(图 2.75 所示为一种闸门启闭设施的传动系统. 已知各齿轮的半径分别为 r_1、r_2、r_3、r_4,鼓轮的半径为 r,闸门重力为 $\boldsymbol{P}$,齿轮的压力角为 θ,不计各齿轮重,求最小的启门力偶矩 M 及轴 O_3 处的约束力.)

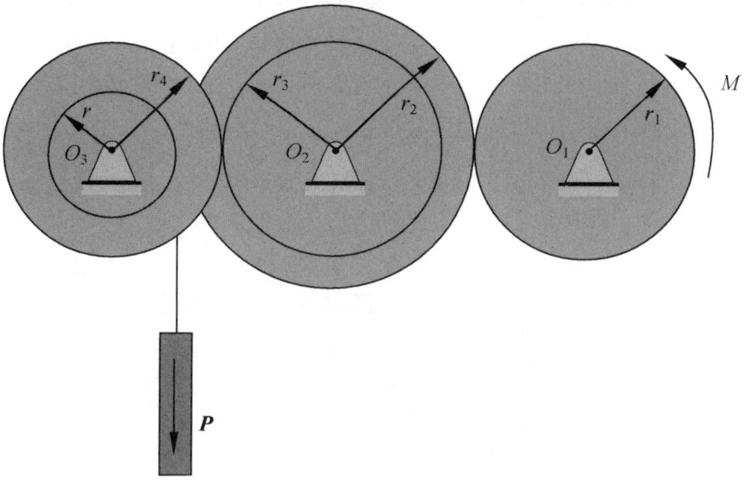

Fig. 2.75 Diagram of Exercise 14(图 2.75 习题 14 图)

15. The member consists of rods AB, AC, DF hinged without regard to their self-weight, as shown in Fig. 2.76. A force couple of moment M is applied to rod DF. Find the forces applied at rods A, D, B. (构件由不计自重的杆 AB、AC、DF 铰接而成,如图 2.76 所示. 在杆 DF 上作用一矩为 M 的力偶,求杆 A、D、B 处所受的力.)

16. Fig. 2.77 is a structure in which an object $\boldsymbol{P}$ weighs 1 200 N and is tied to the wall horizontally by a thin rope across a pulley E with the dimensions shown. Without regard to the weight of the rod and pulley, find the binding force at supports A and B, and the internal force $\boldsymbol{F}_{BC}$ of rod BC. (图 2.77 所示构架中,物体 $\boldsymbol{P}$ 重 1 200 N,由细绳跨过滑轮 E 而水平系与墙上,尺寸如图. 不计杆和滑轮的质量,求支撑 A 和 B 处的约束力,以及杆 BC 的内力 $\boldsymbol{F}_{BC}$.)

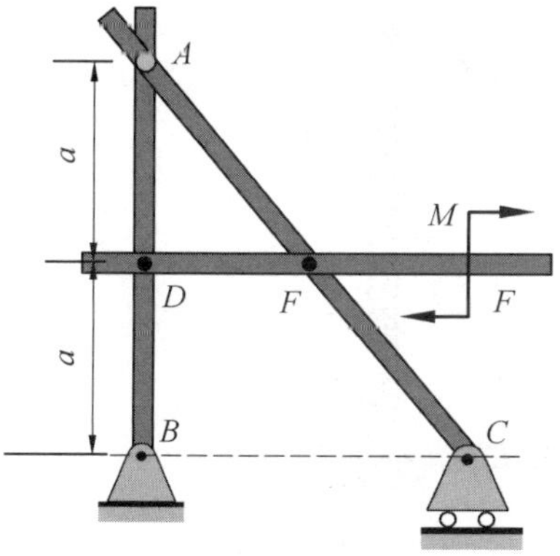

Fig. 2.76 Diagram of Exercise 15
(图 2.76 习题 15 图)

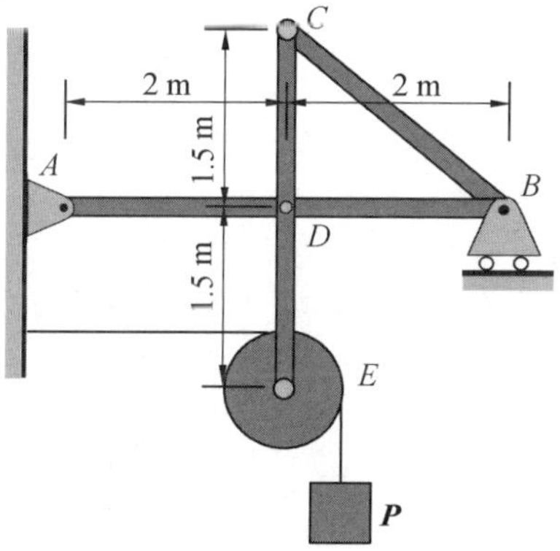

Fig. 2.77 Diagram of Exercise 16
(图 2.77 习题 16 图)

17. In the configuration of Fig. 2.78, with various unit lengths of 300 N/m and a load $\boldsymbol{P}$ = 10 kN, find the restraint at the fixed end A and at the hinge B, C. (图 2.78 的构架中,各种单位长度的重力为 300 N/m,载荷 P = 10 kN,求固定端 A 处与 B、C 铰链处的约束.)

18. The structure in Fig. 2.79 consists of a right-angled bent rod DAB and a straight rod BC,

CD hinged together. The rod DC is subjected to a uniform load q and the rod BC is subjected to a force couple with moment $M=qa^2$. The self-weight of the member is not taken into account. Find the force applied to the hinge D. (图 2.79 所示结构由直角弯杆 DAB 和直杆 BC、CD 铰链而成,杆 DC 受均布载荷 q 作用,杆 BC 受矩为 $M=qa^2$ 的力偶作用. 不计构件自重,求铰链 D 所受的力.)

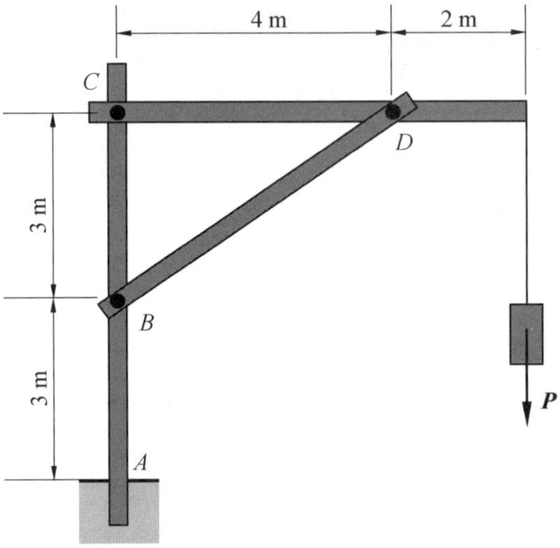

Fig. 2.78 Diagram of Exercise 16(图 2.78 习题 16 图)

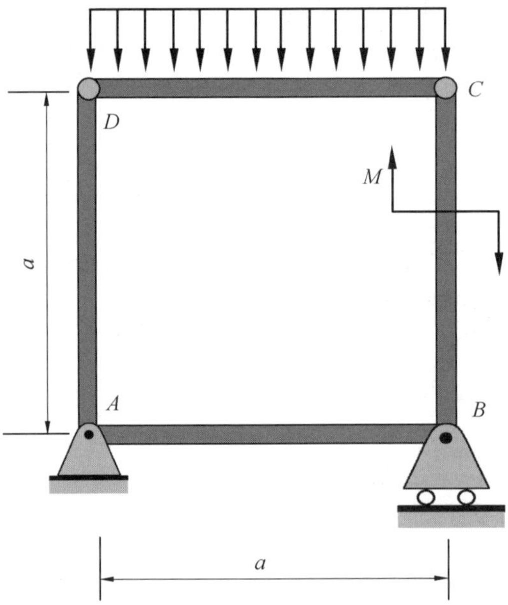

Fig. 2.79 Diagram of Exercise 17(图 2.79 习题 18 图)

Chapter 3 Spatial Force Systems (空间力系)

A force system with arbitrary distribution of action lines in space is called a spatial force system, including spatial concurrent force systems, spatial force couple systems, spatial arbitrary force systems and spatial parallel force systems.（力作用线在空间任意分布的力系称为空间力系，包括空间汇交力系、空间力偶系、空间任意力系和空间平行力系.）

The main content of this chapter includes the followings.（本章主要内容如下.）

(1) Composition and equilibrium of spatial concurrent force systems.（空间汇交力系的合成与平衡.）

(2) Moment of force about a point and moment of force about an axis in space.（空间力对点的矩和力对轴的矩.）

(3) Theory of spatial force couples.（空间力偶理论.）

(4) Simplification of arbitrary spatial force systems.（空间任意力系的简化.）

(5) Equilibrium equations of arbitrary spatial force systems.（空间任意力系的平衡方程.）

(6) Center of gravity.（重心.）

3.1 Spatial Concurrent Force Systems (空间汇交力系)

3.1.1 Projection and Resolution of Forces in Space (空间力的投影和分解)

In Chapter 2, we have already introduced the resolution and projection of forces in space along coordinate axes. The projection of spatial forces can be divided into direct projection method (also known as the first projection method) and indirect projection method (also known as the second projection method).（第2章中已经介绍过力沿空间坐标轴的分解与投影，空间力的投影可分为直接投影法（又称一次投影法）和间接投影法（又称二次投影法）.）

1. Direct Projection Method（直接投影法）

As shown in Fig. 3.1, given the angle α, β, γ between force $\boldsymbol{F}$ and the positive direction of each coordinate axis, we can obtain the projections F_x, F_y, F_z of the force on each axis, those are（如图3.1所示，已知力 $\boldsymbol{F}$ 与三个坐标轴正向的夹角 α、β、γ，可得到该力在坐标轴上的投影 F_x、F_y、F_z 分别为）

$$\begin{cases} F_x = F\cos\alpha \\ F_y = F\cos\beta \\ F_z = F\cos\gamma \end{cases} \tag{3.1}$$

2. Indirect projection method (间接投影法)

If the angle between force $\boldsymbol{F}$ and the coordinate axes Ox and Oy is uncertain, the projection $\boldsymbol{F}_{xy}$ of force $\boldsymbol{F}$ on the plane Oxy can be calculated first, and then projected onto the x and y axes. This method is called indirect projection. As shown in Fig. 3. 2, given the angle γ, φ, the projections of force $\boldsymbol{F}$ on the coordinate axis F_x, F_y, F_z are(当力 $\boldsymbol{F}$ 与坐标轴 Ox、Oy 的夹角不好确定时,可先求出力 $\boldsymbol{F}$ 在平面 Oxy 上的投影 $\boldsymbol{F}_{xy}$,再把这个投影投影到 x、y 轴上,这种方法称为间接投影法. 如图 3.2 所示,已知角度 γ、φ,则力 $\boldsymbol{F}$ 在坐标轴上的投影 F_x、F_y、F_z 为)

$$\begin{cases} F_{xy} = F\sin\gamma \begin{cases} F_x = F_{xy}\cos\varphi = F\sin\gamma\cos\varphi \\ F_y = F_{xy}\sin\varphi = F\sin\gamma\sin\varphi \end{cases} \\ F_z = F\cos\gamma \end{cases}$$

That is (即)

$$\begin{cases} F_x = F\sin\gamma\cos\varphi \\ F_y = F\sin\gamma\sin\varphi \\ F_z = F\cos\gamma \end{cases} \tag{3.2}$$

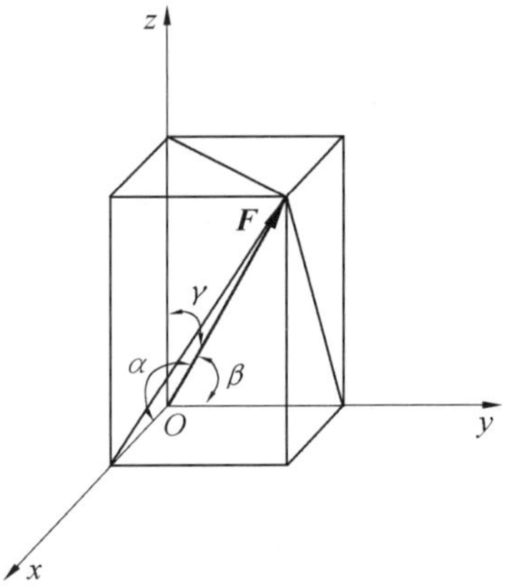

Fig. 3. 1 Direct projection method

(图 3. 1 直接投影法)

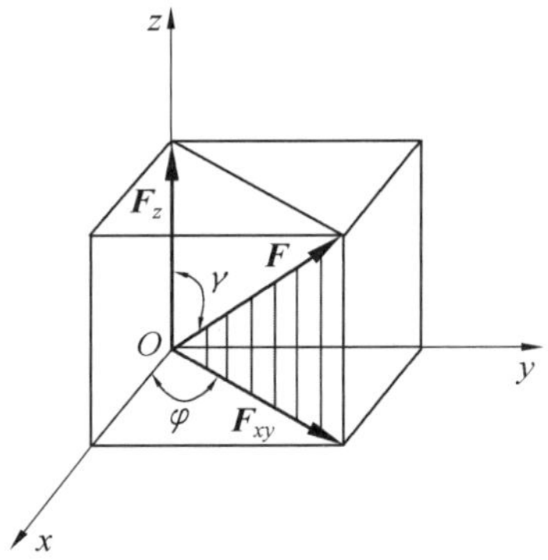

Fig. 3. 2 Indirect projection method

(图 3. 2 间接投影法)

When the projection F_x, F_y, F_z of the force $\boldsymbol{F}$ on the coordinate axis is known, the magnitude of the force $\boldsymbol{F}$ and the direction cosine are(当已知力 $\boldsymbol{F}$ 在坐标轴上的投影 F_x、F_y、F_z 时,力 $\boldsymbol{F}$ 的大小及方向余弦分别为)

$$\begin{cases} F = \sqrt{F_x^2 + F_y^2 + F_z^2} \\ \cos < \boldsymbol{F},\mathbf{i} > = \dfrac{F_x}{F}, \cos < \boldsymbol{F},\mathbf{j} > = \dfrac{F_y}{F}, \cos < \boldsymbol{F},\mathbf{k} > = \dfrac{F_z}{F} \end{cases} \tag{3.3}$$

The component force $\boldsymbol{F}_x$, $\boldsymbol{F}_y$, $\boldsymbol{F}_z$ of the force $\boldsymbol{F}$ on the three coordinate axes can be expressed by projection as(则力 $\boldsymbol{F}$ 在三个坐标轴上的分力 $\boldsymbol{F}_x$、$\boldsymbol{F}_y$、$\boldsymbol{F}_z$ 可以用投影表示为)

$$\boldsymbol{F} = \boldsymbol{F}_x + \boldsymbol{F}_y + \boldsymbol{F}_z = F_x\mathbf{i} + F_y\mathbf{j} + F_z\mathbf{k} \tag{3.4}$$

Eq. (3.4) is called the analytical expression of force. (式(3.4)称为力的解析表达式.)

3.1.2 Composition and Equilibrium of Spatial Concurrent Force System (空间汇交力系的合成与平衡条件)

The resultant force of a spatial confluence system is equal to the vector sum of each component force, and the line of action of the resultant force passes through the intersection point. (空间汇交力系的合力等于各分力的矢量和,合力的作用线通过汇交点.)

Geometric method is(几何法为)

$$\boldsymbol{F}_{\mathrm{R}} = \boldsymbol{F}_1 + \boldsymbol{F}_2 + \cdots + \boldsymbol{F}_n = \sum_{i=1}^{n} \boldsymbol{F}_i = \sum F_x\mathbf{i} + \sum F_y\mathbf{j} + \sum F_z\mathbf{k} \tag{3.5}$$

The closed edge of a force polygon can be used to represent the magnitude and direction of the resultant force. (可用力多边形的封闭边来表示合力的大小和方向.)

Analytic method is(解析法为)

$$\left\{\begin{aligned} &F_{\mathrm{R}x} = \sum F_x,\ F_{\mathrm{R}y} = \sum F_y,\ F_{\mathrm{R}z} = \sum F_z \\ &F_{\mathrm{R}} = \sqrt{\left(\sum F_x\right)^2 + \left(\sum F_y\right)^2 + \left(\sum F_z\right)^2} \\ &\cos < \boldsymbol{F}_{\mathrm{R}}, \mathbf{i} > = \frac{\sum F_x}{F_{\mathrm{R}}}, \cos < \boldsymbol{F}_{\mathrm{R}}, \mathbf{j} > = \frac{\sum F_y}{F_{\mathrm{R}}},\ \cos < \boldsymbol{F}_{\mathrm{R}}, \mathbf{k} > = \frac{\sum F_z}{F_{\mathrm{R}}} \end{aligned}\right. \tag{3.6}$$

Balance condition is(平衡条件为)

$$\boldsymbol{F}_{\mathrm{R}} = \sum_{i=1}^{n} \boldsymbol{F}_i = 0$$

(1) The geometric conditions for equilibrium: The length of the closed edge is zero, and the force polygon closes itself. (平衡时的几何条件:封闭边的长度为零,力多边形自行封闭.)

(2) Analytical conditions for equilibrium: The necessary and sufficient condition for the equilibrium of a spatial intersecting force system is that the algebraicsum of all forces projected on the three coordinate axes in the force system is equal to zero, that is(平衡时的解析条件:空间汇交力系平衡的充要条件是该力系中所有力在三个坐标轴上投影的代数和均等于零,即)

$$\left\{\begin{aligned} &\sum F_x = 0 \\ &\sum F_y = 0 \\ &\sum F_z = 0 \end{aligned}\right. \tag{3.7}$$

These conditions are referred to as the analytical conditions for the equilibrium of spatial intersecting force systems. (这些条件称为空间汇交力系平衡的解析条件.)

3.2 Vector Representation of the Moment of Force on a Point (力对于点之矩的矢量表示)

The moment of force $\boldsymbol{F}$ to a point O can be represented by a vector called the moment vector, represented by $\boldsymbol{M}_O(\boldsymbol{F})$, and the arm of force is represented by h, which passes through the center of moment O, perpendicular to the action surface of the moment. The length of a vector represents the magnitude of the moment (Fig. 3.3), that is(力 $\boldsymbol{F}$ 对于点之矩可以用一个矢量表示,称为力矩矢,用 $\boldsymbol{M}_O(\boldsymbol{F})$ 表示,力臂用 h 表示,该力矩矢量通过矩心 O ,垂直于力矩作用面. 矢量的长度表示力矩大小(图 3.3),即)

$$|\boldsymbol{M}_O(\boldsymbol{F})| = |\boldsymbol{F}|h$$

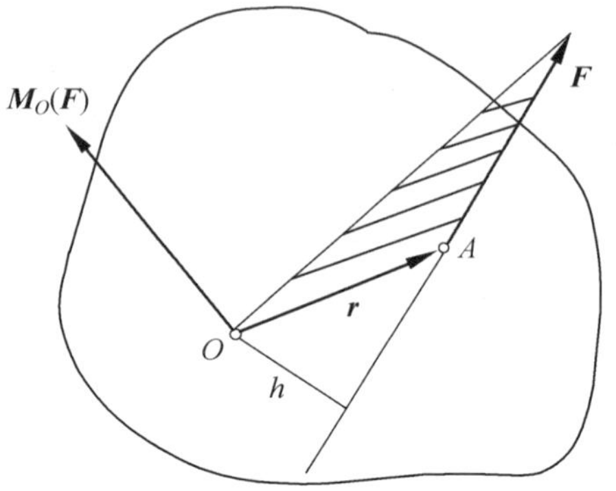

Fig. 3.3 Moment of spatial force to point(图 3.3 空间力对点之矩)

The torque vector relative to a certain center is called the vector product of the radial vector passing through that center and the applied force (Fig. 3.3). (相对于某个中心的力矩矢量称为通过该中心的径向矢量与作用力的矢量积(图 3.3).)

By definition, we have(按照定义,有)

$$\boldsymbol{M}_O(\boldsymbol{F}) = \boldsymbol{r} \times \boldsymbol{F} \tag{3.8}$$

As is well known, any vector can be decomposed according to the coordinate axis, we have (众所周知,任何矢量都可以根据坐标轴进行分解,有)

$$\boldsymbol{M}_O(\boldsymbol{F}) = M_x\mathbf{i} + M_y\mathbf{j} + M_z\mathbf{k} \tag{3.9}$$

Similarly, vector diameter $\boldsymbol{r}$ and force $\boldsymbol{F}$ can be decomposed along the coordinate axis, we have(同样,可以把矢径 $\boldsymbol{r}$ 及力 $\boldsymbol{F}$ 沿坐标轴分解,有)

$$\boldsymbol{r} = x\mathbf{i} + y\mathbf{j} + z\mathbf{k} \text{ , } \boldsymbol{F} = F_x\mathbf{i} + F_y\mathbf{j} + F_z\mathbf{k} \tag{3.10}$$

Execute $\boldsymbol{r} \times \boldsymbol{F}$, we obtain(执行 $\boldsymbol{r} \times \boldsymbol{F}$,得到)

$$\begin{aligned}\boldsymbol{M}_O(\boldsymbol{F}) = \boldsymbol{r} \times \boldsymbol{F} &= \begin{vmatrix} \mathbf{i} & \mathbf{j} & \mathbf{k} \\ x & y & z \\ F_x & F_y & F_z \end{vmatrix} \\ &= (yF_z - zF_y)\mathbf{i} + (zF_x - xF_z)\mathbf{j} + (xF_y - yF_x)\mathbf{k}\end{aligned} \tag{3.11}$$

In this way, the projection of the moment vector $\boldsymbol{M}_O(\boldsymbol{F})$ on the coordinate axis will be(这样,力矩矢量 $\boldsymbol{M}_O(\boldsymbol{F})$ 在坐标轴上的投影将是)

$$\begin{cases} [\boldsymbol{M}_O(\boldsymbol{F})]_x = M_x = yF_z - zF_y \\ [\boldsymbol{M}_O(\boldsymbol{F})]_y = M_y = zF_x - xF_z \\ [\boldsymbol{M}_O(\boldsymbol{F})]_z = M_z = xF_y - yF_x \end{cases} \tag{3.12}$$

The magnitude M_O of the torque vector is(力矩矢量的大小 M_O 为)

$$M_O = |\boldsymbol{M}_O(\boldsymbol{F})| = \sqrt{M_x^2 + M_y^2 + M_z^2} \tag{3.13}$$

The direction cosine of the moment vector determines its direction in space, that is(力矩矢量的方向余弦决定着它在空间的方向,即)

$$\cos < \boldsymbol{M}_O, \mathbf{i} > = \frac{M_x}{M_O}, \cos < \boldsymbol{M}_O, \mathbf{j} > = \frac{M_y}{M_O}, \cos < \boldsymbol{M}_O, \mathbf{k} > = \frac{M_z}{M_O} \tag{3.14}$$

3.3 Moment of Force Relative to an Axis (力对于轴之矩)

The projection of vector $\boldsymbol{M}_O(\boldsymbol{F})$, which is the moment of force $\boldsymbol{F}$ relative to center O , on any z-axis passing through that center is called the moment of force $\boldsymbol{F}$ relative to the z-axis (Fig. 3.4). (矢量 $\boldsymbol{M}_O(\boldsymbol{F})$,即力 $\boldsymbol{F}$ 相对于中心 O 的矩,在任何经过该中心的 z 轴上的投影称为力 $\boldsymbol{F}$ 相对于 z 轴的矩(图 3.4).)

The moment of force relative to an axis is used to measure the rotational effect of force on a rigid body around a certain axis. If force $\boldsymbol{F}$ can cause an object to rotate around the z-axis, take a plane xy perpendicular to the z-axis and intersect with the z-axis at point O_1. $\boldsymbol{F}_{xy}$ represents the projection of force $\boldsymbol{F}$ on this plane, and h represents the vertical distancefrom point O_1 to the line of action of force $\boldsymbol{F}_{xy}$, then the moment of force $\boldsymbol{F}$ on the z-axis is (Fig. 3.4)(力对轴之矩用来度量力使刚体绕某轴的转动效应. 若力 $\boldsymbol{F}$ 作用可使物体绕 z 轴转动,则取一平面 xy 垂直于 z 轴,并与 z 轴交于 O_1点. $\boldsymbol{F}_{xy}$ 表示力 $\boldsymbol{F}$ 在该平面上的投影,h 表示 O_1点至力 $\boldsymbol{F}_{xy}$ 作用线的垂直距离,则力 $\boldsymbol{F}$ 对 z 轴之矩为(图 3.4))

$$M_Z(\boldsymbol{F}) = M_{O_1}(\boldsymbol{F}_{xy}) = \pm |\boldsymbol{F}_{xy}| \cdot h \tag{3.15}$$

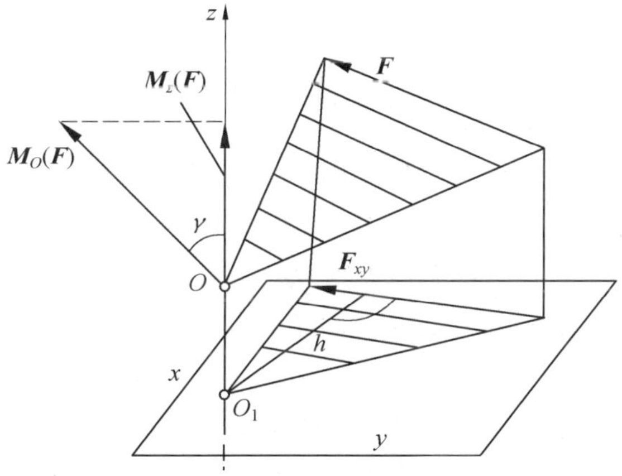

Fig. 3.4 Moment of force relative to axis (图 3.4 力对轴之矩)

The moment of force $\boldsymbol{F}$ relative to the axis is equal to the algebraic moment of force projection on a plane perpendicular to that axis to the inter section point O_1 between the z-axis and the plane. (力 $\boldsymbol{F}$ 对轴的矩等于力在垂直于该轴的平面上的投影对 z 轴与平面交点 O_1 的代数矩.)

The moment of force relative to the z-axis is an algebraic value, and its sign is also defined as the sign of the algebraic moment of force. (力相对于 z 轴的力矩是代数值,其符号也被定义为

力的代数矩的符号.)

1. Steps for calculating the moment of force relative to an axis

(计算力对于轴之矩的步骤)

(1) Draw a plane perpendicular to the axis.(绘制垂直于轴的平面.)

(2) Project force onto the plane.(将力在平面上投影.)

(3) Find the algebraic moment of the projection of force on a plane relative to the intersection of the axis and the plane.(找到力在平面上的投影相对于轴线与平面交点的代数矩.)

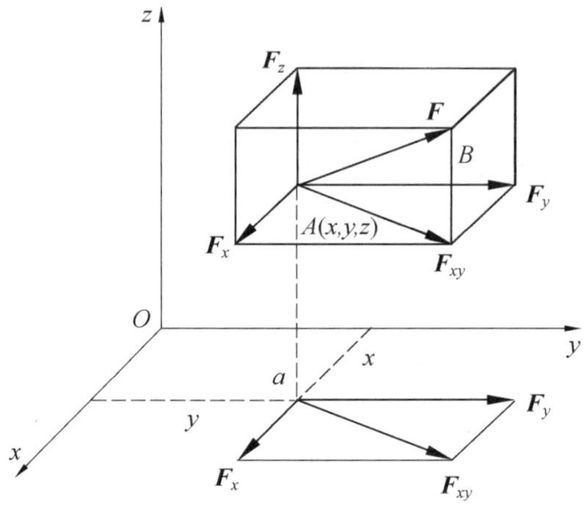

Fig. 3.5 Calculating the moment of force on an axis based on the theorem of resultant moment

(图 3.5 根据合力矩定理计算力对轴之矩)

It is also possible to further calculate the moment of force relative to the axis based on the resultant moment theorem, as shown in Fig. 3.5, to obtain(还可以进一步根据合力矩定理,计算力对轴之矩,如图 3.5 所示,得到)

$$M_z(\boldsymbol{F}) = M_O(\boldsymbol{F}_{xy}) = M_O(\boldsymbol{F}_x) + M_O(\boldsymbol{F}_y) = xF_y - yF_x$$

Similarly, we obtain(同理,得到)

$$\begin{cases} M_x(\boldsymbol{F}) = yF_z - zF_y \\ M_y(\boldsymbol{F}) = zF_x - xF_z \\ M_z(\boldsymbol{F}) = xF_y - yF_x \end{cases} \tag{3.16}$$

x. y, z are the three coordinates of the point of force application.(x、y、z 为力作用点的三个坐标.)

Compare Eq. (3.12) with Eq. (3.16) to obtain(比较式(3.12)和式(3.16),得到)

$$\begin{cases} [\boldsymbol{M}_O(\boldsymbol{F})]_x = M_x = (yF_z - zF_y) = M_x(\boldsymbol{F}) \\ [\boldsymbol{M}_O(\boldsymbol{F})]_y = M_y = (zF_x - xF_z) = M_y(\boldsymbol{F}) \\ [\boldsymbol{M}_O(\boldsymbol{F})]_z = M_z = (xF_y - yF_x) = M_z(\boldsymbol{F}) \end{cases}$$

That is(即)

$$\begin{cases} [\boldsymbol{M}_O(\boldsymbol{F})]_x = M_x(\boldsymbol{F}) \\ [\boldsymbol{M}_O(\boldsymbol{F})]_y = M_y(\boldsymbol{F}) \\ [\boldsymbol{M}_O(\boldsymbol{F})]_z = M_z(\boldsymbol{F}) \end{cases} \tag{3.17}$$

The projection of amoment vector on a certain axis is numerically equal to the moment of the force on that axis. (力矩矢量在某轴上的投影在数值上等于力对该轴的矩.)

2. When the moment of force relative to the axis is zero(力对于轴之矩为零的情况)

When the force is in the same plane with the axis, the moment of the force relative to that axis is equal to zero (Fig. 3. 6). (当力与轴位于同一平面时,力对于该轴之矩等于零(图3.6).)

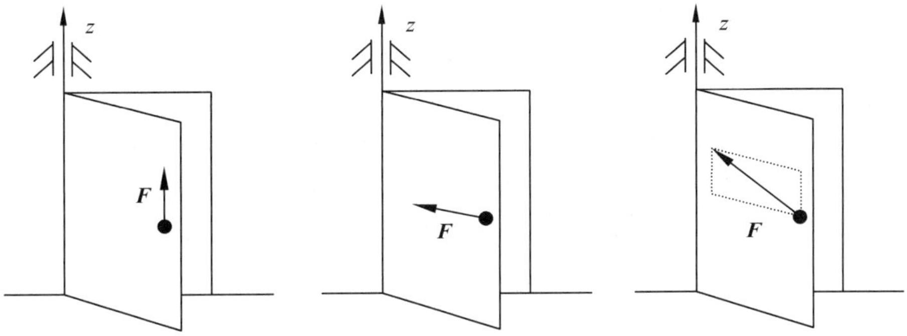

Fig. 3. 6 The moment of force relative to the axis is zero(图3.6 力对轴之矩为零)

3.4 Couple Moment Vector(力偶矩矢)

1. Definition of couple moment vector(力偶矩矢的定义)

The couple moment vector is represented by vector $\boldsymbol{m}$, which is equal to the product of force F and force arm d, and is perpendicular to the action surface of the couple. From the end of the vector along the vector, the direction of rotation of the couple is counterclockwise, is defined as positive (Fig. 3.7). (力偶矩矢用矢量 $\boldsymbol{M}$ 表示,其大小等于力 F 与力臂 d 的乘积,并且垂直于该力偶的作用面. 从矢量的末端沿矢量看去,力偶的转向是以逆时针为正(图3.7).)

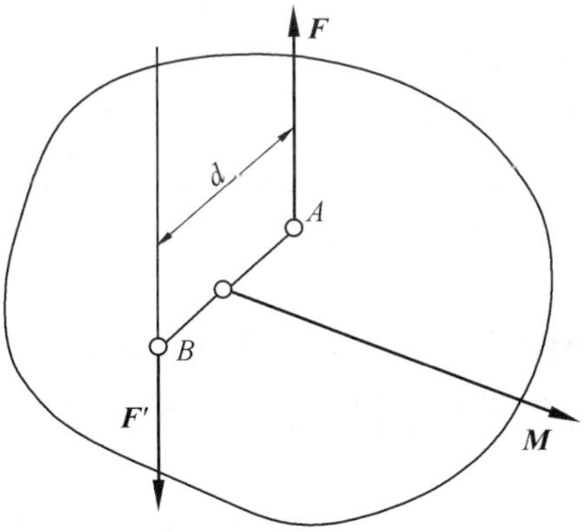

Fig. 3.7 Spatial couple(图3.7 空间力偶)

2. The properties of the couple moment vector(力偶矩矢的性质)

(1)The rotational effect of a couple acting on a rigid body is fully represented by its moment vector. (作用在刚体上的力偶的转动效果完全以其矩矢来表现.)

(2) A couple with equal moment vectors is an equivalent couple.(矩矢相等的力偶为等效力偶.)

(3) The beginning of the couple moment vector can move freely in space, and the couple moment vector is a free vector.(力偶矩矢的始端可在空间内任意移动,力偶矩矢是一自由矢量.)

3. The composition theorem of couple not in the same plane

(不在同一平面上的力偶合成定理)

Theorem 1: The couple system composed of two couple $\boldsymbol{M}_1$ and $\boldsymbol{M}_2$ acting on a rigid body can be replaced by a resultant couple $\boldsymbol{M}$, where the moment vector of the resultant couple is equal to the vector sum of all the moment vectors of each couple in the couple system, i. e. $\boldsymbol{M}=\boldsymbol{M}_1+\boldsymbol{M}_2$ (Fig. 3.8). (**定理 1**:作用在刚体上的两个力偶 $\boldsymbol{M}_1$ 和 $\boldsymbol{M}_2$ 构成的力偶系,可以由一合力偶 $\boldsymbol{M}$ 替代,合力偶矩矢等于力偶系中所有各力偶矩矢的矢量和,即 $\boldsymbol{M}=\boldsymbol{M}_1+\boldsymbol{M}_2$(图 3.8).)

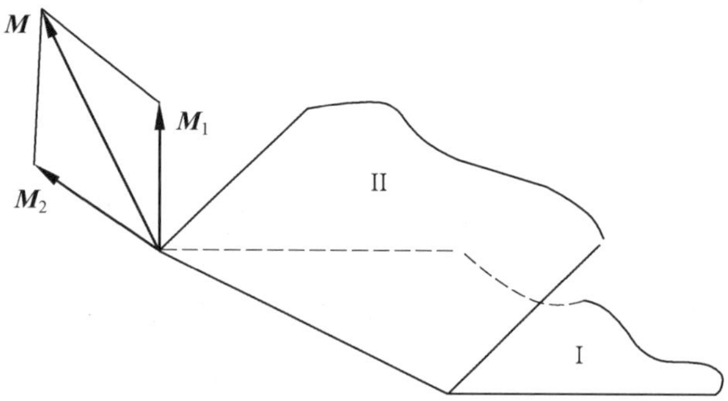

Fig. 3.8 Composition of couples not in the same plane (图 3.8 不在同一平面上的力偶合成)

Theorem 2: A spatial couple system can be synthesized as a resultant couple, and the vector of moments of the resultant couple is equal to the vector sum of all the moment vectors of each couple in the couple system, that is (**定理 2**:空间力偶系可合成一合力偶,合力偶矩矢等于力偶系中所有各力偶矩矢的矢量和,即)

$$\boldsymbol{M}=\boldsymbol{M}_1+\boldsymbol{M}_2+\cdots+\boldsymbol{M}_n=\sum \boldsymbol{M}_k \tag{3.18}$$

Among them, $\boldsymbol{M}_1$, $\boldsymbol{M}_2$, ..., $\boldsymbol{M}_n$ is the moment of each component couple, and $\boldsymbol{M}$ is the moment of the resultant couple. (其中,$\boldsymbol{M}_1$,$\boldsymbol{M}_2$,…,$\boldsymbol{M}_n$ 为各分力偶矩,而 $\boldsymbol{M}$ 为合力偶矩.)

3.5 Equilibrium Conditions for Space Couple Systems (空间力偶系的平衡条件)

When the spatial couple system is in equilibrium, the vector sum of all the moment vectors of each couple in this couple system is equal to zero, we have(空间力偶系平衡时,该力偶系中所有各力偶矩矢的矢量和等于零,有)

$$\boldsymbol{M}=\sum \boldsymbol{M}_k=0 \tag{3.19}$$

The analytical conditions for equilibrium of couple systems (projection form) is as follows.

(力偶系平衡的解析条件(投影形式)如下.)

Project the vectors on both sides onto the same axis to obtain(将两边矢量向同一轴上投影,得到)

$$M_x = \sum \boldsymbol{M}_{kx} = 0, M_y = \sum \boldsymbol{M}_{ky} = 0, M_z = \sum \boldsymbol{M}_{kz} = 0 \tag{3.20}$$

where M_x, M_y, M_z are the algebraic sum of the projection of the moment vectors of each couple in the couple system on three coordinate axes; M_{kx}, M_{ky}, M_{kz} are the projection of the moment vectors of each couple in the couple system on three coordinate axes. (式中,M_x、M_y、M_z为该力偶系中各力偶矩矢分别在三个坐标轴上投影的代数和;M_{kx}、M_{ky}、M_{kz}为该力偶系中各力偶矩矢在分别三个坐标轴上的投影.)

This is the analytical condition for the equilibrium of spatial couple systems. (这就是空间力偶系平衡的解析条件.)

3.6 Theorem of Simplifying a Space Arbitrary Force System based on Any Point (空间任意力系向任一点的简化定理)

For the spatial arbitrary force system shown in Fig. 3.9(a), referring to the simplified method of planar arbitrary force system, selecting point O as the simplified center can translate the forces in the force system towards point O. According to the translation theorem of forces, a spatial intersection force system and an additional couple system acting on point O can be obtained. (对于图3.9(a)所示的空间任意力系,参照平面任意力系的简化方法,选取 O 点作为简化中心,可将力系中各力向 O 点平移. 根据力的平移定理,可得到一个作用在 O 点的空间汇交力系及一个附加力偶系.)

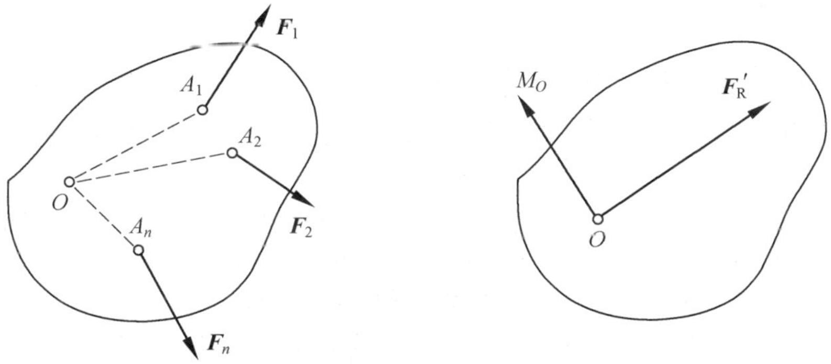

(a) Original force system(原力系)　　(b) Simplification result(简化结果)

Fig. 3.9 Simplification of arbitrary force system towards a point in space
(图3.9 空间任意力系向一点简化)

Theorem: Any spatial arbitrary force system acted on an arbitrarypoint O in a rigid body can be combined to a force $\boldsymbol{F}'_{\mathrm{R}}$ acted on point O, which is equal to the vector sum of the forces of the original force system, is called the principal vector of the original force system. The additional space couple system can be combined into a couple, with a couple moment vector of $\boldsymbol{M}_O$. (**定理**:

任何作用在刚体上 O 点的空间任意力系可合成为作用于 O 点的一个力 $\boldsymbol{F}'_{\mathrm{R}}$,它等于原力系中各力的矢量和,称为原力系的主矢. 附加的空间力偶系可合成为一力偶,其力偶矩矢为 $\boldsymbol{M}_O$.)

Vector $\boldsymbol{M}_O$ is equal to the vector sum of moments of each force in the original force system to the simplified center O , which is called the principal moment of the original force system to point O,we have(矢量 $\boldsymbol{M}_O$ 等于原力系中各力对于简化中心 O 之矩的矢量和,称为原力系对于 O 点的主矩,有)

$$\boldsymbol{M}_O = \sum \boldsymbol{M}_O(\boldsymbol{F}) = \sum M_k \tag{3.21}$$

Conclusion: Two force systems with the same principal vector and principal moment relative to the same center are equivalent. (**结论**:相对于同一中心具有同样主矢和主矩的两个力系等效.)

Carry out project to force $\boldsymbol{F}_1,\boldsymbol{F}_2,\ldots,\boldsymbol{F}_n$, that is, the projection of the force on the coordinate axis is known: F_{1x}, F_{2x}, ... , F_{nx}; F_{1y},F_{2y}, ... ,F_{ny}; F_{1z}, F_{2z}, ... , F_{nz}. Then, the projection of the main vector $\boldsymbol{F}'_{\mathrm{R}}$ on the coordinate axis is(对力 $\boldsymbol{F}_1,\boldsymbol{F}_2,\cdots,\boldsymbol{F}_n$ 进行投影,即已知力在坐标轴上的投影:F_{1x}, F_{2x}, ⋯, F_{nx}; F_{1y},F_{2y}, ⋯,F_{ny}; F_{1z}, F_{2z}, ⋯, F_{nz}. 则主矢量 $\boldsymbol{F}'_{\mathrm{R}}$ 在坐标轴上的投影为)

$$F_{\mathrm{R}x} = \sum F_{kx},\ F_{\mathrm{R}y} = \sum F_{ky},\ F_{\mathrm{R}z} = \sum F_{kz} \tag{3.22}$$

The projections of principal moment $\boldsymbol{M}_O$ on the three coordinate axes are (主矩 $\boldsymbol{M}_O$ 在三个坐标轴上的投影为)

$$M_{Ox} = \sum M_{Ox}(\boldsymbol{F}_k),\ M_{Oy} = \sum M_{Oy}(\boldsymbol{F}_k),\ M_{Oz} = \sum M_{Oz}(\boldsymbol{F}_k) \tag{3.23}$$

The size of the principal vectorand principal moment is(主矢与主矩的大小为)

$$|\boldsymbol{F}_{\mathrm{R}}| = \sqrt{F_{\mathrm{R}x}^2 + F_{\mathrm{R}y}^2 + F_{\mathrm{R}z}^2}\ ,\ |\boldsymbol{M}_O| = \sqrt{M_{Ox}^2 + M_{Oy}^2 + M_{Oz}^2} \tag{3.24}$$

3.7 Special Case of Simplifying a Space Arbitrary Force System Towards Any Point (空间任意力系向任一点简化的特殊情况)

Any spatial force system can be replaced by the principal vector and principal moment. Next, we will investigate several special situations that may arise. (任何空间任意力系都可以用主矢和主矩来代替. 接下来研究一下可能出现几种特殊情况.)

(1) $F'_{\mathrm{R}}=0$, $M_O=0$. Force system is in balance situation. ($F'_{\mathrm{R}}=0$, $M_O=0$. 力系平衡.)

(2) $F'_{\mathrm{R}}=0$, $M_O\neq0$. Simplify the force system into an equivalent resultant couple (rigid body rotation). ($F'_{\mathrm{R}}=0$,$M_D=0$. 力系简化为等效的合力偶 (刚体转动).)

(3) $F'_{\mathrm{R}}\neq0$, $M_O=0$. The force system is simplified as an equivalent resultant force, and the action line of the resultant force pass through the simplified center O. ($F'_{\mathrm{R}}=0$,$M_D=0$. 力系简化为等效的合力,合力的作用线通过简化中心 O.)

(4) $F'_{\mathrm{R}}\neq0$,$M_O\neq0$. The principle vector $\boldsymbol{F}_{\mathrm{R}}$ and the couple vector $\boldsymbol{M}_O$ corresponding to the principle moment vector are in the same plane, and $\boldsymbol{F}_{\mathrm{R}}\perp\boldsymbol{M}_O$, is a special case of a planar force system. As previously mentioned, in this case, the combined force does not act on the simplified

center, but at another point at a distance of M_O/F'_R (Fig. 3.10), we have ($F'_R \neq 0, M_D = 0$. 主矢 $\boldsymbol{F}_R$ 和主矩 $\boldsymbol{M}_O$ 的力偶在同一平面内,且 $\boldsymbol{F}_R \perp \boldsymbol{M}_O$,这是平面力系的一个特殊情况. 之前已经讲过,这种情况合力不作用在简化中心,而是在距离为 M_O/F'_R 的另一点 O_1(图 3.10),有)

$$F_R d = M_O$$

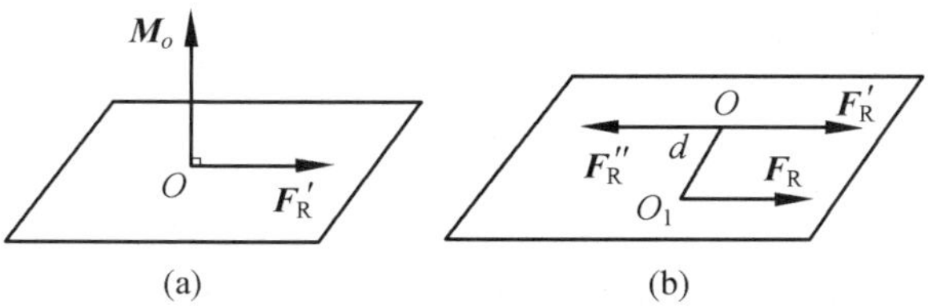

Fig. 3.10 Force and couple is in the same plane (图 3.10 力和力偶在同一平面内)

(5) $F'_R \neq 0$, $M_O \neq 0$, $\boldsymbol{F}_R \,/\!/\, \boldsymbol{M}_O$. Spatial arbitrary force system in space is simplified as a force spiral. ($F'_R \neq 0$, $M_O \neq 0$, $\boldsymbol{F}_R \,/\!/\, \boldsymbol{M}_O$. 空间任意力系被简化为力螺旋.)

① $F'_R \neq 0, M_O \neq 0$, $\boldsymbol{F}_R \,/\!/\, \boldsymbol{M}_O$. As shown in Fig. 3.11(a). ($F'_R \neq 0, M_O \neq 0$, $\boldsymbol{F}_R \,/\!/\, \boldsymbol{M}_O$. 如图 3.11(a)所示.)

At this point, even if the couple moment vector is reduced to two forces perpendicular to the main vector $\boldsymbol{F}_R$ (Fig. 3.11(b)), the three forces that are not in the same plane can no longer be simplified. This situation is called a force spiral. When the main vector is in the same direction as the main moment, it is a right helix, and when it is in the opposite direction, it is a left helix. (此时,即使把力偶矩矢量还原为垂直于主矢量 $\boldsymbol{F}_R$ 的两个力(图 3.11(b)),不在同一平面内的三个力也已不能再简化,这种情况称为力螺旋. 当主矢量与主矩同向时为右螺旋,反向时为左螺旋.)

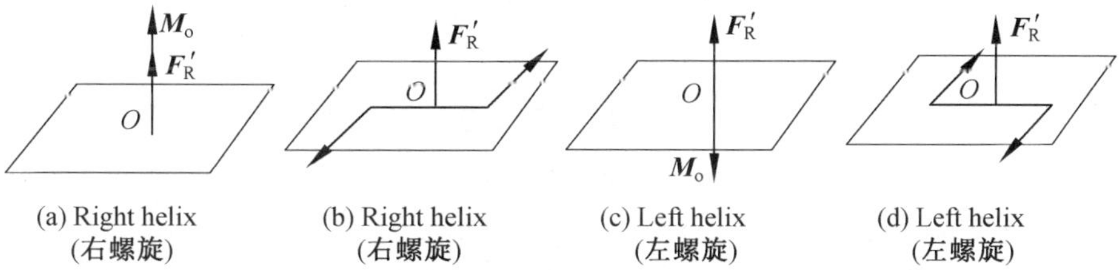

Fig. 3.11 Force spiral (图 3.11 力螺旋)

② $F'_R \neq 0, M_O \neq 0$, included angle α is a general state (Fig. 3.12(a)). The principal moment vector can be decomposed (Fig. 3.12(b)), to obtain ($F'_R \neq 0, M_O \neq 0$,夹角 α 为一般状态(图 3.12(a)). 可以将主矩矢量进行分解(图 3.12(b)),得到)

$$|\boldsymbol{M}'_O| = |\boldsymbol{M}_O| \cos\alpha, \quad |\boldsymbol{M}''_O| = |\boldsymbol{M}_O| \sin\alpha$$

Restore $\boldsymbol{F}_O$ into two forces $\boldsymbol{F}''_R$ and $\boldsymbol{F}_R$ equal in magnitude to the main vector (Fig. 3.12(c)). Finally, remove a pair of equilibrium forces located at point O, translate the moment vector $\boldsymbol{F}_O$ to a new action point O_1, and obtain a force spiral at action point O_1. The distance d is determined by (再将 $\boldsymbol{F}_O$ 还原为与主矢量大小相等的两个力 $\boldsymbol{F}''_R$ 与 $\boldsymbol{F}_R$(图 3.12(c)). 最后去掉位于 O 点的一对平衡力,平移力矩矢量 $\boldsymbol{F}_O$ 到新作用点 O_1,得到一个在作用点 O_1 的力螺旋. 距离 d 为)

$$d = \frac{|\boldsymbol{F}_O|}{|\boldsymbol{F}_R|} = \frac{|\boldsymbol{M}_O|\sin\alpha}{|\boldsymbol{F}_R|}$$

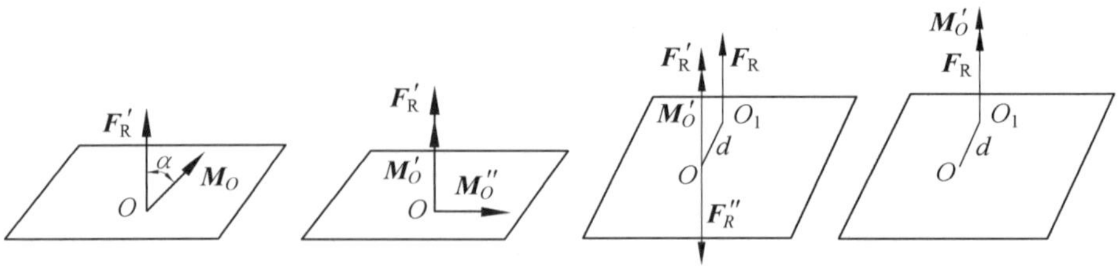

(a) (b) (c) (d)

Fig. 3.12 When the angle between $\boldsymbol{F}_R$ and $\boldsymbol{M}_O$ is in a general state

(图 3.12 $\boldsymbol{F}_R$ 与 $\boldsymbol{M}_O$ 夹角处于一般状态时)

Conclusion: In general, any spatial force system can be simplified as a force spiral. (**结论**：一般情形下空间任意力系可简化为力螺旋.)

3.8 Equilibrium Conditions for Spatial Arbitrary Force Systems (空间任意力系的平衡条件)

1. Geometric Conditions for Equilibrium of Spatial Arbitrary Force Systems (空间任意力系平衡的几何条件)

The necessary and sufficient condition for the equilibrium of any force system in space is that the principal vector of the force system and the following principal moment of the force system for any point are equal to zero, and the following condition must be met, that is (空间任意力系平衡的必要与充分条件是力系的主矢和力系对于任一点的主矩都等于零,要满足以下条件,即)

$$\boldsymbol{F}_R = 0,\ \boldsymbol{M}_O = 0$$

When in balance, the force vector polygon closes itself, while the moment vector polygon closes itself. (平衡时力矢量多边形自行封闭,力矩矢量多边形自行封闭.)

2. Analytical Conditions for Equilibrium of Spatial Arbitrary Force Systems (空间任意力系平衡的解析条件)

When any force system in space is in equilibrium, both the principal vector and principal moment are equal to zero, that is(空间任意力系平衡时,主矢和主矩均等于零,即)

$$|\boldsymbol{F}_R| = \sqrt{F_{Rx}^2 + F_{Ry}^2 + F_{Rz}^2},\ |\boldsymbol{M}_O| = \sqrt{M_{Ox}^2 + M_{Oy}^2 + M_{Oz}^2} = 0$$

Therefore(因此)

$$\begin{cases} F_{Rx} = \sum F_{kx} = 0,\ F_{Ry} = \sum F_{ky} = 0,\ F_{Rz} = \sum F_{kz} = 0 \\ M_{Ox} = \sum M_{Ox}(\boldsymbol{F}_k) = 0,\ M_{Oy} = \sum M_{Oy}(\boldsymbol{F}_k) = 0,\ M_{Oz} = \sum M_{Oz}(\boldsymbol{F}_k) = 0 \end{cases} \tag{3.25}$$

Conclusion: The necessary and sufficient conditions for the equilibrium of spatial arbitrary force system are that the algebraic sum of the projections of all forces in the force system on each of the three mutually perpendicular coordinate axes is equal to zero, and the algebraic sum of the moments of the force system on these three coordinate axes is equal to zero, respectively. (**结论:** 空间任意力系平衡的必要与充分条件是力系中所有力在任意相互垂直的三个坐标轴的每一个轴上之投影的代数和等于零,且力系对于这三个坐标轴之矩的代数和分别等于零.)

This is the analytical condition for the equilibrium of spatial arbitrary force system. (这是空间任意力系平衡的解析条件.)

3. The Theorem of Resultant Moments of Space Forces System
(空间力系的合力矩定理)

If any spatial force system (Fig. 3. 13) can be combined into a combined force, then the moment of the combined force at any point is equal to the vector sum of the moments of each force in the force system to the same point, that is(若空间任意力系(图 3. 13)可以合成为一个合力,则其合力对于任一点之矩等于力系中各力对于同一点之矩的矢量和,即)

$$\boldsymbol{M}_O(\boldsymbol{F}_R) = \sum \boldsymbol{M}_O(\boldsymbol{F}_k) \tag{3.26}$$

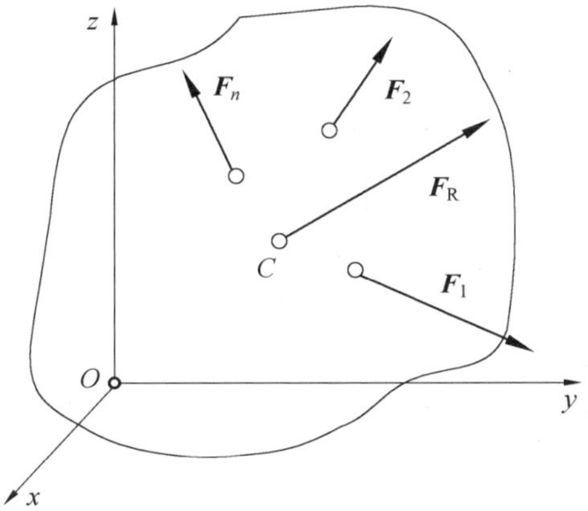

Fig. 3. 13 Resultant moment theorem for spatial force systems
(图 3. 13 空间力系的合力矩定理)

4. Spatial parallel force system(空间平行力系)

If all forces are parallel to each other, the coordinate axis can be selected, so that the O_z axis is parallel to the force (Fig. 3. 14). (如果所有作用力彼此平行,则可以选择坐标轴,使得 O_z 轴与力平行(图 3. 14).)

So there are only three remaining equilibrium equations for the parallel force system in space, those are (那么空间平行力系的平衡方程则只剩下三个,即)

$$\sum F_{kz} = 0, \sum M_{Ox}(\boldsymbol{F}_k) = 0, \sum M_{Oy}(\boldsymbol{F}_k) = 0 \tag{3.27}$$

This is the condition for the equilibrium of spatial parallel force systems. (这是空间平行力系平衡的条件.)

Example 1:If the action points of forces $\boldsymbol{Q}$, $\boldsymbol{T}$, $\boldsymbol{P}$ are known, determine the moments of these forces on the coordinate axis (Fig. 3. 15). (**例题 1**:如果已知 $\boldsymbol{Q}$ 、$\boldsymbol{T}$ 、$\boldsymbol{P}$ 这些力的作用点,请确定这些力对坐标轴的矩(图 3. 15).)

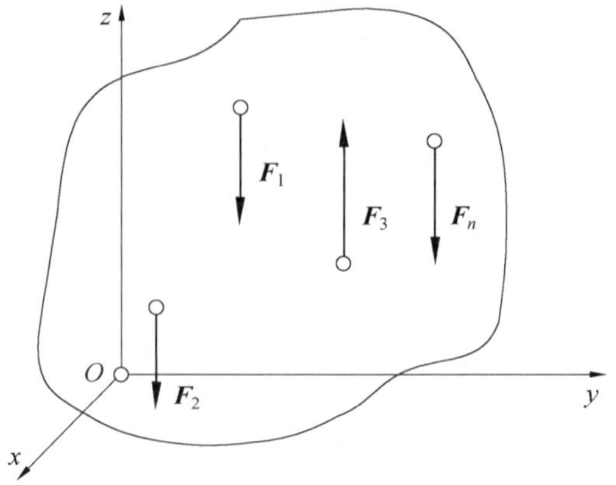

Fig. 3. 14 Space parallel force system(图 3. 14 空间平行力系)

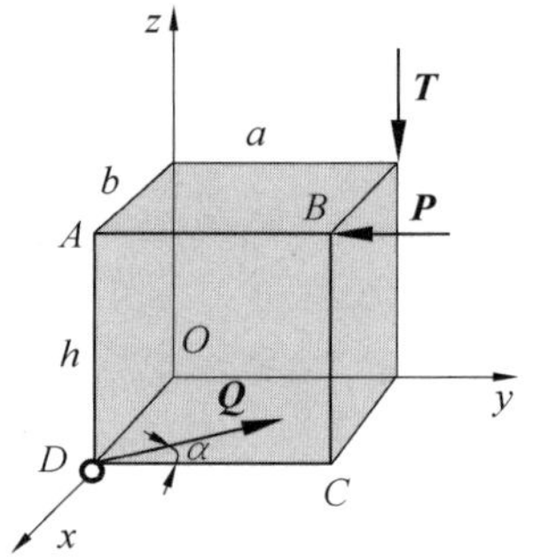

Fig. 3. 15 Diagram of Example 1 (图 3. 15 例题 1 图)

Solution:(解:)

Determine the moment of force $\boldsymbol{T}$ on the shaft, that is(确定力 $\boldsymbol{T}$ 对轴的力矩,即)

$$M_x(\boldsymbol{T}) = -Ta$$

Because force $\boldsymbol{T}$ intersect with axis Oy, so we have(因为力 $\boldsymbol{T}$ 与轴 Oy 相交,故有)

$$M_y(\boldsymbol{T}) = 0$$

Because force $\boldsymbol{T}$ is parallel to axis Oz,so we have(因为力 $\boldsymbol{T}$ 与轴 Oz 平行,故有)

$$M_z(\boldsymbol{T}) = 0$$

Determine the moment of force $\boldsymbol{P}$ relative to the axis, we have(确定力 $\boldsymbol{P}$ 相对于轴的力矩,有)

$$M_x(\boldsymbol{P}) = +Ph$$

Because force $\boldsymbol{P}$ is parallel to axis Oy, so we have(因为力 $\boldsymbol{P}$ 与轴 Oy 平行,故有)

$$M_y(\boldsymbol{P}) = 0$$

$$M_z(\boldsymbol{P}) = -Pb$$

Calculate the moment of force $\boldsymbol{Q}$ relative to the axis. (计算力 $\boldsymbol{Q}$ 相对于轴的力矩.)

Because the force intersects with the axis Ox,so we have(因为力与轴 Ox 相交,故有)

$$M_x(\boldsymbol{Q}) = 0$$

$$M_y(\boldsymbol{Q}) = -(Q\sin\alpha)b$$

$$M_z(\boldsymbol{Q}) = +(Q\cos\alpha)b$$

Example 2: As shown in Fig. 3.16(a), F, a, b, α, β are known, solve $M_O(\boldsymbol{F})$. (**例题 2**:如图 3.16(a)所示,已知 F、a、b、α、β,求 $M_O(\boldsymbol{F})$.)

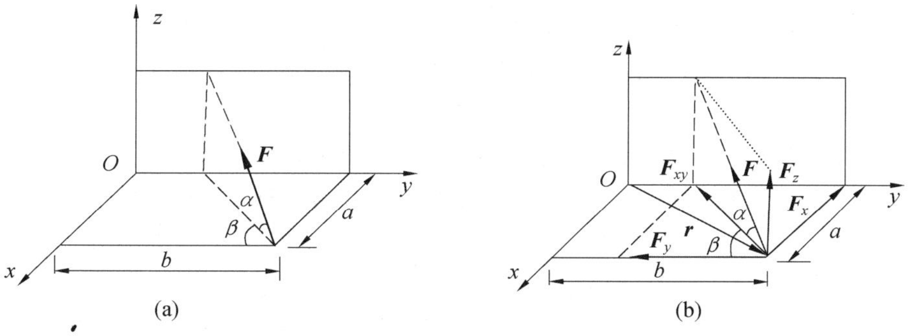

Fig. 3.16 Diagram of Example 2 (图 3.16 例题 2 图)

Solution: (解:)

(1) Direct calculation. (直接计算.)

According to (根据)

$$\boldsymbol{r} = x\mathbf{i} + y\mathbf{j} + z\mathbf{k}, \quad \boldsymbol{M}_O(\boldsymbol{F}) = \boldsymbol{r} \times \boldsymbol{F} = \begin{vmatrix} \mathbf{i} & \mathbf{j} & \mathbf{k} \\ x & y & z \\ F_x & F_y & F_z \end{vmatrix}$$

Among which(其中)

$$\begin{cases} F_x = -F\cos\alpha\sin\beta, & x = a \\ F_y = -F\cos\alpha\cos\beta, & y = b \\ F_z = F\sin\alpha, & z = 0 \end{cases}$$

We obtain(得到)

$$\boldsymbol{M}_O(\boldsymbol{F}) = Fb\sin\alpha\mathbf{i} - Fa\sin\alpha\mathbf{j} + (Fb\sin\alpha\sin\beta - Fa\sin\alpha\cos\beta)\mathbf{k}$$

(2) Using the torque relationship(Fig. 3.17), first calculate the moment of force on the axis to obtain(利用力矩关系(图 3.17),先计算力对轴之矩,得到)

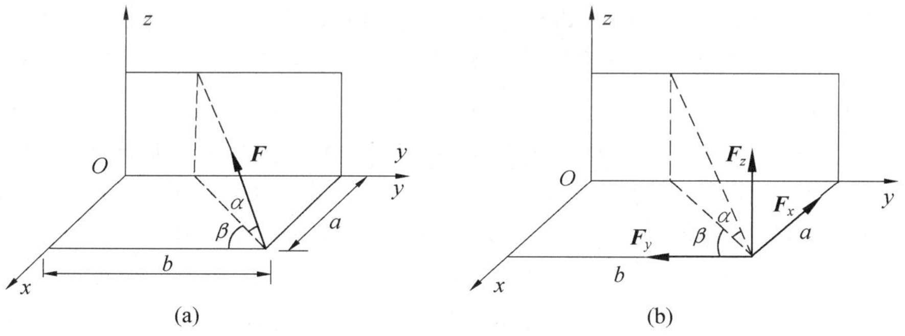

Fig. 3.17 Solving Example 2 using the theorem of resultant moments
(图 3.17 用合力矩定理求解例题 2)

$$\begin{cases} M_x(\boldsymbol{F}) = yF_z - zF_y = F_z b = Fb\sin\alpha \\ M_y(\boldsymbol{F}) = zF_x - xF_z = -F_z a = -Fa\sin\alpha \\ M_z(\boldsymbol{F}) = xF_y - yF_x = aF_y - F_x b = Fb\sin\alpha\sin\beta - Fa\sin\alpha\cos\beta \\ M_O(\boldsymbol{F}) = M_x(\boldsymbol{F})\mathbf{i} + M_y(\boldsymbol{F})\mathbf{j} + M_z(\boldsymbol{F})\mathbf{k} \end{cases}$$

$$= Fb\sin\alpha\mathbf{i} - Fa\sin\alpha\mathbf{j} + (Fb\sin\alpha\sin\beta - Fa\sin\alpha\cos\beta)\mathbf{k}$$

Example 3: Given that the force acting on the rope is $P=60$ N (Fig. 3. 18), calculate how much load can be lifted and calculate the reaction force of the bearing. (**例题** 3:已知作用在绳索上的力为 $P=60$ N(图 3. 18),计算可以提起多少载重,并计算轴承的反力).

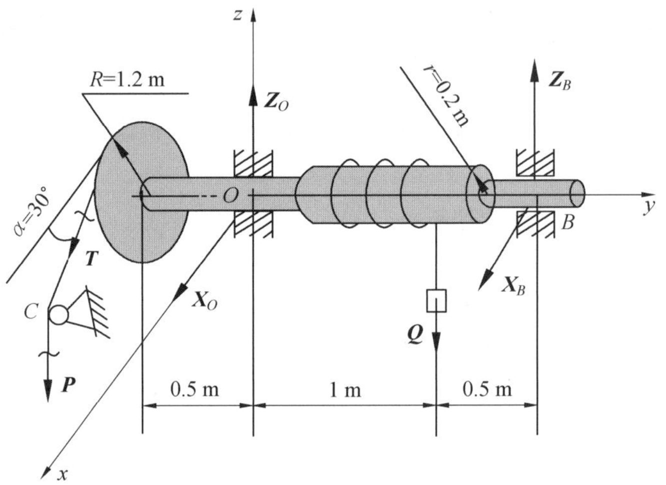

Fig. 3. 18 Diagram of Example 3 (图 3. 18 例题 3 图)

Solution:(解:)

(1) Select axis *AB* as the research object. (选择轴 *AB* 作为研究对象.)

(2) Release the constraint on the axis and replace it with a constraint. Bearings *O* and *B* are cylindrical hinges that only prevent movement in the radial direction, so there are reactions $\boldsymbol{X}_O$, $\boldsymbol{Z}_O$, $\boldsymbol{X}_B$, $\boldsymbol{Z}_B$ in the radial direction of the constraint force at points *O* and *B*. Cut the rope at axis C and use $\boldsymbol{T}$ to indicate the tension of the rope. (解除轴的约束,并将其替换为约束力. 轴承 *O* 和 *B* 是圆柱形铰链,仅在径向方向上防止移动,因此在点 *O* 和 *B* 约束力的径向方向上有反力 $\boldsymbol{X}_O$、$\boldsymbol{Z}_O$、$\boldsymbol{X}_B$、$\boldsymbol{Z}_B$. 在轴 *C* 处将绳子"切"断,并用 $\boldsymbol{T}$ 表示绳索张力.)

(3) Now we can consider the balance of a free body under the action of active and passive forces. Because the sum of the projections of each force on the axis *Oy* is equal to zero, so there are only five remaining equilibrium equations for any force system in space, and because there are also five unknown values, which are $\boldsymbol{X}_O$, $\boldsymbol{X}_B$, $\boldsymbol{Q}$, $\boldsymbol{Z}_O$, $\boldsymbol{Z}_B$, the problem is statically determined. (现在可以在主动力和被动力的作用下考虑自由体的平衡. 因为各力在 *Oy* 轴上的投影总和等于零,所以空间任意力系的六个平衡方程只剩下五个,又因为未知值也是五个,即 $\boldsymbol{X}_O$、$\boldsymbol{X}_B$、$\boldsymbol{Q}$、$\boldsymbol{Z}_O$、$\boldsymbol{Z}_B$,所以问题是静定的.)

(4) Construct the equilibrium equation of the spatial force system, we have(构建空间力系的平衡方程,有)

$$\sum F_x = X_O + X_B + T\cos 30° = 0 \tag{3.28}$$

$$\sum F_y = 0 \tag{3.29}$$

$$\sum F_z = -Q + Z_O + Z_B - T\cos 60° = 0 \tag{3.30}$$

$$\sum M_x = +Z_B \cdot 1.5 - Q \cdot 1 + T\cos 60° \cdot 0.5 = 0 \tag{3.31}$$

$$\sum M_y = -Q \cdot r + T \cdot R = 0 \tag{3.32}$$

$$\sum M_z = -X_B \cdot 1.5 + T\cos 30° \cdot 0.5 = 0 \tag{3.33}$$

(5) Set $r=0.2$ m, $R=1.2$ m, $T=60$ N into Eq. (3.31) to obtain the load weight $Q=360$ N. (将 $r=0.2$ m、$R=1.2$ m、$T=60$ N 代入式(3.31)中,得出载荷重力 $Q=360$ N.)

Derive the constraint reaction force X_B from Eq. (3.33), we have(由式(3.33)得出约束反力 X_B,有)

$$X_B = \frac{60 \cdot \cos 30° \cdot 0.5}{1.5} = 17\ (\text{N})$$

Substitute the obtained $Q=360$ N, $X_B=17$ N into the remaining equations to obtain Z_B, Z_O, X_O, we have (将得到的 $Q=360$ N, $X_B=17$ N 代入剩下的方程中,得出 Z_B、Z_O、X_O,有)

$$Z_B = \frac{360 \cdot 1 - 60 \cdot 0.5 \cdot 0.5}{1.5} = 230\ (\text{N})$$

$$Z_O = 360 - 230 + 60 \cdot 0.5 = 160\ (\text{N})$$

$$X_O = -(17 + 60 \cdot 0.85) = -68\ (\text{N})$$

The negative sign of the constraint reaction X_O indicates that it is opposite to the direction specified in the figure. (约束反力 X_O的负号代表它与图中指定的方向相反.)

Example 4: As shown in Fig. 3.19(a), given the couple moments M_2 and M_3, calculate M_1 and the reaction force at supports A and D at equilibrium. (**例题 4**:如图 3.19(a)所示,已知力偶矩 M_2和 M_3,求平衡时 M_1和支座 A、D 的反力.)

Solution: (**解**:)

Take the curved rod as the research object, draw its force diagram as shown in Fig. 3.19(b), and list the equilibrium equation, we have(取曲杆为研究对象,画其受力图,如图 3.19(b),列平衡方程,有)

$$\begin{cases} \sum F_x = 0, F_{Dx} = 0 \\ \sum F_y = 0, F_{Ay} + F_{Dy} = 0 \\ \sum F_z = 0, F_{Az} + F_{Dz} = 0 \\ \sum M_x(F) = 0, M_1 - bF_{Az} - cF_{Ay} = 0 \\ \sum M_y(F) = 0, aF_{Az} - M_2 = 0 \\ \sum M_z(F) = 0, M_3 - aF_{Ay} = 0 \end{cases}$$

$$F_{Dx} = 0, F_{Az} = -F_{Dz} = \frac{M_2}{a},\ F_{Ay} = -F_{Dy} = \frac{M_3}{a},\ M_1 = \frac{bM_2 + cM_3}{a}$$

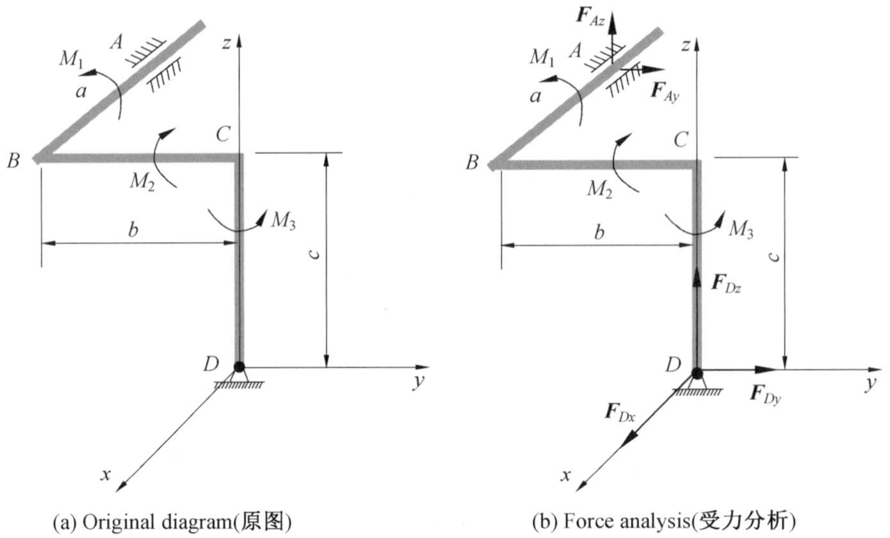

Fig. 3. 19　Diagram of Example 4(图 3. 19　例题 4 图)

3.9　Center of Gravity (重心)

3.9.1　Definition of center of gravity(重心的定义)

The center of gravity of a rigid body is always connected to the point C of the object, and regardless of how the object is placed, the action line of the combined force of gravity acting on the particle always passes through point C (Fig. 3. 20). (刚体的重心是始终与该物体相连接的一点 C,无论物体如何放置,作用在该质点上重力的合力作用线总是通过 C 点(图 3. 20).)

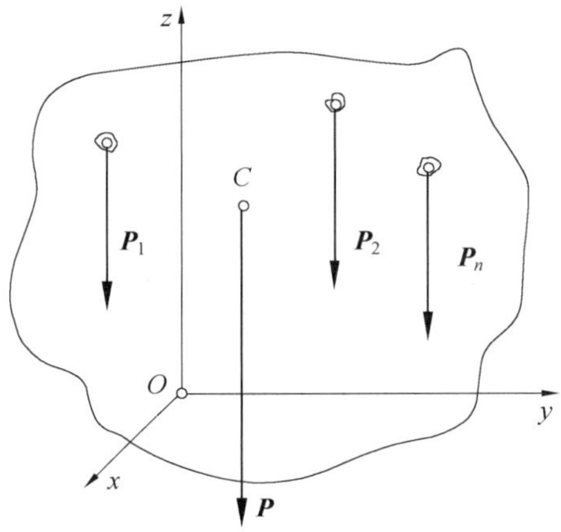

Fig. 3. 20　Center of gravity C(图 3. 20　重心 C)

The coordinates of gravity center of non-uniform bodies are(不均匀体的重心坐标为)

$$X_C = \frac{\sum P_k x_k}{P},\ Y_C = \frac{\sum P_k y_k}{P},\ Z_C = \frac{\sum P_k z_k}{P} \tag{3.34}$$

where the gravity of each small part of an object is P_k; the gravity exerted on the entire object is P; the coordinates of gravity center of each small part are x_k, y_k, z_k.(式中,物体各微小部分的重力为 P_k;整个物体所受的重力为 P;各微小部分的重心坐标为 x_k、y_k、z_k.)

3.9.2 Coordinate of the center of gravity of a homogeneous body (均质体重心坐标)

1. The concept of homogeneous bodies(均质体的概念)

If an object is homogeneous, the weight P_k of any part of is proportional to the volume V_k of that part, $P_k=\gamma \cdot V_k$, and the weight P of the entire object is proportional to the volume V, i. e $P=\gamma \cdot V$,γ is the weight per unit volume. (如果物体是均质的,则其任何部分的重力 P_k 与该部分的体积 V_k 成比例,$P_k=\gamma \cdot V_k$, 而整个物体重力 P 与体积 V 成比例,即 $P=\gamma \cdot V$,γ 为每单位体积的重力.)

2. The center of gravity of volume V(体积 V 的重心)

The coordinates of gravity center of volume V are determined by(体积 V 的重心坐标为)

$$X_C = \frac{\sum V_k x_k}{V},\ Y_C = \frac{\sum V_k y_k}{V},\ Z_C = \frac{\sum V_k z_k}{V} \tag{3.35}$$

where V_k is the volume of each part of an object; V is the volume of the entire object; x_k,y_k,z_k are the coordinates of the center of gravity of each part of an object. (式中,V_k 为物体各部分的体积;V 为整个物体的体积;x_k、y_k、z_k 为物体各部分的重心坐标.)

3. Center of gravity of area S(面积 S 的重心)

The coordinates of gravity center for the area S are determined by (面积 S 重心坐标为)

$$X_C = \frac{\sum S_k x_k}{S},\ Y_C = \frac{\sum S_k y_k}{S} \tag{3.36}$$

where S_k is the area of each part of an object;S the area of the entire object (S 为物体总面积) x_k, y_k are the coordinates of the center of gravity of each part area. (式中,S_k 为各部分物体面积;S 为物体总面积;x_k、y_k 为各分面积重心坐标.)

4. Center of gravity of line(线段重心)

The coordinates of gravity center of the line L are determined by (线 L 重心坐标为)

$$X_C = \frac{\sum l_k x_k}{L},\ Y_C = \frac{\sum l_k y_k}{L},\ Z_C = \frac{\sum l_k z_k}{L}$$

where l_k is the length of each sub section, L is the total length of the object; x_k, y_k,z_k are coordinates of the center of gravity of each component. (式中,l_k 为各分部分长度;l 为物体总长度;x_k、y_k、z_k 为各分部分重心坐标.)

3.9.3 Method for determining the coordinates of the center of gravity of a homogeneous body(确定均质体重心坐标的方法)

1. Symmetry method(对称法)

A homogeneous body with a simple shape and a plane, axis, or center of symmetry must have its center of gravity on itssymmetrical plane, symmetrical axis, or symmetrical center. (凡具有对称面、对称轴或对称中心的简单形状的均质体,其重心一定在它的对称面、对称轴或对称中心上.)

It can be seen from its symmetry that the center of gravity of torus, rectangle, parallelepiped, sphere and other homogeneous bodies with symmetry center are located on the geometric center (symmetry center) of these bodies. (根据其对称性可以看出,圆环形、矩形、平行六面体、球和其他具有对称中心的均质体的重心位于这些形体的几何中心(对称中心)上.)

2. Combination method(组合法)

The essence of the combination method is that if an object can be divided into a finite number of parts, and for each part, the position of its center of gravity is known, then the coordinates of the total center of gravity can be calculated as(组合法的本质是如果一个物体可以被分成有限数量的部分,对于每个部分来说,其重心的位置是已知的,那么总重心的坐标可以用公式计算出来,即)

$$X_C = \frac{\sum P_k x_k}{P},\ Y_C = \frac{\sum P_k y_k}{P},\ Z_C = \frac{\sum P_k z_k}{P}$$

3. Superposition method (negative area method)(叠加法(负面积法))

This method is a special case of the combination method. It is suitable for objects with cuts if the center of gravity of the object without cutting or cutting is known. In this case, the area (volume) of the cut out part is considered as negative area (volume). (该方法是组合法的特殊情况. 如果已知没有切割和切割部分物体的重心,则它适用于有切口的物体. 在这种情况下,切出部分的面积(体积)视为负面积(体积).)

4. Integration method(积分法)

In this method, the sum in the above formula is replaced by the corresponding integral. (在该方法中,上述公式中的求和被相应的积分代替.)

(1) The coordinates of gravity center of volume V are determined by (体积 V 的重心坐标为)

$$X_C = \frac{\int_{(V)} x\mathrm{d}V}{V},\ Y_C = \frac{\int_{(V)} y\mathrm{d}V}{V},\ Z_C = \frac{\int_{(V)} z\mathrm{d}V}{V}$$

where $\mathrm{d}V$ the volume of small parts; V is the entire volume of the object; x,y,z are the coordinate of the center of gravity of a small volume. (式中,$\mathrm{d}V$ 为微小部分的体积;V 为整个物体体积;x、y、z 为微小部分体积的重心坐标.)

(2) The center of gravity coordinates of area S are determined by (面积 S 的重心坐标为)

$$X_C = \frac{\int_{(S)} x\mathrm{d}S}{V},\ Y_C = \frac{\int_{(S)} y\mathrm{d}S}{V}$$

where dS is the area of a small part; S is the entire area of the object; x, y are the coordinate of the center of gravity of a small part of the body area. (式中,dS 为微小部分的面积;S 为整个物体面积;x、y 为微小部分面积的重心坐标.)

(3) The coordinates of the center of gravity of the line are determined by (线段重心坐标为)

$$X_C = \frac{\int_{(L)} x\mathrm{d}l}{L},\ Y_C = \frac{\int_{(L)} y\mathrm{d}l}{L},\ Z_C = \frac{\int_{(L)} z\mathrm{d}l}{L}$$

where dl is tiny line segments; L is the overall length of the line; x, y, z are the center of gravity coordinates of small line segments. (式中,dl 为微小线段;L 线段的整体长度;x、y、z 为微小线段的重心坐标.)

5. Experimental method(实验法)

There are two main experimental methods for determining the position of gravity center of uniform and non-uniform objects of any shape. (确定任意形状的均匀和不均匀物体的重心位置有两种主要的实验方法.)

(1) Suspension method. (悬挂法.)

The suspension method refers to the use of wires or cables to suspend an object from different points. According to the balance condition of two forces, the center of gravity must be on the vertical line passing through the suspension point, and then the object is suspended at any other point, and mark the corresponding vertical line. The intersection of these two vertical lines is the center of gravity of the object (Fig. 3.21). (悬挂法是指使用线或电缆从物体上的不同点将其悬挂起来,根据二力平衡条件,重心必在过悬挂点的铅垂线上,再将物体悬挂在另一任一点,标出铅垂线,这两条铅垂线的交点即为该物体的重心(图 3.21).)

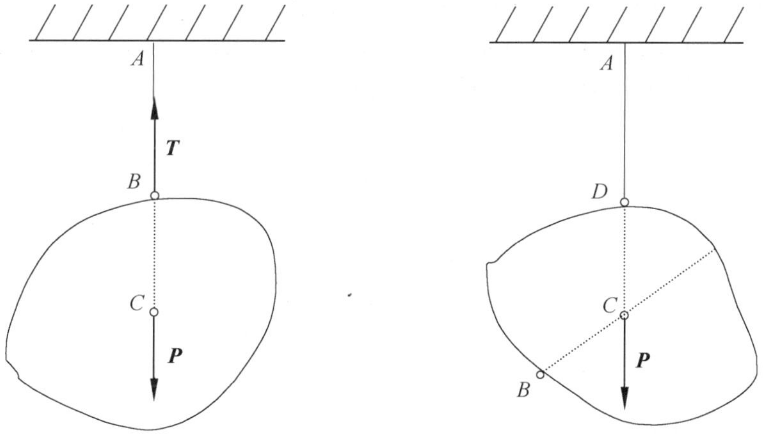

Fig. 3.21 Determination of center of gravity by suspension method (图 3.21 悬挂法确定重心)

(2) Weighing method.(称重法.)

For non-uniform body with complex shape or large volume (such as locomotive, airplane, etc.), one of the supporting reaction forces can be determined by weighing with a special scale, while the other reaction forces can be obtained by the balance equations. (对于形状复杂或体积较大的非均质体(如机车、飞机等),其支撑反力之一可以通过特殊秤称重来确定,其他反力从平衡方程中得出.)

3.9.4 The center of gravity of some simple shaped homogeneous bodies (一些简单形状均质体的重心)

1. The center of gravity of an arc(圆弧的重心)

The coordinates of the center of gravity of the arc with the radius R, and the central angle 2α are(半径为 R,圆心角为 2α 的圆弧的重心坐标为)

$$x_C = (R \cdot \sin \alpha)/\alpha,\ y_C = 0$$

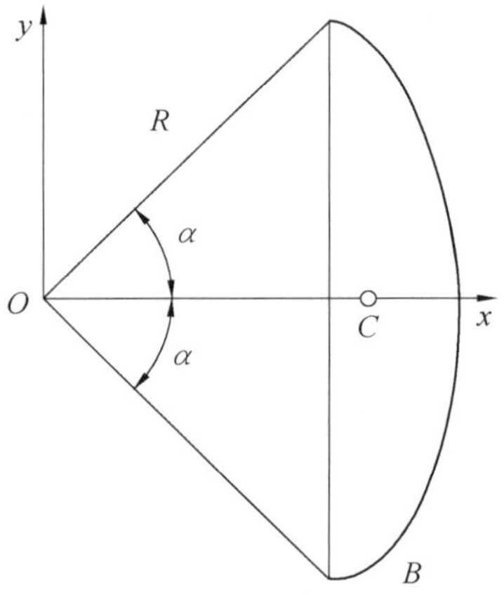

Fig. 3.22 Center of gravity of arc(图 3.22 圆弧的重心)

2. The center of gravity of a triangle(三角形的重心)

The center of gravity of the triangle is located at the intersection of the center line (Fig. 3.23). (三角形的重心位于其中线的交叉点(图 3.23).)

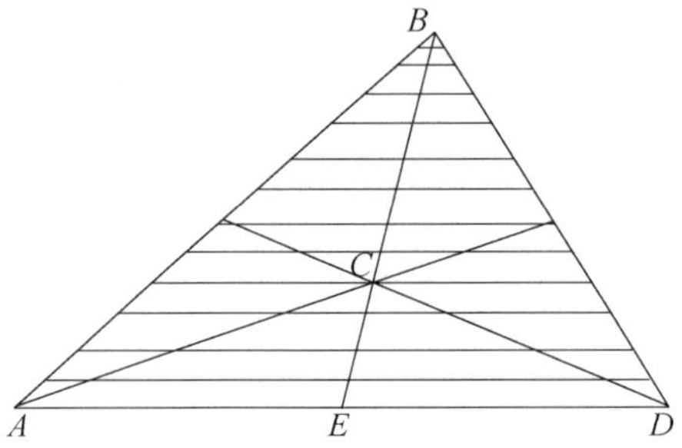

Fig. 3.23 Center of gravity of a triangle (图 3.23 三角形的重心)

3. The center of gravity of the sector(扇形的重心)

The center of gravity of the sector (Fig. 3.24) is located on its axis of symmetry, and the distance from center O is (扇形的重心(图 3.24)位于其对称轴上,与中心 O 的距离为)

$$x_C = (2 \cdot R \cdot \sin\alpha)/(3 \cdot \alpha)$$

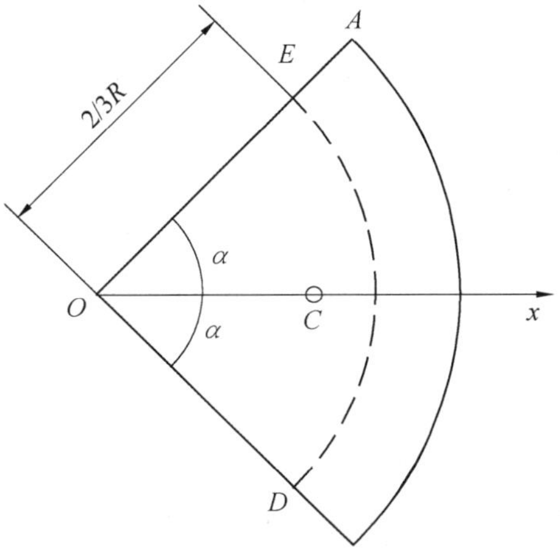

Fig. 3.24　Center of gravity of asector(图 3.24　扇形的重心)

4. The center of gravity of the cone(锥体的重心)

The center of gravity C of the cone is located on line C_1E, where E is the vertex and C_1 is the center of gravity of the bottom surface of the cone (Fig. 3.25), we have(锥体的重心 C 位于线 C_1E 上,其中 E 是顶点,C_1是锥体底面的重心(图 3.25),有)

$$CC_1 = C_1E/4$$

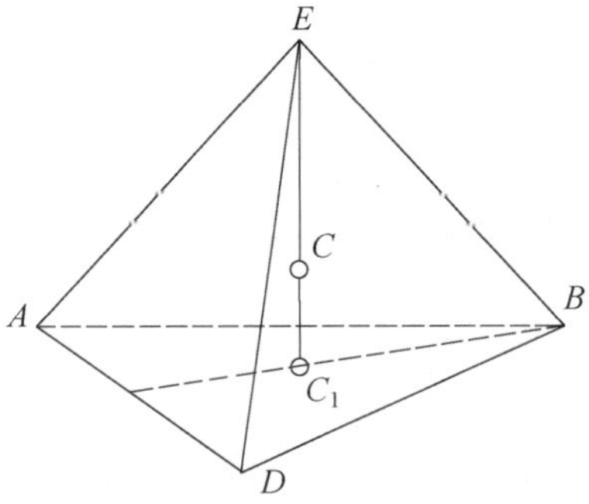

Fig. 3.25　Center of gravity of cone (图 3.25　锥体的重心)

Example 5: Find the coordinates of gravity center of the homogeneous plate in Fig. 3.26(a), with a known fixed plate thickness. (**例题** 5:求图 3.26(a)中均质板的重心坐标,已知板厚度固定.)

Solution:(解:)

(1) Because the uniform plate has a fixed thickness, a formula can be used to determine the center of gravity position in the plan view. (由于均匀板有固定厚度,因此可以使用公式来确定平面图的重心位置.)

(2) Divide the board into three simplest geometric shapes (Fig. 3.26(b)) with known center of gravity positions. (把板分成三个最简单的几何图形(图 3.26(b)),其重心位置已知.)

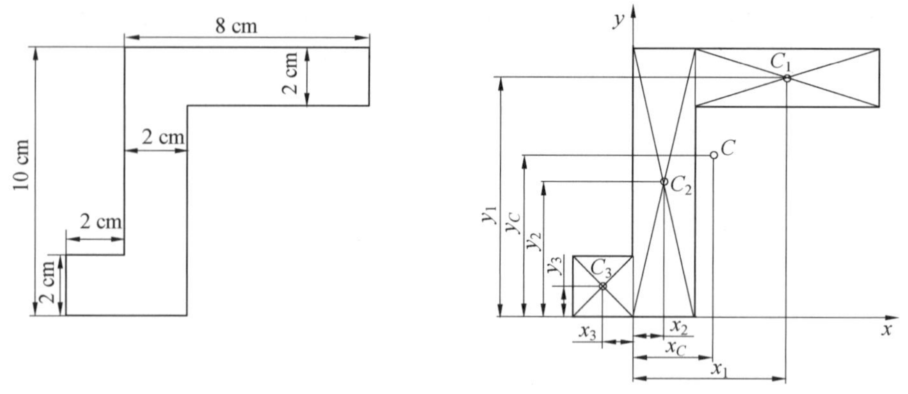

(a) (b)

Fig. 3.26　Diagram of Example 5(图 3.26　例题 5 图)

(3) Select the coordinate system as shown in Fig. 3.26(b).(选择图 3.26(b)所示的坐标系.)

(4) Fill the calculation results in Table 3.1, each rectangle corresponds to one row in the table.(计算结果填入表 3.1 中,每个矩形对应于表格中的一行.)

Table 3.1　Form of Example 5(表 3.1 例题 5 表)

Serial Number (序号)	S_i/cm^2	x_i/cm	y_i/cm	S_ix_i/cm^3	S_iy_i/cm^3
1	12	5	9	60	108
2	20	1	5	20	100
3	4	−1	1	−4	4
$\sum$	36	—	—	76	212

(5) Calculate the sum of S_i, S_ix_i, S_iy_i, and the result is written on the last line.(求和得出 S_i、S_ix_i、S_iy_i 的值,并将结果写在最后一行.)

(6) Obtain the coordinates of the center of gravity of the plate, we have(得出板的重心坐标,即)

$$x_C = \frac{\sum S_ix_i}{S} = \frac{76}{36} = \frac{19}{9}(\text{cm}),\ y_C = \frac{\sum S_iy_i}{S} = \frac{112}{36} = \frac{28}{9}(\text{cm})$$

(7) Based on the calculated coordinates, the center of gravity C of the plate can be determined.(根据计算出的坐标,可以确定板的重心 C.)

5. Methods for determining the position of the center of gravity(确定重心位置的方法)

(1) Graphic segmentation method.(图形分别法.)

For shapes with known positions of center of gravity, the graphic segmentation method is suitable for situations where a rigid body can be divided into a finite number of simple objects.(对于重心位置已知的图形,图形分割法适用于可以将刚体分成有限个简单物体的情况.)

(2) Supplementary method. (补充法.)

The supplementary method is a special case of the segmentation method. This method is used when an object can be divided into simple shapes and its center of gravity is known, but some geometric shapes are missing. (补充法是分割法的特例. 在物体可以被分成简单的形状,其重心的位置已知,但是一些几何形状空缺时使用该方法.)

Example 6: Determine the position of gravity center of a circular cross-section with a diameter of 12 cm, where the drilling diameter is 2 cm (Fig. 3.27). (**例题 6**:确定一个直径为 12 cm 的圆形截面的重心位置,其中钻孔直径为 2 cm(图 3.27).)

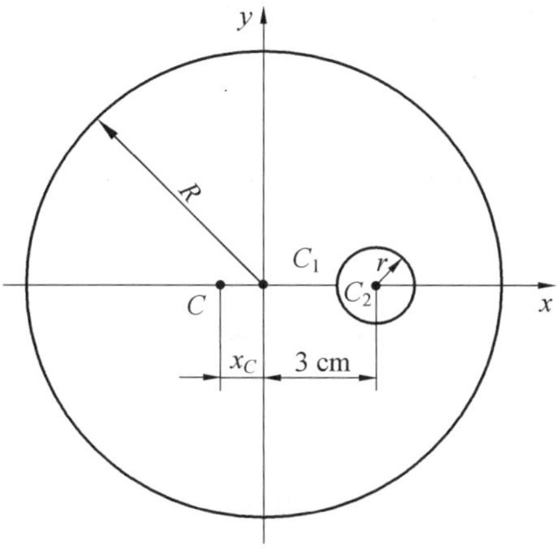

Fig. 3.27 Diagram of Example 6 (图 3.27 例题 6 图)

Solution: (**解**:)

①Since we need to find the center of gravity of the cross-section, we use the formula to determine the center of gravity of the plane figure. (由于需要找到横截面的重心,因此使用确定平面图形重心的公式.)

②Supplement the cross-sectional area of the drilling area (as this part does not actually exist, in the formula, the area of the drilling area is a negative sign), we have(补充钻孔区域的横截面积(因为实际上这个部分不存在,所以在公式中,孔区面积为负号),有)

$$S_2 = -\pi r^2 = -\pi \cdot 1^2 = -\pi \ (\text{cm}^2)$$

③The origin of the coordinate system is located at the center of a circle with a radius of R, i. e. at point C_1. (坐标系的原点位于半径为 R 的圆的中心,即在点 C_1 处.)

④Fill out Table 3.2. (填写表 3.2.)

Table 3.2 Form of Example 6(表 3.2 例题 6 表)

Serial Number (序号)	S_i/cm^2	x_i/cm	y_i/cm	$S_i x_i/\text{cm}^3$	$S_i y_i/\text{cm}^3$
1	36π	0	0	0	0
2	$-\pi$	3	0	-3π	0
$\sum$	35π	—	—	-3π	0

⑤After summing to obtain S_i and $S_i x_i$, write the result on the last line.(求和得出S_i和$S_i x_i$之后,将结果写在最后一行.)

⑥Calculating the coordinates of the center of gravity of the cross-section, we have (计算得到横截面重心的坐标为)

$$x_c = \frac{\sum S_i x_i}{S} = \frac{-3\pi}{35\pi} = -\frac{3}{35}\ (\text{cm})$$

And $y_C = 0$, because axis C_1x is the symmetry axis of that part.(且$y_C = 0$,因为轴C_1x是该部分的对称轴.)

⑦Based on the calculated cross-sectional coordinates, the center of gravity C was obtained.(根据计算出的横截面坐标,得到重心C.)

(3) Integration method.(积分法.)

When the above two methods cannot determine the position of the center of gravity, use the integration method.(在前两种方法不能确定重心位置的情况下,使用积分法.)

(4) Experimental method.(实验法.)

The experiment was conducted using two methods: suspension method and weighing.(实验通过两种方法进行:悬挂法和称重法.)

①Suspension method.(悬挂法.)

The suspension method refers to hanging a flat body that cannot be divided into the simplest known shape of the center of gravity on a rope. Draw a line along the rope on the surface of the object. Then hang the plane from another point, and then draw a vertical line again (along the hanging line). The center of gravity is at the intersection of these two lines.(悬挂法是指将不能被分割为重心已知的最简单图形的平面体悬挂在绳子上,沿着绳子在物体的表面上画一条线,然后将该平面体从另一点悬挂,再次(沿着悬挂线)绘制垂直线,重心在这两条线的交叉点处.)

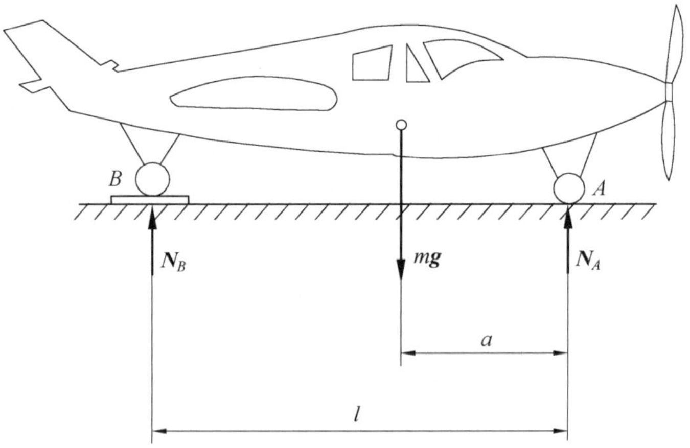

Fig. 3.28 weighing method (图 3.28 称重法)

② Weighing method.(称重法.)

The weighing method is usually used for large items such as airplanes, helicopters and other machines. If the mass is known, such as for an aircraft, place the rear wheels on a scale (Fig. 3.28) and determine the reaction force $\boldsymbol{N}_B$ according to the scale's instructions. Then write an equilibrium equation. It is more convenient to use the moment equation for point A, we have (称重法通常用于大型物品,如飞机、直升机和其他机器. 如果质量已知,如飞机,则将后轮放在秤上(图 3.28),并根据秤的指示确定反力 $\boldsymbol{N}_B$. 然后写一个平衡方程,使用关于点 A 合力矩等式更方便,有)

$$\sum_{i=1}^{3} M_A(\boldsymbol{F}_i) = 0, mga - N_B l = 0$$

From this, the required value a can be found, which is the position of the aircraft's center of gravity, we have(由此可以找到所需的值 a,即飞机重心的位置,有)

$$a = \frac{N_B l}{mg}$$

3.10 Questions(思考题)

1. Calculate the moment of force $\boldsymbol{F}$ to point O using vector product $\boldsymbol{r}_A \times \boldsymbol{F}$. When the force moves along its action line which changes the coordinates x, y, z of the force action point (Fig. 3.29), how does the calculation result change? (用矢量积 $\boldsymbol{r}_A \times \boldsymbol{F}$ 计算力 $\boldsymbol{F}$ 对点 O 之矩,当力沿其作用线移动,改变了力作用点的坐标 x、y、z 时(图 3.29),其计算结果有何变化?)

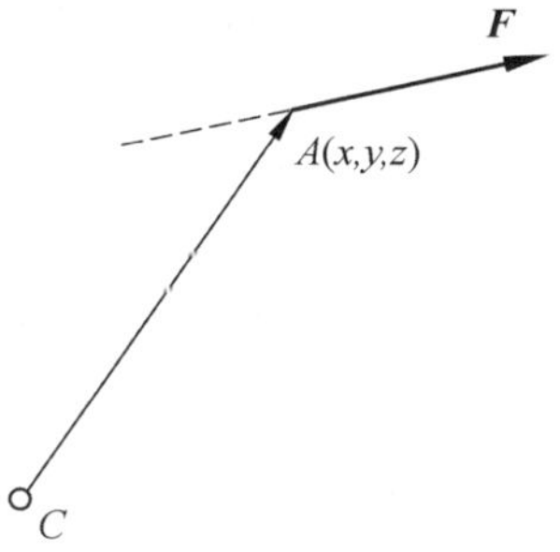

Fig. 3.29 Diagram of Question 1(图 3.29 思考题 1 图)

2. A driving couple is applied on shaft AB, with a torque of M_1 and a meshing radius of $R_2 = 2R_1$ for the gear(Fig. 3.30). When studying the balance problem of axis AB and CD: (1) Can the couple moment acting on axis AB be moved to axis CD on the grounds that it is a free vector? (2) If a couple with a moment of M_2 is applied on the axis CD to balance the two axes, is the moment of the two couples equal? Should the steering be reversed? (轴 AB 上作用一主动力偶,矩为 M_1,齿轮的啮合半径 $R_2 = 2R_1$,如图 3.30 所示. 当研究轴 AB 和 CD 的平衡问题时:(1)能否以力偶矩是自由矢量为由,将作用在轴 AB 上的力偶搬移到轴 CD 上?(2)若在轴 CD 上作用矩为 M_2 的力偶,使两轴平衡,两力偶的矩的大小是否相等? 转向是否应相反?)

3. What is the final result of the spatial equilibrium force system? Is it possible to synthesize into a force spiral? (空间平衡力系的最终结果是什么? 可能合成为力螺旋吗?)

4. (1) The action lines of each force in a spatial force system are parallel to a fixed plane;

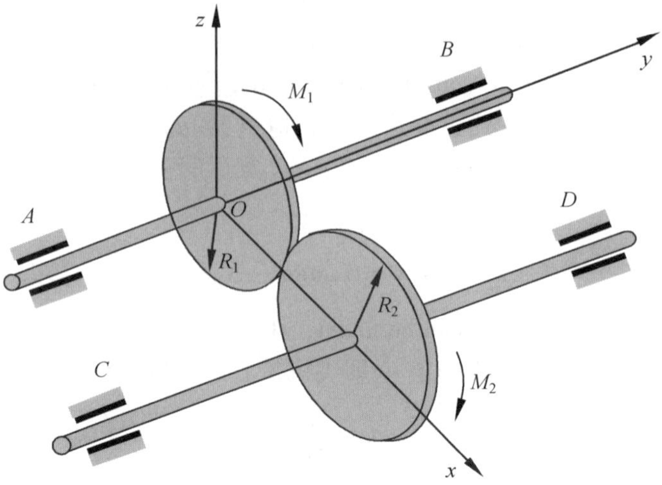

Fig. 3. 30　Diagram of Question 2(图 3. 30　思考题 2 图)

(2) The action lines of each force in a spatial force system converge at two fixed points. How many independent equilibrium equations can each of these two force systems have at most? ((1)空间力系中各力的作用线都平行于某一固定平面;(2)空间力系中各力的作用线分别汇交于两个固定点. 这两种力系最多各有几个独立平衡方程?)

5. The transmission shaft is supported by two thrust bearings, each bearing having three unknown forces and a total of six unknown quantities. And there are exactly six equilibrium equations for spatial arbitrarily force system, is it a statically determined problem? (传动轴用两个止推轴承支撑,每个轴承有三个未知力,共六个未知量. 而空间任意力系的平衡方程恰好有六个,是否为静定问题?)

6. Spatial arbitrarily force system can always be balanced by two forces, why? (空间任意力系总可以用两个力来平衡,为什么?)

7. A certain spatial force system is simplified to three points that are not collinear, is it possible to the following questions: (1) the principal vectors are equal and the principal moments are also equal; (2) the principal vectors are not equal, and the principal moments are equal; (3) the principal vectors are equal, but the principal moments are not equal; (4) the principal vector and principal moment are not equal. (某一空间力系对不共线的三个点简化,试问下述情况是否可能:(1)主矢相等,主矩也相等;(2)主矢不相等,主矩相等;(3)主矢相等,主矩不相等;(4)主矢、主矩都不相等.)

8. As shown in Fig. 3. 31, a force of $\boldsymbol{F}$ is exerted on point A of the cube along an edge direction, questions are as follows: (在图 3. 31 所示正方体上点 A 作用一力 $\boldsymbol{F}$,沿一棱方向,求):

(1) Can we add a non-zero force on point B so that the simplified principal moment of the force system towards point A is zero? (能否在点 B 加上一个不为零的力,使力系向点 A 简化的主矩为零?)

(2) Can we add a nonzero force to point B, so that the simplified principal moment of the

force system towards point B is zero?(能否在点 B 加上一个不为零的力,使力系向点 B 简化的主矩为零?)

(3) Can we add a non- zero force to each point B and C to balance the force system?(能否在点 B、C 各加上一个不为零的力,使力系平衡?)

(4) Can we add a force spiral at point B to balance the force system?(能否在点 B 加上一个力螺旋,使力系平衡?)

(5) Can we add a couple of forces at points B and C to balance the force system?(能否在点 B、C 两处各加上一个力偶,使力系平衡?)

(6) Can we add a force at point B and a couple at point C to balance the force system?(能否在点 B 加上一个力,在点 C 处加一个力偶,使力系平衡?)

9. A cube with side length a is shown in Fig. 3.32. It is known that a certain force system is simplified towards point B to obtain a resultant force, and simplification towards point C' also results in a resultant force. Questions:(图 3.32 所示为一个边长为 a 的正方体,已知某力系向点 B 简化得到一合力,向点 C' 简化也得到一合力. 问:)

(1) Is the principal moment obtained by simplifying the force system towards points A and A' equal?(力系向点 A 和点 A' 简化所得主矩是否相等?)

(2) Is the principal moment obtained by simplifying the force system towards points A and O' equal?(力系向点 A 和点 O' 简化所得主矩是否相等?)

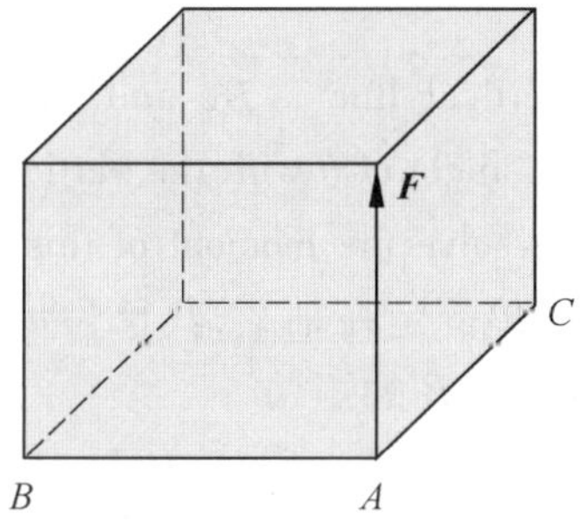

Fig. 3.31 Diagram of Question 8
(图 3.31 思考题 8 图)

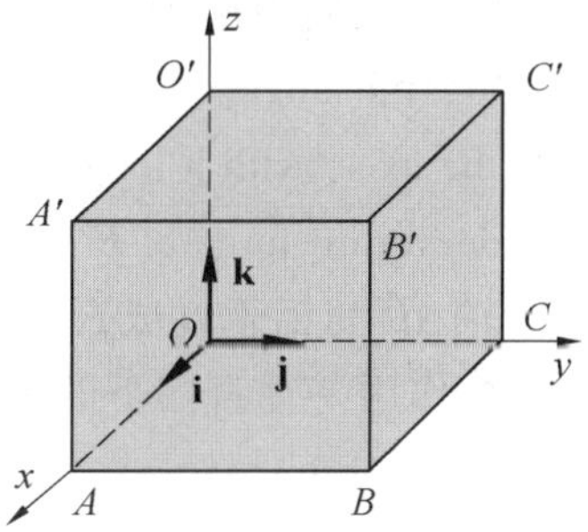

Fig. 3.32 Diagram of Question 9
(图 3.32 思考题 9 图)

3.11 Exercises(习题)

1. The hanging rack is shown in Fig. 3.33, without considering the weight of the three rods. It is connected to point O with a ball hinge, and the plane BOC is a horizontal plane with $OB = OC$, the angle is shown in the Fig. 3.33. Hang a heavy object G with a weight of $P = 10$ kN at point O, please calculate the force on the three rods.(挂物架如图 3.33 所示,三杆的质量不计,用球铰链连接于点 O,平面 BOC 是水平面,且 $OB = OC$,角度如图 3.33 所示. 在点 O 挂一重 $P = 10$ kN 的重物 G,求三根杆所受的力.)

2. The spatial truss shown in Fig. 3.34 consists of six rods 1, 2, 3, 4, 5, 6. Apply a force $\boldsymbol{F}$ on node A, which acts in the rectangular $ABCD$ plane at a 45° angle to the plumb line. Isosceles

triangle $\triangle AEK = \triangle BFM$, and $\triangle AEK$, $\triangle BFM$, $\triangle BND$ are right angles at fixed points A, B, D, and $EC = CK = FD = DM$, force F = 10 kN . Find the force on each rod. (图 3.34 所示空间桁架由六根杆 1、2、3、4、5、6 组成. 在节点 A 上作用一力 $\boldsymbol{F}$,此力作用在矩形 $ABCD$ 平面内,与铅垂线成 45°. 等腰三角形 $\triangle AEK = \triangle BFM$,且 $\triangle AEK$、$\triangle BFM$、$\triangle BND$ 在定点 A、B、D 处均为直角,又有 $EC = CK = FD = DM$,力 F = 10 kN ,求各杆受力.)

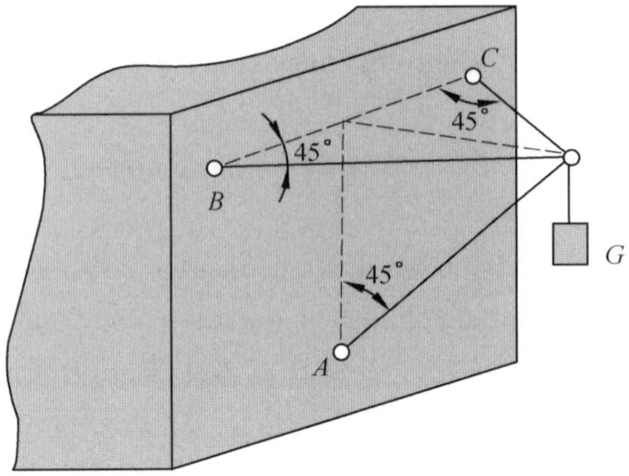

Fig. 3.33 Diagram of Exercise 1

(图 3.33 习题 1 图)

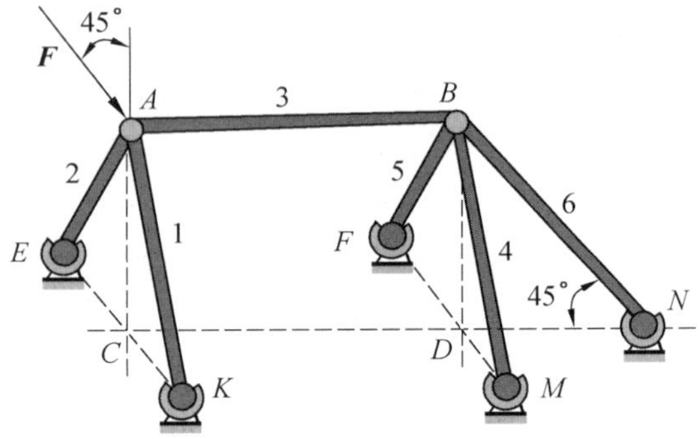

Fig. 3.34 Diagram of Exercise 2

(图 3.34 习题 1 图)

3. Calculate the moment M_z of force $\boldsymbol{F}$ = 1 000 N on the z-axis in Fig. 3.35. (求图 3.35 中力 F = 1 000 N 对于 z 轴的力矩 M_z .)

4. In Fig. 3.36, the angle between axis AB and vertical line is β, and cantilever CD is vertically fixed on the axis, with a length of a and forms a angle of θ with the vertical plane zAB. Apply a vertically downward force of $\boldsymbol{F}$ at point D, please solve the moment of this force on axis AB. (图 3.36 中,轴 AB 与铅垂线成 β 角,悬臂 CD 垂直地固定在轴上,其长为 a,并与铅垂面 zAB 成 θ 角,在点 D 作用一垂直向下的力 $\boldsymbol{F}$,求此力对轴 AB 的矩.)

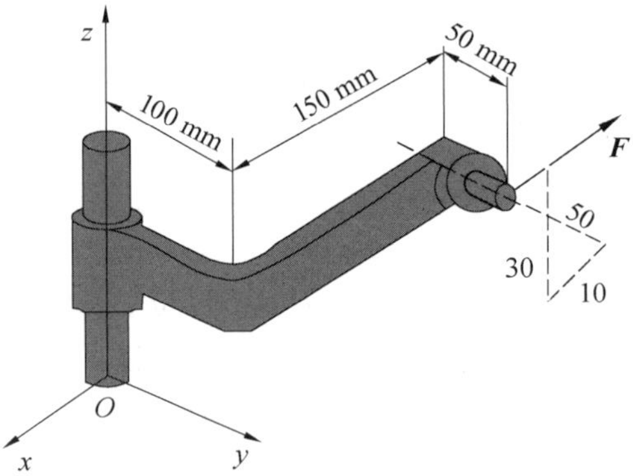

Fig. 3.35 Diagram of Exercise 3

(图 3.35 习题 3 图)

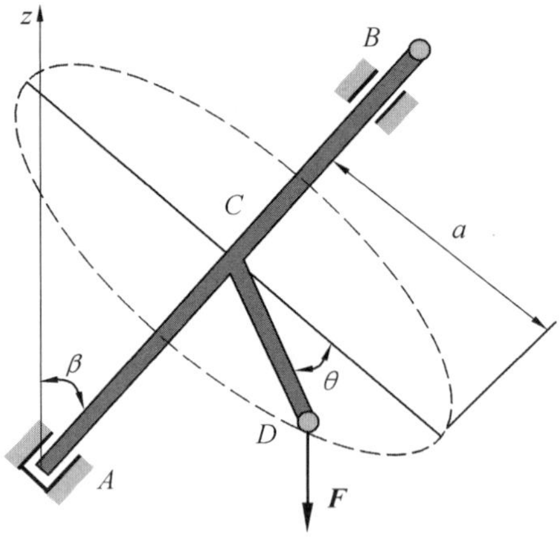

Fig. 3.36 Diagram of Exercise 4

(图 3.36 习题 4 图)

5. The radius of the horizontal disk is R, and there is a force $\boldsymbol{F}$ acting at the outer edge C. The force $\boldsymbol{F}$ is located in the tangent plane of the disk C and has an angle of 60 ° with the tangent

line of the disk C. Other dimensions are shown in Fig. 3.37. Find the moment of force $\boldsymbol{F}$ on the x, y, z axes. (水平圆盘的半径为 R,外缘 C 处作用有力 $\boldsymbol{F}$,力 $\boldsymbol{F}$ 位于圆盘 C 处的切平面内,且与 C 处圆盘切线夹角为 60°,其他尺寸如图 3.37 所示,求力 $\boldsymbol{F}$ 对 x、y、z 轴之矩.)

6. The force acting on a column with an I-shaped cross-section is shown in Fig. 3.38. Please calculate the simplified result of this force towards the centroid of the cross-section. (截面为工字形的立柱受力如图 3.38 所示,求此力向截面形心简化的结果.)

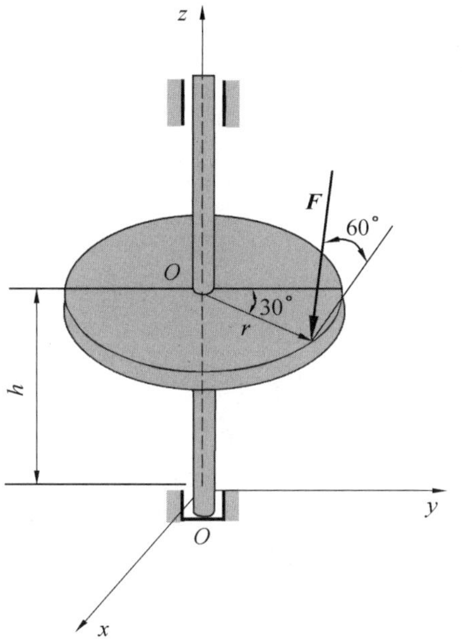

Fig. 3.37 Diagram of Exercise 5
(图 3.37 习题 5 图)

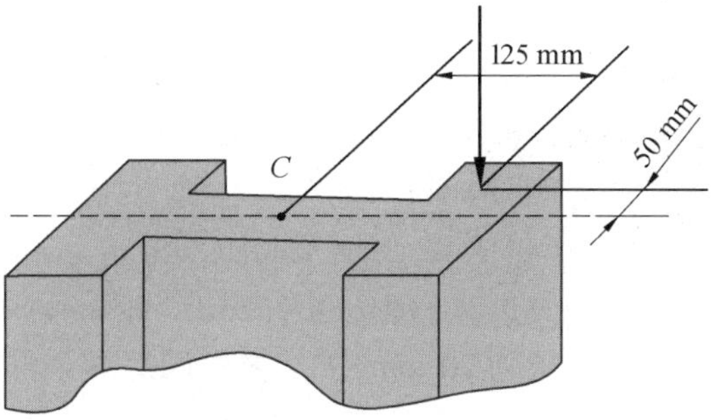

Fig. 3.38 Diagram of Exercise 6
(图 3.38 习题 6 图)

7. The side length of the square is $a = 0.2$ m, and there are six forces equal to 100 N acting along each edge at fixed points A and B, as shown in Fig. 3.39. Find the simplification result of the force system towards point O. (正方形边长 $a = 0.2$ m,在定点 A 和 B 处沿各棱边作用有六个大小都等于 100 N 的力,方向如图 3.39 所示,求力系向点 O 的简化结果.)

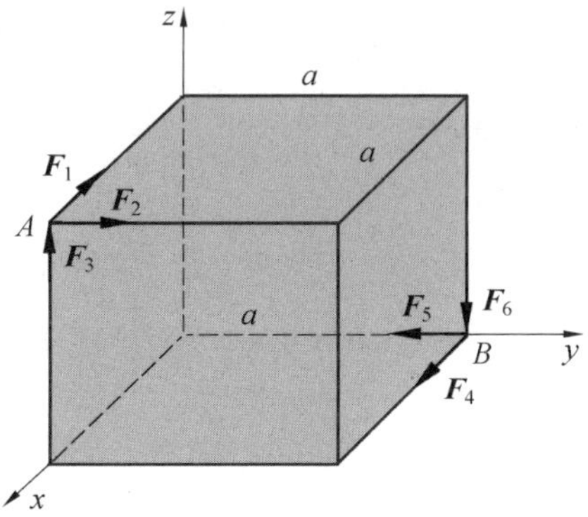

Fig. 3.39 Diagram of Exercise 7
(图 3.39 习题 7 图)

8. There are six forces acting on the vertices A, B, C of the triangular prism in the direction shown in Fig. 3.40, and $AB=300$ mm, $BC=400$ mm, $AC=500$ mm. Simplify this force system towards point A. (在三棱柱的顶点 A、B、C 上作用有六个力,方向如图 3.40 所示,$AB=300$ mm,$BC=400$ mm,$AC=500$ mm,请向点 A 简化此力系.)

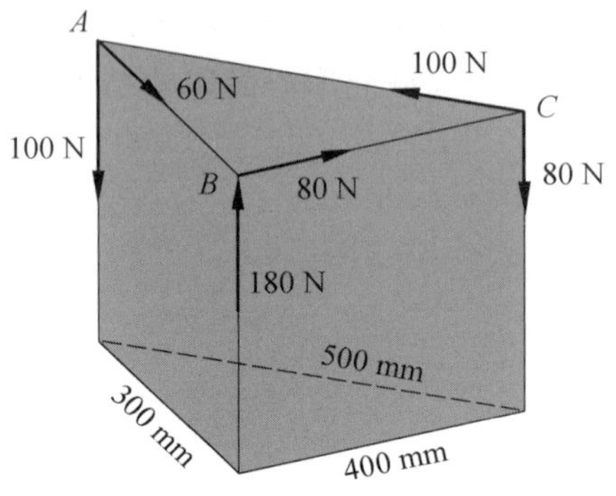

Fig. 3.40 Diagram of Exercise 8
(图 3.40 习题 8 图)

9. The cross-sectional dimensions of the I-beam are shown in Fig. 3.41, please solve the center of gravity (geometric center) of this section. (工字形钢截面尺寸如图 3.41 所示,求此截面的重心(几何中心).)

10. In Fig. 3.42, the side length of each square in the plane graph is 20 mm. Find the position of gravity center of the remaining area after digging out a circle. (图 3.42 平面图形中每一方格的边长为 20 mm,求挖去一圆后剩余部分面积的重心位置.)

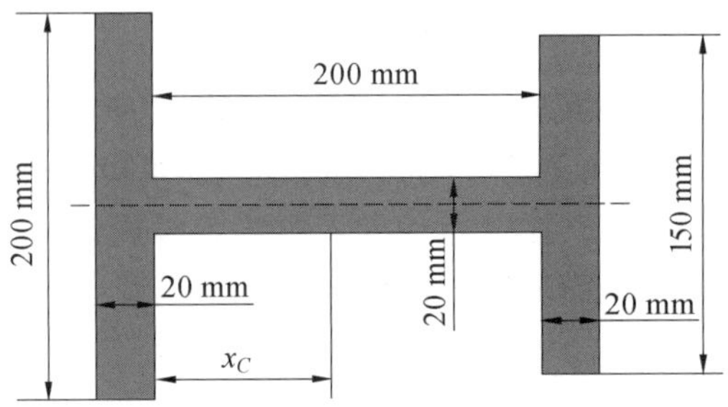

Fig. 3.41 Diagram of Exercise 9
(图 3.41 习题 9 图)

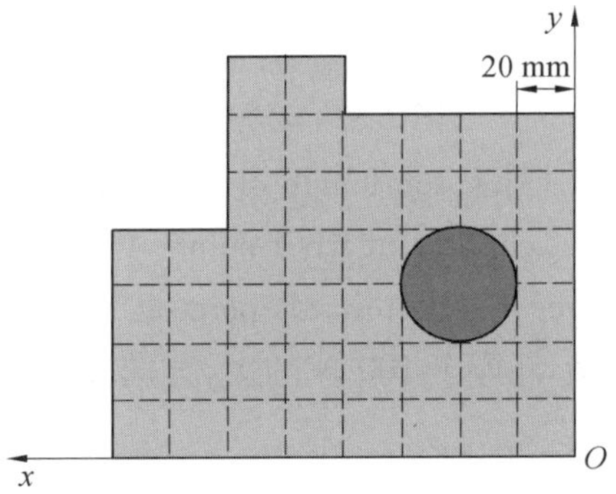

Fig. 3.42 Diagram of Exercise 10
(图 3.42 习题 10 图)

Chapter 4 Friction
(摩擦)

In the previous three chapters, we ignored the influence of friction when we analyzed the force of objects, and regarded the contact surfaces between objects as smooth. But in engineering practice, friction sometimes plays a very important role, and its influence must be taken into account. In the future, unless otherwise specified, the influence of friction will be considered. (在前面 3 章中,在对物体进行受力分析时,忽略了摩擦的影响,把物体之间的接触表面看成光滑的. 但在工程实际中,摩擦有时起到很重要的作用,必须计入其影响. 今后,在没有特别说明的情况,都要考虑摩擦的影响.)

According to the possible relative motion between two objects, friction can be divided into sliding friction and rolling friction. (根据两物体间可能会发生的相对运动形式,摩擦可分为滑动摩擦和滚动摩擦.)

According to the lubrication conditions between two objects, friction can be divided into dry friction and wet friction. (根据两物体间的润滑条件,摩擦又可分为干摩擦和湿摩擦.)

In this chapter, we only consider the balance problems of objects under dry friction, and learn the representation of sliding friction and rolling friction reaction force, as well as the solution of related balance problems. (本章只考虑干摩擦时物体的平衡问题,学习滑动摩擦与滚动摩擦反力的表示及相关平衡问题的求解.)

4.1 Sliding Friction and Sliding Friction Force
(滑动摩擦和滑动摩擦力)

When two objects with rough surfaces have a relative sliding tendency or sliding between their contact surfaces, they act on each other with resistance that hinders relative sliding—sliding friction force. (两个表面粗糙的物体,当其接触表面之间有相对滑动趋势或相对滑动时,彼此作用有阻碍相对滑动的阻力——滑动摩擦力.)

1. Static sliding friction(静滑动摩擦)

The frictional force (or cohesive force) generated when an object has a relative sliding tendency on the surface of another object is called static sliding frictional force. (一个物体在另一个物体表面上有相对滑动趋势时产生的摩擦力(或内聚力)称为静滑动摩擦力.)

In Fig. 4. 1, $\boldsymbol{F}_s$ is the static sliding friction force, $\boldsymbol{F}_N$ is the normal constraint reaction force or normal pressure. The magnitude of $\boldsymbol{F}_s$ must be determined by the balance equation, that is(图 4. 1 中, $\boldsymbol{F}_s$ 为静滑动摩擦力, $\boldsymbol{F}_N$ 为法向约束反力或正压力. $\boldsymbol{F}_s$ 的大小必须由平衡方程确定,即)

$$\sum F_x = 0, F_s = F \tag{4.1}$$

It can be seen from Eq. (4. 1) that the magnitude of static friction force increases with the increase of external force, but it cannot increase without limit. When it increases to a certain extent, the object will slide relative to each other, and the friction is no longer static friction. Therefore, there is a limit value F_{max} for static friction force, which is called the maximum static friction force. (由式(4. 1)可知,静摩擦力的大小随着外力的增加而增加,但是不能无限制地增加. 当增加到一定程度时,物体相对滑动,此时摩擦力就不再是静摩擦力了. 因此,静摩擦力存在一个极限值 F_{max},这个极限值称为最大静摩擦力.)

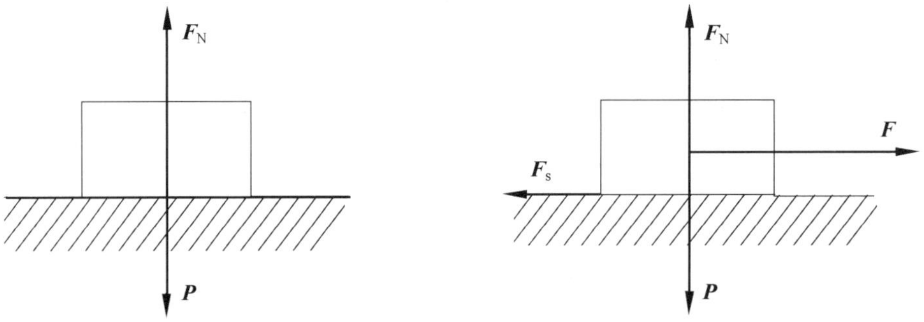

Fig. 4. 1 Static sliding friction force(图 4. 1 静滑动摩擦力)

F_s can take any value from zero to the F_{max} value. (F_s可以取从零到 F_{max}的任何值.)

F_{max} is called the maximum static friction force, we have (F_{max}称为最大静摩擦力,有)

$$0 \leqslant F_s \leqslant F_{max}$$

The direction of static frictional force acting on an object is opposite to the direction of its sliding tendency .(作用在物体上的静摩擦力方向与物体滑动趋势的方向相反.)

2. Law of Sliding Friction(滑动摩擦定律)

The maximum static friction force is numerically equal to the product of the coefficient of static friction and the normal pressure (or normal constraint) between two bodies in contact with each other (Fig. 4. 2), that is(最大静摩擦力在数值上等于静摩擦系数与两个相互接触物体间的正压力或法向约束力)的乘积(图 4. 2),即)

$$F_{max} = f_s F_N \tag{4.2}$$

The dimensionless proportional factor f_s is called as the coefficient of static friction, which is determined by experiments and mainly depends on the material properties and surface conditions of the objects in contact with each other. (无量纲比例因素 f_s 称为静摩擦因数,经实验确定,该因数主要取决于相互接触物体表面的材料性质和表面状况.)

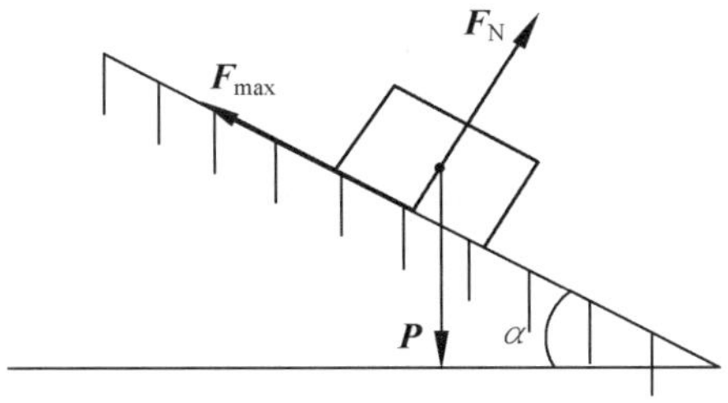

Fig. 4.2 Critical slip state(图 4.2 临界滑动状态)

Some common static friction coefficients are listed in Table 4.1 .(表 4.1 列出了部分常见的静摩擦系数.)

Table 4.1 Friction coefficients of commonly used materials(常用材料的摩擦系数)

Material name(材料名称)	Static friction coefficient (静摩擦系数)		Kinetic friction factor (动摩擦系数)	
	No lubrication (无润滑)	With lubrication (有润滑)	No lubrication (无润滑)	With lubrication (有润滑)
Steel-steel(钢-钢)	0.15	0.1-0.2	0.15	0.05-0.1
Steel-mild steel(钢-软钢)			0.2	0.1-0.2
Steel-cast iron(钢-铸铁)	0.3		0.18	0.05-0.15
Steel-bronze(钢-青铜)	0.15	0.1-0.15	0.15	0.1-0.15
Mild steel-cast iron(软钢-铸铁)	0.2		0.18	0.05-0.15
Mild steel-bronze(软钢-青铜)	0.2		0.18	0.07-0.15
Cast iron-cast iron(铸铁-铸铁)		0.18	0.15	0.07-0.12
Cast iron-bronze(铸铁-青铜)			0.15-0.2	0.07-0.15
Bronze-bronze(青铜-青铜)		0.1	0.2	0.07-0.1
Leather-cast iron(皮革-铸铁)	0.3-0.5	0.15	0.6	0.15
Rubber-cast iron(橡皮-铸铁)			0.8	0.5
Wood-wood(木材-木材)	0.4-0.6	0.1	0.2-0.5	0.07-0.15

The magnitude of the maximum frictional force F_{max} is generally independent of the size of the frictional contact surface. (最大摩擦力 F_{max}的大小一般与摩擦接触表面的尺寸无关.)

In equilibrium, the friction force $0 \leqslant F_s \leqslant F_{max}$. The equilibrium that occurs when the friction force is F_{max} is called limit equilibrium. (在平衡时,摩擦力 $0 \leqslant F_s \leqslant F_{max}$. 在摩擦力为 F_{max}时发生的平衡称为极限平衡.)

3. Dynamic sliding friction (动滑动摩擦)

Sliding friction is the force generated on the contact surface between two objects with relative sliding that hinders their relative sliding. Generally speaking, it is the friction generated when an object slides on the surface of another object. (滑动摩擦力是两个有相对滑动的物体间在接触面上产生的阻碍它们相对滑动的力,通俗来说就是一个物体在另一个物体表面上滑动时产

生的摩擦力.)

The magnitude of kinetic friction is equal to the product of the coefficient of kinetic friction and the normal pressure (or normal constraint) between two objects in contact with each other, that is (动摩擦力的大小等于动摩擦系数与两个相互接触物体间的正压力(或法向约束力)的乘积,即)

$$F_d = fF_N \tag{4.3}$$

where F_d is the kinetic friction force; f is the kinetic friction coefficient, and $f < f_s$. (式中,F_d为动摩擦力;f为动摩擦系数,且$f < f_s$.)

The coefficient of kinetic friction f and the coefficient f_s of static friction are dimensionless proportional coefficients, and both are determined by experiments. (动摩擦系数f与静摩擦系数f_s是无量纲比例系数,且都是通过实验确定的.)

4. Types of balancing problems concerning sliding friction objects (关于滑动摩擦物体平衡问题的类型)

There are two types of problems. (有两种类型的问题.)

(1) The limit critical equilibrium problem when the frictional force $F_{max} = f_s F_N$ (摩擦力$F_{max} = f_s F_N$时的极限临界平衡问题)

This problem is solved in a general way. The critical friction force F_{max} is applied to the object, and its value is represented by the constraint reaction force F_N. That is, there are no new unknown forces when considering the limit equilibrium problem. (这种问题以一般方法解决. 对物体施加临界摩擦力F_{max},其值由约束反力F_N表示. 也就是说,在考虑极限平衡问题时,没有新的未知力.)

(2) Non limit equilibrium problem, $F_s < F_{max}$. (非极限平衡问题,$F_s < F_{max}$.)

In this case, the static friction force F_s is the unknown quantity and is determined together with other unknown forces from the equilibrium equation. (在这种情况下,静摩擦力F_s是未知量,并且与来自平衡方程的其他未知力一起确定.)

4.2 Friction Angle and Self-lock(摩擦角和自锁)

1. Friction angle (摩擦角)

In Fig. 4.3, the resultant force of the normal constraint force $\boldsymbol{F}_N$ and the tangential static friction force $\boldsymbol{F}_s$ is called the full constraint reaction force $\boldsymbol{F}_{RA}$ of the bearing surface, and the maximum angle between $\boldsymbol{F}_{RA}$ and the normal line of the bearing surface is called the friction angle φ. (图 4.3 中,法向的约束力$\boldsymbol{F}_N$和切向的静摩擦力$\boldsymbol{F}_s$的合力称为支承面的全约束反力$\boldsymbol{F}_{RA}$,$\boldsymbol{F}_{RA}$与支承面法线间的最大夹角称为摩擦角φ.)

$\boldsymbol{F}_{RA} = \boldsymbol{F}_N + \boldsymbol{F}_s$, $\boldsymbol{F}_{RA}$ is the fully restrained reaction force, and the angle α between $\boldsymbol{F}_{RA}$ and the normal of the contact surface satisfies ($\boldsymbol{F}_{RA} = \boldsymbol{F}_N + \boldsymbol{F}_s$, $\boldsymbol{F}_{RA}$为全约束反力,$\boldsymbol{F}_{RA}$与接触面法线的夹角α满足)

$$\tan \alpha = \frac{F_s}{F_N}$$

Friction angle φ is the maximum value of the angle between the fully constrained reaction force $\boldsymbol{F}_{RA}$ and the normal line, we have (摩擦角 φ 为全约束反力 $\boldsymbol{F}_{RA}$ 与法线间夹角的最大值，有)

$$\tan\varphi = \frac{F_{max}}{F_N} = \frac{f_s F_N}{F_N} = f_s \tag{4.4}$$

That is, the tangent of the friction angle φ is equal to the coefficient of static friction. (也就是说，摩擦角 φ 的正切等于静摩擦系数.)

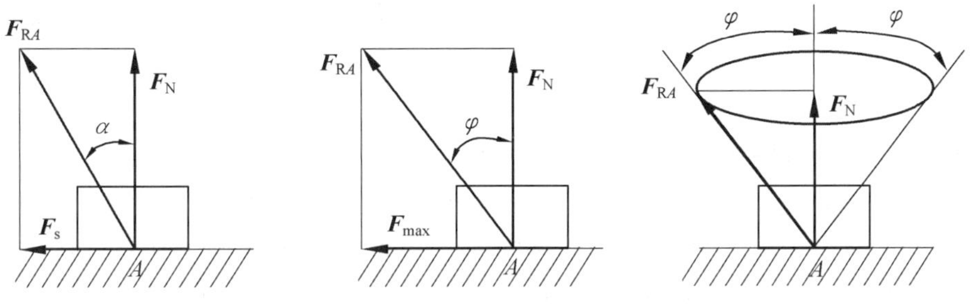

图 4.3 Friction angle(图 4.3 摩擦角)

2. Self-lock phenomenon(自锁现象)

When the block is balanced, $0 \leqslant F \leqslant F_{max}$, so $0 \leqslant \alpha \leqslant \varphi$. (当物块平衡时，$0 \leqslant F \leqslant F_{max}$，所以 $0 \leqslant \alpha \leqslant \varphi$.)

As shown in Fig. 4.4(a), if the action line of the resultant force F_Q of all the active forces acting on the block is within the friction angle, then no matter how large the force is, the block will remain in balance. This phenomenon is called self-lock. (如图 4.4(a)所示，如果作用于物块的全部主动力的合力 F_Q 的作用线在摩擦角之内，则无论这个力多大，物块都必保持平衡，这种现象称为自锁.)

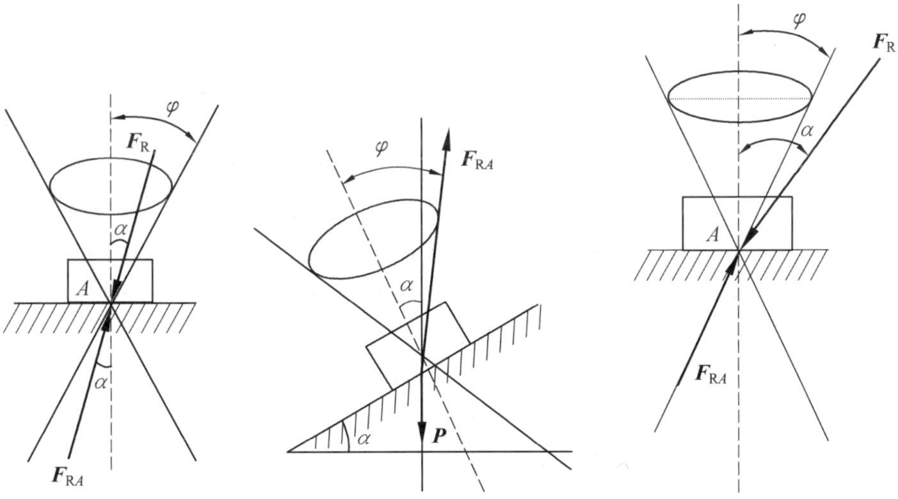

(a) Self-locking(自锁) (b) Self-locking conditions on slopes (斜面自锁条件) (c) Not self-locking(不自锁)

Fig. 4.4 Self-locking and non-self-locking(图 4.4 自锁与非自锁)

When the angle between the resultant force F_Q of the active force and the normal line is smaller than the friction angle φ, the object cannot slide along the surface (self-lock phenomenon).（当主动力的合力与法线夹角小于摩擦角 φ 时，物体沿此表面不会滑动（自锁现象）.）

According to the definition and properties of self-lock, it can be obtained that the self-lock condition of the object on the slope in Fig. 4.4(b) that is only affected by gravity is that the slope angle α of the slope is smaller than the friction angle φ, that is（根据自锁的定义及性质，可以得到图 4.4(b)中斜面上只受重力作用的物体其自锁条件为斜面倾角 α 小于摩擦角 φ，即）

$$\alpha \leqslant \varphi \rightarrow \tan \alpha \leqslant \tan \varphi = f_s$$

$$\tan \alpha \leqslant f_s$$

3. Non-self-lock phenomenon（非自锁现象）

As shown in Fig. 4.4(c), if the action line of the resultant force F_Q of all the active forces acting on the block is outside the friction angle φ, then no matter how small the force is, the block will definitely slide.（如图 4.4(c)所示，如果作用于物块的全部主动力的合力 F_Q 的作用线在摩擦角 φ 之外，则无论这个力怎样小，物块都一定会滑动.）

Questions: As shown in Fig. 4.5, given that the friction angle $\varphi = 20°$, $F = P$, does the object move? Why?（如图 4.5 所示，已知摩擦角 $\varphi = 20°$，$F = P$，则物块动不动，为什么？）

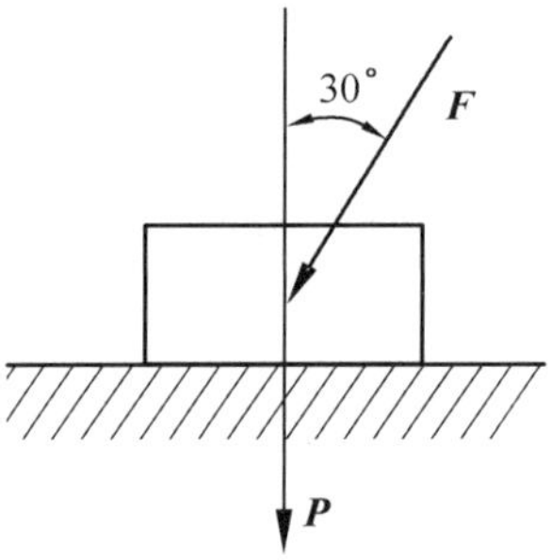

Fig. 4.5 Self-locking condition application（图 4.5 自锁条件应用）

4.3 Rolling Friction（滚动摩擦）

Rolling friction refers to the friction when an object rolls relative to the surface of another object or has a tendency to roll.（滚动摩擦是指一物体沿另一物体表面作相对滚动或具有滚动趋势时的摩擦.）

Rolling friction is caused by the deformation of the contacting objects, resultin that the rolling object does not actually rest on a point on the fixed surface, but on an area AB, as shown in Fig. 4.6.（滚动摩擦是由接触物体的变形引起的，结果导致滚动物体实际上不是停留在固定表面的某一点上，而是在某一区域 AB 上，如图 4.6 所示.）

(1) With a rolling trend.（存在滚动趋势时.）

As shown in Fig. 4.6, when the roller tends to roll to the right relative to the contact surface, the actual contact point is not at point C, but in the area AB on the right side. The distributed force acting on the area AB is simplified towards point A to obtain a force $\boldsymbol{F}_R$ and a force couple M, and the force $\boldsymbol{F}_R$ is decomposed into a tangential static friction force $\boldsymbol{F}_s$ and a normal constraint force $\boldsymbol{F}_N$. Among them, the force restricts the movement of the object, and the couple M restricts the rotation of the object. This couple M is called rolling friction resistance couple.(如图4.6所示,当滚子相对于接触面有向右的滚动趋势时,实际的接触点不在 C 点,而是在右侧的 AB 区域.作用在 AB 区域上的分布力向 A 点简化得到一个力 $\boldsymbol{F}_R$ 和一个力偶 M,力 $\boldsymbol{F}_R$ 分解为一个切向的静摩擦力 $\boldsymbol{F}_s$ 和一个法向的约束力 $\boldsymbol{F}_N$.其中,力限制物体的移动,力偶限制物体的转动.力偶 M 称为滚动摩擦阻力偶.)

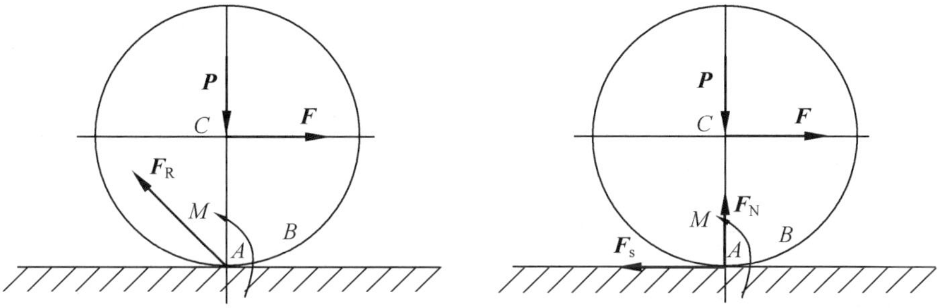

Fig. 4.6 There is a rolling trend(图4.6 存在滚动趋势)

According to the rotational balance equation, we have(根据转动平衡方程得到)

$$\sum M_A(F) = 0, M - F \cdot R = 0 \tag{4.5}$$

It can be seen from the above formula that the magnitude of the couple M increases with the increase of the external force, but it cannot increase without limit. When it increases to a certain extent, the object will roll relative to each other. Therefore, there is a limit value M_{max} in the rolling friction resistance couple M, which is called as the maximum rolling friction resistance couple. Therefore, we have(由上式可知,力偶 M 的大小随着外力的增加而增加,但是不能无限制地增加,增加到一定程度时,物体相对滚动.因此,滚动摩擦阻力偶 M 存在着一个极限值 M_{max},这个极限值称为最大滚动摩擦阻力偶.因此,有)

$$0 \leqslant M \leqslant M_{max}$$

(2) At rolling critical state.(滚动临界状态时.)

$$M_{max} = f_k F_N \tag{4.6}$$

where, f_k is called the rolling friction coefficient, which has the dimension of length and is generally expressed in mm. Table 4.2 is the rolling friction coefficient of several materials.(式中,f_k称为滚动摩阻系数,具有长度的量纲,一般用 mm 表示.表4.2是几种材料的滚动摩阻系数.)

Table 4.2 Rolling friction coefficient(滚动摩阻系数)

Material name(材料名称)	f_k /mm	Material name(材料名称)	f_k /mm
Cast iron vs cast iron(铸铁与铸铁)	0.5	Mild steel vs steel(软钢与钢)	0.5
Steel wheels and rails(质车轮与钢轨)	0.05	Feeder wagons and rails with ball bearings(有滚珠轴承的料车与钢轨)	0.09
Wood and steel(木与钢)	0.3–0.4	Feeder wagons and rails without ball bearings(无滚珠轴承的料车与钢轨)	0.21
Wood with wood(木与木)	0.5–0.8	Steel wooden wheels with wooden surfaces(钢质木轮与木面)	1.5–2.5
Cork vs cork(软木与软木)	1.5	Tires and roads(钢质木轮与木面)	2–10
Hardened steel balls and steel(淬火钢珠与钢)	0.01		

(3) At rolling state.(滚动状态时.)

At rolling state, the frictional resistance moment M is equal to M_{max}, we have(处于滚动状态时,摩擦阻力偶矩 M 等于 M_{max},有)

$$M = M_{max}$$

Physical meaning of rolling friction coefficient is as follows.(滚动摩阻系数物理意义如下.)

The equivalent transformation is performed according to the force diagram corresponding to the rolling and unrolling critical state, as shown in Fig. 4.7. We can restore the force couple M into two forces equal to F_N, and further obtain the force diagram shown in Fig. 4.7(c).(根据将滚未滚临界状态的受力图进行等效变换,如图 4.7 所示,将力偶 M 还原为两个大小等于 F_N 的力,进一步得到图 4.7(c)所示的受力图.)

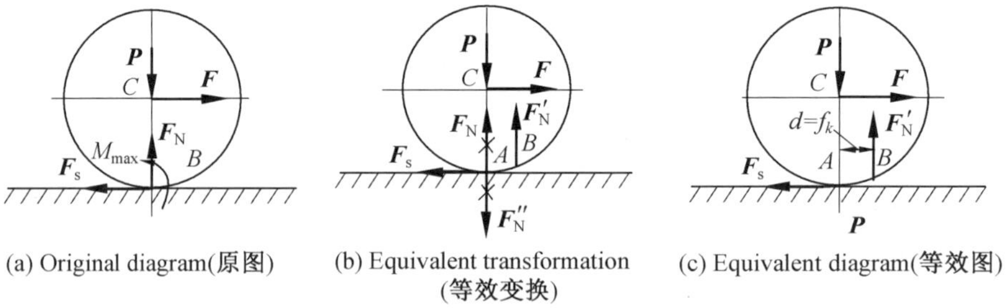

Fig. 4.7 Schematic diagram of the physical meaning of rolling friction coefficient
(图 4.7 滚动摩阻系数物理意义的示意图)

According to Figure 4.7, it can be obtained that(根据图 4.7 可以得到)

$$f_k = \frac{M_{max}}{F'_N} = d$$

In Fig. 4.7, the force couple $(\boldsymbol{F}, \boldsymbol{F}_s)$ is in balance with the force couple $(\boldsymbol{P}, \boldsymbol{F}'_N)$. The offset distance d between $\boldsymbol{F}'_N$ and original $\boldsymbol{F}_N$ is the rolling friction coefficient f_k. (图 4.7 中,力偶 $(\boldsymbol{F}, \boldsymbol{F}_s)$ 与力偶 $(\boldsymbol{P}, \boldsymbol{F}'_N)$ 平衡. $\boldsymbol{F}'_N$ 与原 $\boldsymbol{F}_N$ 偏移的距离 d 即为滚动摩阻系数 f_k.)

The normal reaction force $\boldsymbol{F}_N$ moves a certain distance to the rolling side, and its limit value is equal to f_k, that is, the rolling friction coefficient. In order to consider rolling friction when studying the motion of objects, a moment $M = F_N f_k$ must be added to the acting force. The coefficient f_k has length dimension, i. e. $[f_k] = \mathrm{m}$ (法向反力 $\boldsymbol{F}_N$ 向滚动侧移动了一段距离,其极限值等于 f_k,即滚动摩阻系数,为了在研究物体运动问题时考虑滚动摩擦,必须在作用力上加上力矩 $M_k = F_N f_k$. 系数 f_k 具有长度量纲,即 $[f_k] = \mathrm{m}$).

4.4 Considering the Balance of Objects When Friction (考虑摩擦时物体的平衡问题)

Characteristics of system equilibrium problems when friction is considered is as follows. (考虑摩擦时系统平衡问题的特点如下.)

(1) In addition to active force and constraint force, friction force appears in the balance equation, so the number of unknowns increases. (平衡方程式中除主动力、约束力外还出现了摩擦力,因此未知数增多.)

(2) In addition to the balance equation, the physical equation $F_s \leqslant f_s F_N$ about friction can be supplemented. (除平衡方程外,还可补充关于摩擦力的物理方程 $F_s \leqslant f_s F_N$.)

(3) Avoid solving inequalities, it is possible to solve the critical case, namely the supplementary equation $F_{max} = f_s F_N$. (为避免解不等式,可以解临界情况,即补充方程 $F_{max} - f_s F_N$.)

Common questions are as follows. (常见的问题如下.)

(1) Check whether the object is balanced. (检验物体是否平衡.)

(2) Critical balance problem. (临界平衡问题.)

(3) Find the balance range problem. (求平衡范围问题.)

Example 1: In Fig. 4.8(a), it is known that $Q = 400$ N, $P = 1\ 500$ N, static friction coefficient $f_s = 0.2$, kinetic friction coefficient $f = 0.18$. If the block is at rest on the inclined plane, find the magnitude and direction of the friction force at this time. (**例题** 1:图 4.8(a)中,已知 $Q = 400$ N, $P = 1\ 500$ N,静摩擦系数 $f_s = 0.2$,动摩擦系数 $f = 0.18$. 求物块在斜面上是否静止,并求此时摩擦力的大小和方向.)

Solution: (**解:**)

Take the block as the research object, and assume it is static balance, draw the force diagram, as shown in Fig. 4.8(b), and list the balance equations, we have(取物块为研究对象,并假定其静平衡,画受力图,如图 4.8(b)所示,列出平衡方程,有)

$$\sum F_x = 0,\ Q\cos 30° - P\sin 30° - F_s = 0$$

$$\sum F_y = 0,\ F_N - P\cos 30° - Q\sin 30° = 0$$

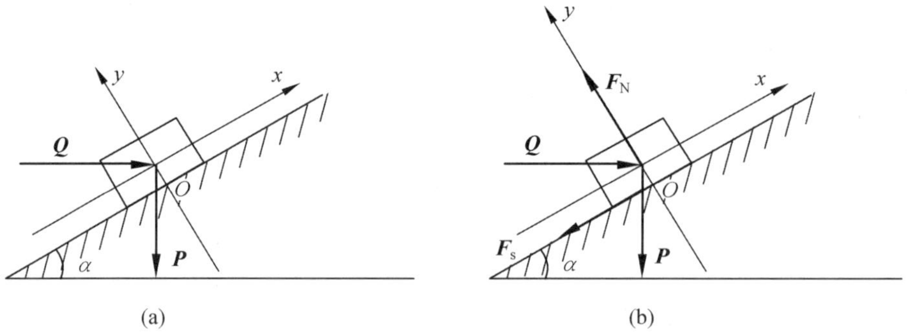

Fig. 4.8　Diagram of Example 1(图 4.8　例题 1 图)

Solutions are(解得)

$$F_s = -403.6\ \text{N}, F_N = 1\ 499\ \text{N}$$

And the maximum static friction force is (而最大静摩擦力为)

$$F_{max} = f_s F_N = 299.8\ \text{N}$$

Because $|F_s| > F_{max}$, so the block cannot stand still, but slide down. (因为$|F_s| > F_{max}$,所以物块不可能静止,而是向下滑动).

At this time, the friction force should be kinetic sliding friction force, the direction is upward along the slope, and the magnitude is(此时的摩擦力应为动滑动摩擦力,方向沿斜面向上,大小为)

$$F_{max} = fF_N = 269.8\ \text{N}$$

Example 2:As shown in Fig. 4.9(a), it is known that $P = 10$ N, the friction coefficient between the two books is $f_{s1} = 0.1$, and the friction coefficient of the outermost layer is $f_{s2} = 0.25$. What is the minimum pressure F required to lift these four books? (**例题** 2:如图 4.9(a)所示,已知 $P = 10$ N,两书间的摩擦系数为 $f_{s1} = 0.1$,最外层摩擦系数为 $f_{s2} = 0.25$. 要提起这四本书需加的最小压力 F 是多少?)

Solution:(解:)

(1) Take the whole as the research object, the force diagram is shown in Fig 4.9(b), and consider the balance in the y direction, we have ((1)取整体为研究对象,受力图如图 4.9(b)所示,考虑 y 方向的平衡,有)

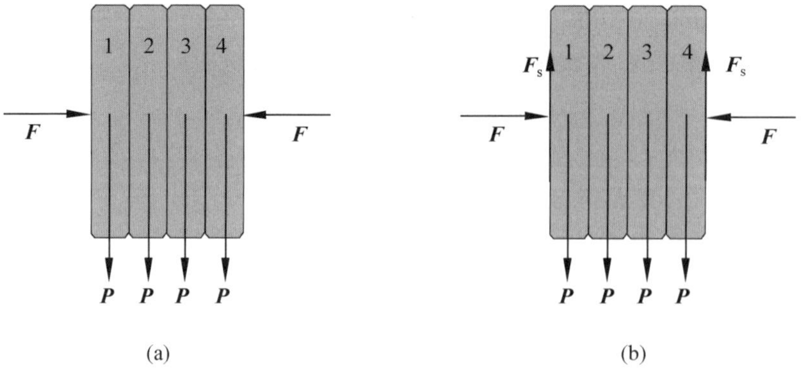

Fig. 4.9　Diagram of Example 2(图 4.9　例题 2 图)

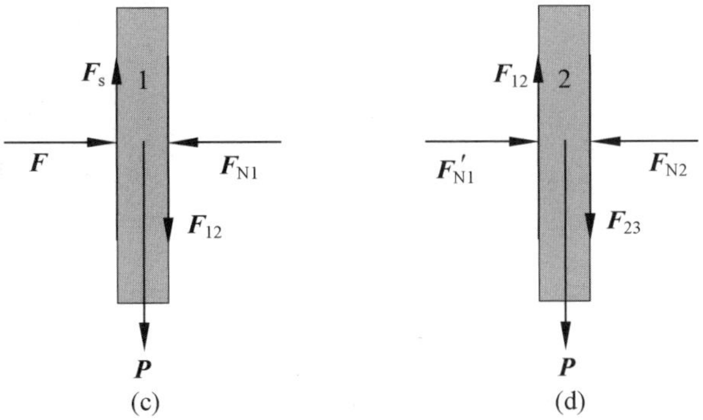

Fig. 4.9(continued)(图 4.9(续))

$$\sum F_y = 0,\ 2F_s - 4P = 0$$

Solutions are(解得)

$$F_s = 20\ \text{N}$$

If the book needs to be lifted, the static friction force must be less than the maximum static friction force, then(如需提起书本,则静摩擦力需小于最大静摩擦力,则有)

$$F_s \leqslant Ff_{s2},\ F \geqslant 80\ \text{N}$$

(2)Take book 1 as the research object, the force diagram is shown in Fig. 4.9(c), and the balance equation is listed as(取书 1 为研究对象,受力图如图 4.9(c)所示,列平衡方程,有)

$$\sum F_y = 0,\ F_s - P - F_{12} = 0$$

We get(得到)

$$F_{12} = 10\ \text{N}$$

If the book needs to be lifted, the sliding friction between book 1 and book 2 must be less than the maximum static friction, then(如需提起书本,则书本 1 和书本 2 之间的滑动摩擦力需小于最大静摩擦力,则有)

$$F_{12} \leqslant Ff_{s1},\ F \geqslant 100\ \text{N}$$

(3)Taking book 2 as the research object, the force diagram is shown in Fig. 4.9(d), and the balance equation is listed as(取书 2 为研究对象,受力图如图 4.9(d)所示,列平衡方程,有)

$$\sum F_y = 0, F'_{12} - P + F_{23} = 0$$

We get(得到)

$$F_{23} = 0\ \text{N}$$

Based on the above analysis, we have(综合上述分析,得到)

$$F_{min} = 100\ \text{N}$$

Example 3: As shown in Fig. 4.10(a), P and α are given, and the coefficient of static friction is f_s, find the magnitude of the horizontal force Q in equilibrium. (**例题** 3:如图 4.10(a)所示,已知 P 和 α,静摩擦系数为 f_s,求平衡时水平力 Q 的大小.)

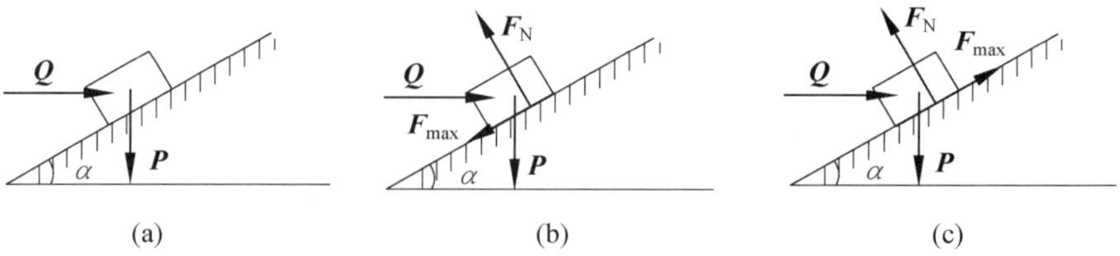

Fig. 4. 10 Diagram of Example 3(图 4. 10 例题 3 图)

Solution:(**解**:)

If the force $\boldsymbol{Q}$ is too large, the block will slide up; if the force $\boldsymbol{Q}$ is too small, the block will slide down. So $\boldsymbol{Q}$ should be between the maximum and minimum values. (若力 $\boldsymbol{Q}$ 太大,则物块上滑;若力 $\boldsymbol{Q}$ 太小,则物块下滑. 因此, $\boldsymbol{Q}$ 应在最大值与最小值之间.)

Take the block as the research object, first find its maximum value, the object is in a critical state about to slide upward, and the friction force $\boldsymbol{F}_s$ goes downward along the slope, as shown in Fig. 4. 10(b), we have(取物块为研究对象,先求其最大值,物体处于将要向上滑动的临界状态,摩擦力 $\boldsymbol{F}_s$ 沿斜面向下,如图 4. 10(b),有)

$$\begin{cases}\sum F_x = 0, Q_{max}\cos\alpha - P\sin\alpha - F_{max} = 0\\ \sum F_y = 0, F_N - P\cos\alpha - Q_{max}\sin\alpha = 0\\ F_{max} = f_s F_N\end{cases}$$

So we have(解得)

$$Q_{max} = P\frac{\sin\alpha + f_s\cos\alpha}{\cos\alpha - f_s\sin\alpha}$$

Then solve its minimum value, the object is in a critical state about to slide down, and the friction force $\boldsymbol{F}_s$ goes up along the slope, as shown in Fig. 4. 10(c), we have (再求其最小值,物体处于将要向下滑动的临界状态,摩擦力 $\boldsymbol{F}_s$ 沿斜面向上, 如图 4. 10(c),有)

$$\begin{cases}\sum F_x = 0, Q_{min}\cos\alpha - P\sin\alpha + F_{max} = 0\\ \sum F_y = 0, F_N - P\cos\alpha - Q_{min}\sin\alpha = 0\\ F_{max} = f_s F_N\end{cases}$$

So we have(解得)

$$Q_{min} = P\frac{\sin\alpha - f_s\cos\alpha}{\cos\alpha + f_s\sin\alpha}$$

To make a block stand still, the following conditions must be met, that is(欲使物块静止,需满足如下条件,即)

$$P\frac{\sin\alpha - f_s\cos\alpha}{\cos\alpha + f_s\sin\alpha} \leqslant Q \leqslant P\frac{\sin\alpha + f_s\cos\alpha}{\cos\alpha - f_s\sin\alpha}$$

Example 4: As shown in Fig. 4. 11 (a), it is a cam mechanism. It is known that the coefficient of static friction between the tappet and the slide-way is f_s , and the width of the slideway is b. The self-weight of the tappet is ignored, and the friction at the contact point between the cam and the rod is ignored. How much is the size of a so that the tappet will not be stuck? (图 4.11(a)所示为一凸轮机构,已知挺杆与滑道间的静摩擦系数为 f_s,滑道宽度为 b. 挺杆自重不计,凸轮与杆件接触处的摩擦忽略不计. 则 a 为多大,挺杆才不致被卡?)

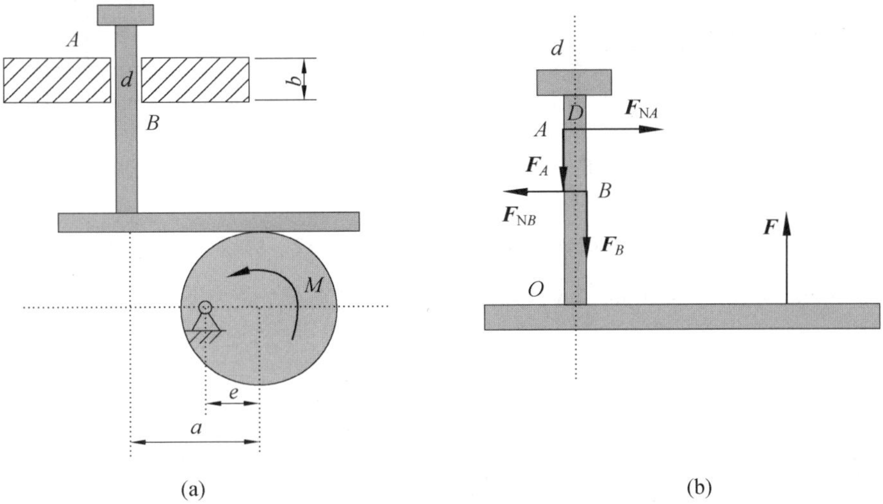

Fig. 4.11 Diagram of Example 4(图 4.11 例题 4 图)

Solution:(解:)

Take the tappet as the research object, if the tappet has an upward sliding tendency, the friction force $\boldsymbol{F}_A$ and $\boldsymbol{F}_B$ are downward. The force diagram is shown in Fig. 4.11(b), and the balance equations are(取挺杆为研究对象,挺杆有向上滑动趋势,则摩擦力 $\boldsymbol{F}_A$ 和 $\boldsymbol{F}_B$ 方向向下. 受力图如图 4.11(b)所示,平衡方程为)

$$\begin{cases} \sum F_x = 0, F_{NA} - F_{NB} = 0 \\ \sum F_y = 0, -F_A - F_B + F = 0 \\ \sum M_D(\boldsymbol{F}) = 0, Fa - F_{NB}b - F_B \dfrac{d}{2} + F_A \dfrac{d}{2} = 0 \end{cases}$$

Considering the critical situation of balance, the friction force reaches the maximum value, then the supplementary equation can be obtained as(考虑平衡的临界情况,摩擦力都达到最大值,则可得补充方程,即)

$$\begin{cases} F_A = f_s F_{NA} \\ F_B = f_s F_{NB} \end{cases}$$

We get(得到)

$$a \leqslant \frac{b}{2f_s}$$

Example 5: As shown in Fig. 4.12(a), it is known that $P = 1\ 000$ N, the coefficient of static friction between the block and the ground is $f_s = 0.52$. Solve Q_{max} when the system balance will not be disturbed. (**例题 5**:如图 4.12(a)所示,已知 $P = 1\ 000$ N, 物块与地面静摩擦系数 $f_s = 0.52$. 求不致破坏系统平衡时的 Q_{max}.)

Solution:(解:)

(1) Take the pin B as the research object, the force diagram is shown in Fig. 4.12(b), and

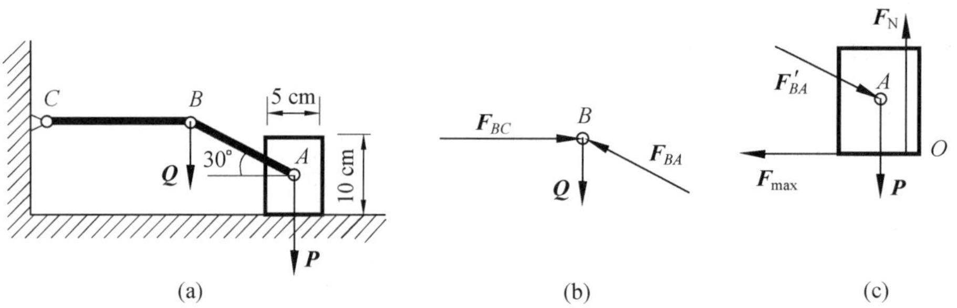

Fig. 4. 12　Diagram of Example 5(图 4. 12　例题 5 图)

the balance equation are listed as ((1)取销钉 B 为研究对象,受力图如图 4. 12(b)所示,列平衡方程,有)

$$\sum F_y = 0, F_{BA}\sin 30° - Q = 0$$

We get(得到)

$$F_{BA} = 2Q$$

(2) Take the block A as the research object. (取物块 A 为研究对象.)

①At the critical equilibrium state of sliding, the force diagram is shown in Fig. 4. 12(c), we have (处于滑动的临界平衡状态时, 受力图如图 4. 12(c)所示,有)

$$\begin{cases} \sum F_x = 0, F'_{BA}\cos 30° - F_{max} = 0 \\ \sum F_y = 0, F_N - P - F'_{BA}\sin 30° = 0 \\ F_{max} = f_s F_N \end{cases}$$

We get(得到)

$$Q_{1max} = \frac{f_s}{\sqrt{3} - f_s}P = 429.03\ \text{N}$$

②When it is in the critical equilibrium state of overturning, the bottom is in the critical state of breaking away from the ground, and $\boldsymbol{F}_N$, $\boldsymbol{F}_{max}$ act on point O in Fig. 4. 13(c), then (处于翻倒的临界平衡状态时,底面处于脱离地面的临界状态,图 4. 13(c)中 $\boldsymbol{F}_N$ 和 $\boldsymbol{F}_{max}$ 作用在 O 点,则有)

$$\begin{cases} \sum M_O(F) = 0 \\ P \cdot 2.5 + F'_{BA}\sin 30° \cdot 2.5 - F'_{BA}\cos 30° \cdot 5 = 0 \end{cases}$$

We get(得到)

$$Q_{2max} = \frac{P}{2(\sqrt{3} - 0.5)} = 405.83\ \text{N}$$

So(所以)

$$Q_{max} = 405.83\ \text{N}$$

Conclusion and discussion: (结论与讨论:)

(1) There are two types of friction phenomena: sliding friction and rolling friction. (摩擦现象分为两类:滑动摩擦和滚动摩阻.)

(2) Sliding friction is the tangential resistance that occurs when there is a relative sliding tendency or relative sliding between the surfaces of two objects in contact with each other. The former is naturally called static sliding friction, and the latter is called kinetic sliding friction. (滑动摩擦力是在两个物体相互接触的表面之间有相对滑动趋势或有相对滑动时出现的切向阻力. 前者自然称为静滑动摩擦力,后者称为动滑动摩擦力.)

① The direction of static friction is opposite to the direction of relative sliding tendency between contact surfaces, and its magnitude should be determined according to the balance equation. When the object is in the critical state of balance, the static friction reaches the maximum value, so the range of static friction with the active force is between zero and the maximum value, that is(静摩擦力的方向与接触面间相对滑动趋势的方向相反,它的大小应根据平衡方程确定. 当物体处于平衡的临界状态时,静摩擦力达到最大值,因此静摩擦力随主动力变化的范围在零与最大值之间,即)

$$0 \leqslant F_s \leqslant F_{max}$$

Law of static friction: The magnitude of the maximum static friction is proportional to the normal pressure between two objects, we have(静摩擦定律:最大静摩擦力的大小与两物体间的正压力成正比,有)

$$F_{max} = f_s F_N, f_s \rightarrow \text{static friction coefficient(静摩擦系数)}$$

② The direction of kinetic friction force is opposite to the speed direction of relative sliding between contact surfaces, and its magnitude is (动摩擦力的方向与接触面间相对滑动的速度方向相反,它的大小为)

$$F = fF_N, f \rightarrow \text{kinetic friction coefficient(动摩擦系数)}$$

③ The friction angle is the maximum value of the angle between the fully restrained reaction force and the normal line, and $0 \leqslant F_s \leqslant F_{max}$. (摩擦角为全约束反力与法线间夹角的最大值,且有 $0 \leqslant F_s \leqslant F_{max}$.)

Self-lock occurs when the action line of the resultant force of the active force is within the friction angle. (当主动力的合力作用线在摩擦角之内时,发生自锁现象.)

(3) When an object rolls, it is affected by rolling resistance couple, which hinders rolling, we have (物体滚动时,会受到阻碍滚动的滚动阻力偶作用,有)

$$0 \leqslant M \leqslant M_{max}, M_{max} = f_k F_N$$

4.5 Questions(思考题)

1. As shown in Fig. 4.13, it is known that an object with a weight of $P = 100$ N is placed on a horizontal surface, and its static friction coefficient $f_s = 0.3$. When the horizontal thrust F acting on the object is 10 N, 20 N, 40 N, analyze that in these three situations, is the object balanced? What is the friction force? (如图 4.13 所示,已知一重为 $P = 100$ N 的物体放在水平面上,其静摩擦因数 $f_s = 0.3$,当作用在物体上的水平推力 F 分别为 10 N、20 N、40 N 时,分析这三种情况下,物体是否平衡,摩擦力等于多少.)

2. It is known that a block with a weight of $P = 100$ N is pressed on a vertical surface with a

horizontal force of $F=500$ N, as shown in Fig. 4.14, and its static friction coefficient is $f_s=0.3$. What is the frictional force of the block at this time? (已知一物块重 $P=100$ N,用水平力 $F=500$ N 的力压在一铅垂表面上,如图 4.14 所示,其静摩擦因数 $f_s=0.3$,则此时物块的摩擦力等于多少?)

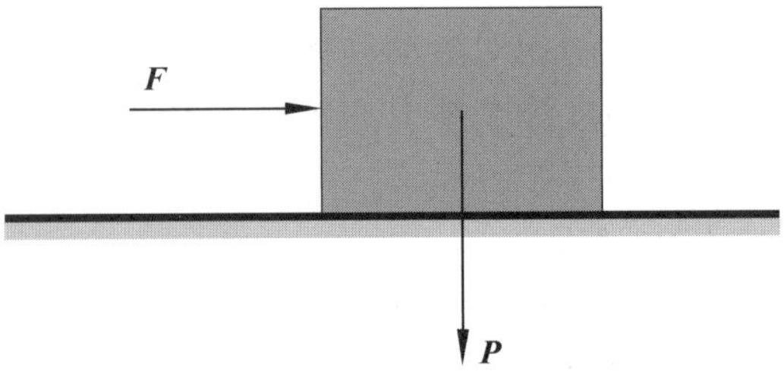

Fig. 4.13 Diagram of Question 1
(图 4.13 思考题 1 图)

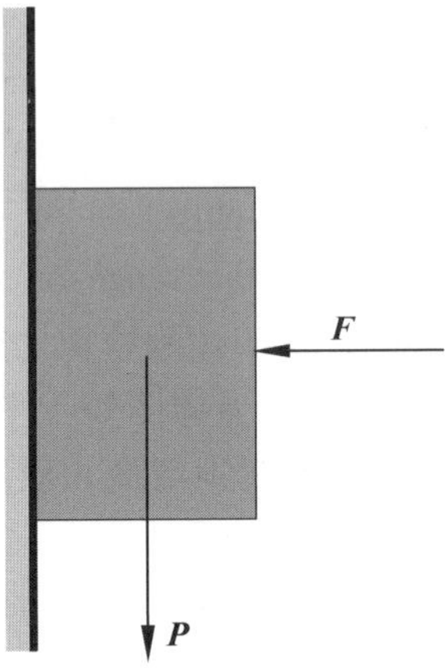

Fig. 4.14 Diagram of Question 2
(图 4.14 思考题 2 图)

3. A block of weight $\boldsymbol{P}$ is placed on a rough horizontal surface, and the coefficient of static friction of contact surface is $f_s=0.3$. To make the block slide to the right in the horizontal direction, pull force $\boldsymbol{F}_1$ can be applied along the direction of OA (Fig. 4.15(a)) or push force $\boldsymbol{F}_2$ can be applied along the direction of OB (Fig. 4.15(b)). Which method of force application saves effort? (物块重 $\boldsymbol{P}$,放置在粗糙的水平面上,接触的静摩擦因数 $f_s=0.3$. 要使物块沿水平方向向右滑动,可沿 OA 方向施加拉力 $\boldsymbol{F}_1$(图 4.15(a)),也可沿 OB 方向施加推力 $\boldsymbol{F}_2$(图 4.15(b)),哪种施力方法省力?)

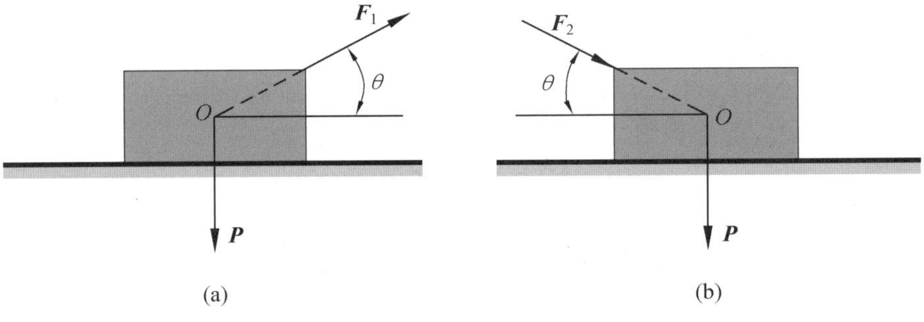

Fig. 4.15 Diagram of Question 3(图 4.15 思考题 3 图)

4. It is known that the weight of the block is $P=100$ N, the inclination angle of the slope is $\theta=30°$, and the coefficient of static friction between the block and the slope is $f_s=0.38$. Solve the friction force between the block and the inclined plane. And, is the object stationary or sliding on the slope at this time (Fig. 4.16(a))? If the object is to move upward along the inclined plane, what is the minimum force applied to the object parallel to the inclined plane (Fig. 4.16(b))? (已知物块重 $P=100$ N,斜面的倾斜角 $\theta=30°$,物块与斜面的静摩擦因数

$f_s = 0.38$. 求物块与斜面之间的摩擦力. 此时物块在斜面上是静止还是下滑(图4.16(a))? 如果要使物块沿斜面向上运动,则加于物块并与斜面平行的力 $\boldsymbol{F}$ 至少应为多大(图4.16(b))?)

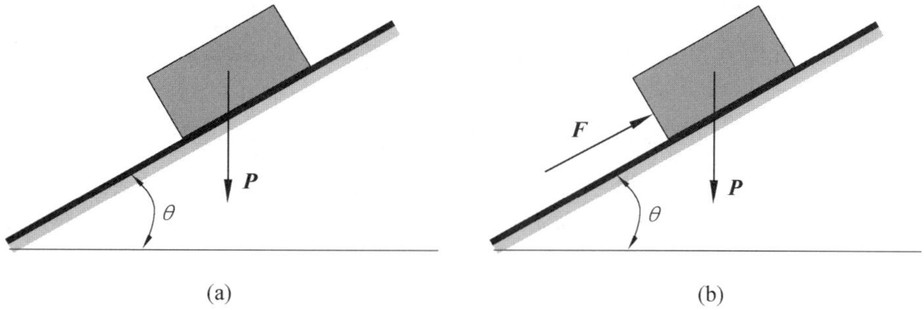

Fig. 4. 16 Diagram of Question 4(图 4. 16 思考题 4 图)

5. The weight of the block is $\boldsymbol{P}$, and a force $\boldsymbol{F}$ is applied outside the friction angle, as shown in Fig. 4. 17. It is known that $\theta = 25°$, friction angle $\varphi_f = 20°$, and force $F = P$, then the question is if the object is stationary and why. (物块重 $\boldsymbol{P}$,一力 $\boldsymbol{F}$ 作用在摩擦角之外,如图 4. 17 所示. 已知 $\theta = 25°$,摩擦角 $\varphi_f = 20°$,且力 $F = P$,则物块是否静止,为什么?)

6. As shown in Fig. 4. 18, if a steel wedge is used to split an object, the friction angle between the contact angles is φ_f, and if the wedge does not slip out after splitting, wedge weight is ignored, what is the included angle between the two planes of the steel wedge? (如图 4. 18 所示,用钢楔劈物,接触角间的摩擦角为 φ_f,劈入后欲使楔不滑出,楔重不计,则钢楔两个平面间的夹角应为多大?)

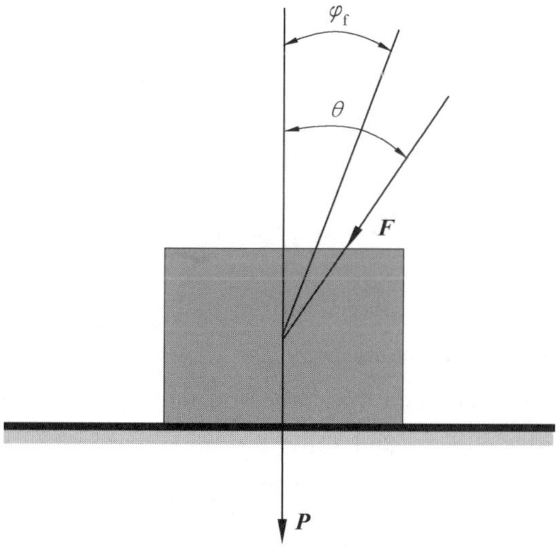

Fig. 4. 17 Diagram of Question 5
(图 4. 17 思考题 5 图)

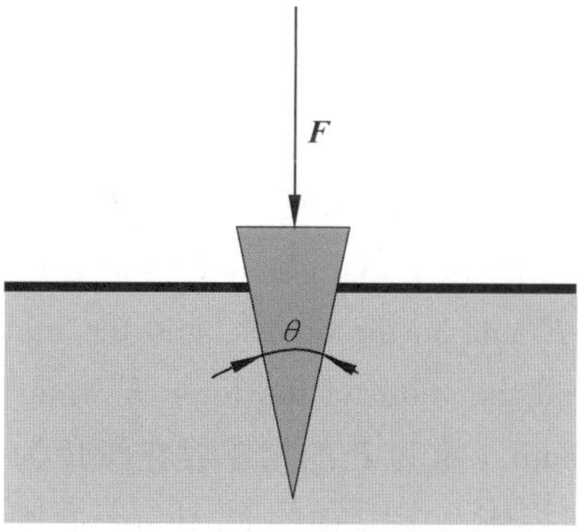

Fig. 4. 18 Diagram of Question 6
(图 4. 18 思考题 6 图)

7. It is known that the weight of an object is $\boldsymbol{P}$, and its size is shown in Fig. 4. 19. Now pull this object with a horizontal force $\boldsymbol{F}$, when you first pull it move, does the friction force at point A, B both reach the maximum value? For example, if the friction coefficients at point A, B are

both f_s , is the maximum static friction force of these two places equal? Another example is that when the force $\boldsymbol{F}$ is too small to pull the object, can the static friction forces at A and B be calculated separately? (已知物体重为 $\boldsymbol{P}$,尺寸如图 4.19 所示. 现以水平力 $\boldsymbol{F}$ 拉此物体,当刚开始拉动时,A、B 两处的摩擦力是否都达到最大值? 若 A、B 两处的摩擦因数均为 f_s ,此二处最大静摩擦力是否相等? 又如力 $\boldsymbol{F}$ 较小而未能拉动物体时,能否分别求出 A、B 两处的静摩擦力?)

8. The car is running horizontally at a constant speed, the ground has sliding friction and rolling friction resistance against the wheels, and the wheels only roll but not slip. The front wheels of the car receive a forward thrust F_1 from the body (Fig. 4.20(a)), while the rear wheels receive a driving couple M and are acted on by a backward force F_2 from the body (Fig. 4.20(b)). Try to draw the force diagram of the front and rear wheels. And how to find its sliding friction? Is it equal to its kinetic friction? Is it equal to its maximum static friction? (汽车匀速水平行驶,地面对车轮有滑动摩擦,也有滚动摩擦阻力,车轮只滚不滑. 汽车前轮收车身施加的一个向前的推力 F_1 作用(图 4.20(a)),而后轮收到驱动力偶 M,并受车身向后的力 F_2 作用(图 4.20(b)). 试画全前、后轮的受力图. 如何求其滑动摩擦力? 是否等于其动摩擦力? 是否等于其最大静摩擦力?)

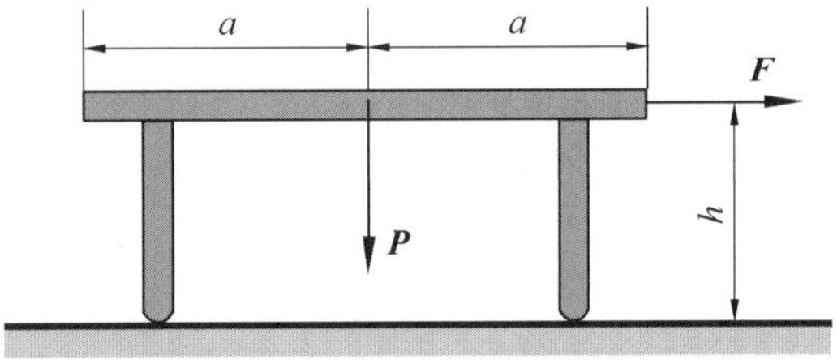

Fig. 4.19　Diagram of Question 7
(图 4.19　思考题 7 图)

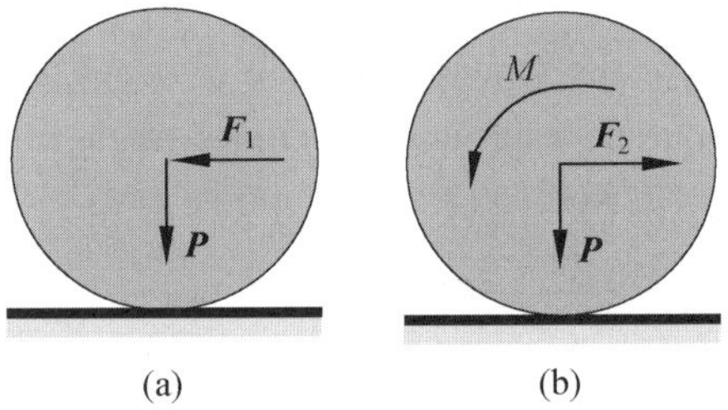

Fig. 4.20　Diagram of Question 8
(图 4.20　思考题 8 图)

4.6　Exercises(习题)

1. An object with a weight $\boldsymbol{P}$ is placed on an inclined plane with an inclination angle of β, and the friction angle between the object and the inclined plane is φ_f, as shown in Fig. 4.21. A force $\boldsymbol{F}$ acts on the object, and the inclination angle between this force and the inclined plane is θ. Find the value of the force F when pulling the object, and solve θ when this force is at its minimum. (重为 $\boldsymbol{P}$ 的物体放在倾角为 β 的斜面上,物体与斜面的摩擦角为 φ_f,如图 4.21 所示. 在物体上作用力 $\boldsymbol{F}$,此力与斜面倾角为 θ. 求拉动物体时力 $\boldsymbol{F}$ 的值,并求当 θ 为何值时此力最小.)

2. As shown in Fig. 4.22, a couple acts on the bar placed in the V-shaped groove. When the moment of the couple $M = 15$ N · m, the bar can just rotate. Known bar weight $P = 400$ N, diameter $D = 0.25$ m, ignoring rolling friction. Find the coefficient of static friction f_s between the bar and the V-groove. (如图 4.22 所示,置于 V 形槽中的棒料上作用一力偶,当力偶的矩 $M =$

15 N · m 时，刚好能转动此棒料. 已知棒料重 $P=400$ N，直径 $D=0.25$ m，不计滚动摩阻. 求棒料与 V 形槽间的静摩擦因数 f_s. ）

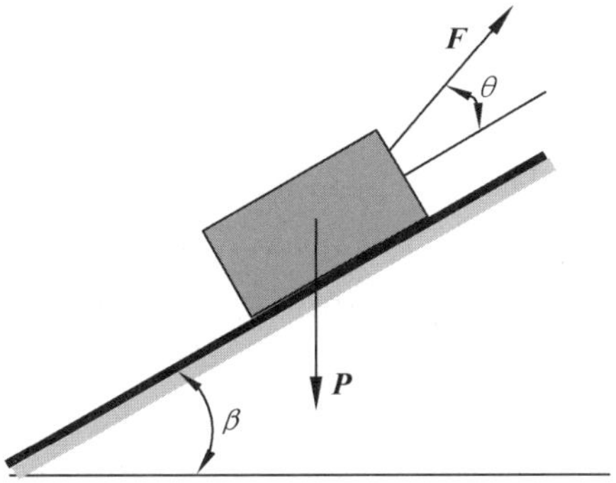

Fig. 4.21　Diagram of Exercise 3
（图 4.21　习题 1 图）

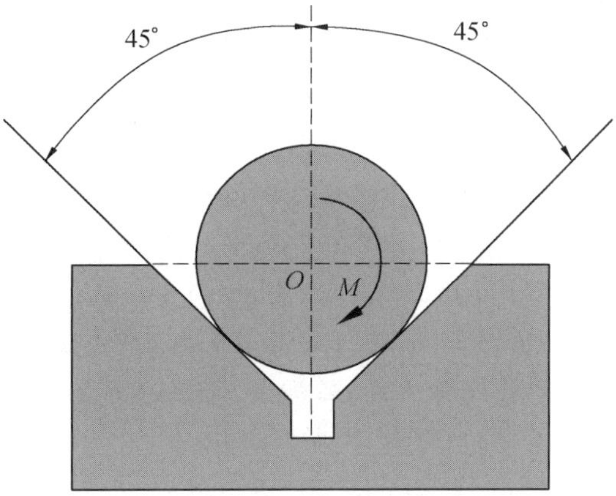

Fig. 4.22　Diagram of Exercise 4
（图 4.22　习题 2 图）

3. Two identical homogeneous rods AB and BC are connected by a smooth hinge at the end point B, and the two points A and C are placed on a non-smooth horizontal plane, as shown in Fig. 4.23. When ABC forms an equilateral triangle, the system is in a critical equilibrium state in the vertical plane. Find the coefficient of friction between the rod end and the horizontal plane. （两根相同的均质杆 AB 和 BC，在端点 B 用光滑铰链连接，A、C 两点放在非光滑的水平面上，如图 4.23 所示. 当 ABC 成等边三角形时，系统在铅垂面内处于临界平衡状态. 求杆端与水平面间的摩擦因数. ）

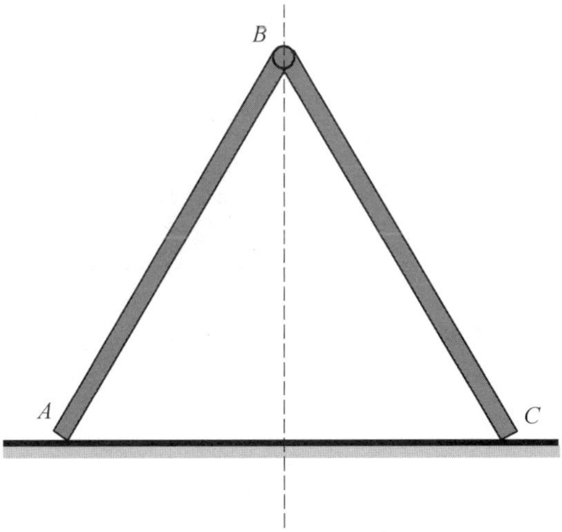

Fig. 4.23　Diagram of Exercise 5
（图 4.23　习题 3 图）

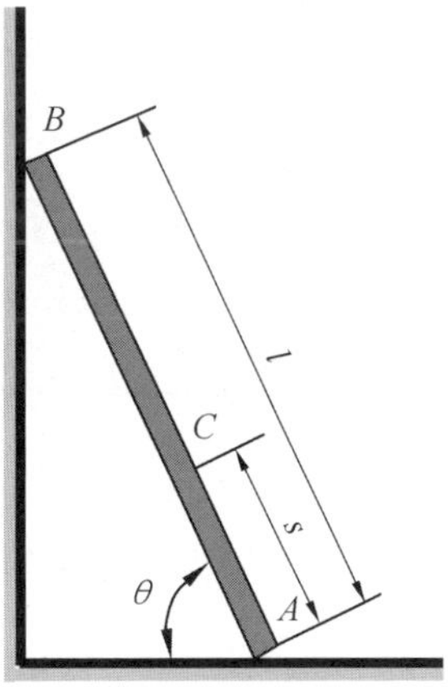

Fig. 4.24　Diagram of Exercise 6
（图 4.24　习题 4 图）

4. The ladder AB leans against the wall, and its weight is $P=200$ N, as shown in Fig. 4.24. The length of the ladder is l, and the angle of intersection with the horizontal plane is $\theta=60°$.

The coefficient of friction with the contact surface is 0.25. If a person with a weight of 650 N climbs up a ladder, what is the distance s from the highest point C to point A that the person can reach? (梯子 AB 靠在墙上,其重 $P=200$ N,如图 4.24 所示. 梯长为 l,并与水平面交角 $\theta=60°$. 与接触面间的摩擦因数均为 0.25. 如有一重为 650 N 的人沿梯子向上爬,则人所能达到的最高点 C 到点 A 的距离 s 是多少?)

5. As shown in Fig. 4.25, the coefficient of static friction between the sliding door and the upper-lower slide-ways is f_s regardless of its own weight, and the door height is h. If a horizontal force $\boldsymbol{F}$ is used to pull the door at $2/3h$ above the door without being stuck, find the minimum value of the door width b. Whether the self-weight of the door has any influence on the minimum value of the door width that is not stuck? (如图 4.25 所示,不计自重的拉门与上下滑道之间的静摩擦因数均为 f_s,门高为 h. 若在门上 $2/3h$ 处用水平力 $\boldsymbol{F}$ 拉门而不会被卡住,求门宽 b 的最小值. 门的自重对不被卡住的门宽最小值是否有影响?)

6. Drum B weighs 500 N and is placed in the corner, as shown in Fig. 4.26. It is known that the coefficient of static friction between the drum and the horizontal floor is 0.25, while the vertical wall is absolutely smooth. A weight hangs from the lower end of the rope on the drum. Radius $R=200$ mm, $r=100$ mm, ignoring rolling friction, find the maximum weight of weight A in balance. (鼓轮 B 重 500 N,放在墙角里,如图 4.26 所示. 已知鼓轮与水平地板间的静摩擦因数为 0.25,而铅垂墙壁绝对光滑. 鼓轮上的绳索下端挂着重物. 半径 $R=200$ mm,$r=100$ mm,不计滚动摩阻,求平衡时重物 A 的最大重力.)

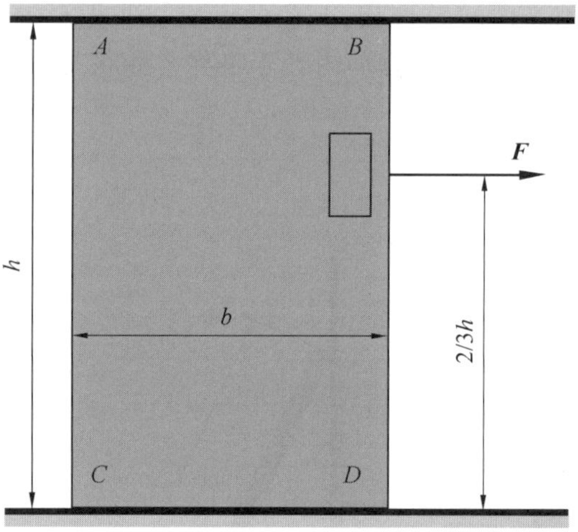

Fig. 4.25 Diagram of Exercise 7
(图 4.25 习题 5 图)

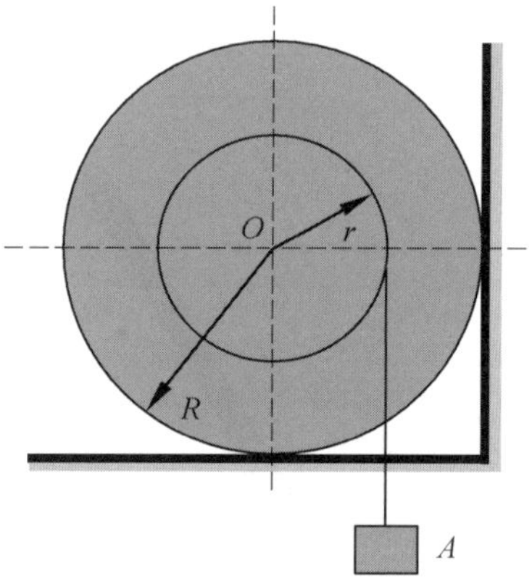

Fig. 4.26 Diagram of Exercise 8
(图 4.26 习题 6 图)

7. The rolling machine is composed of two wheels, the diameter of both wheels is $d=500$ mm, the gap between the wheels $a=5$ mm, and the two wheels rotate in opposite directions, as shown in Fig. 4.27. It is known that the coefficient of static friction between the red-hot iron plate and the cast iron wheel is $f_s=0.1$, what is the thickness b of the iron plate that can be pressed? Tip: To make the machine work, the iron plate must be driven by two runners, that is, the resultant force of the normal force and friction acting on the iron plates A and B must be

horizontal to the right.（轧压机由两轮组成，两轮的直径均为 $d=500$ mm，轮间的间隙 $a=5$ mm，两轮反向转动，如图 4.27 所示. 已知烧红的铁板与铸铁轮间的静摩擦因数 $f_s=0.1$，则能扎压的铁板的厚度 b 是多少？提示：欲使机器工作，则铁板必须被两转轮带动，即作用在铁板 A、B 处的法向作用力和摩擦力的合力必须水平向右.）

8. The planar crank-link slider mechanism is shown in Fig. 4.28. $OA=l$, there is a couple with moment M acting on the crank OA, and OA is horizontal. The angle between the connecting rod AB and the vertical line is θ, the static friction coefficient between the slider and the horizontal plane is f_s, regardless of the weight of each component, $\tan\theta>f_s$. Find the value of F force when the mechanism is in balance at the position shown in the Fig. 4.28.（平面曲柄连杆滑块机构如图 4.28 所示. $OA=l$，在曲柄 OA 上作用有一矩为 M 的力偶，OA 水平. 连杆 AB 与铅垂线的夹角为 θ，滑块与水平面之间的静摩擦因数为 f_s，不计各构件质量，$\tan\theta>f_s$. 求机构在图 4.28 所示位置保持平衡时 F 力的值.）

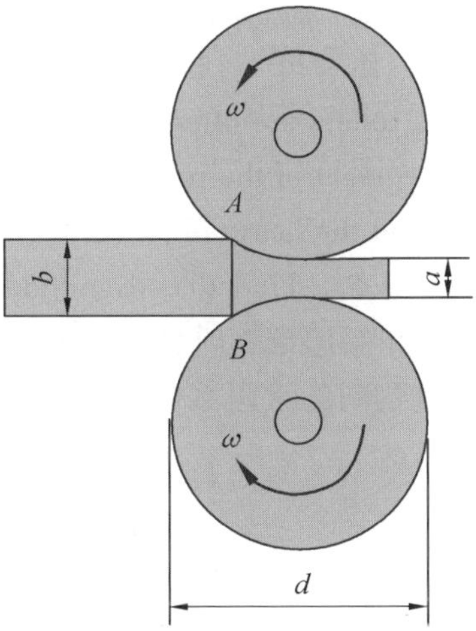

Fig. 4.27 Diagram of Exercise 7
（图 4.27 习题 7 图）

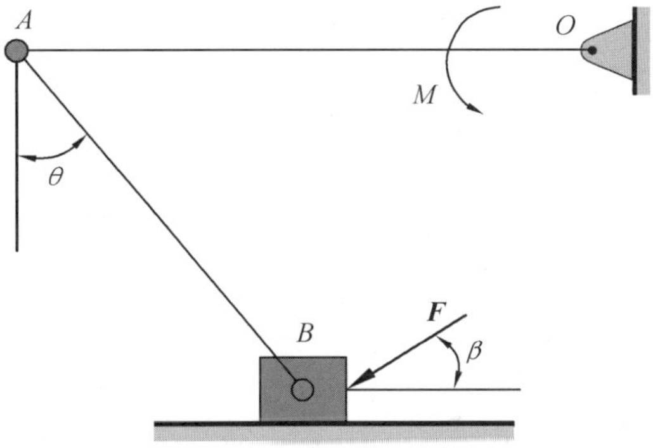

Fig. 4.28 Diagram of Exercise 8
（图 4.28 习题 8 图）

9. As shown in Fig. 4.29, for a homogeneous beam AB with a weight of $P_1=450$ N, the A end of the beam is a fixed hinge support, and the other end rests on the mandrel of the coil frame with a weight of $P_2=343$ N, and the wheel center C is the coil frame center of gravity. The static friction coefficients between the coil frame and the AB beam and the ground are $f_{s1}=0.4$, $f_{s2}=0.2$, ignoring rolling friction, the radius of the coil frame is $R=0.3$ m, and the radius of the mandrel is $r=0.1$ m. Now wind a soft rope regardless of weight on the mandrel of the bobbin, find the minimum value of the horizontal tension $\boldsymbol{F}$ that makes the bobbin move from rest.（如图 4.29 所示，重力 $P_1=450$ N 的均质梁 AB，梁的 A 端为固定铰支座，另一端搁置在重力 $P_2=343$ N 的线圈架的芯轴上，轮心 C 为线圈架的重心. 线圈架与 AB 梁和地面间的静摩擦因数分别为 $f_{s1}=0.4$，$f_{s2}=0.2$，不计滚动摩阻，线圈架的半径 $R=0.3$ m，芯轴的半径 $r=0.1$ m. 在线圈架的芯轴上绕一不计质量的软绳，求使线圈架由静止而开始运动的水平拉力 $\boldsymbol{F}$ 的最小值.）

10. The homogeneous long slab AD has a weight $\boldsymbol{P}$ and a length of 4 m, supported by a short

slab BC ignoring its own weight, as shown in Fig. 4.30, $AC = BC = AB = 3$ m. Find the friction angles at A, B, C to be large enough to balance them.（均质长板 AD 重为 $\boldsymbol{P}$，长为 4 m，用一不计自重的短板 BC 支撑，如图 4.30 所示，$AC = BC = AB = 3$ m. 求 A、B、C 处的摩擦角各为多大才能使之平衡.）

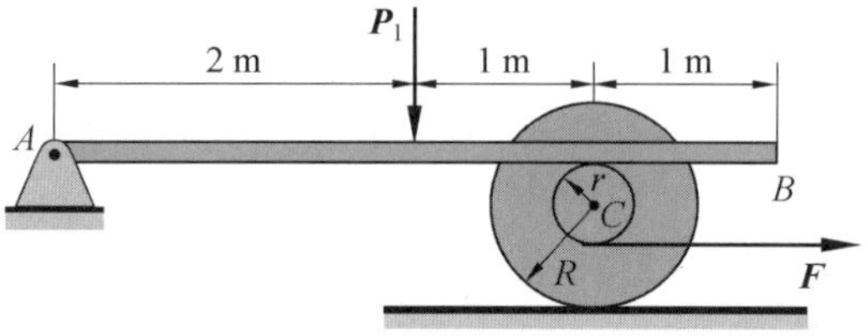

Fig. 4.29 Diagram of Exercise 9
（图 4.29 习题 9 图）

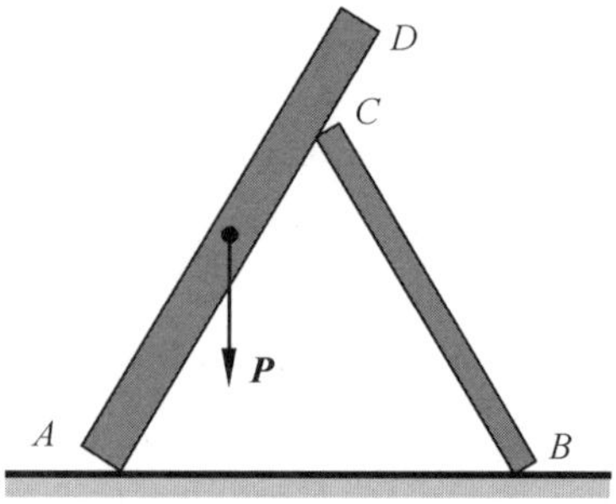

Fig. 4.30 Diagram of Exercise 10
（图 4.30 习题 10 图）

11. As shown in Fig. 4.31, when carrying heavy objects, rollers are often placed under the board. It is known that the weight of the heavy object is $\boldsymbol{P}$, the weight of the roller is $P_1 = P_2$, the radius is r, the rolling resistance coefficient between the roller and the heavy object is δ_1, and the rolling resistance coefficient between the roller and the ground is δ_2. Find the magnitude of the horizontal force $\boldsymbol{F}$ when pulling the weight.（如图 4.31 所示，在搬运重物时，常在板下面垫滚子. 已知重物重力为 $\boldsymbol{P}$，滚子重力 $P_1 = P_2$，半径为 r，滚子与重物间的滚阻系数为 δ_1，与地面间的滚阻系数为 δ_2. 求拉动重物时水平力 $\boldsymbol{F}$ 的大小.）

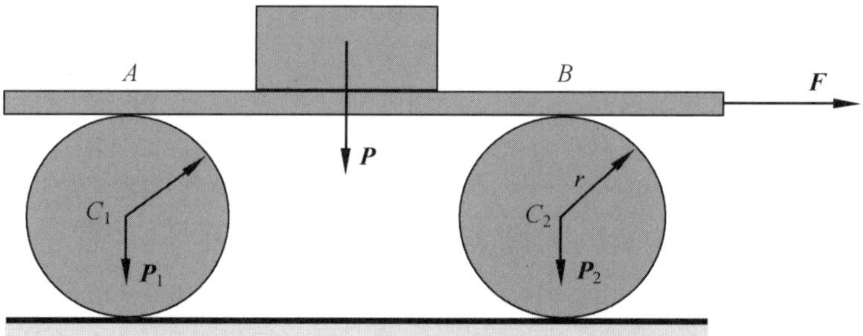

Fig. 4.31 Diagram of Exercise 11
（图 4.31 习题 11 图）

Part 2 Kinematics(第 2 篇 运动学)

Preface（引言）

Statics studies the simplification of the force system acting on the object, the equilibrium condition of the force system and its application. When the force system acting on an object is unbalanced, the state of motion of the object will change. The law of motion of an object is not only related to the force, but also related to the inertia of the object itself and the original state of motion. In order to study the motion law of the object under the action of the force system, this chapter does not consider the "force" and "mass" of the object's motion for the time being, but separately studies the geometric properties of the object's motion (trajectory, motion equation, velocity and acceleration, etc.), this part of the content called kinematics. As for the relationship between the law of motion of an object and force, inertia, etc. , it will be studied in dynamics. Therefore, kinematics is a science that studies the geometric properties of object motion, and is an important basis for the study of dynamics and subsequent professional courses such as mechanism analysis.（静力学研究作用在物体上的力系简化、力系平衡条件及应用. 当作用在物体上的力系不平衡时,物体的运动状态将发生改变. 物体的运动规律不仅与受力情况有关,而且与物体本身的惯性及原来的运动状态有关. 为研究物体在力系作用下的运动规律,本章暂不考虑物体运动的"力"和"质量",而单独研究物体运动的几何性质(轨迹、运动方程、速度和加速度等),这部分内容称为运动学. 物体的运动规律与力、惯性等的关系将在动力学中研究. 因此,运动学是研究物体运动几何性质的科学,是研究动力学及后续专业课程如机构分析等的重要基础.）

Kinematics is a branch of mechanics that describes and studies the change of the position of an object in space over time from a geometric point of view (refers to not involving the physical properties of the object itself and the force exerted on the object). (运动学是从几何的角度(指不涉及物体本身的物理性质和施加在物体上的力)描述和研究物体在空间的位置随时间变化规律的力学分支.)

In kinematics, it is first necessary to determine the spatial position of an object at any moment. To determine the position of an object in space, another object must be selected as a reference, and this reference object is called a reference body. Due to the relativity of motion, the selected reference body is different, and the motion law of the object relative to different reference bodies is also different. Therefore, in mechanics, describing the motion of any object requires specifying a reference body. If the coordinate axes are fixed on the reference body, the reference coordinate system is formed abbreviated as the reference system. In general engineering problems, the coordinate system fixed to the ground is taken as the reference system, which is called the fixed reference system.(在运动学中,首先需要确定物体在任何时刻所在的空间位置.而要确定物体在空间的位置,必须选取另一个物体作为参考,这个参考的物体称为参考体.由于运动的相对性,所选的参考体不同,物体相对于不同参考体的运动规律也不同,因此在力学中,描述任何物体的运动都需要指明参考体.如果将坐标轴固连于参考体上,就构成参考坐标系,简称参考系.一般工程问题中,都取与地面固连的坐标系为参考系,称为定参考系.)

In theoretical mechanics, mechanical motion is the simplest form of motion. When studying mechanical motion, a relative reference coordinate system must always be selected, which can move or be still. For example, when studying the mechanical motion of objects on the earth, a coordinate system fixed to the earth is always selected as a stationary reference system.(在理论力学中,机械运动是最简单的运动形式.在研究机械运动时,总要选定相对参考系,这种参考系可以是移动的,也可以是静止的.例如,在研究地球上物体的机械运动时,总是选择一个与地球相固连的坐标系作为静止的参考系.)

Chapter 5　Kinematics of Points
(点的运动学)

The kinematics of points is the basis of studying the motion of objects. This chapter studies the simple motion of a point, and studies the change of the geometric position of a point relative to a certain reference system with time, including the motion equation, motion trajectory, velocity and acceleration of the point, etc.(点的运动学是研究物体运动运动的基础.本章研究点的简单运动,研究点相对于某一参考系的几何位置随时间的变化规律,包括点的运动方程、运动轨迹、速度和加速度等.)

5.1　Methods of Determining the Motion of Particles (确定质点运动的方法)

Before studying the motion of a point and determining the characteristics of this motion, we must learn how to determine the position of a point in space at a certain instant.(在研究点的运动并确定这种运动的特征之前,必须学习如何确定点在某瞬时在空间中的位置.)

There are several ways to determine motion for this.(为此,有几种确定运动的方法.)

(1) Natural coordinate method.(自然坐标法.)

(2) Cartesian coordinate method.(笛卡儿坐标法.)

(3) Radius method.(矢径法.)

A point moves along a certain curve or trajectory in space.(点在空间沿某条曲线或轨迹运动.)

Introduce the methods of kinematic research on points: natural method (Fig. 5.1(a)). We need to know: ① the trajectory of the point; ② the relationship of the arc length with time ($\widehat{OM}=S=f(t)$), this relationship is called is the motion equation of the point; ③ the starting point of motion; ④ the origin; ⑤ the reference direction.(介绍点的运动学研究的方法:自然法(图 5.1(a)).需要已知:①点的轨迹;②弧长随时间的变化关系($\widehat{OM}=S=f(t)$),这种关系式称为点的运动方程;③运动的起始点;④原点;⑤参考方向.)

The position of a point in space is uniquely determined by the radius $\boldsymbol{r}$ drawn from a fixed center to the point M (Fig. 5.1(b)). This way of defining motion is called the sagittal method, we have (空间中某点的位置由从某个固定中心到该点 M 绘制的矢径 $\boldsymbol{r}$ 唯一确定(图 5.1(b)).这种定义运动的方式称为矢径法,有)

$$\boldsymbol{r}=\boldsymbol{r}(t) \tag{5.1}$$

In this case, the location of the point in space will be determined by the geometric location of

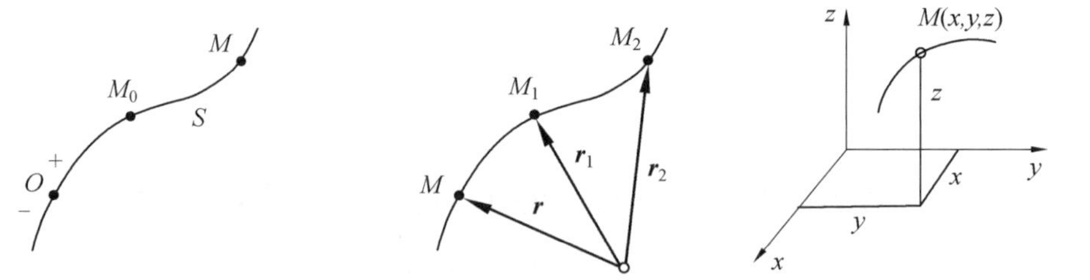

(a) Natural method(自然法)　(b) Radius method(矢径法)　(c) Cartesian coordinate method (笛卡儿坐标法)

Fig. 5. 1　method of describing the point motion (图 5. 1　描述点运动的方法)

the endpoints of the vector, i. e. the endpoints of its radius. (在这种情况下, 空间点的位置将由矢量 $\boldsymbol{r}$ 端点的几何位置,即其矢径的终点确定.)

The Cartesian coordinate method of point motion (Fig. 5. 1(c)) should know the following functional relationship, by which it can be determined how the coordinates of a point in space change with time, that is (点运动的笛卡儿坐标法 (图 5. 1(c)) 应该知道如下函数关系, 通过它可以确定空间中某个点的坐标是如何随时间变化的,即)

$$x = f_1(t), y = f_2(t), z = f_3(t) \tag{5.2}$$

These equations are called the equations of motion in the Cartesian coordinate system, and from these equations the position of a point in space can be determined according to time. If a point moves on a plane, its position is described by two equations which are(这些方程式称为笛卡儿坐标系下的运动方程, 根据时间, 由这些方程可以确定空间中某一点的位置. 如果点在平面上运动, 则其位置由两个方程式描述,即)

$$x = f_1(t), y = f_2(t) \tag{5.3}$$

If a point moves along a straight line, then only one equation is sufficient, that is(如果一个点沿直线运动,那么只有一个方程就足够了,即)

$$x = f(t) \tag{5.4}$$

Example 1: As shown in Fig. 5. 2, the equation of motion of a point on a known plane is(如图 5. 2,已知平面上点的运动方程为)

$$x = 2 + 4t, y = -3 + 8t$$

where x and y are in cm; t is in s. Determine the motion trajectory of the point. (式中,x 和 y 以 cm 为单位;t 以 s 为单位. 确定点的运动轨迹.)

Solution: (解:)

We eliminate t from the given equation of motion to obtain the trajectory equation. From the first equation, we get $t=(x-2)/4$, and from the second equation, we get $t=(y+3)/8$. Equating the right-hand sides of these equations, we get(从给定运动方程中消除 t,得到轨迹方程. 从第一个等式中得到 $t=(x-2)/4$, 从第二个等式中得到 $t=(y+3)/8$. 令这些等式的右边部分相等,得到)

$$\frac{x-2}{4} = \frac{y+3}{8}$$

The trajectory of motion is a straight line, we draw this straight line. Assuming $x=0$, we find

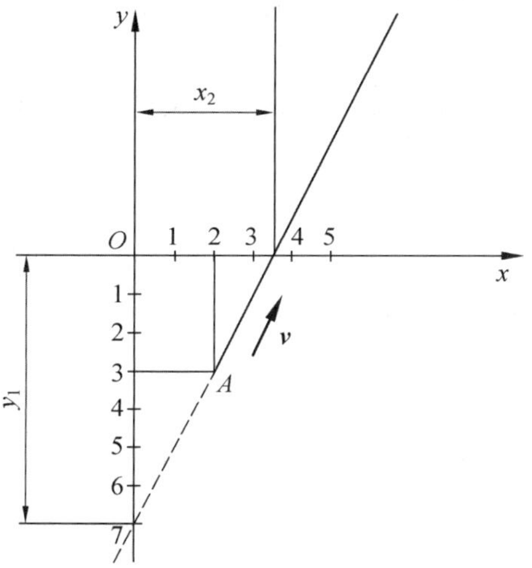

Fig. 5. 2 Diagram of Example 1(图 5. 2 例题 1 图)

the intersection of the trajectory with the Oy axis: $y_1=-7$.(运动的轨迹是一条直线,画出这条直线.假设 $x=0$,可以找到轨迹与 Oy 轴的交点:$y_1=-7$.)

Assuming $y=0$, we will find the intersection of the trajectory with the Ox axis: $x_2=3.5$ cm. After drawing straight lines through these points, we will obtain the trajectory of a mass point in a straight line (Fig. 5. 2). On this line, we need to find the starting point of the movement.(假设 $y=0$, 可以找到轨迹与 Ox 轴的交点: $x_2=3.5$ cm.通过这些点绘制直线后,将得到一个质点的直线运动轨迹(图 5. 2).在这一直线上,需要找到运动的起始点.)

At the moment the movement starts, i. e. when $t=0$, the point has coordinates $x_A=2$ cm, $y_A=-3$ cm. We only need to determine the direction of the mass point A. Over time, the x and y coordinates will increase. So the particle will start moving from point A and will continue to move up the arrow to infinity.(在运动开始的那一刻,即当 $t=0$ 时,该点具有坐标 $x_A=2$ cm,$y_A=-3$ cm,只需确定质点 A 的方向.随着时间的推移,x 和 y 坐标将增加. 因此,质点将从 A 点开始运动,并将继续沿箭头向上运动到无穷大.)

Therefore, the motion trajectory of the particle is obtained, which is determined by the natural method: its starting point is point A, the direction of motion is along the direction of the arrow, and $S=S(t)$ is called the motion equation of the moving point along the known trajectory.(因此, 得到了质点的运动轨迹, 它是通过自然法确定的:其起点是 A 点,运动方向沿着箭头方向,$S=S(t)$ 称为动点沿着已知轨迹的运动方程.)

5.2 Speed of Point(点的速度)

As we all know, when a point moves along a straight line at a constant speed, its speed is determined by dividing the distance s passed by the time, that is $v=s/t$. This formula does not apply to variable speed motion. The method of determining speed depends on determining motion methods.(众所周知,当一个点匀速地沿直线移动时,它的速度是用通过的距离 s 除以时间

来确定的,即 $v=s/t$. 这个公式不适用于变速运动,确定速度的方法取决于确定运动的方法.)

1. The speed of a point when the motion is described by the vector method (矢量法描述运动时点的速度)

The speed of the point determines the speed and direction of the point's motion. Using the vector method to describe motion, the position of each moment point is determined by the vector $\boldsymbol{r}=\boldsymbol{r}(t)$.(点的速度决定点的运动快慢和方向. 利用描述运动的矢量方法,每个时刻点的位置由矢径 $\boldsymbol{r}=\boldsymbol{r}(t)$ 确定.)

At time t, the particle is at point M, and its position is determined by the vector $\boldsymbol{r}=\boldsymbol{r}(t)$ (Fig. 5.3(a)). At time $t+\Delta t$, the point will be at the position of M_1, which is determined by the vector $\boldsymbol{r}_1=\boldsymbol{r}+\Delta\boldsymbol{r}$. The ratio $\frac{\Delta\boldsymbol{r}}{\Delta t}$ is the mean velocity vector $\boldsymbol{v}_{CP}$, and the derivative of vector $\boldsymbol{r}$ with respect to time t is the velocity vector at a given instant, we have(在 t 时刻质点在 M 点,其位置由矢径 $\boldsymbol{r}=\boldsymbol{r}(t)$ 确定(图 5.3(a)). 在 $t+\Delta t$ 时刻时,点将在 M_1 位置,由矢径 $\boldsymbol{r}_1=\boldsymbol{r}+\Delta\boldsymbol{r}$ 确定. 比值 $\frac{\Delta\boldsymbol{r}}{\Delta t}$ 是平均速度矢量 $\boldsymbol{v}_{CP}$,矢量 $\boldsymbol{r}$ 对时间 t 的导数是给定时刻的速度矢量,有)

$$\boldsymbol{v}=\frac{d\boldsymbol{r}}{dt}$$

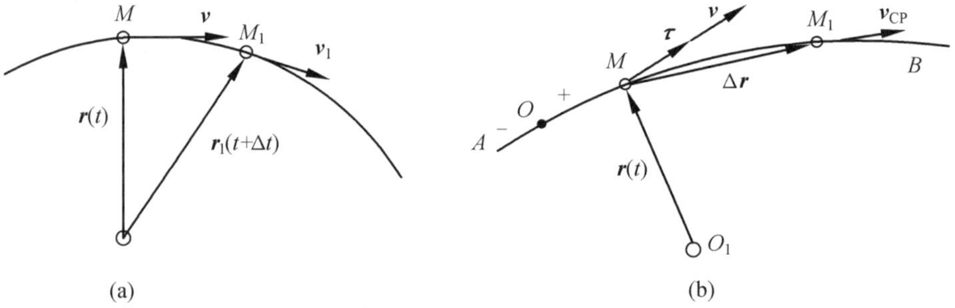

Fig. 5.3 Velocity calculation by vector method(图 5.3 矢径法求速度)

Since $\boldsymbol{v}$ is the derivative of $\boldsymbol{r}=\boldsymbol{r}(t)$, the velocity vector $\boldsymbol{v}$ is always tangent to the trajectory of the particle.(因为 $\boldsymbol{v}$ 是 $\boldsymbol{r}=\boldsymbol{r}(t)$ 的导数,所以速度向量 $\boldsymbol{v}$ 始终与质点的轨迹相切.)

2. Velocity of a point in natural coordinate method (自然坐标法中点的速度)

If the motion of a point is determined by the natural method, its trajectory AB, the starting position of the motion, the direction of motion and the equation $S=S(t)$ are all known. In the resulting expression $\boldsymbol{v}=\frac{d\boldsymbol{r}}{dt}$, we introduce an intermediate variable — the arc coordinate S,and we have(如果通过自然法确定点的运动,则其轨迹 AB、运动起始位置、运动方向和方程 $S=S(t)$ 都是已知的. 在结果表达式 $\boldsymbol{v}=\frac{d\boldsymbol{r}}{dt}$ 中,引入一个中间变量——弧坐标 S,则有)

$$\boldsymbol{v}=\frac{d\boldsymbol{r}}{dS}\cdot\frac{dS}{dt}$$

Since dS is a scalar value, $\frac{d\boldsymbol{r}}{dS}$ will follow the tangential direction of the path at point M, this vector is denoted by $\boldsymbol{\tau}$ (Fig. 5.3(b)), and along the direction of motion, its magnitude is equal to 1, always pointing in the direction of increasing S. (因为 $\mathrm{d}S$ 是标量值，所以 $\frac{\mathrm{d}\boldsymbol{r}}{\mathrm{d}S}$ 将沿点 M 处路径的切线方向,这个向量用 $\boldsymbol{\tau}$ (图5.3(b)) 表示，并且顺着运动的方向，其大小等于1,始终指向 S 增加的方向.)

Therefore, the velocity vector is determined in the form of natural coordinates, that is(因此，用自然坐标的形式确定速度矢量,即)

$$\boldsymbol{v} = \frac{\mathrm{d}S}{\mathrm{d}t}\boldsymbol{\tau}$$

The derivative $\mathrm{d}S/\mathrm{d}t$ is the algebraic value of velocity. If $\mathrm{d}S/\mathrm{d}t > 0$, then at the moment under consideration, the point is moving in the direction of increasing arc coordinate S, therefore, the direction of its velocity coincides with the direction of the unit vector $\boldsymbol{\tau}$. If $\mathrm{d}S/\mathrm{d}t<0$, the arc coordinate S decreases, therefore, the velocity vector points in the opposite direction to the vector. (导数 $\mathrm{d}S/\mathrm{d}t$ 是速度的代数值. 如果 $\mathrm{d}S/\mathrm{d}t > 0$,那么在所考虑的时刻,该点在弧坐标 S 增加的方向上运动. 因此,其速度的方向与单位矢量 $\boldsymbol{\tau}$ 的方向一致. 如果 $\mathrm{d}S/\mathrm{d}t<0$,则弧坐标 S 减小. 因此,速度矢量指向与矢量 $\boldsymbol{\tau}$ 相反的方向.)

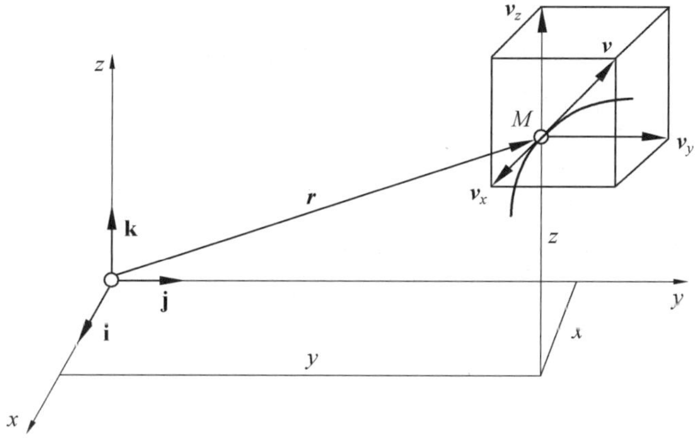

Fig. 5.4 Cartesian coordinate method to find speed(图 5.4 笛卡儿坐标法求速度)

3. Velocity of a point under the equation of motion in Cartesian coordinates (笛卡儿坐标运动方程下点的速度)

The velocity of a point is defined under the Cartesian coordinate method that describes the point's motion. The equation of motion for point M (Fig. 5.4) is(在描述点运动的笛卡儿坐标法下定义点的速度. 设点 M 的运动方程 (图 5.4) 为)

$$x = f_1(t), y = f_2(t), z = f_3(t)$$

Its position in space is determined by the vector, that is(它在空间中的位置由矢径决定,即)

$$\boldsymbol{r} = \mathbf{i}x + \mathbf{j}y + \mathbf{k}z$$

According to previous research results, the velocity of the moving point is equal to the first derivative of the vector of the moving point with respect to time, and the velocity vector can be written as (根据之前的研究结果,动点的速度等于动点的矢径对于时间的一阶导数,速度矢

量可以写成）

$$\boldsymbol{v} = \lim_{\Delta t \to 0} \frac{\Delta \boldsymbol{r}}{\Delta t} = \frac{\mathrm{d}\boldsymbol{r}}{\mathrm{d}t} = \frac{\mathrm{d}}{\mathrm{d}t}(\mathbf{i}x + \mathbf{j}y + \mathbf{k}z) = \mathbf{i}\frac{\mathrm{d}x}{\mathrm{d}t} + \mathbf{j}\frac{\mathrm{d}y}{\mathrm{d}t} + \mathbf{k}\frac{\mathrm{d}z}{\mathrm{d}t}$$

Therefore, $\boldsymbol{v} = \mathbf{i}v_x + \mathbf{j}v_y + \mathbf{k}v_z$, we establish a projected coordinate system on x, y, z (Fig. 5.4) and define the velocity vector, we have（因此，$\boldsymbol{v} = \mathbf{i}v_x + \mathbf{j}v_y + \mathbf{k}v_z$，在 x、y、z 上建立投影坐标系（图 5.4）并定义速度向量，有）

$$v = \sqrt{v_x^2 + v_y^2 + v_z^2}$$

5.3 Acceleration of Point（点的加速度）

Acceleration of the point is a vector, which describes the speed and direction of velocity vector change with time: $\boldsymbol{a} = \frac{\mathrm{d}\boldsymbol{v}}{\mathrm{d}t} = \frac{\mathrm{d}^2\boldsymbol{r}}{\mathrm{d}t^2}$. An expression for the projection of the acceleration vector on the axis can be written, which are（点的加速度是一个矢量，它描述速度矢量随时间变化的快慢和方向：$\boldsymbol{a} = \frac{\mathrm{d}\boldsymbol{v}}{\mathrm{d}t} = \frac{\mathrm{d}^2\boldsymbol{r}}{\mathrm{d}t^2}$. 可以在轴上写出加速度矢量投影的表达式，即）

$$a_x = \mathrm{d}v_x/\mathrm{d}t, a_y = \mathrm{d}v_y/\mathrm{d}t, a_z = \mathrm{d}v_z/\mathrm{d}t$$

The acceleration magnitude can be determined if the projections a_x, a_y, a_z are known, that is（如果已知投影 a_x、a_y、a_z，则可以确定加速度大小，即）

$$a = \sqrt{a_x^2 + a_y^2 + a_z^2}$$

In the natural method where the motion trajectory of a particle is known, its acceleration vector can be decomposed along the natural axes $\boldsymbol{\tau}$ and $\boldsymbol{n}$ (Fig. 5.5), that is（在已知质点运动轨迹的自然方法中，其加速度矢量可沿自然轴 $\boldsymbol{\tau}$ 和 $\boldsymbol{n}$（图 5.5）分解，即）

$$\boldsymbol{a} = a_\tau \boldsymbol{\tau} + a_n \boldsymbol{n}$$

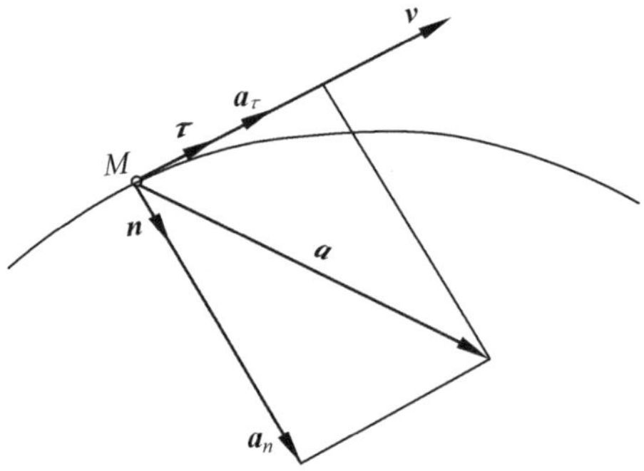

Fig. 5.5 Full acceleration（图 5.5 全加速度）

The projection of acceleration on vector $\boldsymbol{\tau}$ is called tangential acceleration, which characterizes the rate of change of velocity: $a_\tau = \mathrm{d}v/\mathrm{d}t$. Tangential acceleration only exists when moving in a non-uniform curve.（加速度在矢量 $\boldsymbol{\tau}$ 上的投影称为切向加速度，其表征速度大小的变化率为 $a_\tau = \mathrm{d}v/\mathrm{d}t$. 切向加速度仅在非匀速曲线运动时存在.）

The normal acceleration $a_\mathrm{n} = v^2/\rho$ represents the change in the direction of the velocity $\boldsymbol{v}$

vector when the point moves along the curved trajectory (ρ is the radius of curvature of the trajectory at a point).（法向加速度 $a_n = v^2/\rho$ 表示当点沿曲线轨迹运动时速度 $\boldsymbol{v}$ 矢量方向的变化（ρ 是一点处轨迹的曲率半径）.）

Example 2: A ellipse gauge mechanism, $\theta = \omega t$, $OA = AC = AB = l$, $BM = d$, find the motion equation, velocity and acceleration of point M.（**例题 2**：一个椭圆规机构，$\theta = \omega t$，$OA = AC = AB = l$，$BM = d$，求 M 点的运动方程、速度和加速度.）

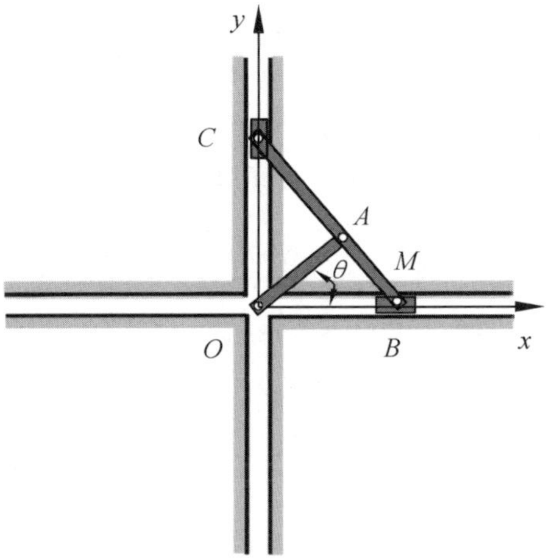

Fig. 5.6　Diagram of Example 2（图 5.6　例题 2 图）

(1) Establish a fixed reference frame Oxy.（建立固定参考系 Oxy.）

(2) Place the point under consideration at its general position in the coordinate system.（将所考查的点置于坐标系中的一般位置.）

(3) Write the equation of motion of the point according to the known constraints.（根据已知的约束条件列写点的运动方程.）

The equation of motion of point M is（M 点的运动方程为）

$$\begin{cases} x = (2l - d)\cos\theta = (2l - d)\cos \omega t \\ y = d\sin \theta = d\sin \omega t \end{cases}$$

Eliminate t from it to get the trajectory equation of point M, we have（从中消去 t，得到 M 点的轨迹方程，有）

$$\left(\frac{x}{2l - d}\right)^2 + \left(\frac{y}{d}\right)^2 = 1$$

The speed of point M is（M 点的速度为）

$$v_x = \dot{x} = -\omega(2l - d)\sin \omega t\ ,\ \dot{y} = \omega d\cos \omega t$$

The acceleration at point M is（M 点的加速度为）

$$a_x = \ddot{x} = -\omega^2(2l - d)\cos \omega t\ ,\ \ddot{y} = -\omega^2 d\sin \omega t$$

5.4 Special Circumstances of Point Motion(点运动的特殊情况)

(1) $a_n=0$, $a_\tau=0$.

Therefore, the total acceleration $a=0$. The point moves in a straight line with uniform velocity. The equation of motion at this time is $S=S_0+v_0t$, where S_0 is the arc coordinate at the initial moment; v_0 is the velocity at the point at the initial moment of motion, at any other point in time t, the velocity will not change because the motion will not accelerate.(因此,总加速度 $a=0$. 该点做匀速直线运动. 此时的运动方程是 $S=S_0+v_0t$. 其中,S_0是初始时刻的弧坐标; v_0是初始运动时刻点的速度. 在任何其他时间点 t,速度不会改变,因为运动不会加速.)

(2) $a_n\neq0$, $a_\tau=0$.

Uniform velocity curve motion. The velocity vector of a point changes only in direction, the law of motion of the curve is similar to the first case, we have (匀速率曲线运动. 点的速度矢量仅在方向上变化,曲线运动定律与第一种情况类似,有)

$$S=S_0+v_0t$$

(3) $a_n=0$, $a_\tau\neq0$.

Variable speed linear motion. (变速直线运动.)

(4) $a_n\neq0$, $a_\tau\neq0$.

Variable speed curve movement. (变速曲线运动.)

If $a_\tau=a=\text{const}$ in (3) and $a_\tau=\text{const}$ in(4), the points will move in a straight line with constant speed (uniform change in speed) respectively(如果在(3)的情况下 $a_\tau=a=\text{const}$,在(4)的情况下 $a_\tau=\text{const}$, 则点将分别做匀变速(速度均匀变化)直线运动)

$$S=S_0+v_0t\pm\frac{1}{2}at^2$$

and uniform speed change (uniform change in speed) curved motion(和匀变速(速率均匀变化)曲线运动)

$$S=S_0+v_0t\pm\frac{1}{2}a_\tau t^2$$

Example 3: The car decelerates uniformly within a radius of $R=1$ km. At the beginning of the segment, the speed of the train is 36 km/h with a full acceleration $a_0=0.125$ m/s^2. If the arc length is 560 m, determine the speed and acceleration of the train at the end of the curved region. (**例题** 3:火车在 $R=1$ km 的半径范围内匀减速运动. 在该段开始时,列车的速度为 36 km/h,全加速度 $a_0=0.125$ m/s^2. 如果弧长为 560 m,则确定曲线区域末端的列车速度和加速度.)

Solution:(解:)

(1) Consider the motion of one of the train's points, such as its center of gravity. The origin O of arc coordinates coincides with the initial position of point M_O, and the direction of movement is positive (Fig. 5.7). At this point, the value of S_0 is zero. (考虑列车其中一个点的运动,如它的重心. 将弧坐标的原点 O 与点 M_O的初始位置相重合,运动方向为正(图 5.7). 此时,S_0的值为零.)

(2) Write down the equation of motion of a point's uniform velocity curve, that is(写出点的匀变速曲线运动方程,即)

$$S = v_0 t - \frac{1}{2} a_\tau t^2$$

A formula for determining the velocity of this motion is(一个确定这个运动速度的公式为)

$$v = v_0 - a_\tau t$$

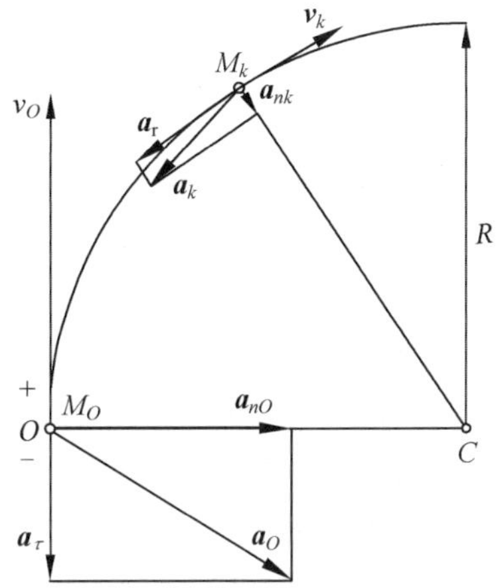

Fig. 5.7 Diagram of Example 3 (图 5.7 例题 3 图)

(3) Determine the normal acceleration at the start, we have(确定开始处的法向加速度,有)

$$a_{n0} = \frac{v_0^2}{R} = \frac{100}{1\ 000} = 0.1\ \mathrm{m/s^2}$$

$$(v_0 = 36\ \mathrm{km/h} = 10\ \mathrm{m/s},\ R = 1\ \mathrm{km} = 1\ 000\ \mathrm{m})$$

(4) Knowing the magnitude of the full acceleration at the starting point, we define its tangent component as(知道开始处点的全加速的大小后,可以定义其切线分量为)

$$a_O^2 = a_{nO}^2 + a_\tau^2, a_\tau = \sqrt{a_O^2 - a_{nO}^2} = \sqrt{0.125^2 - 0.1^2} = 0.075\ \mathrm{m/s^2}$$

(5) Substitute the expression for the tangential component of the acceleration into the formula, given the time t during which the train has passed a section of 560 m, we have (用加速度切向分量的表达式代入公式,已知时间 t,在此期间火车通过了 560 m 的路段,有)

$$560 = 10t - \frac{0.075t^2}{2}$$

Thus, we have(由此,有)

$$t = \frac{10 \pm \sqrt{100 - 1\ 120 \cdot 0.075}}{0.075} = \frac{10 \pm 4}{0.075}\ (\mathrm{s})$$

Therefore(因此)

$$t_1 = \frac{14}{0.075}\ \mathrm{s},\ t_2 = \frac{6}{0.075}\ \mathrm{s}$$

The value of t_1 is false because this time exceeds the time the train was at the end of the route. So only the second root of the equation is considered, that is (t_1的值是不真实的,因为这

个时间超过了列车在路径终点的时间.所以只考虑方程的第二个根,即)

$$t_2 = \frac{6}{0.075}\ \text{s}$$

(6) Determine the velocity at the end of the path, that is(确定路径终点的速度,即)

$$v_k = 10 - 0.075t_k = 10 - 0.075 \cdot 80 = 4\ (\text{m/s})$$

(7) Calculate the normal acceleration at the end point of the path, we have(计算路径终点的法向加速度,有)

$$a_{nk} = \frac{v_k^2}{R} = \frac{4^2}{1\ 000} = 0.016\ (\text{m/s}^2)$$

(8) Determine the velocity at the end of the path, we have(定义路径末尾的全加速度,有)

$$a_k = \sqrt{a_{nk}^2 + a_\tau^2} = \sqrt{0.016^2 + 0.075^2} = 0.076\ 7\ (\text{m/s}^2)$$

From the calculations, it can be seen that due to the reduction of the normal acceleration, the total acceleration decreases, while the tangential acceleration remains unchanged. (从计算中可以看出,由于法向加速度降低,因此全加速度减小,而切线加速度保持不变.)

Example 4: It is known that the radius of the circle is R, the rod AB rotates around the axis A, the rotation angle is $\varphi = \omega t$, and ω is a constant. Find: (1) the motion equation, velocity and acceleration of the small ring M; (2) the small ring M relative to the velocity and acceleration of AB bar. (**例题** 4:已知圆半径为 R,AB 杆绕 A 轴转动,转角 $\varphi = \omega t$,ω 为常数.求:(1)小环 M 的运动方程、速度和加速度;(2)小环 M 相对于 AB 杆的速度和加速度.)

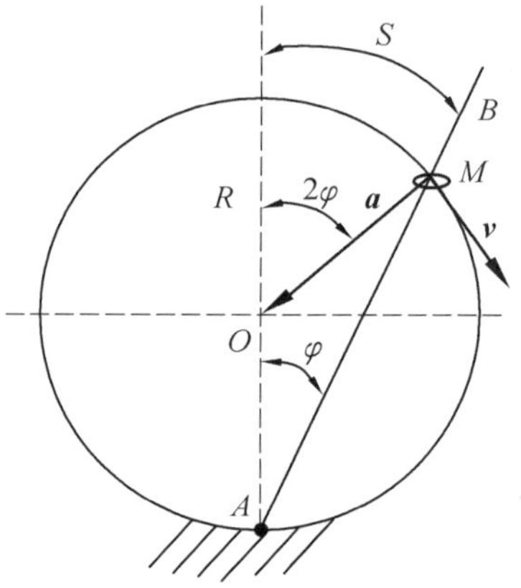

Fig. 5.8 Diagram of Example 4(图 5.8 例题 4 图)

Solution:(解:)

(1) Establish the arc coordinates of the diagram. (建立图示弧坐标.)

The equation of motion is(运动方程为)

$$S = R \cdot 2\varphi = 2R\omega t$$

The speed is(速度为)

$$v = \frac{\mathrm{d}S}{\mathrm{d}t} = 2R\omega$$

The acceleration is(加速度为)

$$a_\tau = \frac{\mathrm{d}v}{\mathrm{d}t} = 0,\ a_n = \frac{v^2}{R} = 4R\omega^2$$

(2) Create a Cartesian coordinate system.（建立笛卡儿坐标系.）

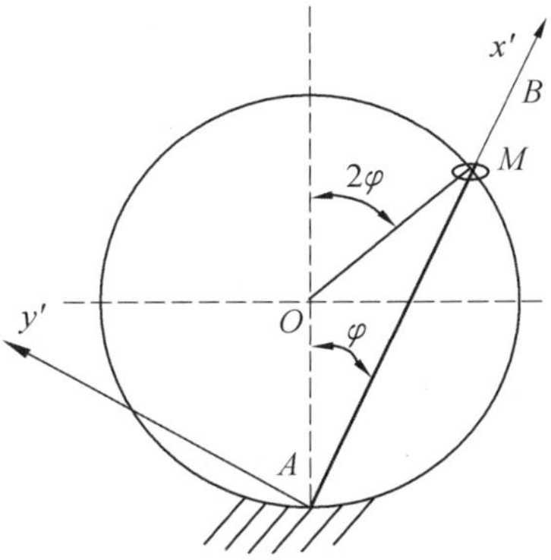

Fig. 5.9 Calculate relative velocity using Cartesian coordinate method
（图 5.9 笛卡儿坐标法求相对速度）

The equation of motion is（运动方程为）

$$x_{1M} = 2R\cos\varphi = 2R\cos\omega t$$

The speed is（速度为）

$$v_{1M} = \frac{\mathrm{d}x_{1M}}{\mathrm{d}t} = -2R\omega\sin\omega t$$

The acceleration is（加速度为）

$$a_{1M} = \frac{\mathrm{d}v_{1M}}{\mathrm{d}t} = -2R\omega^2\cos\omega t$$

5.5 Questions（思考题）

1. Are $\frac{\mathrm{d}v}{\mathrm{d}t}$ and $\frac{\mathrm{d}v}{\mathrm{d}t}$, $\frac{\mathrm{d}r}{\mathrm{d}t}$ and $\frac{\mathrm{d}r}{\mathrm{d}t}$ the same?（$\frac{\mathrm{d}\boldsymbol{v}}{\mathrm{d}t}$ 和 $\frac{\mathrm{d}v}{\mathrm{d}t}$、$\frac{\mathrm{d}\boldsymbol{r}}{\mathrm{d}t}$ 和 $\frac{\mathrm{d}r}{\mathrm{d}t}$ 是否相同?）

2. Point M moves along the helix from outside to inside, as shown in Fig. 5.10. The length of the arc it traveled is proportional to the first power of time. Is the acceleration of the point getting bigger or smaller? Point M to run faster or slower?（点 M 沿螺旋线自外向内运动，如图 5.10所示. 它走过的弧长与时间的一次方成正比，则点的加速度是越来越大，还是越来越小? 点 M 越跑越快，还是越跑越慢?）

3. When the point moves in a curve, the acceleration $\boldsymbol{a}$ of the point is a constant vector, as shown in Fig. 5.11. Does the questioning point move at a uniform speed?（当点做曲线运动时，点的加速度 $\boldsymbol{a}$ 是恒矢量，如图 5.11 所示. 则点是否做匀变速运动?）

4. Two moving points in a curved motion have the same initial velocity, the same trajectory, and the same normal acceleration of the two points in motion. Determine whether the following statements are correct:（做曲线运动的两个动点，初速度相同、运动轨迹相同、运动中两点的法向加速度也相同. 判断下述说法是否正确:）

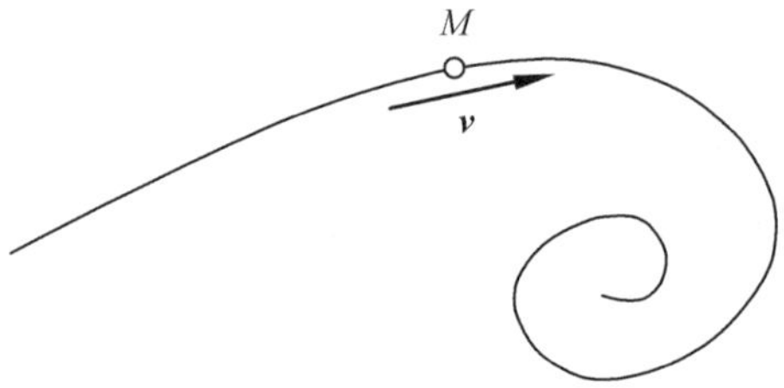

Fig. 5.10 Diagram of Quetion 2
(图 5.10 思考题 2 图)

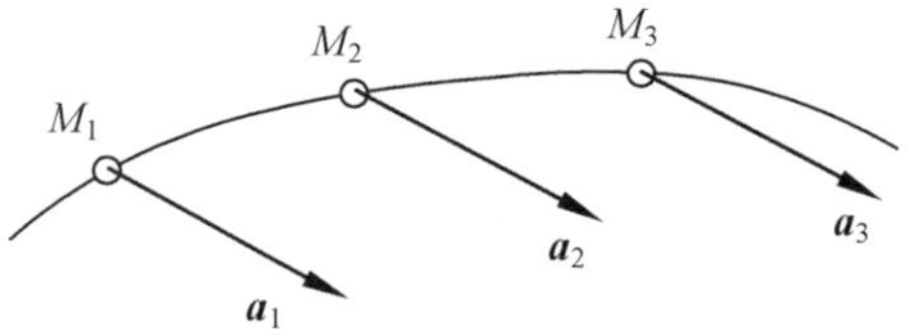

Fig. 5.11 Diagram of Quetion 3
(图 5.11 思考题 3 图)

(1) The tangential acceleration of the two moving points at any moment must be the same; (任一瞬间两动点的切向加速度必须相同;)

(2) The speed of the two moving points at any moment must be the same; (任一瞬间两动点的速度必须相同;)

(3) The equations of motion for the two moving points must be the same. (两动点的运动方程必须相同.)

5. When a point moves in a curve, are the following statements correct? (点做曲线运动时,下列说法是否正确?)

(1) If the tangential acceleration is positive, the point will move with acceleration; (若切向加速度为正,则点做加速运动;)

(2) If the sign of the tangential acceleration is the same as that of the velocity, the point will move with acceleration; (若切向加速度与速度符号相同,则点做加速运动;)

(3) If the tangential acceleration is zero, the velocity is a constant vector. (若切向加速度为零,则速度为常矢量.)

6. A circle of radius r rolls without slipping on a circle of radius $2r$. Does any point P on the small circle move along a straight line? (一半径为 r 的圆周在半径为 $2r$ 的圆周内做无滑动的滚动. 小圆周上任一点 P 是否沿直线运动?)

7. Assuming a projectile is shot horizontally on a target, its speed is equal to n times the speed of sound ($n>1$). When the sound of firing the gun and the sound of the projectile hitting the target are heard at the same time, they are all on a hyperbola with eccentricity n, and when the value of n is close to 1, what kind of curve does this hyperbola approximate to? (设一射弹水平地射在一靶子上,它的速度等于声速的 n 倍($n>1$). 当同时听到放枪与射弹碰靶声的地点皆在一偏心度 n 的双曲线上时,在 n 的数值与 1 相近时,此双曲线与何种曲线近似?)

8. Two points A and B are on the same spoke, if the rotational angular velocity of the wheel is uniform, does point A make constant circular motion around point B? (两点 A 和 B 在同一根轮辐上,若轮的转动角速度时均匀的,则 A 点是否绕 B 点做等速圆周运动?)

5.6 Exercises(习题)

1. The speed of flowing water in a small river is 3 km/h, and the speed of a person rocking a boat in still water is 5 km/h. If the person wants to rock the boat to reach the opposite bank

vertically, what should the direction of the boat be? If he wants to reach the opposite shore in the shortest time, what is the direction of the boat? (一小河中流水的速率为 3 km/h,一人在静止水中摇船的速度为 5 km/h. 若此人欲摇船垂直到达对岸,则船行的方向应该如何? 假如他要在最短的时间到达对岸,则船行的方向是什么?)

2. A train decelerates from 45km/h to 30 km/h in 8 s. The deceleration rate of the train is a constant. How long does the train need to move before it comes to a complete stop? (一火车以 45 km/h 的速度在 8 s 中减至 30 km/h,该火车的减速率是一常数,则火车在完全停止前还需要运动多久?)

3. As shown in Fig. 5. 12, the length of rods of the curved ruler is $OA=AB=200$ mm, $CD=DE=AC=AE=50$ mm. If OA rotates around the O axis with an angular velocity $\omega=\frac{\pi}{5}$ rad/s, and when the movement starts, the rod OA moves horizontally to the right, find the motion equation and trajectory of point D on the ruler. (如图 5. 12 所示,曲线规尺的各杆长为 $OA=AB=200$ mm, $CD=DE=AC=AE=50$ mm. 若 OA 以角速度 $\omega=\frac{\pi}{5}$ rad/s 绕 O 轴转动,并且当运动开始时,杆 OA 水平向右,求尺上点 D 的运动方程和轨迹.)

4. As shown in Fig. 5. 13, the rod AB has a length l and rotates around point B at constant angular velocity ω, and its rotation equation is $\varphi=\omega t$. And the slider B connected with the bar makes harmonic vibration along the horizontal direction according to the law $s=a+b\sin\omega t$, in which a and b are constants. Find the locus of point A. (如图 5. 13 所示,杆 AB 长 l,以等角速度 ω 绕点 B 转动,其转动方程为 $\varphi=\omega t$. 而与杆连接的滑块 B 按 $s=a+b\sin\omega t$ 的规律沿水平方向做谐振运动,其中 a 和 b 均为常数. 求点 A 的轨迹.)

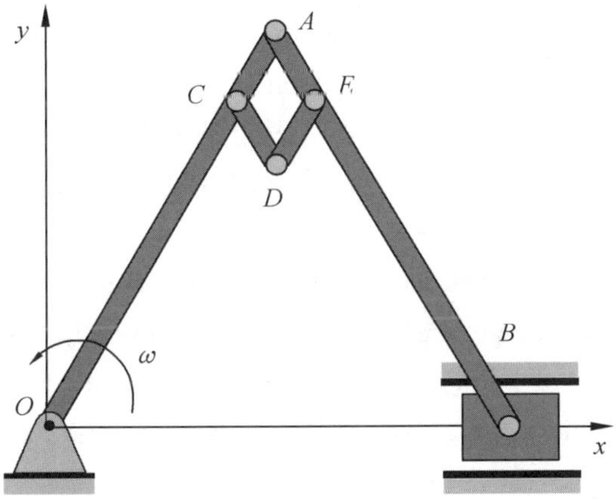

Fig. 5. 12 Diagram of Exercise 3
(图 5. 12 习题 3 图)

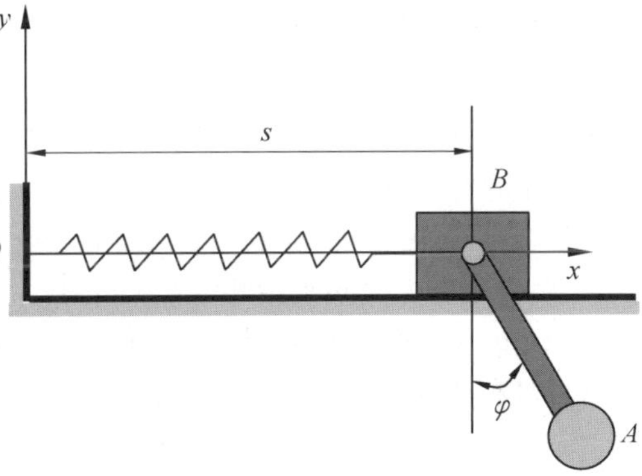

Fig. 5. 13 Diagram of Exercise 4
(图 5. 13 习题 4 图)

5. As shown in Fig. 5. 14, the radius of the eccentric wheel is R, and it rotates around the O axis, the rotation angle $\varphi=\omega t$ (ω is a constant), the eccentricity $OC=e$, and the cam drives the ejector rod AB to reciprocate along the vertical line. Find the equation of motion and velocity of the jack. (如图 5. 14 所示,偏心轮半径为 R,绕 O 轴转动,转角 $\varphi=\omega t$ (ω 为常量),偏心距 $OC=e$,凸轮带动顶杆 AB 沿铅垂线做往复运动. 求顶杆的运动方程和速度.)

6. As shown in Fig. 5. 15, the two rods OA and O_1B rotate around the O and O_1 axes respectively, and the cross slider D is used to connect the two rods. During the movement, the two bars remain intersected at right angles. $OO_1 = a$, $\varphi = kt$, where k is a constant. Find the speed of slider D and the speed relative to OA. (如图 5.15 所示,OA 和 O_1B 两杆分别绕 O 和 O_1 轴转动,用十字形滑块 D 将两杆连接. 在运动过程中,两杆保持相交成直角. $OO_1 = a$, $\varphi = kt$,其中 k 为常数. 求滑块 D 的速度和相对于 OA 的速度.)

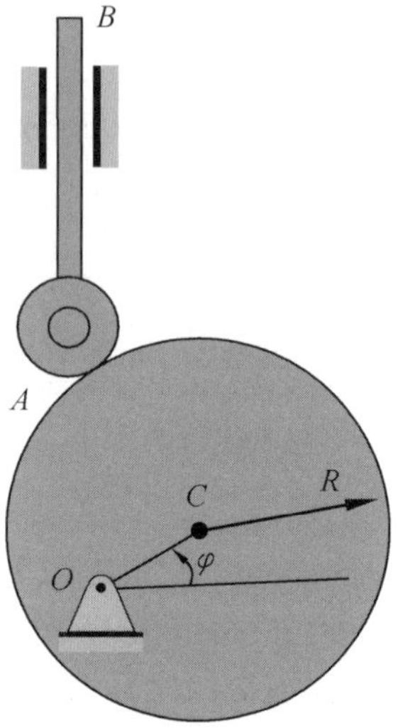

Fig. 5. 14 Diagram of Exercise 5
(图 5.14 习题 5 图)

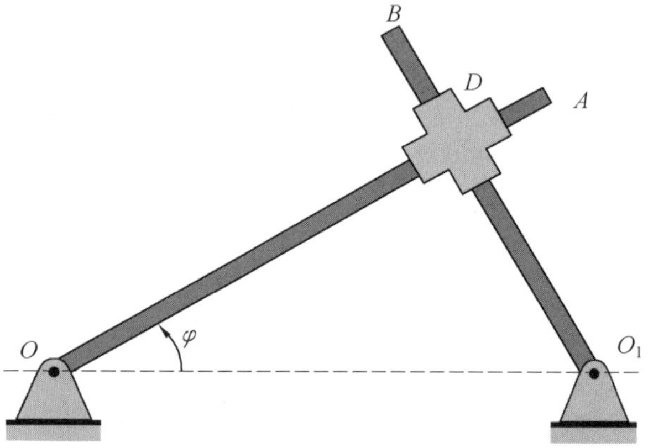

Fig. 5. 15 Diagram of Exercise 6
(图 5.15 习题 6 图)

7. As shown in Fig. 5. 16, a rod rotates around its fixed end O at a uniform angular velocity ω_0, and a slider slides along the rod at a constant speed v_0. Assume that the rod is at the horizontal position and the slider is at point O when the movement starts. Find the trajectory of the slider (in polar coordinates). (如图 5.16 所示,一杆以匀角速度 ω_0 绕其固定端 O 转动,沿此杆有一滑块以匀速 v_0 滑动. 设运动开始时,杆在水平位置,滑块在点 O. 求滑块的轨迹(以极坐标表示).)

8. As shown in Fig. 5. 17, the agitator moves up and down periodically along the z-axis, $z = z_0 \sin 2\pi ft$, and rotates around the z-axis with a rotation angle of $\varphi = \omega t$. Assuming the radius of the stirring wheel is r, find the maximum acceleration of point A on the rim of the wheel. (如图 5.17 所示,搅拌器沿 z 轴周期性上下运动, $z = z_0 \sin 2\pi ft$,并绕 z 轴转动,转角 $\varphi = \omega t$. 设搅拌轮半径为 r,求轮缘上点 A 的最大加速度.)

9. The arc with radius R is tangent to the wall AB, and there is a light source at the center O of the circle. The point M moves along the arc at the same speed v_0 from the tangent point C, as shown in Fig. 5. 18. Find the velocity and acceleration of the shadow M' of point M on the wall. (半径为 R 的圆弧与 AB 墙相切,在圆心 O 处有一光源,点 M 从切点 C 处开始以等速度 v_0 沿圆弧运动,如图 5.18 所示. 求 M 点在墙上影子 M' 的速度大小和加速度大小.)

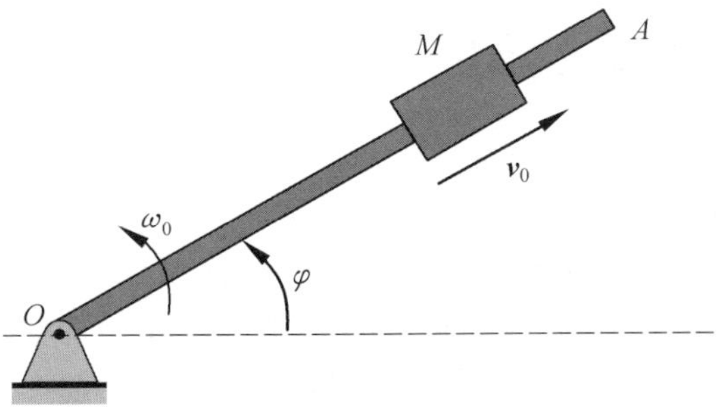

Fig. 5.16　Diagram of Exercise 7
(图 5.16　习题 7 图)

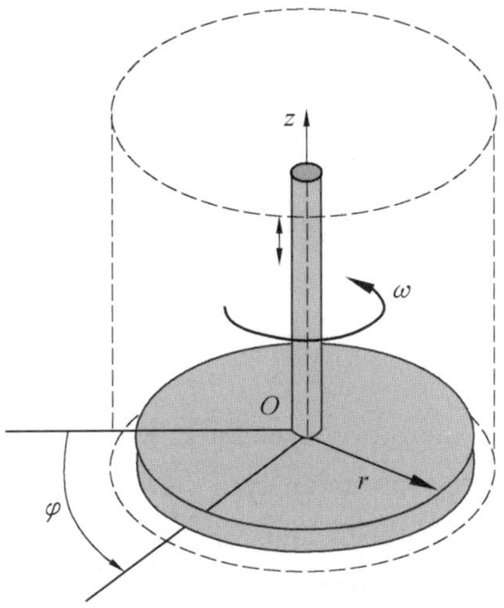

Fig. 5.17　Diagram of Exercise 8
(图 5.17　习题 8 图)

10. As shown in Fig. 5.19, two particles A and B move along two concentric circles with radii r and R respectively. The ratio of the velocity of these two particles is inversely proportional to the radius of the circle where they are located. Prove that: if the relative velocity of the particle is the same as that of AB When the lines are parallel, $\angle AOB = \arccos\dfrac{2rR}{r^2+R^2}$. (如图 5.19 所示,两个质点 A 和 B 分别沿着两个半径分别为 r 和 R 的同心圆做圆周运动,这两个质点的速率之比与所在圆周半径成反比. 求证:若质点的相对速度与 AB 连线平行,则有 $\angle AOB = \arccos\dfrac{2rR}{r^2+R^2}$.)

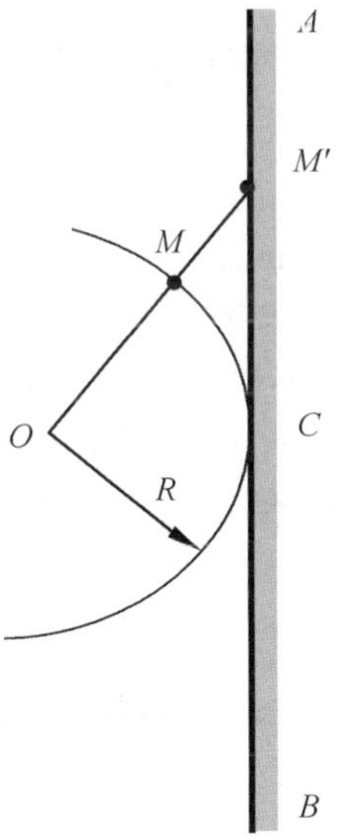

Fig. 5.18　Diagram of Exercise 9
(图 5.18　习题 9 图)

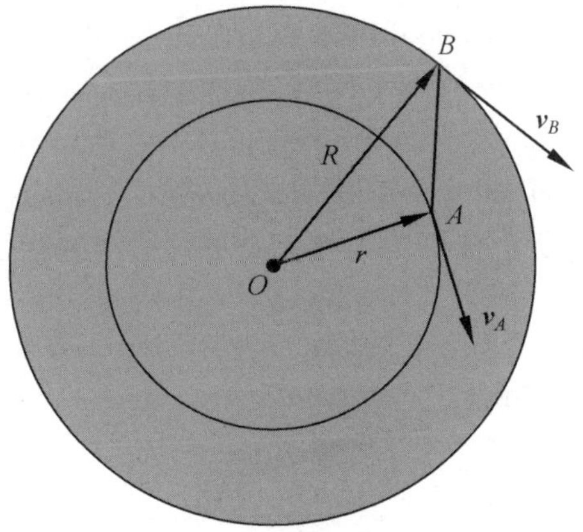

Fig. 5.19　Diagram of Exercise 10
(图 5.19　习题 10 图)

Chapter 6 Simple Motion of Rigid Bodies (刚体的简单运动)

In the previous chapter, we studied the motion of points, but in engineering practice, it is often necessary to analyze the motion of rigid bodies, such as the rotation of gears, the motion of locomotive wheels and carriages, and the motion of vibrating screens, etc. They are all motion of rigid bodies and cannot be abstracted as the motion of points. To study the motion of a rigid body, it is necessary to study not only the motion of the entire rigid body, but also the geometric properties and motion laws of each point on the rigid body. (上一章研究了点的运动,但在工程实际中常常需要分析刚体的运动,如齿轮的转动、机车车轮及车厢的运动、振动筛筛子的运动等. 它们都是刚体的运动,一般不能抽象为点的运动. 研究刚体的运动,既要研究整体刚体的运动,又要研究刚体上各点的几何性质及运动规律.)

The forms of rigid body motion are diverse, and this chapter studies two basic movements of rigid bodies: parallel movement and fixed axis rotation. These two types of motion are also the two simplest types of motion for rigid bodies, which not only widely exist in engineering practice, but also serve as the basis for studying complex motion of rigid bodies. (刚体运动的形式是多样的,本章研究刚体的两种基本运动:平行移动和定轴转动. 这两种运动也是刚体两种最简单的运动,不仅在工程实际中广泛存在,而且也是研究刚体复杂运动的基础.)

6.1 Parallel Movement of Rigid Bodies(刚体的平行移动)

Translational motion is a kind of rigid body motion. When a rigid body is in translational motion, any straight line on the rigid body is always parallel to its initial position. (平移运动是刚体运动中的一种,刚体平移时,刚体上任何一条直线都始终与其初始位置平行.)

Each mechanism shown in Fig. 6.1 has rigid bodies that perform translational motion, such as slider B, push rod C and AB rigid bodies. Translational motion is called translational motion for short. (图 6.1 所示各机构中都有做平移运动的刚体,如滑块 B、推杆 C 及 AB 刚体. 平移运动简称平动.)

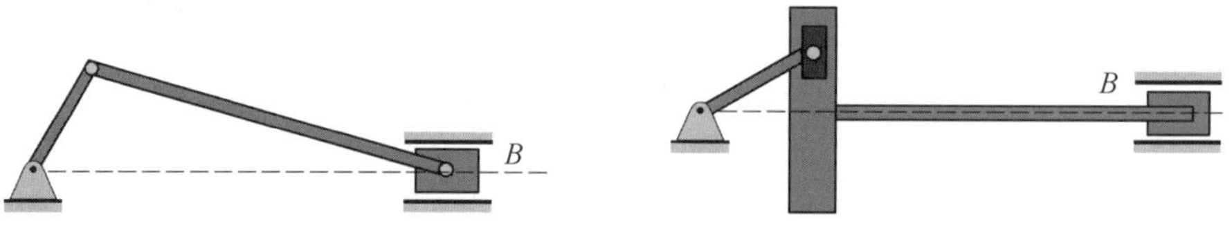

(a) Rectilinear translation motion(直线平动)

Fig. 6.1 Example of a translational rigid body(图 6.1 平动刚体的实例)

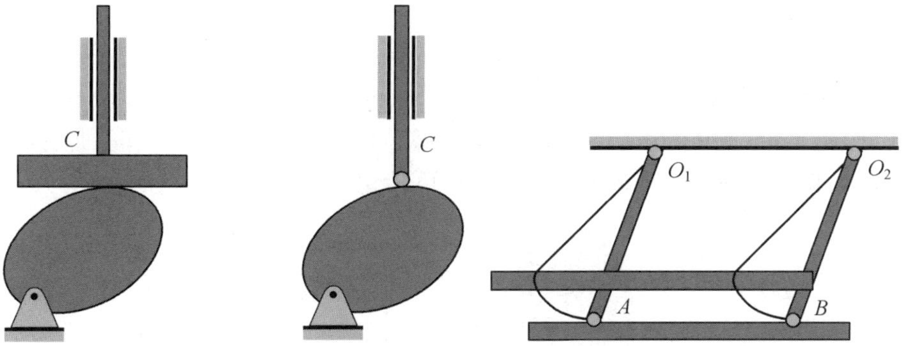

(b) Rectilinear translation motion(**直线平动**)   (c) Curve translation motion(**曲线平动**)

Fig. 6. 1(continued)(图 6. 1(续))

(1) Rectilinear translation. (直线平动.)

If the motion trajectories of each point on a rigid body are straight lines, this motion is called rectilinear translation. (如果刚体上各点的运动轨迹为直线,则称这种运动为直线平动.)

(2) Curve translation. (曲线平动.)

If the trajectory of each point on a rigid body is a curve, then this motion is called curve translation. (如果刚体上各点的运动轨迹为曲线,则称这种运动为曲线平动.)

In translational motion, all points on the rigid body have the same trajectory, and each time has the same speed and acceleration(in size and direction). According to the basic properties of translational motion, the translational motion of a rigid body can be represented by the motion of its previous point. Examples of translational motion include the movement of the steam engine piston, the movement of the slider with the cutter on the cross planer, etc. In these cases, the trajectory of the points on the rigid body is a straight line. In the two wheel connecting rod (Fig. 6.2), the trajectories of these points are circular. Linkage AA_1 translational motion, the wheel rotates. The trajectory of each point may become more complex as it rolls forward. (在平移运动中，刚体上的所有点都有相同的轨迹，每个时刻（在大小和方向上)都有相同的速度和加速度. 根据平移运动的基本性质,刚体的平移运动可以用其上一个点的运动来代表. 平移运动的例子有蒸汽机活塞的运动、带有刀具的滑块在十字刨床上的运动等. 在这些情况下，刚体上点的轨迹是直线. 而在二轮连杆中(图 6. 2),这些点的轨迹是圆形，连杆 AA_1 平移运动，轮子转动. 当其在向前滚动时,各点的轨迹可能更加复杂.)

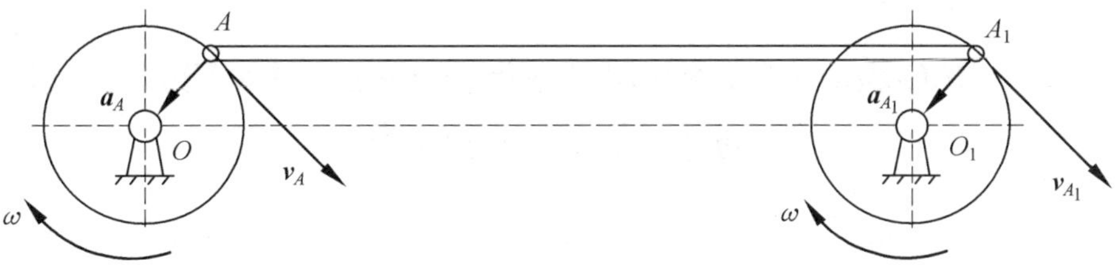

Fig. 6. 2 Linkage curve translation(图 6. 2 连杆曲线平动)

The nature of rigid body translational motion is as follows. (刚体平移运动的性质如下.)

(1) When a rigid body is in translation, the trajectory shape of each point on it is the same.

(当刚体平动时,其上各点的轨迹形状相同.)

(2) At each moment, the velocity and acceleration of each point are the same, as shown in Fig. 6.3. That is $\boldsymbol{v}_B = \boldsymbol{v}_A, \boldsymbol{a}_B = \boldsymbol{a}_A$. The translational motion of a rigid body can be transformed into the motion of any point within the rigid body. (在每一瞬间,各点的速度相同,加速度也相同,如图 6.3 所示,即 $\boldsymbol{v}_B = \boldsymbol{v}_A, \boldsymbol{a}_B = \boldsymbol{a}_A$. 刚体的平动可转化为刚体内任一点的运动.)

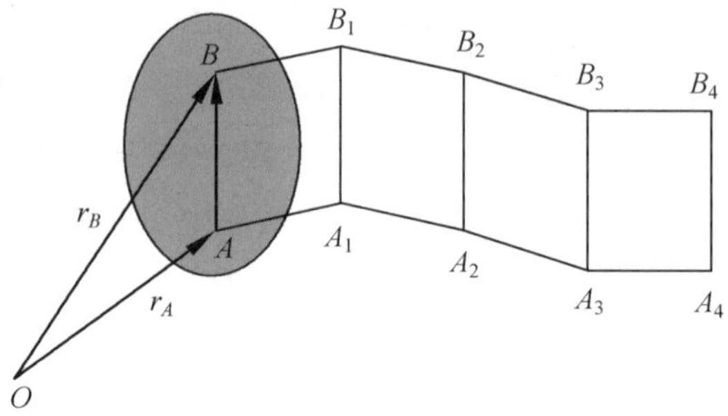

Fig. 6.3 Properties of rigid bodytranslation (图 6.3 刚体平移时的性质)

Example 1: As shown in Fig. 6.4, it is known that $OA = l$, $\varphi = \omega t$. Find the velocity and acceleration of the T-shaped rod. (**例题** 1:如图 6.4 所示,已知 $OA = l, \varphi = \omega t$. 求 T 形杆的速度和加速度.)

Solution: (**解**:)

Considering the T-shaped rod is in translational motion, we establish a graphical coordinate system, and take point M as the research object, we can get (T 形杆平动,建立图示坐标系,取 M 点为研究对象,可以得到)

$$x_M = l\sin\varphi = l\sin\omega t,\quad v_M = \frac{\mathrm{d}x_M}{\mathrm{d}t} = l\omega\cos\omega t,\quad a_M = \frac{\mathrm{d}v_M}{\mathrm{d}t} = -l\omega^2\sin\omega t$$

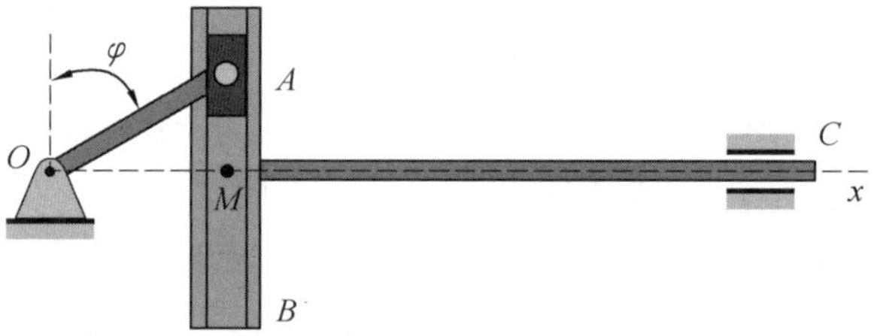

Fig. 6.4 Diagram of Example 1(图 6.4 例题 1 图)

6.2 Rotation of a Rigid Body Around the Fixed Axis (刚体绕定轴的转动)

When a rigid body rotates on the fixed axis, its points move in a circular motion on a plane perpendicular to a fixed straight line. The fixed straight line is called the rigid body's axis of rotation, and the center of the circle is located on that axis of rotation. (刚体定轴转动时,其上各点在垂直于固定直线的平面上做圆周运动,不动的直线称为刚体的转轴,圆周的中心位于该转轴上.)

To achieve this motion, points A and B on the rigid body should be fixed (Fig. 6.5). At this point, the straight line passing through these two points is the axis of rotation. The rotation angle of a rigid body changes over time, we have (为实现这一运动,刚体上的 A、B 两点应固定不动(图 6.5). 此时,通过这两点的直线是转轴. 刚体的转角随时间变化,有)

$$\varphi = f(t)$$

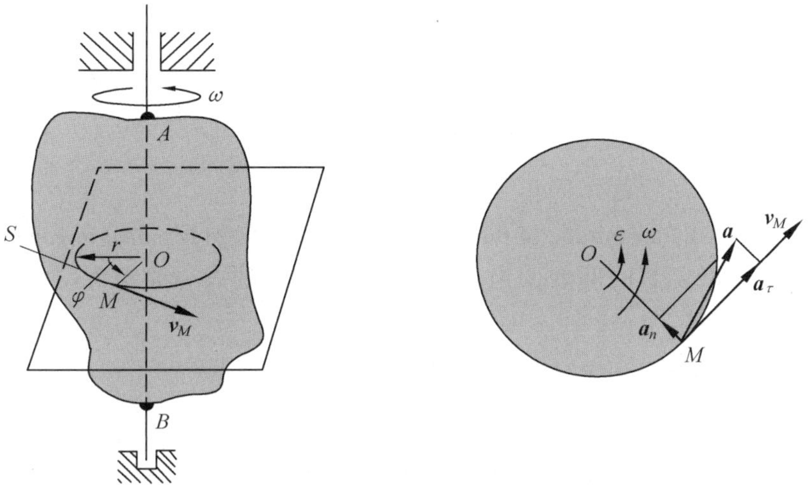

Fig. 6.5 Rigid body rotation around a fixed axis(图 6.5 刚体绕定轴转动)

This equation is called the rotational equation of a rigid body. The turning angle (expressed in radians) is usually represented by the number of revolutions N, we have(该式称为刚体的转动方程. 转角(以弧度表示)通常用转数 N 表示,有)

$$\varphi = 2\pi N$$

The numerical value of the speed at which the rotation angle changes over time is called the angular velocity of the rigid body, which is 1/s and determined by the equation, that is(随着时间的变化,转角变化快慢的数值大小称为刚体的角速度,其值为 1/s, 其数值由公式确定,即)

$$\omega = \frac{\mathrm{d}\varphi}{\mathrm{d}t} \tag{6.1}$$

Considering the arc length $S = r \times \varphi$, therefore $\varphi = S/r$, we have(考虑到弧长 $S = r \times \varphi$, 因此 $\varphi = S/r$, 得到)

$$\omega = \frac{\mathrm{d}S}{r\mathrm{d}t} = \frac{v_M}{r} \tag{6.2}$$

From here, we can obtain the linear velocity of a point on a rotating rigid body $v_M = \omega \times r$. The relationship between angular velocity ω and speed n($\mathrm{r \cdot min^{-1}}$) is (可以得到转动刚体上点的线速度 $v_M = \omega \times r$. 角速度 ω 与转速 n($\mathrm{r \cdot min^{-1}}$)之间的关系为)

$$\omega = \frac{2\pi n}{60} = \frac{\pi n}{30} \tag{6.3}$$

In this case, the linear velocity of the point on the rigid body can also be expressed as rotational speed, that is(在这种情况下,刚体上点的线速度也可以用转速表示,即)

$$v = \frac{\pi n}{30}\frac{d}{2} \tag{6.4}$$

The unit of speed will depend on the unit of diameter d. If d is measured in mm, then v will be expressed in m/s, that is(速度的单位将取决于直径 d 的单位. 如果 d 以 mm 为单位, 那么 v 将以 m/s 表示,即)

$$v = \frac{\pi d n}{60 \cdot 1\ 000} \tag{6.5}$$

In engineering, speed is usually expressed in m/min, we have(在工程学中,速度通常以 m/min 表示,即)

$$v = \frac{\pi d n}{1\ 000} \tag{6.6}$$

As time goes by, the magnitude of the angular velocity change is called angular acceleration, which is s^{-2}, we have(随着时间的变化, 角速度变化快慢的大小称为角加速度, 其单位为 s^{-2},有)

$$\varepsilon = \frac{d\omega}{dt} = \frac{d^2\varphi}{dt^2} \tag{6.7}$$

If $\varepsilon = d\omega/dt>0$ and $\omega = d\varphi/dt>0$, it indicates accelerated motion; if $d\omega/dt<0$, $d\varphi/dt>0$, it indicates decelerated motion. (如果 $\varepsilon = d\omega/dt>0$ 及 $\omega = d\varphi/dt>0$, 则做加速运动;如果 $d\omega/dt<0$, $\omega = d\varphi/dt>0$, 则做减速运动.)

6.3 The Velocity and Acceleration of Each Point Within the Rotating Rigid Body(转动刚体内各点的速度和加速度)

When the rigid body rotates on the fixed axis, any point within the rigid body undergoes circular motion. Study the motion of each point using the natural coordinate method. As shown in Fig. 6. 6, the circular radius of point M rotation is R, and M_O is selected as the arc coordinate of the circular point. The positive direction of the rotation angle φ is specified as the positive direction of the arc coordinate S. So, the arc length that point M passes through is(当刚体定轴转动时,刚体内任一点都做圆周运动. 采用自然坐标法研究各点的运动. 如图 6.6 所示,点 M 转动的圆周半径为 R,选 M_O 为弧坐标的圆点,按转角 φ 的正向规定为弧坐标 S 的正向. 于是,点 M 走过的弧长为)

$$S = R\varphi$$

The linear velocity v is(线速度 v 为)

$$v = \frac{dS}{dt} = R\frac{d\varphi}{dt} = R\omega \tag{6.8}$$

That is, the speed of rotating any point within the rigid body is equal to the product of the angular velocity of the rigid body and the vertical distance from the point to the axis, and its direction is the direction of rotation along the tangent of the circumference. The larger the turning radius, the greater the linear velocity, as shown in Fig. 6. 6. (即转动刚体内任一点的速度的大小等于刚体的角速度与该点到轴线的垂直距离的乘积,它的方向沿圆周的切线向转动的一方. 转动半径

越大,线速度越大,如图 6.6 所示.)

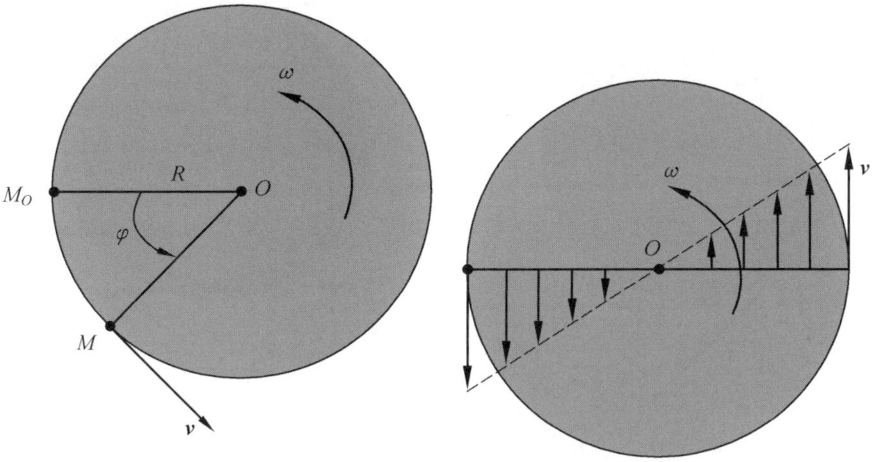

Fig. 6.6 Relationship between linear velocity and angular velocity on a rotating rigid body
(图 6.6 转动刚体上的线速度与角速度关系)

The point M on the rigid body also participates in circular motion with a radius of $OM=R$ (Fig. 6.6). Due to its trajectory being a curve, the acceleration can be expressed as(刚体上 M 点也参与沿半径 $OM=R$ 的圆周运动(图 6.6). 由于它的轨迹是曲线, 因此加速度可以表示为)

$$\boldsymbol{a} = \boldsymbol{a}_\tau \boldsymbol{\tau} + \boldsymbol{a}_n \boldsymbol{n} \tag{6.9}$$

Tangential component of acceleration is(加速度的切向分量为)

$$a_\tau = \mathrm{d}v/\mathrm{d}t = \mathrm{d}\omega \times R/\mathrm{d}t = R\mathrm{d}\omega/\mathrm{d}t = R\varepsilon \tag{6.10}$$

The direction of ε determines the direction of acceleration $\boldsymbol{a}_\tau$ (Fig. 6.7(a)). (ε 方向决定了加速度 $\boldsymbol{a}_\tau$ 的方向(图 6.7(a)).)

The normal acceleration is(法向加速度为)

$$a_n = v^2/R = (\omega R)^2/R = \omega^2 R \tag{6.11}$$

The direction of this acceleration always points to the center of curvature, so it is called centripetal acceleration.(这种加速度的方向一直指向曲率中心, 因此称为向心加速度.)

The total acceleration of a point when a rigid body rotates around a fixed axis is(刚体绕定轴旋转时点的全加速度为)

$$\boldsymbol{a} = \boldsymbol{a}_\tau + \boldsymbol{a}_n$$

The magnitude of the total acceleration a is(全加速度 a 的大小为)

$$a = \sqrt{a_\tau^2 + a_n^2} = R\sqrt{\varepsilon^2 + \omega^4} \tag{6.12}$$

The angle θ between the total acceleration a and the normal line in Fig. 6.7(b) satisfies (图 6.7(b)中全加速度 a 与法线夹角 θ 满足)

$$\tan\theta = \frac{a_\tau}{a_n} = \frac{\varepsilon}{\omega^2} \tag{6.13}$$

Conclusion:(结论:)

(1) At every moment, the velocity and acceleration of all points within the rigid body are

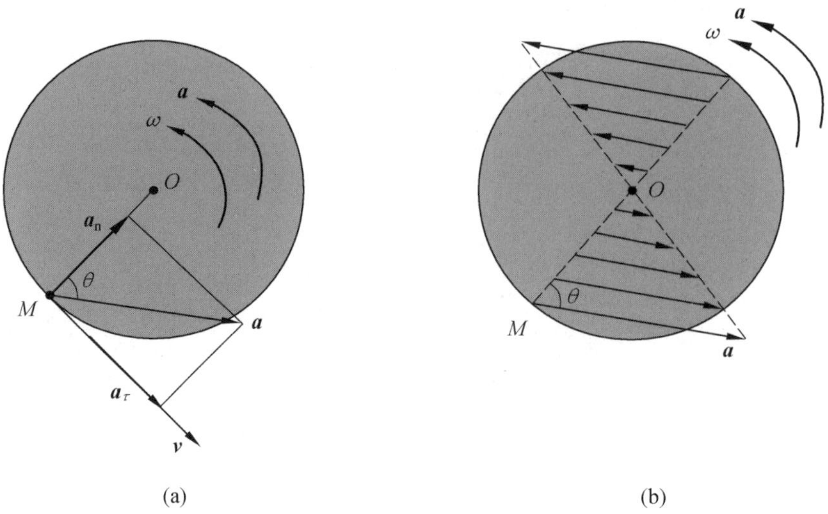

Fig. 6.7 Acceleration of a point on a rotating rigid body(图 6.7 转动刚体上点的加速度)

directly proportional to the vertical distance from these points to the axis. (每一瞬时,刚体内所有点的速度和加速度的大小分别与这些点到轴线的垂直距离成正比.)

(2) At every moment, the angle between the acceleration and radius of all points within the rigid body is the same. ((2)每一瞬时,刚体内所有点的加速度与半径间的夹角都相同.)

6.4 Special Cases of Fixed Axis Rotation of Rigid Bodies (刚体定轴转动的特殊情况)

(1) $\omega=$const.

Already known $\omega=|\mathrm{d}\varphi/\mathrm{d}t|=$const, rewriting the relationship and apply the equation at the corresponding initial time t_0(respectively φ_0) and any time t, we have(已知 $\omega=|\mathrm{d}\varphi/\mathrm{d}t|=$const, 重写该关系式并将该方程在对应于初始时间 t_0(分别为 φ_0)和任意时间 t 的时间内积分,有)

$$\int_{\varphi_0}^{\varphi}\mathrm{d}\varphi=\omega\int_{t_0}^{t}\mathrm{d}t$$

$$\varphi=\varphi_0+\omega t$$

$$\omega=\frac{2\pi n}{60}=\frac{\pi n}{30}$$

Then $\varphi=\varphi_0+\omega t$. This result follows the law of uniform rotation of the rigid body. (则有 $\varphi=\varphi_0+\omega t$. 该结果遵循刚体的匀速转动定律.)

(2) $\varepsilon=$const.

It is the constant speed rotation of the rigid body (constant acceleration or constant deceleration). The laws of motion are similar, we have(其是刚体的匀变速转动(匀加速或匀减速). 其运动定律是相似的,有)

$$\begin{cases}\omega = \omega_0 + \varepsilon t \\ \varphi = \varphi_0 + \omega_0 t + \dfrac{1}{2}\varepsilon t^2 \\ \omega^2 - \omega_0^2 = 2\varepsilon(\varphi - \varphi_0)\end{cases}$$

Example 2: As shown in Fig. 6. 8(a), calculating the velocity and acceleration of point A and point C at the rod end, and drawing their directions. (**例题**2:如图6.8(a)所示,试计算杆端A点和C点的速度、加速度,并画出其方向.)

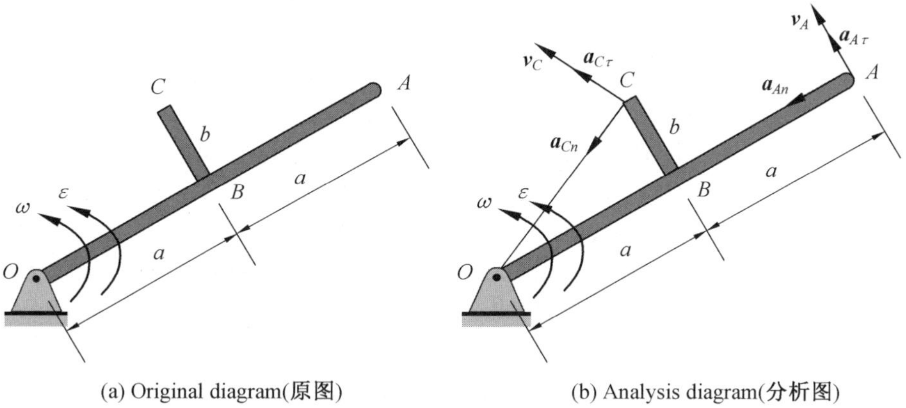

Fig. 6. 8 Diagram of Example 2(图6.8 例题2图)

Solution:(解:)

$$v_A = 2a\omega\ ,\ a_A^\tau = 2a\varepsilon\ ,\ a_A^n = 2a \times \omega^2$$

$$v_C = \sqrt{(a^2 + b^2)} \times \omega\ ,\ a_A^\tau = \sqrt{(a^2 + b^2)} \times \varepsilon\ ,\ a_A^n = \sqrt{(a^2 + b^2)} \times \omega^2$$

The direction is shown in Fig. 6. 8(b). (其方向如图6.8(b)所示.)

Example 3: As shown in Fig. 6. 9, h and $\boldsymbol{v}_0$ are known. Find the rotation equation, angular velocity and angular acceleration of the OA rod. (如图6.9所示,已知h和$\boldsymbol{v}_0$. 求OA杆的转动方程、角速度和角加速度.)

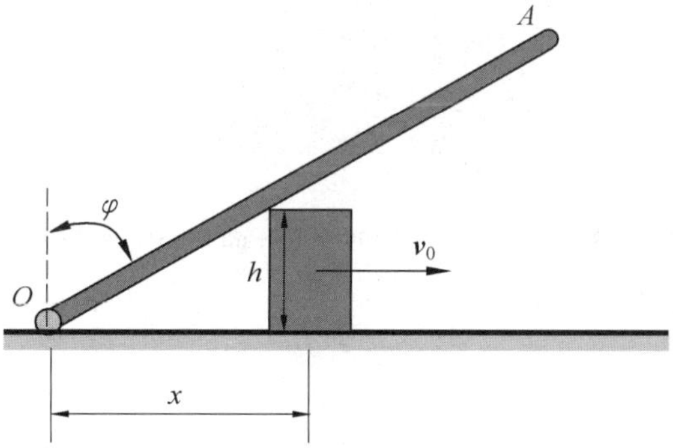

Fig. 6. 9 Diagram of Example 3(图6.9 例题3图)

Solution:(解:)

Establishing a graphical coordinate system, and we get(建立图示坐标系,然后得到)

$$\tan\varphi = \frac{x}{h} = \frac{v_0 t}{h}$$

The angle equation is(转角方程为)

$$\varphi = \arctan\frac{v_0 t}{h}$$

The angular velocity is(角速度为)

$$\omega = \frac{\mathrm{d}\varphi}{\mathrm{d}t} = \frac{hv_0}{h^2 + v_0^2t^2}$$

The angular acceleration is(角加速度为)

$$\varepsilon = \frac{\mathrm{d}\omega}{\mathrm{d}t} = -\frac{2hv_0^3t}{(h^2 + v_0^2t^2)^2}$$

Example 4: As shown in Fig. 6.10, it is known that $O_1A = O_2B = l$, angular velocity ω and angular acceleration ε of the O_1A rod are also known. (**例题**4:如图6.10所示,已知$O_1A = O_2B = l$,又知O_1A杆的角速度ω和角加速度ε.)

Find the motion trajectory, velocity and acceleration of the point C. (求C点的运动轨迹、速度和加速度.)

Solution:(解:)

During the motion of the board, any straight line on it is always parallel to its initial position. Therefore, the board is translated. (板在运动过程中,其上任意直线始终平行于它的初始位置.因此,板平移.)

The motion trajectory of point C has the same shape as that of points A and B, that is, an arc line with point O as the center and l as the radius. (C点的运动轨迹与A、B两点的运动轨迹形状相同,即以O点为圆心,l为半径的圆弧线.)

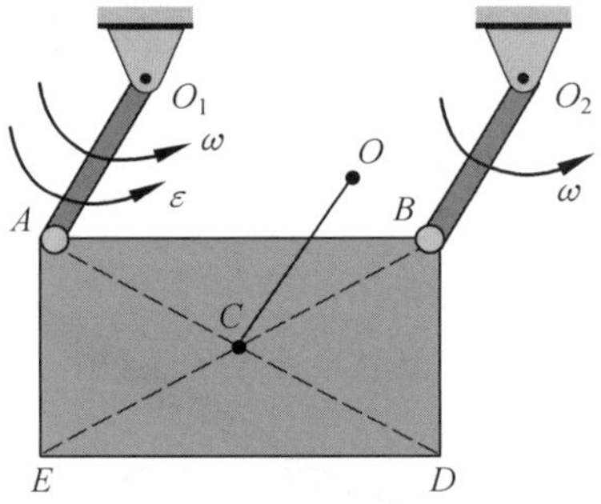

Fig. 6.10 Diagram of Example 4(图6.10 例题4图)

The velocity is(速度为)

$$v_C = v_A = v_B = \omega l$$

The acceleration is(加速度为)

$$\begin{aligned} a_C = a_A &= \sqrt{(a_C^{\tau})^2 + (a_C^n)^2} \\ &= \sqrt{(a_A^{\tau})^2 + (a_A^n)^2} \\ &= \sqrt{(\varepsilon l)^2 + (\omega^2 l)^2} \\ &= l\sqrt{\varepsilon^2 + \omega^4} \end{aligned}$$

6.5 The Transmission Ratio of the Gear Train(轮系的传动比)

Gear trains are commonly used in engineering to increase or decrease the speed of machinery, with the most common being gear trains and belt gear trains. As shown in Fig. 6.11, the two external meshing gears rotate around the O_1 and O_2 axes respectively, the meshing circle radii are R_1 and R_2, the numbers of teeth are z_1 and z_2, and the angular velocities are ω_1 and ω_2. Setting A and B are respectively the contact points of two gear meshing circles, as there is no relative sliding between the two circles, so(工程中常用轮系传动提高或降低机械的转速,最常见的有齿轮系和带轮系. 如图 6.11 所示,两个外啮合齿轮分别绕 O_1 轴和 O_2 轴定轴转动,啮合圆半径分别为 R_1 和 R_2,齿数分别为 z_1 和 z_2,角速度分别为 ω_1 和 ω_2. 设 A 和 B 分别是两个齿轮啮合圆的接触点,因为两圆之间没有相对滑动,故)

$$v_A = v_B$$

$$v_A = R_1\omega_1 = v_B = R_2\omega_2$$

$$\frac{\omega_1}{\omega_2} = \frac{R_2}{R_1}, \frac{\omega_1}{\omega_2} = \frac{R_2}{R_1} = \frac{z_2}{z_1} \tag{6.14}$$

From this, it can be seen that the angular velocity of the two fixed shaft gears in meshing is inversely proportional to the number of teeth of the two gears. (由此可知,处于啮合中的两个定轴齿轮的角速度与两齿轮的齿数成反比.)

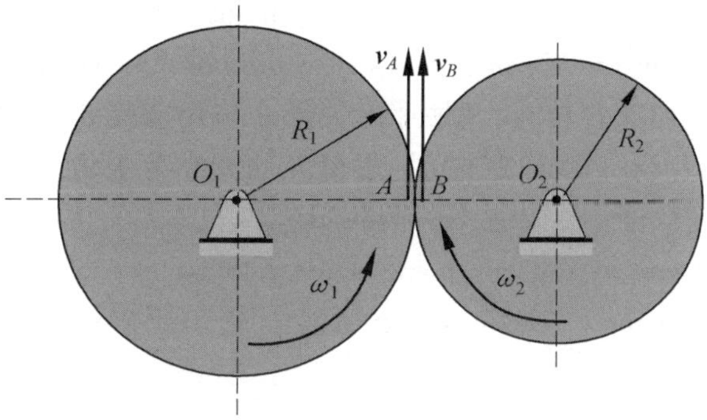

Fig. 6.11 Gear train transmission (图 6.11 轮系传动)

The transmission ratio: Setting wheel 1 as the driving wheel and wheel 2 as the driven wheel. In mechanical engineering, the ratio of the angular velocity of the driving wheel to the driven wheel is often referred to as the transmission ratio, denoted as(传动比:设轮 1 是主动轮,轮 2 是从动轮. 在机械工程中,常常把主动轮与从动轮的角速度之比称为传动比,记为)

$$i_{12} = \frac{\omega_1}{\omega_2} = \frac{R_2}{R_1} = \frac{z_2}{z_1} \tag{6.15}$$

6.6 Vector Representation of Angular Velocity and Angular Acceleration(角速度和角加速度的矢量表示)

The concepts of "angular velocity vector $\boldsymbol{\omega}$" and "angular acceleration vector $\boldsymbol{\varepsilon}$" are introduced. The angular velocity vector $\boldsymbol{\omega}$ is along the axis, and the direction of the curved arrow indicates the turning direction of the rigid body. If you look from the end of the angular velocity vector towards the beginning, you will see the rigid body rotating counterclockwise (Fig. 6.12). The size of a vector $\boldsymbol{\omega}$ is equal to the absolute value of angular velocity, that is(下面介绍"角速度矢量 $\boldsymbol{\omega}$"和"角加速度矢量 $\boldsymbol{\varepsilon}$"的概念. 角速度矢量 $\boldsymbol{\omega}$ 沿轴线,弯曲箭头的指向表示刚体的转向. 如果从角速度矢量的末端向开始端看,则看到刚体做逆时针的转动(图 6.12). 矢量 $\boldsymbol{\omega}$ 的大小等于角速度的绝对值,即)

$$\omega = |\mathrm{d}\varphi/\mathrm{d}t|$$

Notice: The direction is also determined by the right-hand spiral rule, where the four fingers of the right hand represent the rotation of the rigid body around the axis, and the thumb indicates the direction of $\boldsymbol{\omega}$, as shown in Fig. 6.12(a).(注意:指向也由右手螺旋法则确定,即以右手四指表示刚体绕轴的转向,大拇指的指向表示 $\boldsymbol{\omega}$ 的指向,如图 6.12(a)所示.)

When the rigid body rotates with a fixed axis for acceleration, the angular acceleration vector $\boldsymbol{\varepsilon}$ and the angular velocity vector $\boldsymbol{\omega}$ are in the same direction (Fig. 6.12(a)), while when the rigid body rotates with a fixed axis for deceleration, the two directions are opposite (Fig. 6.12(b)). The size of the vector $\boldsymbol{\varepsilon}$ is equal to the absolute value of angular acceleration, that is (刚体定轴转动做加速旋转时,角加速度矢量 $\boldsymbol{\varepsilon}$ 与角速度 $\boldsymbol{\omega}$ 矢量同向(图 6.12(a)),而减速旋转时二者方向相反(图 6.12(b)). 矢量 $\boldsymbol{\varepsilon}$ 的大小等于角加速度的绝对值,即

$$\varepsilon = |\mathrm{d}\omega/\mathrm{d}t|$$

The vectors $\boldsymbol{\omega}$ and $\boldsymbol{\varepsilon}$ can act on any point of the rotation axis of a rigid body, hence these vectors are called sliding vectors.(矢量 $\boldsymbol{\omega}$ 和 $\boldsymbol{\varepsilon}$ 可以作用于刚体转轴上的任意一点,因此称这些矢量为滑动矢量.)

The starting point of the angular velocity vector $\boldsymbol{\omega}$ on the axis can be any position. If the rotation axis z is taken as the positive direction of the axis, it is represented by the direction of the unit vector $\mathbf{k}$ (Fig. 6.12), so the angular velocity vector rotating around the fixed axis can be written as(角速度矢 $\boldsymbol{\omega}$ 在轴上的起点可以是任意位置,如取转轴为 z 轴的正向,用单位矢量 $\mathbf{k}$ 的方向表示(图 6.12),于是绕定轴转动的角速度矢可写为)

$$\boldsymbol{\omega} = \omega\mathbf{k} \tag{6.16}$$

where ω is an algebraic quantity, $\omega = \mathrm{d}\varphi/\mathrm{d}t$.(式中, ω 是代数量, $\omega = \mathrm{d}\varphi/\mathrm{d}t$.)

Similarly, the angular velocity vector $\boldsymbol{\varepsilon}$ can also be represented by a sliding vector along the axis, that is(同理,角速度矢 $\boldsymbol{\varepsilon}$ 也可以用一个沿轴线的滑动矢量表示,即)

$$\boldsymbol{\varepsilon} = \varepsilon\mathbf{k} \tag{6.17}$$

where ε is an algebraic quantity, $\varepsilon = \mathrm{d}\omega/\mathrm{d}t = \mathrm{d}^2\varphi/\mathrm{d}t^2$. So(式中, ε 是代数量, $\varepsilon = \mathrm{d}\omega/\mathrm{d}t = \mathrm{d}^2\varphi/\mathrm{d}t^2$. 于是)

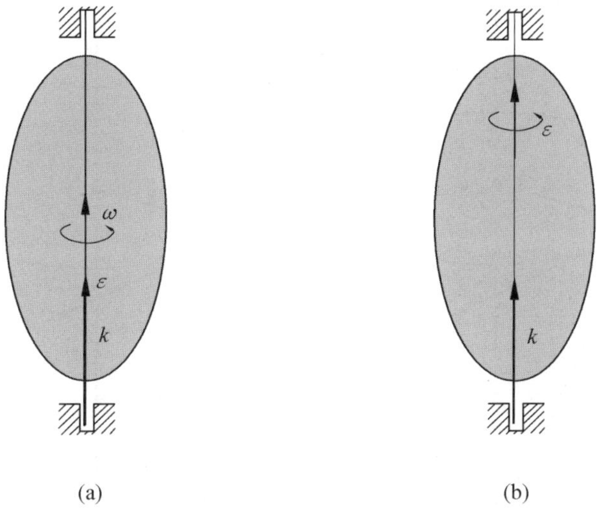

Fig. 6.12 Angular velocity vector and angular acceleration vector
(图 6.12 角速度矢量与角加速度矢量)

$$\boldsymbol{\varepsilon}=\frac{\mathrm{d}\omega}{\mathrm{d}t}\mathbf{k}=\frac{\mathrm{d}(\omega\mathbf{k})}{\mathrm{d}t}=\frac{\mathrm{d}\boldsymbol{\omega}}{\mathrm{d}t} \tag{6.18}$$

That is, the acceleration vector is equal to the first derivative of angular velocity $\boldsymbol{\omega}$ over time. (即加速度矢等于角速度 $\boldsymbol{\omega}$ 对时间的一阶导数.)

The vector product representation of velocity and acceleration is as follows. (速度和加速度的矢积表示如下.)

The vector diameter $\boldsymbol{r}$ of point M is represented by any point A on the rotation axis z, and the angle between the vector $\boldsymbol{\omega}$ and $\boldsymbol{r}$ is represented by θ , as shown in Fig. 6.13. From the geometric relationship shown in the diagram, it can be seen that the velocity of point M is(转轴 z 上任意一点 A 表示点 M 的矢径 $\boldsymbol{r}$,并用 θ 表示矢量 $\boldsymbol{\omega}$ 与 $\boldsymbol{r}$ 之间的夹角,如图 6.13 所示. 由图中的几何关系可知,点 M 的速度大小为)

$$|\boldsymbol{v}|=|\boldsymbol{\omega}|R=|\boldsymbol{\omega}|r\sin\theta=|\boldsymbol{\omega}\times\boldsymbol{r}| \tag{6.19}$$

Its direction is the same as the direction of the vector product $\boldsymbol{\omega}\times\boldsymbol{r}$, so the velocity of point M can be written as(其方向与矢积 $\boldsymbol{\omega}\times\boldsymbol{r}$ 的方向相同,因此点 M 的速度可写为)

$$\boldsymbol{v}=\boldsymbol{\omega}\times\boldsymbol{r} \tag{6.20}$$

That is, the velocity vector of any point on a fixed axis rotating rigid body is equal to the cross product of the angular velocity vector of the rigid body and the vector diameter of that point. (也就是说,定轴转动刚体上任一点的速度矢等于刚体的角速度矢与该点矢径的叉乘.)

The acceleration vector of a point is(点的加速度矢为)

$$\boldsymbol{a}=\frac{\mathrm{d}\boldsymbol{v}}{\mathrm{d}t}$$

Substituting Eq. (6.20) into the above formula, we have(将式(6.20)代入上式,有)

$$\boldsymbol{a}=\frac{\mathrm{d}}{\mathrm{d}t}(\boldsymbol{\omega}\times\boldsymbol{r})=\frac{\mathrm{d}\boldsymbol{\omega}}{\mathrm{d}t}\times\boldsymbol{r}+\boldsymbol{\omega}\times\frac{\mathrm{d}\boldsymbol{r}}{\mathrm{d}t}$$

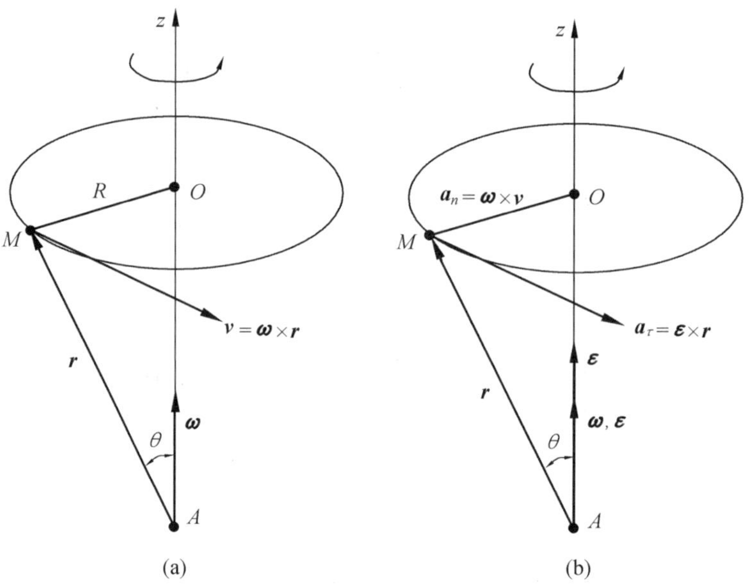

Fig. 6. 13　Vector product representation of velocity and acceleration
(图 6. 13　速度和加速度的矢积表示)

Due to $\frac{d\boldsymbol{\omega}}{dt}=\boldsymbol{\varepsilon}$, $\frac{d\boldsymbol{r}}{dt}=\boldsymbol{v}$,so (由于 $\frac{d\boldsymbol{\omega}}{dt}=\boldsymbol{\varepsilon}$, $\frac{d\boldsymbol{r}}{dt}=\boldsymbol{v}$,因此)

$$\boldsymbol{a}=\boldsymbol{\varepsilon}\times\boldsymbol{r}+\boldsymbol{\omega}\times\boldsymbol{v} \tag{6.21}$$

In the above equation, the sizes of the two items at the right end are(上式中,右端两项的大小分别为)

$$|\boldsymbol{\varepsilon}\times\boldsymbol{r}|=|\boldsymbol{\varepsilon}|\cdot|\boldsymbol{r}|\sin\theta=|\boldsymbol{\varepsilon}|\cdot R$$
$$|\boldsymbol{\omega}\times\boldsymbol{v}|=|\boldsymbol{\omega}|\cdot|\boldsymbol{v}|\sin 90^\circ=R\omega^2$$

Their directions are consistent with tangential acceleration and normal acceleration, therefore (它们的方向分别与切向加速度和法向加速度一致,因此)

$$\boldsymbol{a}_\tau=\boldsymbol{\varepsilon}\times\boldsymbol{r}\ ,\ \boldsymbol{a}_n=\boldsymbol{\omega}\times\boldsymbol{v} \tag{6.22}$$

That is, the tangential acceleration of any point on a fixed axis rotating rigid body is equal to the vector product of the angular acceleration vector of the rigid body and the vector diameter of the point; the normal acceleration vector is equal to the product of the angular velocity vector of the rigid body and the velocity vector at that point. (也就是说,定轴转动刚体上任一点的切向加速度等于刚体的角加速度矢与该点矢径的矢积;法向加速度矢等于刚体的角速度矢与该点速度矢的矢积.)

Example 5: As shown in Fig. 6. 14, a coordinate system is fixed on a rigid body rotating around a fixed axis, the three axes x',y',z' rotate together with the rigid body, the unit vector is $\mathbf{i}'$,$\mathbf{j}'$,$\mathbf{k}'$, the angular velocity of a rigid body is $\boldsymbol{\omega}$, please test the relationship as(**例题 5**: 如图 6. 14所示,在绕定轴转动的刚体上固结一个坐标系,三个轴 x'、y'、z' 随刚体一起转动,单位矢量为 $\mathbf{i}'$、$\mathbf{j}'$、$\mathbf{k}'$,刚体的角速度为 $\boldsymbol{\omega}$,试证关系式,即)

$$\begin{cases}\dfrac{\mathrm{d}\mathbf{i}'}{\mathrm{d}t}=\boldsymbol{\omega}\times\mathbf{i}'\\\dfrac{\mathrm{d}\mathbf{j}'}{\mathrm{d}t}=\boldsymbol{\omega}\times\mathbf{j}'\\\dfrac{\mathrm{d}\mathbf{k}'}{\mathrm{d}t}=\boldsymbol{\omega}\times\mathbf{k}'\end{cases} \tag{6.23}$$

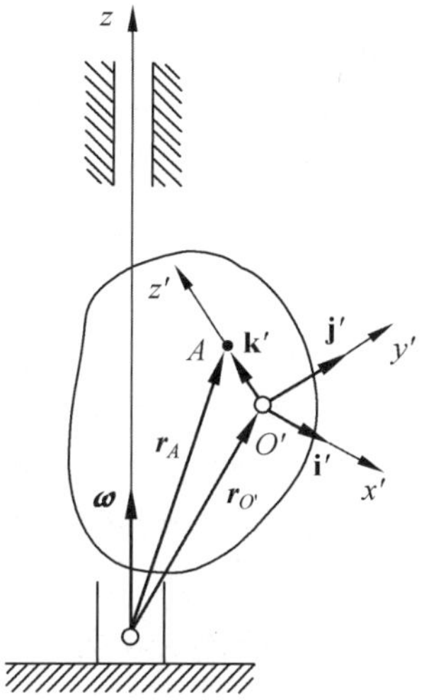

Fig. 6. 14 Diagram of Example 5(图 6. 14 例题 5 图)

Proofing:(证明:)

Analyze $\dfrac{\mathrm{d}\mathbf{k}'}{\mathrm{d}t}$ first. Setting the vector diameter of the vector endpoint A of $\mathbf{k}'$ is $\boldsymbol{r}_A$, the vector diameter of the origin of the motion system O' is $\boldsymbol{r}_{O'}$, both point A and point O' move in a circular motion around the z axis, the speeds are (先分析 $\dfrac{\mathrm{d}\mathbf{k}'}{\mathrm{d}t}$. 设 $\mathbf{k}'$ 的矢端点 A 的矢径为 $\boldsymbol{r}_A$,动系原点 O' 的矢径为 $\boldsymbol{r}_{O'}$,点 A 和点 O' 均绕 z 轴做圆周运动,其速度为)

$$\boldsymbol{v}_A=\frac{\mathrm{d}\boldsymbol{r}_A}{\mathrm{d}t}=\boldsymbol{\omega}\times\boldsymbol{r}_A,\ \boldsymbol{v}_{O'}=\frac{\mathrm{d}\boldsymbol{r}_{O'}}{\mathrm{d}t}=\boldsymbol{\omega}\times\boldsymbol{r}_{O'}$$

As shown in Fig. 6. 14, we have (由图 6. 14 可知)

$$\mathbf{k}'=\boldsymbol{r}_A-\boldsymbol{r}_{O'}$$

Taking the derivative of the above equation, so(对上式求导,得到)

$$\frac{\mathrm{d}\mathbf{k}'}{\mathrm{d}t}=\frac{\mathrm{d}(\boldsymbol{r}_A-\boldsymbol{r}_{O'})}{\mathrm{d}t}=\boldsymbol{\omega}\times\boldsymbol{r}_A-\boldsymbol{\omega}\times\mathbf{k}'$$

The derivative of $\mathbf{i}'$, $\mathbf{j}'$ is similar to the above equation. Combined writing as($\mathbf{i}'$、$\mathbf{j}'$ 的导数与上式相似,合写成)

$$\begin{cases}\dfrac{\mathrm{d}\mathbf{i}'}{\mathrm{d}t}=\boldsymbol{\omega}\times\mathbf{i}'\\\dfrac{\mathrm{d}\mathbf{j}'}{\mathrm{d}t}=\boldsymbol{\omega}\times\mathbf{j}'\\\dfrac{\mathrm{d}\mathbf{k}'}{\mathrm{d}t}=\boldsymbol{\omega}\times\mathbf{k}'\end{cases}$$

Proofing completed. The above equation is commonly referred to as Poisson's formula. (证毕.上式通常称为泊松公式.)

6.7 Questions(思考题)

1."When the rigid body is translating, the track of each point must be a straight line or plane curve; when the rigid body rotates around a fixed axis, the track of each point must be a circle." Is this statement correct? ("刚体平移时,各点的轨迹一定是直线或平面曲线;刚体绕定轴转动时,各点的轨迹一定是圆."这种说法对吗?)

2. Which of the following conditions must be met for rigid body motion to be translational? (满足下述哪些条件的刚体运动一定是平移)?

(1) When a rigid body moves, three points on it that are not in a straight line always move in a straight line;(刚体运动时,其上不在一条直线上的三个点始终做直线运动;)

(2) When a rigid body moves, the distance from all points on it to a fixed plane remains constant;(刚体运动时,其上所有点到某固定平面的距离始终保持不变;)

(3) When a rigid body moves, there are two intersecting lines on it that always remain parallel to their respective initial positions;(刚体运动时,其上有两条相交直线始终与各自初始位置保持平行;)

(4) When a rigid body moves, the velocity and direction of three points on it that are not in a straight line are always the same.(刚体运动时,其上有不在一条直线上的三点的速度大小、方向始终相同.)

3. The rigid body rotates on a fixed axis, and the distance from a certain point A to the axis of rotation is R. Which of the following conditions is sufficient to determine the instantaneous velocity and acceleration of any point on a rigid body? (刚体做定轴转动,其上某点 A 到转轴距离为 R.为求出刚体上任意点在某一瞬时的速度和加速度大小,下述哪组条件是充分的?)

(1) The velocity of point A and the direction of full acceleration at that point are known;(已知点 A 的速度及该点的全加速度方向;)

(2) The tangential and normal accelerations of point A are known;(已知点 A 的切向加速度及法向加速度;)

(3) The tangential velocity of point A and the direction of full acceleration at that point are known;(已知点 A 的切向速度及全速度的方向;)

(4) The normal acceleration and velocity of point A are known;(已知点 A 的法向加速度及该点的速度;)

(5) The normal acceleration and total acceleration direction of point A are known.(已知点 A 的法向加速度及该点全加速度方向.)

4. Try to derive the rotation equations for a rigid body to rotate at uniform speed and acceleration.(试推导刚体做匀速转动和匀加速度转动的转动方程.)

5. The equation of motion for an object to rotate on a fixed axis is $\varphi = 4t - 3t^2$ (φ calculated in rad, t in s.). Try to find the velocity and acceleration of a point within this object with a radius of

rotation of $r=0.5$ m at $t_0=0$ s and $t_1=1$ s, and ask at what instant did the object change direction.(物体做定轴转动的运动方程为 $\varphi=4t-3t^2$(φ 以 rad 计,t 以 s 计).试求此物体内,转动半径 $r=0.5$ m 的一点,在 $t_0=0$ s 和 $t_1=1$ s 时的速度和加速度大小,并求物体在哪一瞬时改变方向.)

6.8 Exercises(习题)

1. As shown in Fig. 6.15, in the planar parallel mechanism, $O_1A=O_2B=150$ mm, $\varphi=\frac{\pi}{6}t^2$ (φ calculated in rad, t in s.), find the velocity and acceleration of point M in plate $ABCD$ when $t=1$ s.(图 6.15 所示的平面平行机构中,$O_1A=O_2B=150$ mm, $\varphi=\frac{\pi}{6}t^2$(φ 以 rad 计,t 以 s 计),求 $t=1$ s 时,板 $ABCD$ 中点 M 的速度和加速度.)

2. The mixer is shown in Fig. 6.16. It is known that $O_1A=O_2B=R$, $O_1O_2=AB$, and the rod O_1A rotates at a constant speed of n (r/min). Analyze the trajectory, velocity and acceleration of the M point on the component BAM.(搅拌机如图 6.16 所示,已知 $O_1A=O_2B=R$,$O_1O_2=AB$,杆 O_1A 以不变转速 n (r/min)转动.试分析构件 BAM 上 M 点的轨迹、速度和加速度.)

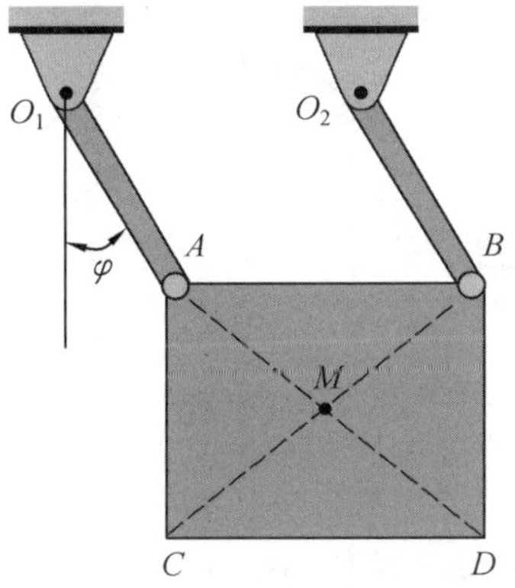

Fig. 6.15 Diagram of Exercise 1
(图 6.15 习题 1 图)

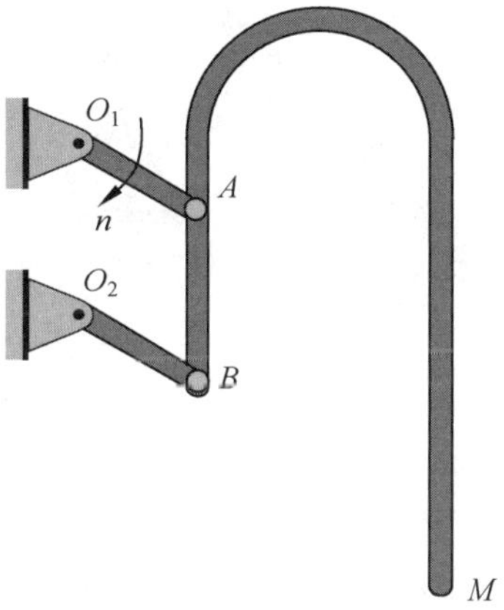

Fig. 6.16 Diagram of Exercise 2
(图 6.16 习题 2 图)

3. During the rotation of a certain flywheel around a fixed axis O, the angle between the total acceleration at any point on the rim and its rotation radius is constant $\alpha=60°$. When the motion begins, its angle of rotation φ_0 is zero, angular velocity is ω_0. Finding the rotation equation of the flywheel and the relationship between its angular velocity and rotation angle.(某飞轮绕固定轴 O 转动的过程中,轮缘上任何一点的全加速度与其转动半径的夹角恒为 $\alpha=60°$. 当运动开始时,其转角 φ_0 为零,角速度为 ω_0,求飞轮的转动方程及其角速度与转角间的关系.)

4. When the gyrocompass is started, the angular acceleration of its rotor increases from zero in direct proportion totime. After 5 min, the angular velocity of the rotor $\omega=600\pi$ rad/s. How many revolutions did the rotor rotate during this period?(当启动陀螺罗盘时,其转子的角加速度从零开始与时间成正比地增大.经过 5 min 后,转子的角速度 $\omega=600\pi$ rad/s. 则转子在这

段时间内转过多少转?)

5. Fig. 6.17 is the mechanism for feeding the workpiece into the drying furnace. The fork rod $OA = 1.5$ m rotates in the vertical plane, with the rod $AB = 0.8$m. The A-end is a hinge, and the B-end has a frame for placing the workpiece. When in motion, the speed of the workpiece is always 0.05 m/s, and the rod AB is always vertical. At the beginning of the motion, the angle $\varphi = 0$. Find the relationship of the angle φ during the motion process with time and the trajectory equation of point B. (图 6.17 为把工件送入干燥炉内的机构,叉杆 $OA = 1.5$ m,在铅垂面内转动,杆 $AB = 0.8$ m,A 端为铰链,B 端有放置工件的框架. 在运动时,工件的速度恒为 0.05 m/s,杆 AB 始终铅垂. 设运动开始时,角 $\varphi = 0$. 求运动过程中角 φ 与时间的关系以及点 B 的轨迹方程.)

6. The mechanism is shown in Fig. 6.18, assuming that the rod AB moves at a uniform velocity v, at the beginning $\varphi = 0$. Seek when $\varphi = \pi/4$, the angular velocity and angular acceleration of rocker OC. (机构如图 6.18 所示,假定杆 AB 以匀速 v 运动,开始时 $\varphi = 0$. 求当 $\varphi = \pi/4$ 时,摇杆 OC 的角速度和角加速度.)

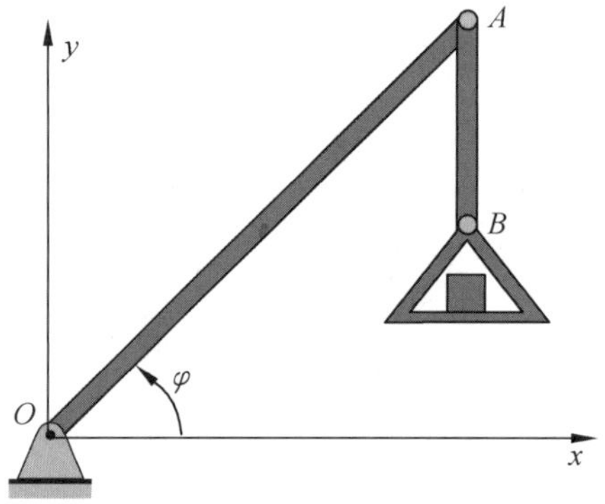

Fig. 6.17 Diagram of Exercise 5
(图 6.17 习题 5 图)

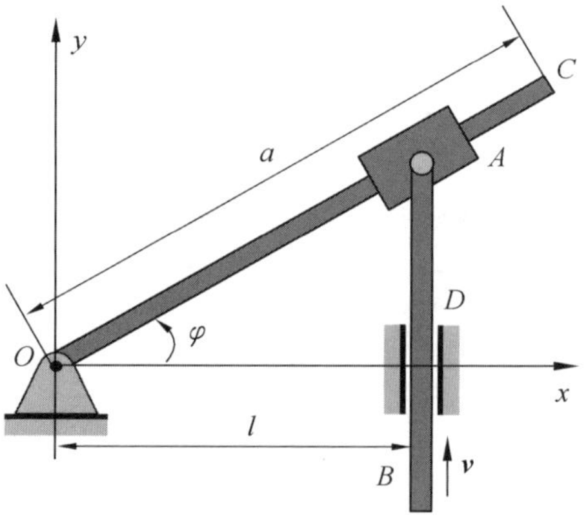

Fig. 6.18 Diagram of Exercise 6
(图 6.18 习题 6 图)

7. The elevator device is driven by a drum with a radius of R, as shown in Fig. 6.19. There is no sliding between the wheel and the rope, and the motion equation of the lifted object is $x = at^2$. Find the magnitude of the total acceleration at point M on the rim of the drum wheel at any instant. (升降机装置由半径为 R 的鼓轮带动,如图 6.19 所示. 轮与绳子之间无滑动,被升降物体的运动方程为 $x = at^2$. 求任意瞬时,鼓轮轮缘上点 M 的全加速度的大小.)

8. A disk with a radius of $R = 100$mm rotates around its center. At the moment of Fig. 6.20, the velocity at point A is $v_A = 200\mathrm{j}$ mm/s, and the tangential acceleration at point B is $a_B^\tau = 150\mathrm{i}$ mm/s^2. Find angular velocity ω and angular acceleration α, and further write the vector expression of the acceleration at point C. (半径 $R = 100$ mm 的圆盘绕其圆心转动,在图 6.20 所示的瞬时,点 A 的速度为 $v_A = 200\mathrm{j}$ mm/s,点 B 的切向加速度 $a_B^\tau = 150\mathrm{i}$ mm/s^2. 求角速度 ω 和角加速度 α,并进一步写出点 C 加速度的矢量表达式.)

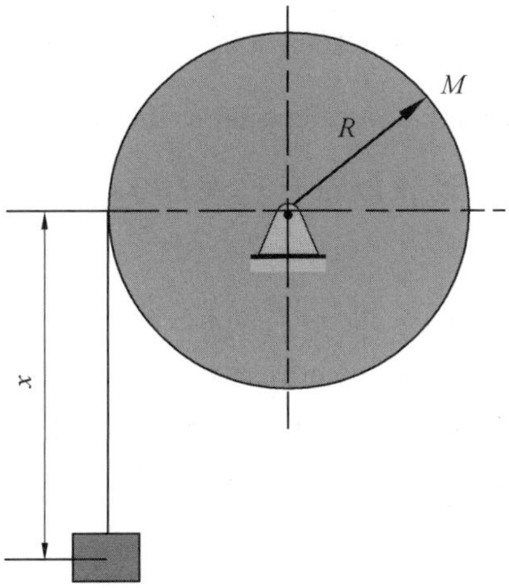

Fig. 6. 19 Diagram of Exercise 7
(图 6. 19 习题 7 图)

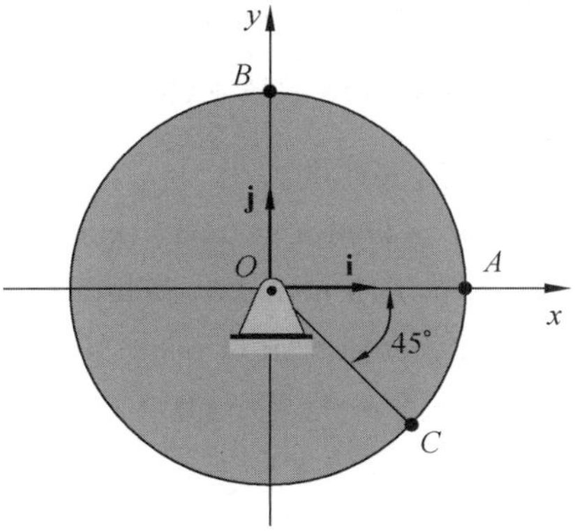

Fig. 6. 20 Diagram of Exercise 8
(图 6. 20 习题 8 图)

9. The instantaneous angular velocity of a rectangular body rotating around a fixed axis AB, instantaneous angular velocity $\omega = 6$ rad/s, angular acceleration $\alpha = 3$ rad/s^2, steering as shown in Fig. 6. 21. Point B is the center of the $CDEF$ on the top surface of the rectangle, $EG = 100$ mm, calculate this instantaneous value: (长方体绕固定轴 AB 转动,某瞬时的角速度 $\omega = 6$ rad/s,角加速度 $\alpha = 3$ rad/s^2,转向如图 6. 21 所示. 点 B 为长方体顶面 $CDEF$ 的中心,$EG = 100$ mm,求此瞬时:)

(1) The vector expression and size of point G velocity; (点 G 速度的矢量表达式及其大小;)

(2) The vector expression and magnitude of the normal acceleration of point G; (点 G 法向加速度的矢量表达式及其大小;)

(3) The vector expression and magnitude of the tangential acceleration at point G; (点 G 切向加速度的矢量表达式及其大小;)

(4) The vector expression of the total acceleration of point G and its magnitude. (点 G 全加速度的矢量表达式及其大小.)

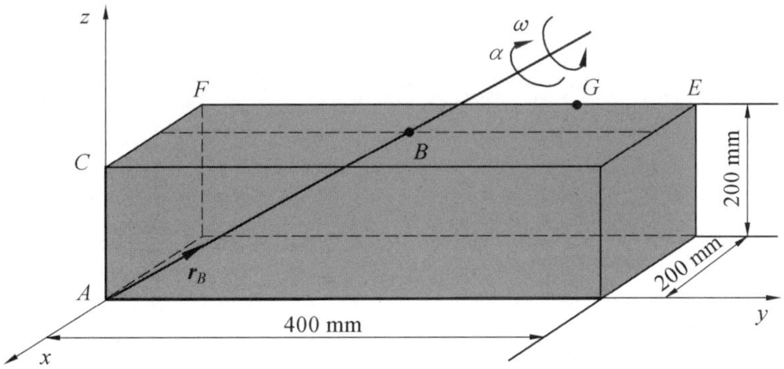

Fig. 6. 21 Diagram of Exercise 9(图 6. 21 习题 9 图)

10. The motion law of a known point is $r = (7t)\mathbf{i} + (3 + t^2)\mathbf{j} + \frac{t^3}{3}\mathbf{k}$ (t calculated in s, r in m). Find the velocity, tangential acceleration and normal acceleration at the time $t=3$ s. (已知点的运动规律为 $r = (7t)\mathbf{i} + (3 + t^2)\mathbf{j} + \frac{t^3}{3}\mathbf{k}$ (t 以 s 计, r 以 m 计). 求 $t=3$ s 时点的速度和切向加速度、法向加速度大小.)

11. Point M is known to travel at a constant speed of $v=20$ m/s, spiraling along a cylinder with a radius of $r=0.5$ m and a pitch of $p=0.2$ m. Please use cylindrical coordinates to represent the velocity and acceleration of point M. (已知点 M 以匀速 $v=20$ m/s 行驶,沿圆柱螺旋运动,圆柱半径为 $r=0.5$ m,螺距 $p=0.2$ m. 试用柱坐标表示点 M 的速度和加速度.)

Chapter 7 Composite Motion of Points
(点的合成运动)

In the previous two chapters, we have studied the motion of a point or rigid body relative to a fixed reference system, which can be called simple motion. The characteristics and laws of the motion of the same object relative to different reference systems are different. In this chapter, we study the relationship between the different kinematic characteristics exhibited by the same moving point relative to two different reference systems, mainly the relationship between velocities and the relationship between accelerations, expressed respectively by the theorem of composition of velocities and the theorem of composition of accelerations. The content of this chapter forms the basis for the study of complex motion of points and rigid bodies.(前两章分析了点或刚体相对于一个定参考系的运动,可称为简单运动.同一物体相对于不同参考系的运动形式及规律各不相同.本章研究同一动点相对于两个不同参考系而表现出的不同运动学特征之间的关系,主要是速度之间的关系和加速度之间的关系,分别用速度合成定理及加速度合成定理来表述.本章内容是研究点和刚体复杂运动的基础.)

7.1 The Concepts of Composite Motion of Points
(点的合成运动概念)

The composite motion of a point is when a point is involved in two or more motions at the same time. For example, a passenger is walking on the deck of a boat that is traveling on a river. If the river passes through the Earth's meridian at a certain angle, what is the trajectory of the passenger and his velocity relative to the Earth's surface? This question can be answered only after learning the concepts of relative motion, transport motion and absolute motion of points.(点的合成运动是指某个点同时参与两个或多个运动.例如,乘客行走在一艘正在河流上行驶的船的甲板上.如果河道以一定角度通过地球的子午线,那么乘客的轨迹及其相对于地球表面的速度是多少?只有在学习了点的相对运动、点的牵连运动和点的绝对运动的概念之后,才能回答这个问题.)

Observe an object moving in space (Fig. 7.1) and point M. Point M which is not part of the object undergoes a certain displacement relative to the object. The axes Ox, Oy, Oz that are solidly connected to the object are plotted by means of any point O on the moving object. The coordinate system is called the moving reference system.(观察一个在空间中运动的物体(图7.1)和点 M,点 M 不属于该物体,但点相对于该物体发生一定的位移.通过运动物体上的任意点 O,绘制与该物体相固连的轴 Ox、Oy、Oz,该坐标系称为动参考系.)

The reference coordinate system O_1x_1，O_1y_1，O_1z_1 which is fixed to the Earth is called the fixed reference system.（把固连于地球上的参考坐标系 O_1x_1、O_1y_1、O_1z_1 称为定参考系.）

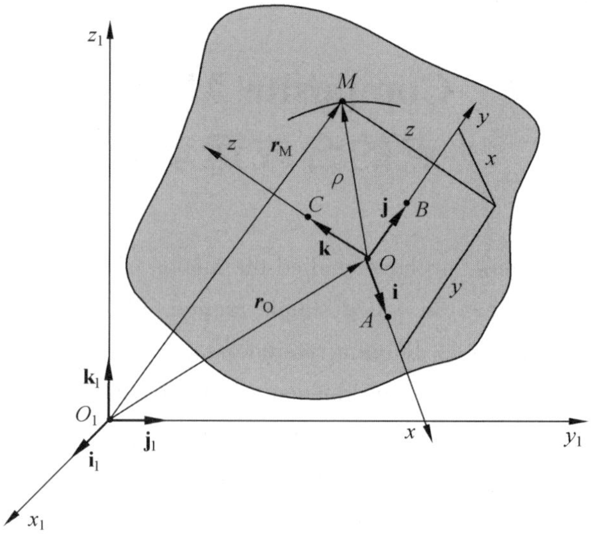

Fig. 7.1　Composite motion of point M（图 7.1　点 M 的合成运动）

The motion of moving point M relative to the fixed reference system is called the absolute motion of the point. In the absolute motion, the velocity and acceleration of the point are called absolute velocity $\boldsymbol{v}$ and absolute acceleration $\boldsymbol{a}$.（动点 M 相对于定参考系的运动称为点的绝对运动. 绝对运动中点的速度和加速度称为绝对速度 $\boldsymbol{v}$ 和绝对加速度 $\boldsymbol{a}$.）

The motion of point M relative to the moving reference system is called the relative motion of the point. In the relative motion, the velocity and acceleration of the point are called relative velocity $\boldsymbol{v}_r$ and relative acceleration $\boldsymbol{a}_r$.（动点 M 相对于动参考系的运动称为点的相对运动，相对运动中点的速度和加速度称为相对速度 $\boldsymbol{v}_r$ 和相对加速度 $\boldsymbol{a}_r$.）

The motion of the moving referencesystem $Oxyz$ and the object fixed to it relative to the fixed reference frame $O_1x_1y_1z_1$ is called transport motion. The velocity and acceleration of the point in the object that coincides with the moving point M along the object at the current moment are called transport velocity $\boldsymbol{v}_e$ and transport acceleration $\boldsymbol{a}_e$.（动参考系 $Oxyz$ 及与其固定相连的物体相对于定参考系 $O_1x_1y_1z_1$ 的运动是牵连运动. 该物体在当前时刻与沿其运动的动点 M 相重合的点的速度和加速度称为牵连速度 $\boldsymbol{v}_e$ 和牵连加速度 $\boldsymbol{a}_e$.）

When studying the complex motion of a point, the "stop motion principle" is preferred. The transport motion needs to be stopped so that a stationary observer can see the relative motion of the point. Only relative motion will be present in this way, and it will be absolute. In contrast, if the relative motion is stopped, the transport motion will become absolute, and the stationary observer will only see the transport motion.（在研究点的复杂运动时，优先使用"止动原则". 需要停止牵连运动，目的是让静止的观察者看到点的相对运动. 这样只会有相对运动，相对运动将是绝对的. 相反，如果停止相对运动，牵连运动将变成绝对的，且静止的观察者将只能看到这种牵连运动.）

In the latter condition, a crucial observation is made when determining the transport motion of a point. The transport motion of the point is contingent upon the moment at which relative motion ceases and the position of the moving point on the moving reference system at that juncture. Given that the motion of all points on a moving object is generally different, it is more appropriate to define the transport motion of the point as the absolute motion of the point on the moving reference system that coincidences with the instantaneous position of the current moving point. (在后一种情况下,当确定点的牵连运动时发现了一个非常重要的情况. 点的牵连运动取决于停止相对运动的时间及此时动点在动参考体上的位置. 一般而言,运动物体上所有点的运动都不同. 因此,更符合逻辑的是将点的牵连运动定义为动系上与当前动点之瞬间位置重合点的绝对运动.)

The fixed coordinate axes x_1, y_1, z_1 and the moving coordinate axes x, y, z are shown in Fig. 7.1. The absolute motion of point M is determined by(定坐标轴 x_1、y_1、z_1 和动坐标轴 x、y、z 如图 7.1 所示. 点 M 的绝对运动为)

$$\begin{cases} x_1 = x_1(t) \\ y_1 = y_1(t) \\ z_1 = z_1(t) \end{cases} \tag{7.1}$$

Relative motion is determined with the moving coordinate axes by (相对运动由动坐标轴下的方程确定,即)

$$\begin{cases} x = x(t) \\ y = y(t) \\ z = z(t) \end{cases} \tag{7.2}$$

The motion of the moving reference system relative to the fixed reference system (transport motion) can be determined by (动系相对于定系的运动(牵连运动)可确定为)

$$\begin{cases} x_{1O} = x_{1O}(t) \\ y_{1O} = y_{1O}(t) \\ z_{1O} = z_{1O}(t) \end{cases} \tag{7.3}$$

The equation for determining the transport motion of a point cannot exist. This is because, by definition, the transport motion of point M is the motion of the moving reference system $Oxyz$ relative to the fixed reference system, and motion point M is coinciding with that point at this moment. However, the motion of points on the moving reference system $Oxyz$ are different. (确定点牵连运动的方程是不可能存在的. 这是因为,根据定义,点 M 的牵连运动是动系 $Oxyz$ 相对于定系的运动,此刻动点 M 与该点重合. 但是动系上 $Oxyz$ 各点的运动都不同.)

The absolute motion of point M is determined by the vector radius $\boldsymbol{r}_M$, while the relative motion is determined by the vector radius $\boldsymbol{\rho}$. The vector radius $\boldsymbol{r}_O$ determines the motion of the moving reference system $Oxyz$ relative to the origin O_1 (but not the transport motion of point M). (点 M 的绝对运动由矢径 $\boldsymbol{r}_M$ 决定, 而相对运动由矢径 $\boldsymbol{\rho}$ 决定. 矢径 $\boldsymbol{r}_O$ 决定动系 $Oxyz$ 相对于原点 O_1 的运动 (但不是点 M 的牵连运动).)

The motion of an object relative to a reference object can be composed of several simple motions relative to other reference objects. This type of motion is called composite motion. The motion of point M in Fig. 7.2(a), the motion of point M on the rim of the wheel in Fig. 7.2(b) and the motion of slider A in Fig. 7.2(c) are all composite motions. The relationships between the moving point, the reference system and the velocity (or acceleration) are shown in Fig. 7.3. (物体相对于某一参考体的运动可由相对于其他参考体的几个简单运动组合而成,这种运动称为合成运动. 图7.2(a)中M点的运动、图7.2(b)中车轮轮缘上M点的运动、图7.2(c)中滑块A的运动都是合成运动. 动点、参考系、速度(或加速度)之间的关系如图7.3所示.)

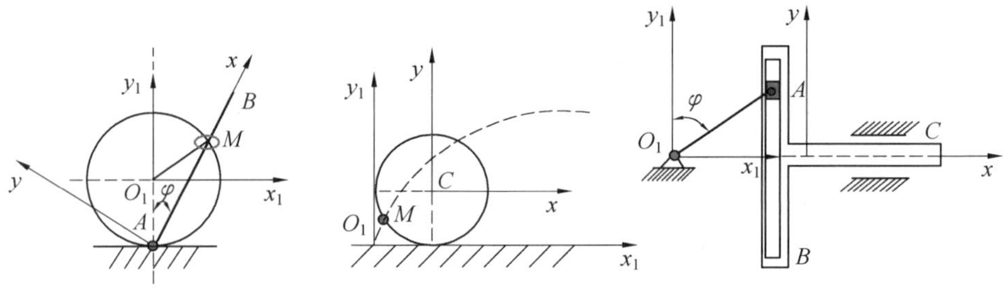

Fig. 7.2 Examples of Composite motion of points(图7.2 点的合成运动实例)

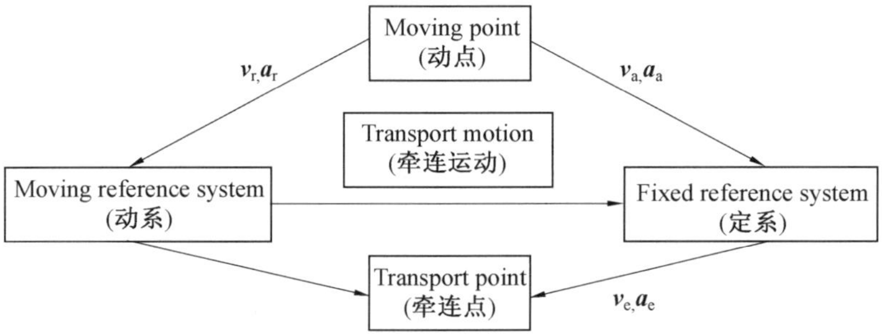

Fig. 7.3 Motion description of moving point relative to the reference system
(图7.3 动点相对于参考系的运动描述)

7.2 The Theorem of Composition of Velocities of Points (点的速度合成定理)

The velocity vector of the particle is(质点的速度矢量为)

$$\boldsymbol{v} = \frac{\mathrm{d}\boldsymbol{r}}{\mathrm{d}t}$$

The relationship between the radius vector $\boldsymbol{r}_M$ of point M with respect to the fixed reference system and the radius vector $\boldsymbol{r}_O$ of the reference origin O in the moving coordinate system is shown as (Fig. 7.1) (点M相对于定系的矢径$\boldsymbol{r}_M$与动坐标系的参考原点O的矢径$\boldsymbol{r}_O$关系为(图7.1))

$$\boldsymbol{r}_M = \boldsymbol{r}_O + \boldsymbol{\rho} \tag{7.4}$$

where $\boldsymbol{\rho}$ represents the radius vector of point M in the moving reference system, it is used to determine the position of a point in its relative motion. (式中, $\boldsymbol{\rho}$ 表示动参考系中点 M 的矢径,用来确定一个点在其相对运动中的位置.)

Calculate the velocity vector of the point M, we have(计算点 M 的速度矢量,有)

$$\boldsymbol{v}_M = \frac{\mathrm{d}}{\mathrm{d}t}(\boldsymbol{r}_M) = \frac{d}{\mathrm{d}t}(\boldsymbol{r}_O + \boldsymbol{\rho}) \tag{7.5}$$

or(或者)

$$\boldsymbol{v}_M = \frac{\mathrm{d}}{\mathrm{d}t}(\boldsymbol{r}_O) + \frac{\mathrm{d}}{\mathrm{d}t}(\boldsymbol{\rho}) \tag{7.6}$$

In the above expression, the first term $\frac{\mathrm{d}}{\mathrm{d}t}(\boldsymbol{r}_O)$ is the velocity $\boldsymbol{v}_O$ of the point O in the moving system relative to the fixed coordinate system. Since the positon of the unit vectors $\mathbf{i}$, $\mathbf{j}$, $\mathbf{k}$ in space together with the object change with time, their time derivatives are not zero. It should also be noted that these unit vectors intersect at point O and do not move relative to point O. Therefore, these unit vectors perform instantaneous rotations around an axis passing through point O. (在上述表达式中,第一项 $\frac{\mathrm{d}}{\mathrm{d}t}(\boldsymbol{r}_O)$ 是动系点 O 相对于定坐标系的速度 $\boldsymbol{v}_O$. 由于单位向量 $\mathbf{i}$, $\mathbf{j}$, $\mathbf{k}$ 在空间中的位置和物体一起随时间发生改变,因此它们的时间导数不等于零. 还应注意的是,这些单位向量相交于 O 点,相对于 O 点不发生移动. 因此,这些单位向量绕通过点 O 的轴做瞬间旋转.)

Calculate the derivative $\frac{\mathrm{d}}{\mathrm{d}t}(\boldsymbol{\rho})$,we have(计算导数 $\frac{\mathrm{d}}{\mathrm{d}t}(\boldsymbol{\rho})$,有)

$$\frac{\mathrm{d}}{\mathrm{d}t}(\boldsymbol{\rho}) = \frac{\mathrm{d}}{\mathrm{d}t}(x\mathbf{i} + y\mathbf{j} + z\mathbf{k}) = \mathbf{i}\frac{\mathrm{d}x}{\mathrm{d}t} + \mathbf{j}\frac{\mathrm{d}y}{\mathrm{d}t} + \mathbf{k}\frac{\mathrm{d}z}{\mathrm{d}t} + x\frac{\mathrm{d}\mathbf{i}}{\mathrm{d}t} + y\frac{\mathrm{d}\mathbf{j}}{\mathrm{d}t} + z\frac{\mathrm{d}\mathbf{k}}{\mathrm{d}t} \tag{7.7}$$

Hence(因此)

$$\boldsymbol{v}_M = \frac{\mathrm{d}}{\mathrm{d}t}(r_O) + \mathbf{i}\frac{\mathrm{d}x}{\mathrm{d}t} + \mathbf{j}\frac{\mathrm{d}y}{\mathrm{d}t} + \mathbf{k}\frac{\mathrm{d}z}{\mathrm{d}t} + x\frac{\mathrm{d}\mathbf{i}}{\mathrm{d}t} + y\frac{\mathrm{d}\mathbf{j}}{\mathrm{d}t} + z\frac{\mathrm{d}\mathbf{k}}{\mathrm{d}t} \tag{7.8}$$

By using the "stop motion rule", Eq. (7.8) can be used to determine the relative velocity the transport velocity of a point. In fact, by stopping the transport motion and the motion of axes x, y, z, we know that $\boldsymbol{r}_O = \text{const}, \mathbf{i} = \text{const}, \mathbf{j} = \text{const}, \mathbf{k} = \text{const}$. From Eq. (7.8), we can obtain that(使用"止动规则", 可以使用式 (7.8) 来确定点的相对速度和点的牵连速度. 事实上,通过停止牵连运动和轴 x、y、z 的运动, 即 $\boldsymbol{r}_O = \text{const}, \mathbf{i} = \text{const}, \mathbf{j} = \text{const}, \mathbf{k} = \text{const}$,由式(7.8)可以得到)

$$\boldsymbol{v}_r = \frac{\mathrm{d}x}{\mathrm{d}t}\mathbf{i} + \frac{\mathrm{d}y}{\mathrm{d}t}\mathbf{j} + \frac{\mathrm{d}z}{\mathrm{d}t}\mathbf{k} = v_x\mathbf{i} + v_y\mathbf{j} + v_z\mathbf{k} \tag{7.9}$$

where $v_x = \frac{\mathrm{d}x}{\mathrm{d}t}, v_y = \frac{\mathrm{d}y}{\mathrm{d}t}, v_z = \frac{\mathrm{d}z}{\mathrm{d}t}$ represent the projection of the vector of the relative velocity $\boldsymbol{v}_\mathrm{r}$ onto the corresponding coordinate axis. (式中, $v_x = \frac{\mathrm{d}x}{\mathrm{d}t}, v_y = \frac{\mathrm{d}y}{\mathrm{d}t}, v_z = \frac{\mathrm{d}z}{\mathrm{d}t}$ 是将相对速度矢量 $\boldsymbol{v}_\mathrm{r}$ 投影到相应的坐标轴上得到的.)

Therefore, substitute Eq. (7.9) into Eq. (7.8), we obtain(因此,将式(7.9)代入式(7.8)中,得到)

$$\boldsymbol{v}_M = \frac{\mathrm{d}}{\mathrm{d}t}(\boldsymbol{r}_O) + x\frac{\mathrm{d}\mathbf{i}}{\mathrm{d}t} + y\frac{\mathrm{d}\mathbf{j}}{\mathrm{d}t} + z\frac{\mathrm{d}\mathbf{k}}{\mathrm{d}t} + \boldsymbol{v}_\mathrm{r} \tag{7.10}$$

By stopping the relative motion of point M (x = const, y = const, z = const), its transport velocity can be obtained as(通过停止点 M 的相对运动(x = const, y = const, z = const), 可得到其牵连速度为)

$$\boldsymbol{v}_\mathrm{e} = \frac{\mathrm{d}}{\mathrm{d}t}(\boldsymbol{r}_O) + x\frac{\mathrm{d}\mathbf{i}}{\mathrm{d}t} + y\frac{\mathrm{d}\mathbf{j}}{\mathrm{d}t} + z\frac{\mathrm{d}\mathbf{k}}{\mathrm{d}t} = \boldsymbol{v}_O + x\frac{\mathrm{d}\mathbf{i}}{\mathrm{d}t} + y\frac{\mathrm{d}\mathbf{j}}{\mathrm{d}t} + z\frac{\mathrm{d}\mathbf{k}}{\mathrm{d}t} \tag{7.11}$$

Therefore, substitute Eq. (7.11) into Eq. (7.10), we obtain (因此,将式(7.11)代入式(7.10)中,得到)

$$\boldsymbol{v}_M = \boldsymbol{v}_\mathrm{a} = \boldsymbol{v}_\mathrm{e} + \boldsymbol{v}_\mathrm{r} \tag{7.12}$$

The proof of equations (from Eq. (7.10) to Eq. (7.12)) can also be proven using another approach. (式(7.10) ~ (7.12)的证明也可采用另外一种思路.)

Consider what the derivative, such as $\frac{\mathrm{d}\mathbf{k}}{\mathrm{d}t}$, represents. In summary, if the unit vector $\mathbf{k}$ can only perform instantaneous rotations around point O, then there will be an instantaneous rotation axis of $O\Omega$ (Fig. 7.4). (思考一下导数表示什么,如 $\frac{\mathrm{d}\mathbf{k}}{\mathrm{d}t}$. 综上所述,如果单位向量 $\mathbf{k}$ 只能绕点 O 进行瞬时转动,那么就会存在一个瞬时转动轴 $O\Omega$(图 7.4).)

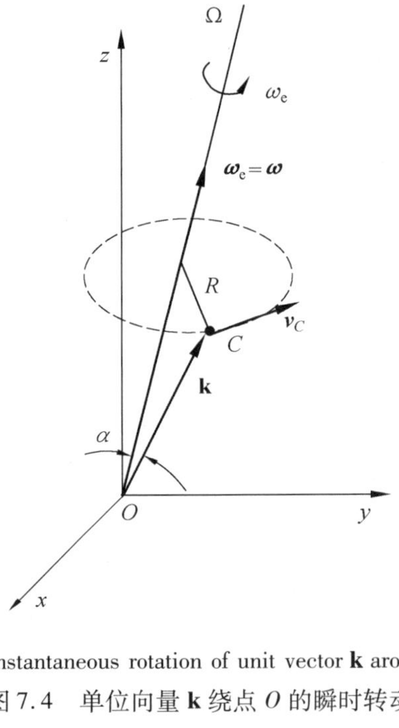

Fig. 7.4 Instantaneous rotation of unit vector **k** around point O
(图 7.4 单位向量 **k** 绕点 O 的瞬时转动)

It is known that the derivative of the radius vector is the linear velocity of the end of that vector. Since the unit vector $\mathbf{k}$ is a vector, then $\frac{\mathrm{d}\mathbf{k}}{\mathrm{d}t} = \boldsymbol{v}_C$, we can obtain(已知矢径的导数是该矢量末端的线速度. 由于单位向量 $\mathbf{k}$ 是矢量,因此 $\frac{\mathrm{d}\mathbf{k}}{\mathrm{d}t} = \boldsymbol{v}_C$, 由此可知)

$$x\frac{d\mathbf{i}}{dt}+y\frac{d\mathbf{j}}{dt}+z\frac{d\mathbf{k}}{dt}=x\boldsymbol{v}_A+y\boldsymbol{v}_B+z\boldsymbol{v}_C \tag{7.13}$$

When calculating the linear velocity of the end of the vector along the circumference, we have $\boldsymbol{v}_C=R\omega=|\mathbf{k}|\omega\sin\alpha=\omega\sin\alpha$. Given that the magnitude of the cross product $\boldsymbol{\omega}\times\mathbf{k}$ is also $\omega\sin\alpha$ and the vectors $\boldsymbol{\omega}$, $\mathbf{k}$, $\boldsymbol{v}_C$ are perpendicular to each other, we can write $\boldsymbol{v}_C=\boldsymbol{\omega}\times\mathbf{k}$. Similarly, we can write $\boldsymbol{v}_B=\boldsymbol{\omega}\times\mathbf{j}$, $\boldsymbol{v}_A=\boldsymbol{\omega}\times\mathbf{i}$. Therefore(计算矢量末端沿圆周切向的线速度时，$v_C=R\omega=|\mathbf{k}|\omega\sin\alpha=\omega\sin\alpha$. 已知矢量积 $\boldsymbol{\omega}\times\mathbf{k}$ 的大小也等于 $\omega\sin\alpha$，且矢量 $\boldsymbol{\omega}$、$\mathbf{k}$、$\boldsymbol{v}_C$ 相互垂直,可以写出 $\boldsymbol{v}_C=\boldsymbol{\omega}\times\mathbf{k}$. 同样,可以写出 $\boldsymbol{v}_B=\boldsymbol{\omega}\times\mathbf{j}$ 和 $\boldsymbol{v}_A=\boldsymbol{\omega}\times\mathbf{i}$. 因此)

$$\begin{cases} x\dfrac{d\mathbf{i}}{dt}=x\boldsymbol{v}_A=x\boldsymbol{\omega}\times\mathbf{i}=\boldsymbol{\omega}\times x\mathbf{i} \\ y\dfrac{d\mathbf{j}}{dt}=y\boldsymbol{v}_B=y\boldsymbol{\omega}\times\mathbf{j}=\boldsymbol{\omega}\times y\mathbf{j} \\ z\dfrac{d\mathbf{k}}{dt}=z\boldsymbol{v}_C=z\boldsymbol{\omega}\times\mathbf{k}=\boldsymbol{\omega}\times z\mathbf{k} \end{cases}$$

or(或者)

$$x\frac{d\mathbf{i}}{dt}+y\frac{d\mathbf{j}}{dt}+z\frac{d\mathbf{k}}{dt}=\boldsymbol{\omega}\times(x\mathbf{i}+y\mathbf{j}+z\mathbf{k})=\boldsymbol{\omega}\times\boldsymbol{\rho} \tag{7.14}$$

Finally, the following relation is(最后得到关系式为)

$$\boldsymbol{v}_M=(\boldsymbol{v}_O+\boldsymbol{\omega}\times\boldsymbol{\rho})+\boldsymbol{v}_r \tag{7.15}$$

The expressions in parentheses represent the velocity of thetransport point, i. e. the velocity of the point on the moving system that instantaneously coincides with the moving point $\boldsymbol{v}_e$ (equal to the sum of the velocity of the origin in the moving system $\boldsymbol{v}_O$ and the linear velocity relative to this origin $\boldsymbol{\omega}\times\boldsymbol{\rho}$), and we obtain (括号中的表达式代表牵连点的速度,如动系上与动点瞬间重合的点的速度 $\boldsymbol{v}_e$ (等于动系原点速度 $\boldsymbol{v}_O$ 与相对于此原点的线速度 $\boldsymbol{\omega}\times\boldsymbol{\rho}$ 之和),于是得到)

$$\boldsymbol{v}_M=\boldsymbol{v}_e+\boldsymbol{v}_r=\boldsymbol{v}_a$$

This equation expresses the theorem of composition of velocities: The absolute velocity of a point is equal to the vector sum of itstransport velocity and relative velocity. (该式表示出速度合成定理:点的绝对速度等于其牵连速度和相对速度的矢量和.)

This theorem is sometimes referred to as the velocity parallelogram theorem. (该定理有时称为速度平行四边形定理.)

In general, the magnitude of the absolute velocity can be calculated by(在一般情况下,绝对速度的大小为)

$$v=\sqrt{v_r^2+v_e^2+2v_rv_e\cos(\boldsymbol{v}_r,\boldsymbol{v}_e)} \tag{7.16}$$

Example 1: Passengers walk in the direction opposite to that of the train at a speed of 0.5 km/h towards the side of the carriage. The train moves along a straight track at a speed of 60 km/h. What is the speed of the passenger's movement relative to the building(Fig. 7.5)? (**例题** 1:乘客沿着与火车方向相反的方向以 0.5 km/h 的速度向着车厢的一边行走. 火车以 60km/h 的速度沿着直线轨道移动. 乘客相对于建筑物的运动速度是多少(图 7.5)?)

Solution:(解:)

(1) Determine the transport speed. As the carriage moves forward along a straight line, the

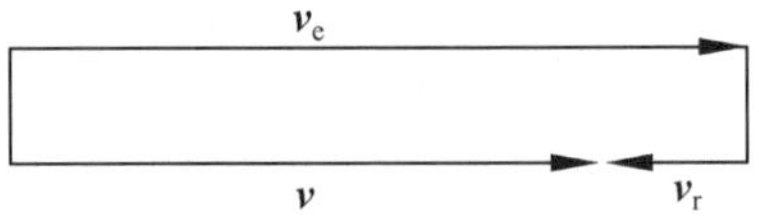

Fig. 7. 5 Diagram of Example 1(图 7. 5 例题 1 图)

velocity is the same at all points, i. e. $v_e = 60$ km/h. (确定牵连速度. 因为车厢沿直线向前运动,所以各点速度相同,即 $v_e = 60$ km/h.)

(2) Determine the absolute velocity of the passenger. Based on the theorem of composition of velocities, when the points move synthetically, we have $\boldsymbol{v}_a = \boldsymbol{v}_e + \boldsymbol{v}_r$ (Fig. 7. 5). Since all vectors are parallel, we can obtain $v = 60 - 0.5 = 59.5$ km/h. (确定乘客的绝对速度. 基于速度合成定理,当点合成运动时,有 $\boldsymbol{v}_a = \boldsymbol{v}_e + \boldsymbol{v}_r$ (图 7. 5). 因为所有矢量是平行的,因此 $v = 60 - 0.5 = 59.5$ km/h.)

Therefore, passengers move relative to the building at an absolute speed of 59. 5 km/h in the direction of train movement. (因此,乘客以 59. 5 km/h 的绝对速度在火车运行的方向上相对于建筑物运动.)

Example 2: As shown in Fig. 7. 6 (a), it is known that $OA = l$, $\varphi = \omega t$. Determine the velocity and acceleration of the T-shaped rod. (**例题** 2:如图 7. 6(a)所示,已知 $OA = l$,$\varphi = \omega t$. 求 T 形杆的速度和加速度.)

Solution:(解:)

(1) Take the slider A on OA as the moving point and the T-shaped rod as the moving system. (取 OA 上的滑块 A 为动点,T 形杆为动系.)

(2) Analyze the three types of motion and determine the direction of velocity. (分析三种运动,确定速度的方向.)

(3) Make a velocity parallelogram as shown in Fig. 7. 6 (b) and solve for the unknown variables, we have (做出速度平行四边形,如图 7. 6(b),求解未知量,有)

$$\boldsymbol{v}_a = \boldsymbol{v}_e + \boldsymbol{v}_r \text{ , } v_a = l\omega \text{ , } v_e = v_a\cos\varphi = l\omega\cos\varphi$$

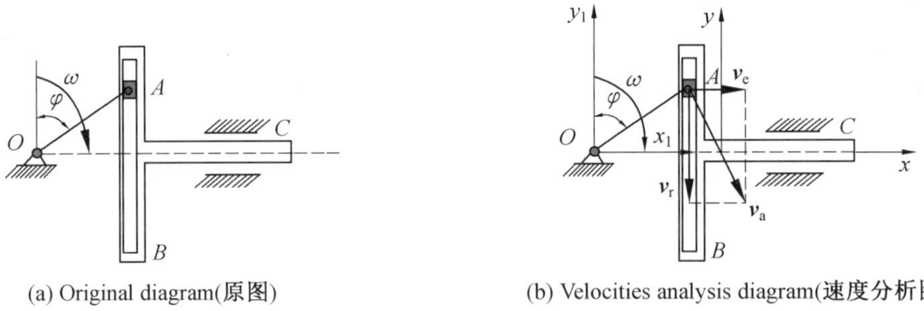

Fig. 7. 6 Diagram of Example 2(图 7. 6 例题 2 图)

Example 3: As shown in Fig. 7. 7 (a), it is known that R and $\varphi = \omega t$ (ω is a constant). (1) Determine the velocity of the small ring M; (2) Determine the velocity of the small ring M with respect to the AB rod. (**例题** 3:如图 7. 7(a)所示,已知 R 且 $\varphi = \omega t$(ω 为常数). 求:(1)小环 M 的速度;(2)小环 M 相对于 AB 杆的速度.)

Solution:(解:)

(1)Take the small ring M as the moving point and the AB rod as the moving system, draw the velocity vector diagram as shown in Fig. 7.7(b), according to(取小环 M 为动点,AB 杆为动系,画速度矢量图如图 7.7(b)所示,根据)

$$\boldsymbol{v}_a = \boldsymbol{v}_e + \boldsymbol{v}_r$$

we can obtain (得到)

$$v_e = \overline{AM} \cdot \omega = 2R\omega\cos\varphi \text{ , } v_a = v_e/\cos\varphi = 2R\omega \text{ , } v_r = v_a\sin\varphi = 2R\omega\sin\varphi$$

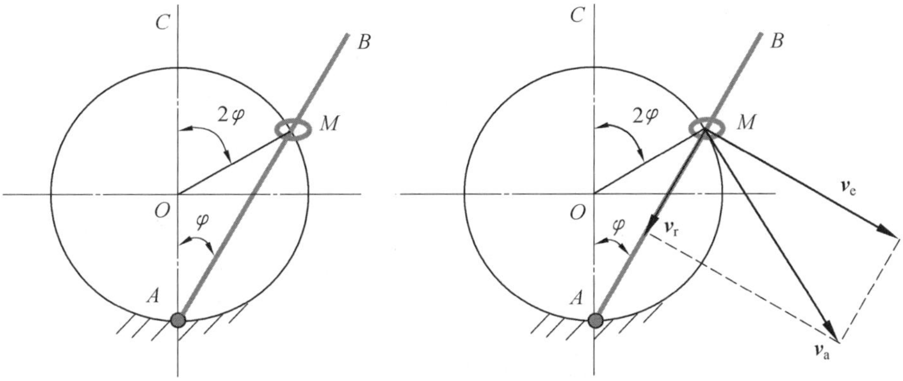

(a) Original diagram(原图) (b) Velocities analysis diagram(速度分析图)

Fig. 7.7 Diagram of Example 3(图 7.7 例题 3 图)

Example 4: As shown in Fig. 7.8(a), it is known that ω and $OO_1 = l, OA = r$. Determine the angular velocity of the rocker when OA is horizontal ω_1. (**例题** 4:如图 7.8(a)所示,已知 ω 和 $OO_1 = l, OA = r$. 求当 OA 水平时摇杆的角速度 ω_1.)

Solution:(解:)

(1)Take the end point of the crank A as the moving point and the rocker as the moving system, and draw the velocity vector diagram as shown in Fig. 7.8(b), we have(取曲柄端点 A 为动点,摇杆为动系, 画速度矢量图,如图 7.8(b)所示,有)

$$\boldsymbol{v}_a = \boldsymbol{v}_e + \boldsymbol{v}_r \text{ , } v_a = r\omega$$

We can obtain(得到)

$$v_e = v_a\sin\varphi = r\omega\sin\varphi$$

$$\omega_1 = v_e/O_1A = v_e\sin\varphi/r = \omega\sin^2\varphi = \frac{r^2\omega}{l^2 + r^2}$$

Example 5: On the basis of Example 4, as shown in Fig. 7.9(a), if the angular velocity of OA is known, determine the velocity of the BC rod. (**例题** 5:在例题 4 基础上,如图 7.9(a)所示,若已知 OA 的角速度,求 BC 杆的速度.)

Solution:(解:)

In Example 4, we have taken the end point of the crank A as the moving point and the rocker as the moving system, and obtain the angular velocity of the rocker ω_1. On this basis, take the slider B on the CB rod as the moving point and the rocker as the moving system, and draw the velocity vector diagram as shown in Fig. 7.9(b). According to(在例题 4 中已经取曲柄端点 A

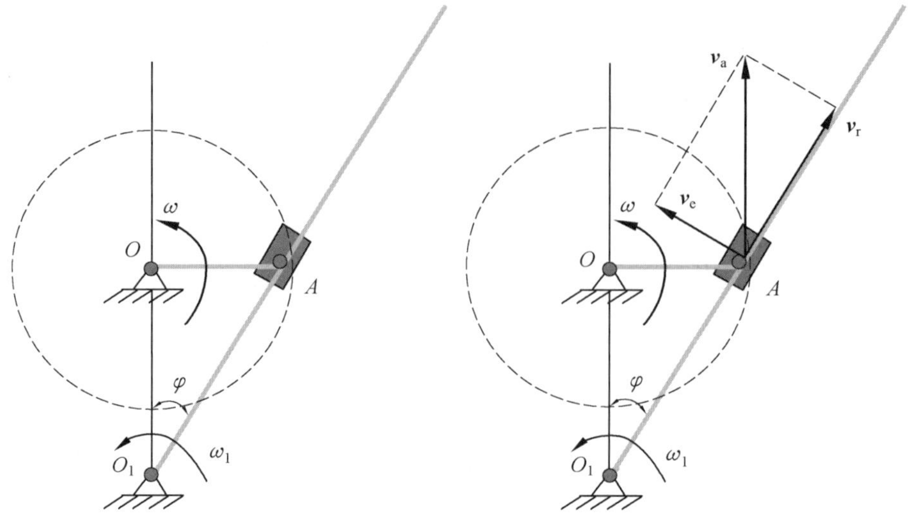

(a) Original diagram(原图)　　(b) Velocities analysis diagram(速度分析图)

Fig. 7.8　Diagran of Example 4(图 7.8　例题 4 图)

为动点,摇杆为动系,得到了摇杆的角速度 ω_1. 在此基础上,再取 *CB* 杆上的滑块 *B* 为动点,摇杆为动系,画速度矢量图,如图 7.9(b)所示. 根据)

$$\boldsymbol{v}_{Ba} = \boldsymbol{v}_{Be} + \boldsymbol{v}_{Br}$$

we can obtain(可求得)

$$v_{Ba} = \frac{v_{Be}}{\cos \varphi} = \frac{\omega_1 O_1 B}{\cos \varphi}$$

Since the *BC* bar makes translation motion, the speed of the *BC* bar is v_{Ba} . (由于 *BC* 杆做平动,因此 *BC* 杆的速度为 v_{Ba} .)

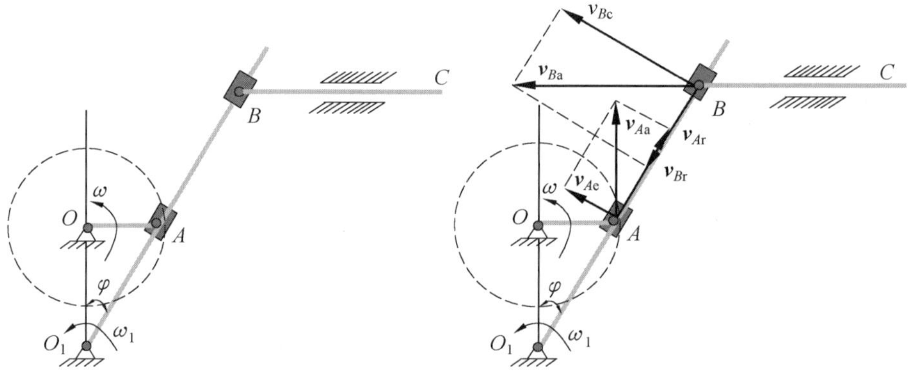

(a) Original diagram(原图)　　(b) Velocities analysis diagram(速度分析图)

Fig. 7.9　Diagram of Example 5(图 7.9　例题 5 图)

Example 6: As shown in Fig. 7. 10(a), h, φ, ω are known. Determine the velocity of the *AB* rod. (**例题 6**:如图 7. 10(a)所示,已知 h、φ、ω. 求 *AB* 杆的速度.)

Solution:(解:)

Take the end point *A* of the *AB* rod as the moving point and the cam as the moving system,

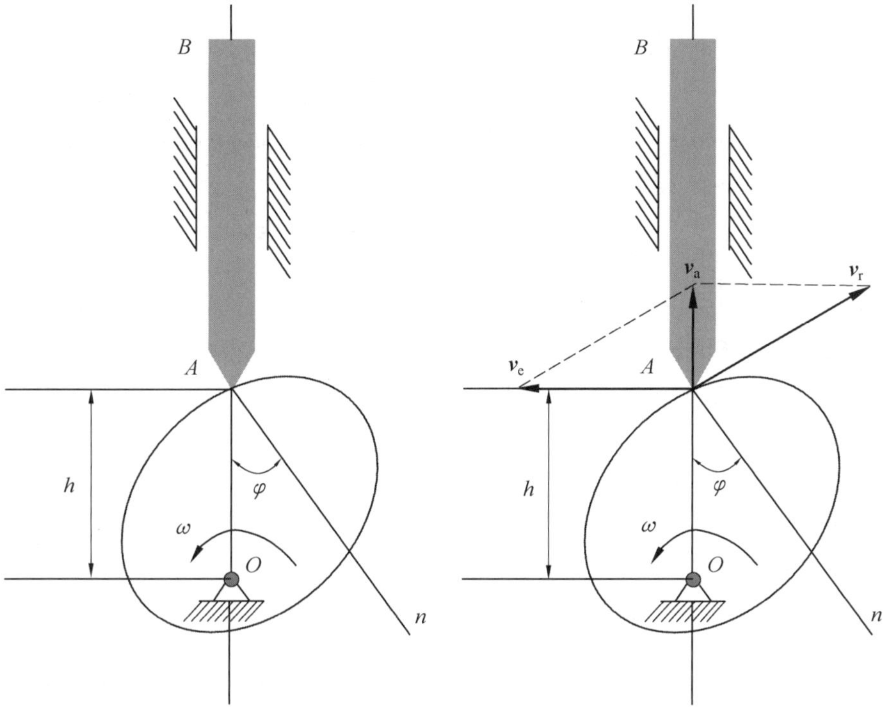

(a) Original diagram(原图) (b) Velocities analysis diagram(速度分析图)

Fig. 7.10 Diagram of Example 6(图 7.10 例题 6 图)

draw the velocity vector diagram as Fig. 7.10(b). According to(取 AB 杆端点 A 为动点,凸轮为动系,画速度矢量图,如图 7.10(b)所示. 根据)

$$\boldsymbol{v}_a = \boldsymbol{v}_e + \boldsymbol{v}_r \ , \ v_e = h\omega$$

we can obtain (得到)

$$v_a = v_e \tan\varphi = h\omega \tan\varphi$$

7.3 The Theorem of Acceleration Composition of Points and Coriolis Acceleration(点的加速度合成定理和科里奥利加速度)

The acceleration at the point is the first derivative of the velocity vector. Therefore, using Eq. (7.8), we can obtain the absolute acceleration, that is(点的加速度为速度矢量的一阶导数. 因此,由式(7.8)可得绝对加速度为)

$$\boldsymbol{a}_M = \frac{\mathrm{d}\boldsymbol{v}_M}{\mathrm{d}t} = \frac{\mathrm{d}^2\boldsymbol{r}_O}{\mathrm{d}t^2} + \frac{\mathrm{d}^2 x}{\mathrm{d}t^2}\mathbf{i} + \frac{\mathrm{d}^2 y}{\mathrm{d}t^2}\mathbf{j} + \frac{\mathrm{d}^2 z}{\mathrm{d}t^2}\mathbf{k} + 2\left(\frac{\mathrm{d}x}{\mathrm{d}t}\frac{\mathrm{d}\mathbf{i}}{\mathrm{d}t} + \frac{\mathrm{d}y}{\mathrm{d}t}\frac{\mathrm{d}\mathbf{j}}{\mathrm{d}t} + \frac{\mathrm{d}z}{\mathrm{d}t}\frac{\mathrm{d}\mathbf{k}}{\mathrm{d}t}\right) + x\frac{\mathrm{d}^2\mathbf{i}}{\mathrm{d}t^2} + y\frac{\mathrm{d}^2\mathbf{j}}{\mathrm{d}t^2} + z\frac{\mathrm{d}^2\mathbf{k}}{\mathrm{d}t^2} \tag{7.17}$$

Using the "stopping motion rule", we can obtain the relative and transport acceleration of the points. (使用止动规则, 可以得到点的相对加速度和牵连加速度.)

Bring $\boldsymbol{r}_O$ = const, $\mathbf{i}$ = const, $\mathbf{j}$ = const, $\mathbf{k}$ = const into Eq. (7.17), we obtain the relative acceleration, that is(将 $\boldsymbol{r}_O$ = const, $\mathbf{i}$ = const, $\mathbf{j}$ = const, $\mathbf{k}$ = const 代入式(7.17)中,得到相对加

速度为)

$$\boldsymbol{a}_{\mathrm{r}}=\frac{\mathrm{d}\boldsymbol{v}_M}{\mathrm{d}t}=\frac{\mathrm{d}^2x}{\mathrm{d}t^2}\mathbf{i}+\frac{\mathrm{d}^2y}{\mathrm{d}t^2}\mathbf{j}+\frac{\mathrm{d}^2z}{\mathrm{d}t^2}\mathbf{k} \tag{7.18}$$

Let $x=\text{const}$, $y=\text{const}$, $z=\text{const}$ to obtain the transport acceleration, we can get(令 $x=\text{const}$, $y=\text{const}$, $z=\text{const}$,得到牵连加速度为)

$$\boldsymbol{a}_{\mathrm{e}}=\frac{\mathrm{d}^2\boldsymbol{r}_O}{\mathrm{d}t^2}+x\frac{\mathrm{d}^2\mathbf{i}}{\mathrm{d}t^2}+y\frac{\mathrm{d}^2\mathbf{j}}{\mathrm{d}t^2}+z\frac{\mathrm{d}^2\mathbf{k}}{\mathrm{d}t^2} \tag{7.19}$$

Therefore, we can know from Eq. (7.17) that the absolute acceleration is not composed of two accelerations, but of three, that is(因此,从式(7.17)中可以得出,绝对加速度不是由两个加速度组成,而是由三个加速度组成,即)

$$\boldsymbol{a}=\boldsymbol{a}_{\mathrm{e}}+\boldsymbol{a}_{\mathrm{r}}+\boldsymbol{a}_{\mathrm{C}} \tag{7.20}$$

The additional acceleration is called Coriolis acceleration (named after the scientist who first discovered it) and is equal to(附加的加速度称为科里奥利加速度(以首次发现此加速度的科学家名字命名),其等于)

$$\boldsymbol{a}_{\mathrm{C}}=2\left(\frac{\mathrm{d}x}{\mathrm{d}t}\frac{\mathrm{d}\mathbf{i}}{\mathrm{d}t}+\frac{\mathrm{d}y}{\mathrm{d}t}\frac{\mathrm{d}\mathbf{j}}{\mathrm{d}t}+\frac{\mathrm{d}z}{\mathrm{d}t}\frac{\mathrm{d}\mathbf{k}}{\mathrm{d}t}\right) \tag{7.21}$$

This additional acceleration occurs because the transport velocity depends on the relative motion and the position of the transport point, while the relative velocity changes with the transport motion. (这种附加加速度的出现是因为牵连速度取决于相对运动及牵连点的位置,而相对速度则通过牵连运动而发生改变.)

In two special cases, the way to determine the Coriolis acceleration is the simplest. (在两种特殊情况下确定科里奥利加速度是最简单的方法.)

(1) The transport motion is translational (the moving system x, y, z moves in translation). (牵连运动为平动(动系 x, y, z 平行移动).)

Since the coordinate axes of the moving system does not rotate in motion, we can know the unit vector $\mathbf{i}=\text{const}$, $\mathbf{j}=\text{const}$, $\mathbf{k}=\text{const}$,and therefore according to Eq. (7.21), we can obtain the Coriolis acceleration $\boldsymbol{a}_{\mathrm{C}}=0$ and the absolute acceleration are only the sum of two accelerations, $\boldsymbol{a}=\boldsymbol{a}_{\mathrm{e}}+\boldsymbol{a}_{\mathrm{r}}$. (由于动系坐标轴在运动中不旋转,因此单位矢量 $\mathbf{i}=\text{const}$, $\mathbf{j}=\text{const}$, $\mathbf{k}=\text{const}$. 因此,根据式(7.21)得到科里奥利加速度 $\boldsymbol{a}_{\mathrm{C}}=0$ 和绝对加速度仅为两项加速度之和,有 $\boldsymbol{a}=\boldsymbol{a}_{\mathrm{e}}+\boldsymbol{a}_{\mathrm{r}}$.)

This is understandable, because the transport motion of the point will not depend on the relative motion and the transport motion will not change the direction of the vector of the relative velocity. (这是可以理解的, 因为点的牵连运动将不依赖于相对运动,并且牵连运动不会改变相对速度的矢量方向.)

(2) The transport motion is rotation about fixed axis. (牵连运动为定轴转动.)

Let the moving system $Oxyz$ rotate with angular velocity $\boldsymbol{\omega}_{\mathrm{e}}$ around a fixed axis ξ (Fig. 7.11). (设动系 $Oxyz$ 以角速度 $\boldsymbol{\omega}_{\mathrm{e}}$ 绕固定轴 ξ 旋转(图 7.11).)

We represent the axes of the moving system as radial vectors whose end pass over the endpoints of the unit vectors. The derivative of the unit vector with respect to time can then be

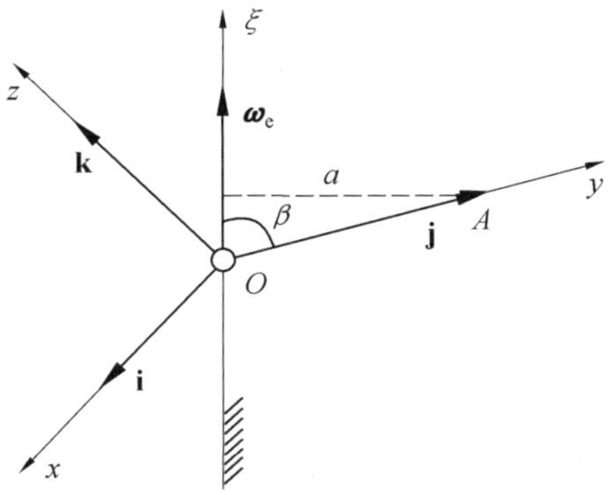

Fig. 7. 11 Moving system rotating around a fixed axis(图 7. 11 围绕固定轴旋转的动系)

considered as the velocity of these endpoints. (把动系坐标轴表示成径向向量,末端过单位向量的终点. 然后可以将单位向量对时间的导数视为这些终点的速度.)

For example, the velocity of the point A at the end of the vector $\mathbf{j}$ is $\boldsymbol{v}_A = \frac{\mathrm{d}\mathbf{j}}{\mathrm{d}t}$. But because its modulus $v_A = \omega_e a = \omega_e \cdot \mathrm{j} \sin\beta$, and the velocity vector $\boldsymbol{v}_A$ is perpendicular to $\boldsymbol{\omega}_e$ and $\mathbf{j}$ in the direction, i. e. , $\boldsymbol{v}_A = \boldsymbol{\omega}_e \times \mathbf{j}$. So similarly we can write $\boldsymbol{v}_A = \boldsymbol{\omega}_e \times \mathbf{j}$, $\frac{\mathrm{d}\mathbf{j}}{\mathrm{d}t} = \boldsymbol{\omega}_e \times \mathbf{j}$ and $\frac{\mathrm{d}\mathbf{k}}{\mathrm{d}t} = \boldsymbol{\omega}_e \times \mathbf{k}$. (例如, 矢量 $\mathbf{j}$ 末端 A 点的速度 $\boldsymbol{v}_A = \frac{\mathrm{d}\mathbf{j}}{\mathrm{d}t}$. 但因为它的模 $v_A = \omega_e a = \omega_e \cdot \mathbf{j} \sin\beta$, 而速度矢量 $\boldsymbol{v}_A$ 在方向上是垂直于 $\boldsymbol{\omega}_e$ 和 $\mathbf{j}$ 的, 即 $\boldsymbol{v}_A = \boldsymbol{\omega}_e \times \mathbf{j}$. 因此,类似地有 $\boldsymbol{v}_A = \boldsymbol{\omega}_e \times \mathbf{j}$, $\frac{\mathrm{d}\mathbf{j}}{\mathrm{d}t} = \boldsymbol{\omega}_e \times \mathbf{j}$ 和 $\frac{\mathrm{d}\mathbf{k}}{\mathrm{d}t} = \boldsymbol{\omega}_e \times \mathbf{k}$.)

To prove this, we can use the formula $\boldsymbol{v} = \boldsymbol{\omega} \times \boldsymbol{r}$. (为证明这一点, 可以使用公式 $\boldsymbol{v} = \boldsymbol{\omega} \times \boldsymbol{r}$.)

Then the Coriolis acceleration is obtained by Eq. (7.21), that is(然后通过式(7.21)得到科里奥利加速度为)

$$\boldsymbol{a}_C = 2\boldsymbol{\omega}_e \times \left(\frac{\mathrm{d}x}{\mathrm{d}t}\mathbf{i} + \frac{\mathrm{d}y}{\mathrm{d}t}\mathbf{j} + \frac{\mathrm{d}z}{\mathrm{d}t}\mathbf{k}\right) \tag{7.22}$$

Taking Eq. (7.9) into account, we obtain(且考虑到式(7.9),可得到)

$$\boldsymbol{a}_C = 2\boldsymbol{\omega}_e \times \boldsymbol{v}_r \tag{7.23}$$

Coriolis acceleration is 2 times the product of the angular velocity vector of the transport motion and the relative velocity vector at the point. (科里奥利加速度是该点的牵连运动角速度矢量与相对速度矢量乘积的 2 倍.)

Its magnitude is(其大小为)

$$a_C = 2\omega_e v_r \sin\alpha \tag{7.24}$$

where α is the acute angle between the vector $\boldsymbol{\omega}_e$ and the vector $\boldsymbol{v}_r$. (式中, α 为矢量 $\boldsymbol{\omega}_e$ 与 $\boldsymbol{v}_r$ 间所夹的锐角.)

Note: It can be proven that this result is true for any transport motion, not only when rotating around a fixed axis. (注: 可以证明,这一结果在任何牵连运动中都是正确的, 而不仅仅是在

绕固定轴旋转时.)

Example 7: As shown in Fig. 7. 12, it is known that $OA = l$, $\varphi = \omega t$ (ω is a constant). Determine the acceleration of the T-shaped rod when $\varphi = 30°$. (**例题** 7:如图 7. 12 所示,已知 $OA = l$, $\varphi = \omega t$ (ω 为常量). 求当 $\varphi = 30°$ 时,T 形杆的加速度.)

Solution:(解:)

Take the slider A as the moving point and the T-shaped rod as the moving system. The analysis of the velocity of the moving point has been carried out in Example 2. As the moving system does translational motion and there is no Coriolis acceleration, then(取滑块 A 为动点,T 形杆为动系. 例题 2 中已经进行了动点的速度分析. 因为动系做平动,没有科里奥利加速度,故有)

$$\boldsymbol{a} = \boldsymbol{a}_{\mathrm{e}} + \boldsymbol{a}_{\mathrm{r}}$$

Its acceleration vector is shown in Fig. 7. 12(b). Moreover, the normal acceleration of the absolute acceleration is $a_{\mathrm{a}}^{n} = l\omega^2$, and the tangential acceleration of the absolute acceleration is $a_{\mathrm{a}}^{\tau} = OA \cdot \varepsilon = 0$, so we can obtain the acceleration of the T-shaped rod, that is(其加速度矢量如图 7. 12(b)所示. 其中,绝对加速度的法向加速度 $a_{\mathrm{a}}^{n} = l\omega^2$, 切向加速度 $a_{\mathrm{a}}^{\tau} = OA \cdot \varepsilon = 0$,得到 T 形杆的加速度,即)

$$a_{\mathrm{e}} = a_{\mathrm{a}} \sin 30° = \frac{1}{2} l\omega^2$$

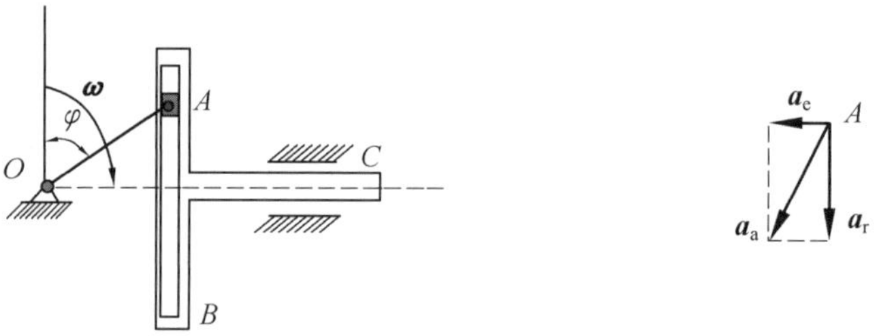

(a) Original diagram(原图)　　(b) Acceleration analysis diagram(加速度分析图)

Fig. 7. 12　Diagram of Example 7 (图 7. 12　例题 7 图)

Example 8: As shown in Fig. 7. 13, it is known that $OA = r$, $\omega = \mathrm{const}$. Determine the velocity and acceleration of the CD rod. (**例题** 8:如图 7. 13 所示,已知 $OA = r$,$\omega = \mathrm{const}$. 求 CD 杆的速度和加速度.)

Solution:(解:)

Take the point C of the CD rod as the moving point and the triangle ABC as the moving system. The velocity vector and acceleration vector diagrams are shown in Fig. 7. 13, we have(取 CD 杆 C 点为动点,三角板 ABC 为动系. 速度矢量及加速度矢量图如图 7. 13 所示,有)

$$\boldsymbol{v}_{\mathrm{a}} = \boldsymbol{v}_{\mathrm{e}} + \boldsymbol{v}_{\mathrm{r}}$$

From Fig. 7. 13, it can be known that(由图 7. 13 可知)

$$v_{\mathrm{e}} = v_A = r\omega$$

Then we have(则有)

$$v_{\mathrm{e}} = v_{\mathrm{r}} = v_{\mathrm{a}} = r\omega$$

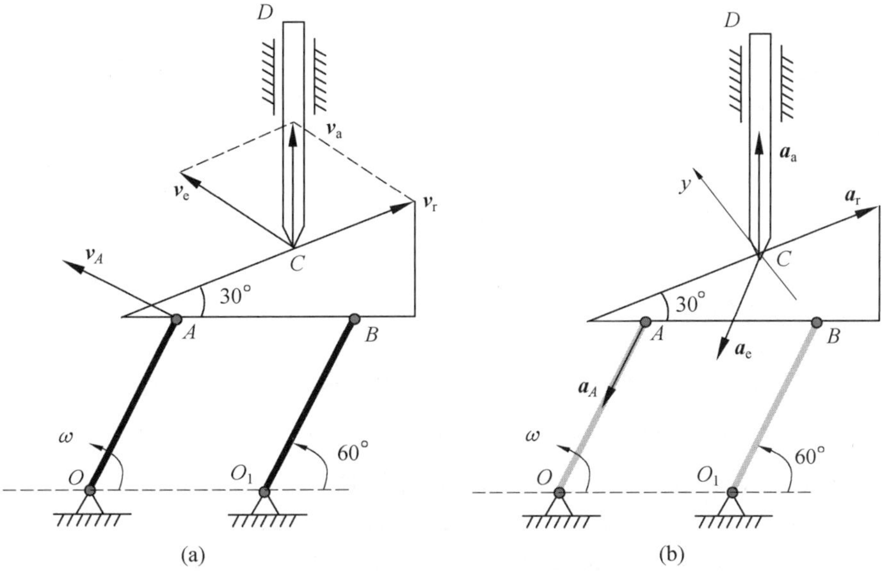

Fig. 7.13 Diagram of Example 8(图 7.13 例题 8 图)

The velocity of the *CD* rod is(*CD* 杆的速度为)

$$v_{CD} = v_a = r\omega$$

Since the moving system makes a curve translation motion, then we have(由于动系做曲线平动,因此有)

$$\boldsymbol{a} = \boldsymbol{a}_e + \boldsymbol{a}_r$$

Make a line *Cy* perpendicular to the relative velocity $\boldsymbol{a}_r$ through point *C*, and project the accelerations in the above equation to the *Cy* axis to obtain(过 *C* 点做与相对速度 $\boldsymbol{a}_r$ 垂直的直线 *Cy*,将上式各加速度向 *Cy* 轴投影,得到)

$$a_a \cos 30° = -a_e \cos 60°$$

where(其中)

$$a_e = a_A = r\omega^2$$

So we can obtain the acceleration of the *CD* rod, that is (因此,可以得到 *CD* 杆的加速度为)

$$a_{CD} = a_a = -a_e \tan 30° = -\frac{\sqrt{3}}{3} r\omega^2$$

Example 9: Determine in Example 3: (1) the acceleration of the small ring *M*; (2) the acceleration of the small ring *M* with respect to the *AB* bar.(求例题 3 中:(1) 小环 *M* 的加速度;(2) 小环 *M* 相对于 *AB* 杆的加速度.)

Solution:(解:)

Take the small ring *M* as the moving point and the *AB* rod as the moving system. The moving system makes a fixed-axis rotation around the *A* axis, and the acceleration analysis of the moving point is carried out based on the results of the velocity analysis in Example 3 (Fig. 7.14(a)).(取小环 *M* 为动点,*AB* 杆为动系. 动系绕 *A* 轴做定轴转动,基于例题 3 中的速度分析结果(图 7.14(a))进行动点的加速度分析.)

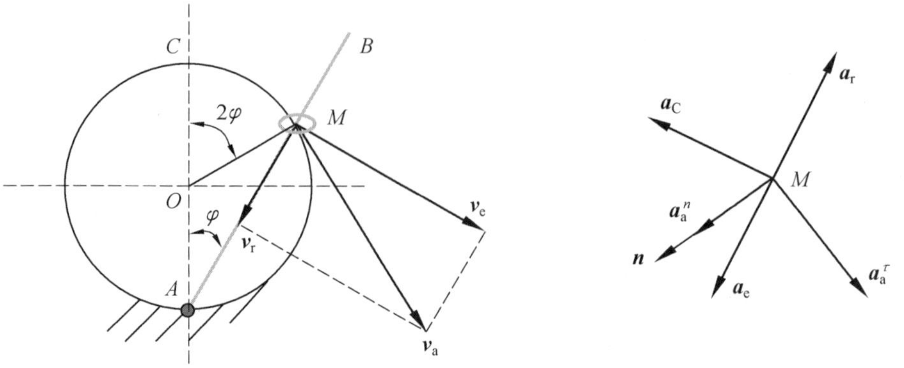

(a) Original diagram(原图)　　(b) Acceleration analysis diagram(加速度分析图)

Fig. 7. 14　Diagram of Example 9(图 7. 14　例题 9 图)

It is known from Example 3 that(由例题 3 已知得到)

$$v_a = v_e/\cos\varphi = 2R\omega\ ,\ v_r = v_a\sin\varphi = 2R\omega\sin\varphi$$

Since the moving system makes a fixed-axis rotation around the A-axis, there is Coriolis acceleration at the moving point. According to $\boldsymbol{a} = \boldsymbol{a}_e + \boldsymbol{a}_r + \boldsymbol{a}_C$, we can get(由于动系绕 A 轴做定轴转动,因此动点存在科里奥利加速度. 根据 $\boldsymbol{a} = \boldsymbol{a}_e + \boldsymbol{a}_r + \boldsymbol{a}_C$,得到)

$$\boldsymbol{a}_a^n + \boldsymbol{a}_a^\tau = \boldsymbol{a}_e + \boldsymbol{a}_r + \boldsymbol{a}_C$$

where(其中)

$$a_a^n = v_a^2/R = 4R\omega^2,\ a_a^\tau = 0,\ a_e = v_e^2/AM = 2R\omega^2\cos\varphi,\ a_C = 2\omega_e v_r = 4R\omega^2\sin\varphi$$

Project the acceleration equation to the Mn axis (Fig. 7. 14(b)), we obtain(将加速度方程向 Mn 轴投影(图 7. 14(b)),得到)

$$a_a^n = a_e\cos\varphi + a_C\sin\varphi - a_r\cos\varphi$$

We can get the acceleration of the small ring M with respect to the AB rod, that is(解得小环 M 相对于 AB 杆的加速度为)

$$a_r = -2R\omega^2\cos\varphi$$

Therefore, the absolute acceleration of the small ring M is (因此,小环 M 的绝对加速度为)

$$a_a^n = 4R\omega^2$$

Example 10: Determine the angular acceleration ε of the rocker in Example 4. (**例题** 10:求例题 4 中摇杆的角加速度 ε .)

Solution:(解:)

Take the end point of the crank A as the moving point and the rocker as the moving system. The angular velocity of the rocker ω_1 has been obtained through the velocity analysis in Example 4 (Fig. 7. 15(a)), $\omega_1 = \dfrac{r^2\omega}{l^2 + r^2}$. According to the theorem of composition of accelerations, we can obtain that(取曲柄端点 A 为动点,摇杆为动系. 例题 4 中通过速度分析已经得到摇杆的角速度 ω_1(图 7. 15(a))为 $\omega_1 = \dfrac{r^2\omega}{l^2 + r^2}$. 根据加速度合成定理,有)

$$\boldsymbol{a}_a = \boldsymbol{a}_e^n + \boldsymbol{a}_e^\tau + \boldsymbol{a}_r + \boldsymbol{a}_C$$

where(其中)

$$a_a = \omega^2 r \text{ , } a_e^n = \omega_1^2 O_1 A = \frac{r^4 \omega^2}{(l^2 + r^2)^{3/2}} \text{ , } v_r = v_a \cos \varphi = \omega r l / \sqrt{l^2 + r^2}$$

$$a_C = 2\omega_1 v_r \sin 90° = \frac{2\omega^2 r^3 l}{(l + r^2)^{3/2}}$$

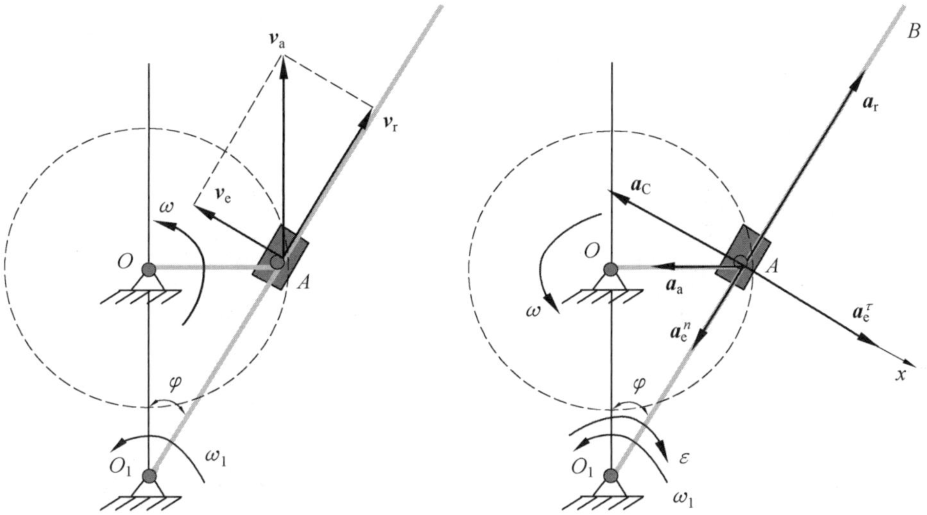

(a) Original diagram(原图)　　(b) Acceleration analysis diagram(加速度分析图)

Fig. 7.15 Diagram of Example 10(图 7.15 例题 10 图)

Project the acceleration equation to the Ax axis (Fig. 7.15(b)), and we have(将加速度方程向 Ax 轴投影(图 7.15(b)),得到)

$$-a_a \cos \varphi = a_e^\tau - a_C$$

we can obtain(解得)

$$a_e^\tau = -\frac{rl(l^2 - r^2)}{(l^2 + r^2)^{3/2}}\omega^2$$

Therefore, we obtain the angular acceleration of the rocker, that is(则摇杆的角加速度为)

$$\varepsilon = \frac{a_e^\tau}{AO_1} = -\frac{rl(l^2 - r^2)}{(l^2 + r^2)^2}\omega^2$$

Example 11: Determine the acceleration of the AB rod in Example 6. (**例题** 11:求例题6 中 AB 杆的加速度.)

Solution:(解:)

Take the end point A of the AB rod as the moving point and the cam as the moving system. The absolute and relative velocities of the AB rod have been obtained through the velocity analysis in Example 6 (Fig. 7.16(a)). We have known that (取 AB 杆端点 A 为动点,凸轮为动系. 例题 6 中通过速度分析已经得到 AB 杆的绝对速度及相对速度(图 7.16(a)),已知)

$$v_a = h\omega \tan \varphi \text{ , } v_r = h\omega \sec \varphi$$

According to the theorem of composition of accelerations, we have(根据加速度合成定理,有)

$$\boldsymbol{a}_a = \boldsymbol{a}_e + \boldsymbol{a}_r^n + \boldsymbol{a}_r^\tau + \boldsymbol{a}_C$$

besides(其中)

$$a_e = \omega^2 h\ ,\ a_r^n = v_r^2/\rho = (h\omega \sec\varphi)^2/\rho\ ,\ a_C = 2\omega_1 v_r \sin 90° = \frac{2h\omega^2 \sec\varphi}{\rho}$$

Project the acceleration equation to the An axis (Fig. 7.16(b)), we obtain (将加速度方程向 An 轴投影 (图 7.16(b)), 得到)

$$-a_a \cos\varphi = a_r^n + a_e \cos\varphi - a_C$$

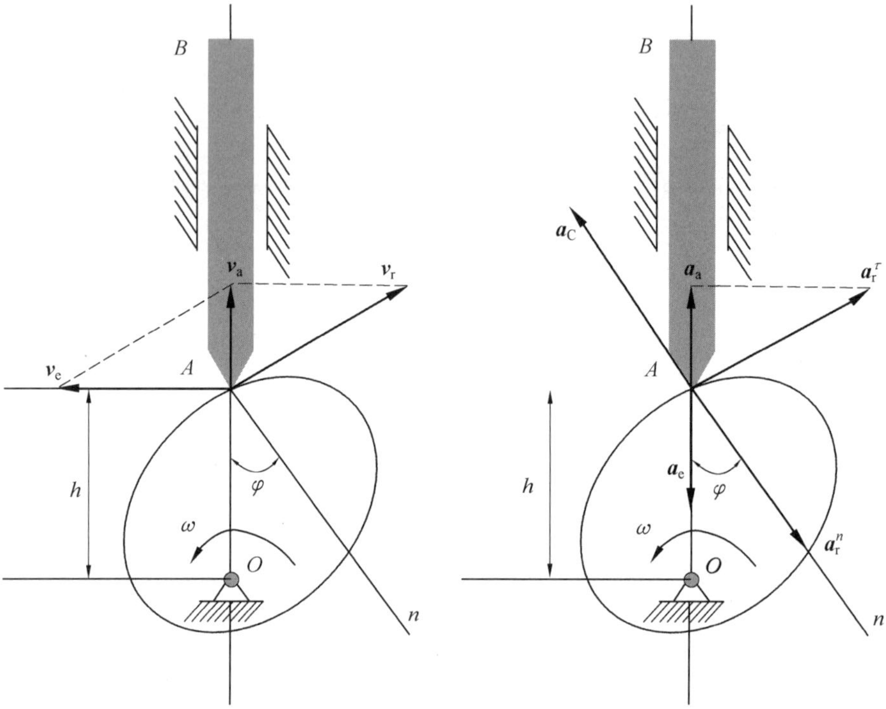

(a) Original diagram(原图) (b) Acceleration analysis diagram(加速度分析图)

Fig. 7.16 Diagram of Example 11(图 7.16 例题 11 图)

Solve equations and we can get (解上述公式,得)

$$a_a = h\omega^2\left(2\sec^2\varphi - \frac{h}{\rho}\sec^3\varphi - 1\right)$$

7.4 Composition of Two Rotation Motions (两个转动运动的合成)

7.4.1 Composition of the Rotation around an Intersecting Axis of Rigid Body(刚体绕相交轴的转动的合成)

Considering the composition of the motion of a rigid body, the composite motion is the resultant of the rotation motion of an object around two axes intersecting at a point. The movement of the disc is shown in Fig. 7.17. The rotation of the disc relative to the axis ON is its relative motion. Therefore, the angular velocity of the rotation motion is represented by ω_r. The rotation of the *ON axis itself around the Oz* axis is the transport motion, so we use ω_e to represent the angular velocity. What is the absolute motion of the object in this situation? (考虑刚体的运动合成,该合成运动为物体绕相交于一点的两个轴的转动运动的合成. 盘的运动如图 7.17 所示. 该盘相对于轴线 ON 的转动是其相对运动. 因此,该转动运动的角速度用 ω_r 来表示. ON 轴本身围绕 Oz 轴的转动是牵连运动,因此用 ω_e 表示角速度. 在这种情况下,物体的绝对运动是什么?)

We constructed parallelograms with vectors $\boldsymbol{\omega}_e$ and $\boldsymbol{\omega}_r$ (Fig. 7.17). Using the diagonal of the parallelogram OC to represent the angular velocity vector synthesized by the rotation of the object around the axis $O\Omega$. The velocity of point O is zero because it is simultaneously rotating on two instantaneous axes ON and Oz. Since point C belongs to the object participating in composite motion, its velocity is determined by the theorem of composition of velocities, we have (以向量 $\boldsymbol{\omega}_e$ 和 $\boldsymbol{\omega}_r$ 构建平行四边形(图 7.17). 用该平行四边形的 OC 对角线来表示物体绕轴 $O\Omega$ 转动合成的角速度矢量. 点 O 的速度等于零,因为它同时在两个瞬时转动轴 ON 和 Oz 上. 由于点 C 属于参与运动合成的物体,因此其速度由速度合成定理确定,有)

$$\boldsymbol{v}_C = \boldsymbol{v}_r + \boldsymbol{v}_e$$

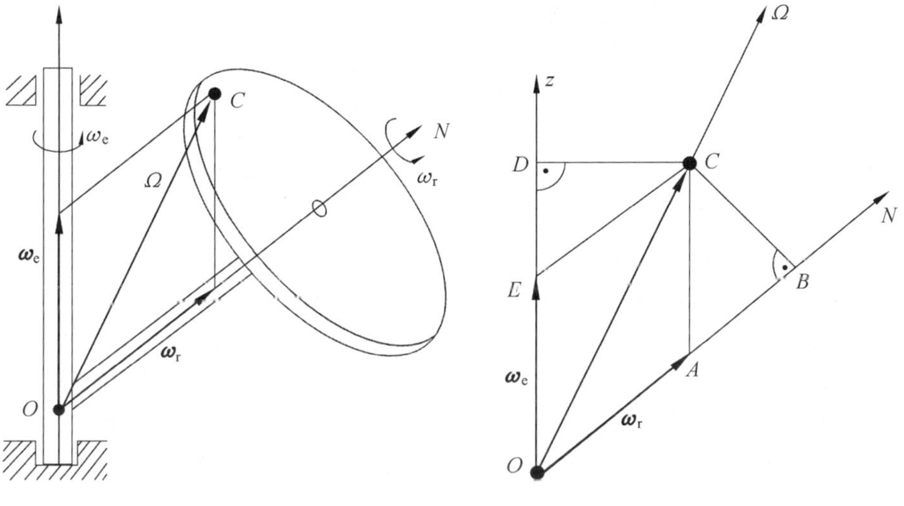

Fig. 7.17 Composition of rotation of an object around two axes intersecting at a point
(图 7.17 物体绕相交于一点的两个轴的转动运动的合成)

Calculate the linear velocity of point C in relative rotation around axis ON, we have (计算点 C 在绕轴 ON 的相对转动中的线速度,有)

$$v_r = BC \times \omega_r = 2 \text{ times the area of } \triangle OAC (\triangle OAC \text{ 面积的 2 倍})$$

The velocity vector $\boldsymbol{v}_r$ is perpendicular to the OAC plane and "rotates in the direction of angular velocity". The magnitude of the linear velocity of the transport motion of point C is (速度矢量 $\boldsymbol{v}_r$ 垂直于 OAC 平面并"顺着角速度的转动方向". C 点在其牵连运动中线速度的大小等于)

$$v_e = DC \times \omega_e = 2 \text{ times the area of } \triangle OEC (\triangle OEC \text{ 面积的 2 倍})$$

The velocity vector is perpendicular to the *OEC* plane and rotates in the direction of angular velocity $\boldsymbol{\omega}_e$. Due to the same area of △*OAC* and △*OEC*, there are two velocity vectors acting at point *C*, with the same magnitude and opposite directions, so the sum of the two vectors is zero (该速度矢量顺着角速度 $\boldsymbol{\omega}_e$ 的转动方向垂直于 *OEC* 平面. 由于△*OAC* 和△*OEC* 的面积相等,因此在点 *C* 处作用有两个速度矢量,数量相等且方向相反,则它们的和为零.)

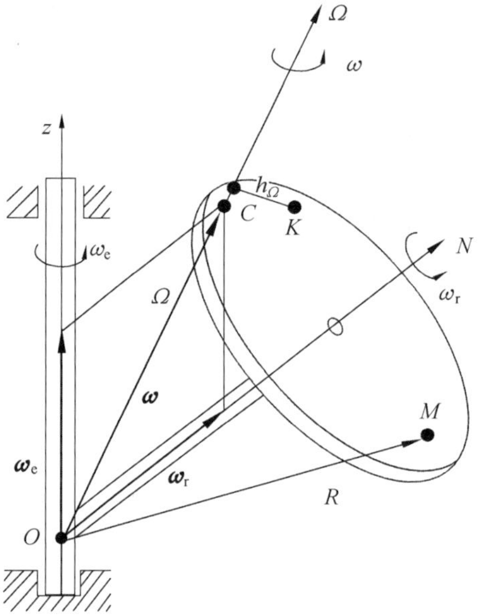

Fig. 7. 18　Calculation of the velocity of a point on a rigid body rotating around two intersecting axes
(图 7. 18　绕两相交轴转动的刚体上点的速度计算)

In this way, the straight line *OΩ* passing through two fixed points *O* and *C* is the instantaneous axis of the rotation of the object. Then it can be assumed that the disc (which can also be a rigid body of any other shape) rotates instantaneously around the axis *OΩ*. In this case, the velocity of any point *M* (Fig. 7. 18) can be defined as (这样,穿过两个固定的点 *O* 和点 *C* 的直线 *OΩ* 是物体的瞬时转动轴. 然后可以假设盘(也可为像任何其他形状的刚体)绕轴 *OΩ* 瞬时转动. 在这种情况下,任何点 *M*(图 7. 18)的速度可以定义为)

$$\boldsymbol{v}_M = \boldsymbol{\omega} \times R$$

On the other hand, this point participates in composite motion, so its velocity can be written in different ways, such as (另外,该点参与了合成运动,因此其速度可以用不同的方式来书写,如)

$$\boldsymbol{v}_M = \boldsymbol{v}_r + \boldsymbol{v}_e$$

where(其中)

$$\boldsymbol{v}_r = \boldsymbol{\omega}_r \times R, \boldsymbol{v}_e = \boldsymbol{\omega}_e \times R$$

So we have(则有)

$$\boldsymbol{\omega} \times R = \boldsymbol{\omega}_r \times R + \boldsymbol{\omega}_e \times R$$

Then(由此可见)

$$\boldsymbol{\omega} \times R = (\boldsymbol{\omega}_r + \boldsymbol{\omega}_e) \times R$$

Therefore (因此)

$$\boldsymbol{\omega} = \boldsymbol{\omega}_r + \boldsymbol{\omega}_e$$

Therefore, the vector sum of the relative rotation angular velocity and the transport rotation angular velocity is equal to the absolute rotation angular velocity vector. (因此,相对转动角速度和牵连转动角速度的矢量和等于绝对转动角速度矢量.)

The established expressions is called the parallelogram theorem of angular velocities. (建立的关系式称为角速度的平行四边形定理.)

By constructing a parallelogram with angular velocities, when two rotation motions are added to the intersecting axis, the velocity of any point of the object can be determined based on the relative instantaneous rotational axis (for example, for point K) (Fig. 7.18), that is(通过构造角速度的平行四边形,当对相交轴附加两个转动运动时,可以根据相对瞬时转动轴确定物体任何点的速度(如对于点 K)(图 7.18),即)

$$v_K = \omega h_\Omega$$

7.4.2 Composition of Rotation Motions around Parallel Axes of Rigid Bodies (刚体绕平行轴转动的合成)

In this case, the vectors of relative angular velocity and transport angular velocity are parallel. There are following several possible situations. (在这种情况下,相对角速度和牵动角速度的矢量是平行的. 有以下几种可能的情况.)

1. The relative rotation and induced rotation directions are the same
(相对转动和牵连转动方向相同)

Assume that Plane Ⅰ (Fig. 7.19(a)) rotates relative to Plane Ⅱ, while Plane Ⅱ rotates relative to the stationary Plane Ⅲ, and then the absolute motion of Plane Ⅰ will be synthesized relative to Plane Ⅲ. In this case, the motion of Plane Ⅱ is the transport motion. Plane Ⅰ and Plane Ⅱ can undergo similar movements in Plane Ⅲ (Fig. 7.19(b)). Due to the fact that both motions are rotation, the velocity at the intersection of the rotation axes ω_e and ω_r with Plane Ⅲ will be zero: at point P_e —the transport velocity, and at point P_r —the relative velocity. As is well known, the absolute velocity of any point in complex motion is equal to the geometric vector sum of relative velocity and transport velocity. As the transport velocity of the point is zero, its absolute velocity is equal to the relative velocity, we have(假设平面Ⅰ(图 7.19(a))相对于平面Ⅱ转动,而平面Ⅱ相对于静止的平面Ⅲ转动,然后平面Ⅰ的绝对运动将通过相对于平面Ⅲ进行合成. 在这种情况下,平面Ⅱ的运动是牵连运动. 平面Ⅰ和平面Ⅱ可以在平面Ⅲ中进行类似的运动(图 7.19(b)). 由于两个运动都是转动的,因此转动轴 ω_e 和 ω_r 与平面Ⅲ的交点处,速度将为零:在 P_e 点是牵连速度,在 P_r 点是相对速度. 众所周知,复杂运动中任何点的绝对速度等于相对速度和牵连速度的几何矢量和. 由于点 P_e 的牵连速度为零,因此其绝对速度将等于相对速度,有)

$$\boldsymbol{v}_{P_e} = \boldsymbol{v}_r + \boldsymbol{v}_e = \boldsymbol{v}_r + 0 = \boldsymbol{v}_r$$

The magnitude of the velocity is determined by the formula $v_{P_e} = P_eP_r\omega_r$. Its direction is perpendicular to its rotational radius (P_eP_r segment) and angular velocity vector, determined by the right-hand spiral rule.(该速度的大小由公式 $v_{P_e} = P_eP_r\omega_r$ 确定. 它的方向将垂直于其转动半径(P_eP_r段)和角速度矢量,并由右手螺旋法则确定.)

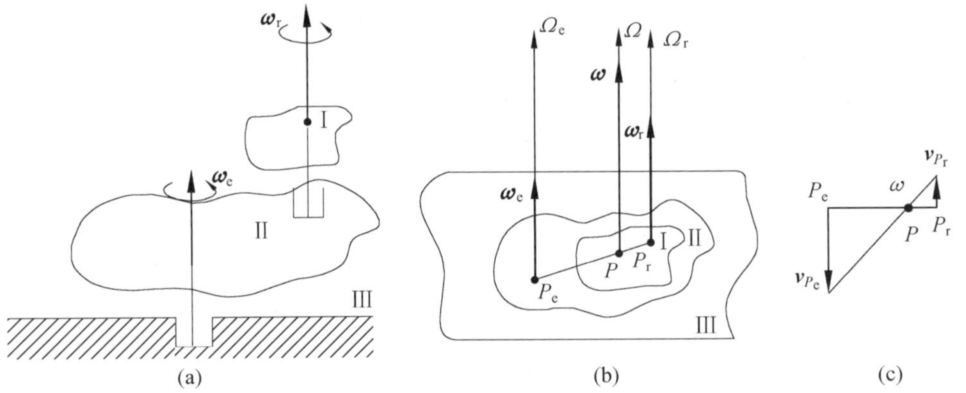

Fig. 7.19 Same Direction of Relative Rotation and Transport Rotation
(图 7.19 相对转动和牵连转动方向相同)

Similarly, for the point P_r, the absolute velocity is $\boldsymbol{v}_{P_r} = \boldsymbol{v}_r + \boldsymbol{v}_e = 0 + \boldsymbol{v}_e = \boldsymbol{v}_e$. The magnitude of the velocity is $v_{P_r} = P_eP_r\omega_e$. The vector is perpendicular to the line segment P_eP_r and the vector of the transport rotation angular velocity, determined by the right-hand spiral rule.(对于 P_r 点,同理,绝对速度 $\boldsymbol{v}_{P_r} = \boldsymbol{v}_r + \boldsymbol{v}_e = 0 + \boldsymbol{v}_e = \boldsymbol{v}_e$. 该速度的大小是 $v_{P_r} = P_eP_r\omega_e$,并且矢量垂直于线段 P_eP_r 及牵连转动角速度矢量,即由右手螺旋法则确定.)

Draw the absolute velocity vectors of points P_e and P_r in the graph, and then we find the instantaneous center of velocity P (Fig. 7.19(c)). This Figure indicates that the motion of Plane Ⅰ is composed of two parallel unidirectional rotation motions, and we can obtain the angular velocity as (在图中绘制点 P_e 和 P_r 的绝对速度矢量,然后找到速度瞬时中心 P(图 7.19(c)). 该图表明平面Ⅰ的运动是由两个平行的单向转动运动组成的,其角速度为)

$$\omega = v_{P_r}/PP_r = v_{P_e}/PP_e$$

From the similarity of triangles, $v_{P_e}/v_{P_r} = PP_e/PP_r$. Substitute the velocities v_{P_r} and v_{P_e} representing angular velocity of relative motion and transport motion and we obtain (从三角形的相似性来看,$v_{P_e}/v_{P_r} = PP_e/PP_r$. 代入代表相对运动和牵连运动角速度的速度 v_{P_r} 和 v_{P_e} 的值,可以得到)

$$\frac{PP_e}{PP_r} = \frac{\omega_r}{\omega_e}$$

From this, it can be seen that the instantaneous rotation axis Ω which is parallel to the axes Ω_e and Ω_r (Fig. 7.19) passes through the instantaneous center point P of velocity and divides the distance between these axes into sections that are inversely proportional to the angular velocity. (由此可见,瞬时转动轴 Ω(图 7.19)通过速度的瞬时中心点 P,平行于轴 Ω_e 和 Ω_r,将这些轴之间的距离分成与角速度成反比的区段.)

Determine the magnitude of the absolute angular velocity. To achieve this, we will replace $v_{P_r} = P_eP_r\omega_e$ with $v_{P_r} = \omega PP_r$. So, we have (确定绝对角速度大小. 为此,将 $v_{P_r} = P_eP_r\omega_e$ 替换为其值 $v_{P_r} = \omega PP_r$,则得到)

$$\omega PP_r = P_eP_r\omega_e$$

Considering $PP_e + PP_r = P_eP_r$, rewrite this equation, we have(考虑到 $PP_e + PP_r = P_eP_r$,重写该等式,有)

$$\omega PP_r = \omega_e(PP_e + PP_r)$$

or(或者)

$$\omega PP_r = \omega_e PP_e + \omega_e PP_r$$

Given $\omega_e PP_e = \omega_r PP_r$, we obtain(已知 $\omega_e PP_e = \omega_r PP_r$,得到)

$$\omega PP_r = \omega_r PP_r + \omega_e PP_r$$

By eliminating the common factor PP_r , we can obtain (在通过消除公共因子 PP_r 后,可以得到)

$$\omega PP_r = P_eP_r\omega_e$$

$$\omega PP_r = \omega_e(PP_e + PP_r)$$

or(或者)

$$\omega PP_r = \omega_e PP_e + \omega_e PP_r$$

$$\omega PP_r = \omega_r PP_r + \omega_e PP_r$$

We have(则有)

$$\omega = \omega_e + \omega_r$$

The absolute angular velocity is equal to the sum of the angular velocities of the components of unidirectional rotational motion. (绝对角速度大小等于单向转动运动分量的角速度大小之和.)

From Fig. 7. 19 (c), it can be known that the absolute rotation of the plane pattern is counterclockwise, and its composite motion is also the same. The direction of the absolute angular velocity vector $\boldsymbol{\omega}$ is along the Ω-axis, and its direction is the same as those of the vectors ω_e and ω_r (Fig. 7. 19). (从图 7. 19(c)中可以看出平面图形的绝对转动也是逆时针方向,其合成运动也是如此. 绝对角速度矢量 $\boldsymbol{\omega}$ 的指向沿着 Ω 轴,其方向与矢量 ω_e 和 ω_r 相同(图 7. 19).)

2. The direction of the relative rotation and the transport rotation are opposite, and their angular velocities are not equal in magnitude(相对转动和牵连转动方向相反,并且它们的角速度大小不相等)

We determine the absolute velocity of the transport velocity center P_e (Fig. 7. 20 (a)), we have(确定牵连速度中心 P_e 的绝对速度(图 7. 20(a)),有)

$$\boldsymbol{v}_{P_e} = \boldsymbol{v}_r + \boldsymbol{v}_e = \boldsymbol{v}_r + 0 = \boldsymbol{v}_r$$

The modulus of the vector is(该向量的模为)

$$v_{P_e} = \omega_r P_e P_r$$

The direction of the vector is determined by the right-hand spiral rule. Similarly, we can get the absolute velocity of the relative velocity center P_r, that is(方向按右手螺旋法则确定. 类似地,可以定义相对速度中心 P_r 的绝对速度,有)

$$\boldsymbol{v}_{P_r} = \boldsymbol{v}_r + \boldsymbol{v}_e = 0 + \boldsymbol{v}_e = \boldsymbol{v}_e$$

When viewed from the end of the vector $\boldsymbol{\omega}_e$, the point P_r relative to the center of motion rotates counterclockwise around the Ω_e axis. Therefore, the direction of vector $\boldsymbol{v}_{P_r}$ is determined by

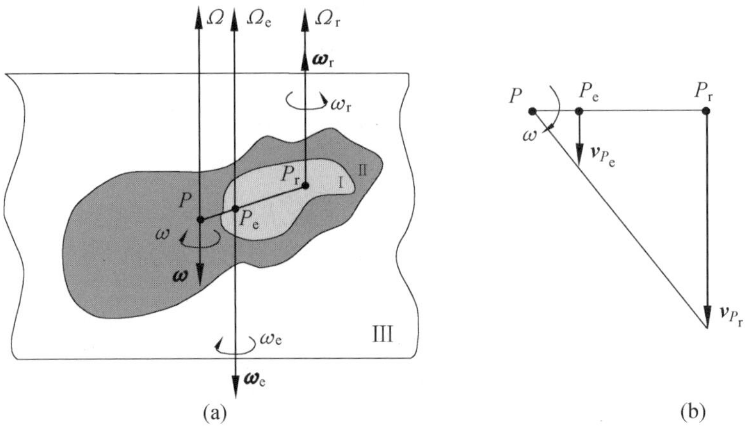

Fig. 7.20 Angular velocity of relative rotation and transport rotation with opposite direction and unequal magnitude

(图 7.20 相对转动和牵连转动角速度方向相反,大小不相等)

the right-hand spiral rule, pointing outward. The magnitude of vector $\boldsymbol{v}_{P_r}$ is(当从 $\boldsymbol{\omega}_e$ 矢量的末端观察时,相对运动中心的 P_r 点逆时针方向绕 Ω_e 轴旋转. 因此,向量 $\boldsymbol{v}_{P_r}$ 指向由右手螺旋法则确定,指向朝外. 向量 $\boldsymbol{v}_{P_r}$ 的大小为)

$$v_{P_r} = \omega_e P_e P_r$$

In Fig. 7.20(a), it shows that(在图中 7.20(a),可知)

$$\omega_e > \omega_r$$

So(因此)

$$v_{P_r} > v_{P_e}$$

We intercept the velocity vector $\boldsymbol{v}_{P_e}$ of point P_e and velocity vector $\boldsymbol{v}_{P_r}$ of point P_r (Fig. 7.20(b)) and use the graph to find the instantaneous velocity center, i. e. point P. If observed from the end of the instantaneous rotation axis Ω, the absolute rotation proceeds in a clockwise direction. (截取点 P_e 和 P_r 的速度向量 $\boldsymbol{v}_{P_e}$ 和 $\boldsymbol{v}_{P_r}$ (图 7.20(b))并利用图形找到瞬时速度中心,即点 P. 当从瞬时旋转轴 Ω 的末端观察时,绝对转动将沿着顺时针方向进行.)

Based on the similarity of triangles (Fig. 7.20(b)), we obtain (根据三角形的相似性(图 7.20(b)), 得出)

$$v_{P_e}/v_{P_r} = PP_e/PP_r$$

Substitute in the values of velocities v_{P_r} and v_{P_e} representing angular velocities of relative motion and transport motion, we obtain(代入表示相对运动和牵连运动角速度的速度 v_{P_r} 和 v_{P_e} 的值,得到)

$$\frac{PP_e}{PP_r} = \frac{\omega_r}{\omega_e}$$

Therefore, the instantaneous absolute rotation axis of a planar figure is parallel to the axis of the transport rotation and relative rotation. The axis passes through a plane, lies in the outside of the axis of the transport rotation and relative rotation, and the distance between the axis and other two axes is inversely proportional to the magnitude of the angular velocity. (因此,平面图形的瞬

时绝对转动轴平行于牵连转动和相对转动的轴. 该轴穿过平面,并在外侧与这些轴之间的距离和角速度大小成反比.)

To determine the angular velocity of absolute rotation, we use the equation (Fig. 7.20(b))(为确定绝对转动的角速度大小,使用关系式)

$$v_{P_r} = \omega PP_r$$

Meanwhile, as mentioned above, we have(与此同时,如上所述,有)

$$v_{P_r} = P_eP_r\omega_e$$

Make the right side of the equation equal and consider $P_eP_r = PP_r - PP_e$, we obtain(令等式右边相等,并考虑到 $P_eP_r = PP_r - PP_e$,得到)

$$\omega PP_r = \omega_e(PP_r - PP_e)$$

or (或者)

$$\omega PP_r = \omega_e PP_r - \omega_e PP_e$$

We have previously proven the equation $\omega_e PP_e = \omega_r PP_r$ and obtain $\omega PP_r = \omega_e PP_r - \omega_r PP_r$. After removing the factor PP_r, we can obtain(前面已证明了 $\omega_e PP_e = \omega_r PP_r$ 这个等式,得到了 $\omega PP_r = \omega_e PP_r - \omega_r PP_r$,在去掉因子 PP_r 后,可得到)

$$\omega = \omega_e - \omega_r$$

That is to say, the magnitude of the absolute angular velocity is equal to the difference in angular velocities of the rotation components in the opposite direction. The vector of the absolute angular velocity is consistent with the larger angular velocity vector and is located on the side of the axis with the larger rotation angular velocity (Fig. 7.20(a)). (也就是说,绝对角速度大小等于相反方向转动分量的角速度之差. 绝对角速度的矢量与较大的角速度矢量一致,并且位于转动角速度较大的轴一侧(图 7.20(a)).)

3. The directions of relative rotation and transport rotation are opposite, and their angular velocities are equal in magnitude (Fig. 7.21(a))(相对转动和牵连转动方向相反,它们的角速度大小相等(图 7.21(a)))

In this case, we can determine the absolute motion of the Plane Ⅰ. Since the magnitude of angular velocity is equal, then (针对这种情况,可以确定平面Ⅰ的绝对运动. 由于角速度的大小相等,因此)

$$\boldsymbol{\omega}_e = -\boldsymbol{\omega}_r$$

Plane Ⅰ makes a relative rotation around axis Ω_r, and Plane Ⅱ makes transport rotation around axis Ω_e. Because for any point in Plane Ⅰ, there are (平面Ⅰ绕轴 Ω_r 做相对转动,平面Ⅱ绕轴 Ω_e 做牵连转动. 因为对于平面Ⅰ中的任何一点,都有)

$$\boldsymbol{v} = \boldsymbol{v}_e + \boldsymbol{v}_r$$

so for point M, this equation is also correct. We determine its relative velocity. The vector $\boldsymbol{v}_r$ is perpendicular to the MP_r segment along the direction of the angular velocity ω_r (Fig. 7.21(b)). We can obtain the magnitude as(所以对于 M 点来说,该等式也是正确的. 下面确定它的相对速度. 矢量 $\boldsymbol{v}_r$ 将顺着角速度 ω_r 的转向垂直于 MP_r 段(图 7.21(b)),计算其大小,有)

$$v_r = \omega_r MP_r$$

Because (因为)

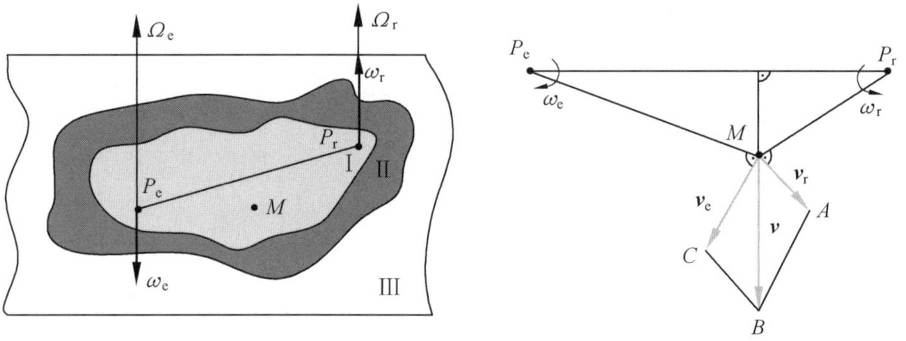

(a) (b)

Fig. 7.21 Angular velocities of relative rotation and transport rotation with opposite directions and equal magnitude

(图 7.21 相对转动和牵连转动角速度方向相反,大小相等)

$$\omega_r = \omega_e$$

so we obtain(得到)

$$v_r = \omega_e MP_r$$

The transport velocity of point M is perpendicular to the line segment MP_e along the direction of the transport angular velocity $\boldsymbol{\omega}_e$. Determine the magnitude of the transport velocity $\boldsymbol{v}_e$, we have (点 M 的牵连速度将顺着牵连角速度 $\boldsymbol{\omega}_e$ 转向垂直于线段 MP_e . 计算牵连速度大小 $\boldsymbol{v}_e$,有)

$$v_e = \omega_e MP_e$$

We construct a parallelogram based on the velocity vectors $\boldsymbol{v}_r$ and $\boldsymbol{v}_e$ (Fig. 7. 21 (b)). $\triangle MBC$ and $\triangle P_eMP_r$ is similar because their edges are proportional and perpendicular to each other. From the similarity of triangles, we obtain(依据速度矢量 $\boldsymbol{v}_r$ 和 $\boldsymbol{v}_e$ 构建一个平行四边形(图 7.21(b)). $\triangle MBC$ 和 $\triangle P_eMP_r$ 是相似的,因为两个三角形的边长成比例且互相垂直. 从三角形的相似性中得到)

$$\frac{v}{P_eP_r} = \frac{v_e}{MP_e} = \frac{v_r}{MP_r} = \omega_e = \omega_r = \omega$$

The MC and CB sides are respectively perpendicular to MP_e and MP_r , so the third side of these triangles is also perpendicular, that is, the vector $\boldsymbol{v}$ is perpendicular to the side P_eP_r . Therefore, the velocity vector of any point must be perpendicular to the P_eP_r segment and its magnitude is equal to(由于 MC 和 CB 边分别垂直于 MP_e 和 MP_r ,因此这些三角形的第三边也是垂直的,即矢量 $\boldsymbol{v}$ 垂直于侧面 P_eP_r . 因此,任意点的速度矢量必垂直于 P_eP_r 段,并且其大小等于)

$$v = \omega P_eP_r$$

If the velocities of all points of an object are the same in magnitude and direction, then the instantaneous velocity center of the object is located at an infinite distance, that is, the object undergoes translation motion. (如果物体所有点的速度在大小和方向上是相同的,那么这个物体的瞬时速度中心位于无限远处,即物体做平动.)

Therefore, when angular velocities of two rotations which is equal in magnitude and opposite in directions are combined, the motion we get is translational. (因此,当两个角速度大小相等、方向相反的转动合成时,所产生的运动是平动.)

The combination of two rotationswhose angular velocity has equal magnitude and opposite directions is called rotation couple. (两个角速度方向相反、大小相等的转动合成称为转动偶.)

Example 12: The mechanism is driven by an *OB* crank, which rotates at the angular velocity ω_0 (Fig. 7. 22 (a)). Use the rotation composition theorem of parallel axes to determine the velocity and instantaneous absolute angular velocity of the connecting rod *AC* relative to the *OB* crank. (**例题** 12:一个机构由一个 *OB* 曲柄驱动,它以角速度 ω_0转动(图 7.22(a)). 运用平行轴的转动合成定理,确定连杆 *AC* 相对于 *OB* 曲柄转动的速度及其瞬时绝对角速度.)

Solution:(解:)

(1) Determine the relative angular velocity of the connecting rod *AC*. The connecting rod *AC* performs a composite motion. Point *B* belongs to both *OB* and *AC* lingkages, so its relative rotation velocity is zero. (确定连杆 *AC* 的相对角速度. 连杆 *AC* 做复合运动. *B* 点同时属于 *OB* 和 *AC* 连杆,因此其相对转动的速度为零.)

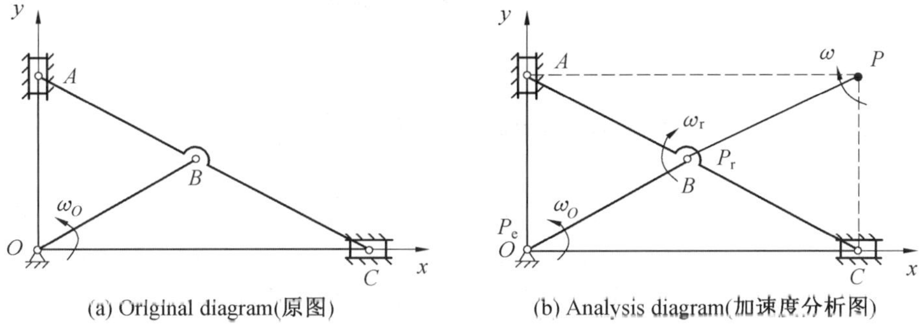

Fig. 7.22 Diagram of Example 12(图 7.22 例题 12 图)

So, this is an instantaneous velocity center P_r (Fig. 7. 22 (b)), which means it is the instantaneous velocity center of the connecting rod *AC* in its relative rotation motion. The transport rotation is the rotation of the crank *OB* relative to the fixed point *O*, denoted by P_e for rotation center. The axes of transport rotation and relative rotation are perpendicular to the graph surface, i. e. parallel to each other. Therefore, the parallel axis rotation motion composition theorem can be applied. (因此,这是一个瞬时的速度中心,这意味着它是 *AC* 连杆在其相对转动运动中的瞬时速度中心,用 P_r 表示(图 7.22(b)). 牵连转动是曲柄 *OB* 相对于固定点 *O* 的转动,用 P_e 表示该中心. 牵连转动和相对转动的轴垂直于图面,即彼此平行. 因此,可以应用平行轴转动运动合成定理.)

Since the directions of the velocities of point *A* and point *C* on the connecting rod *AC* are known (respectively along the axes *Ox* and *Oy*), the instantaneous center of velocity of the connecting rod can be found, represented by point *P* (Fig. 7.22(b)),we have(由于连杆 *AC* 的点 *A* 和点 *C* 的速度方向是已知的(分别沿着轴 *Ox* 和 *Oy*),因此可以找到该连杆的速度瞬心,即点 *P*,有(图 7.22(b)))

$$\frac{\omega_r}{\omega_e} = \frac{PP_e}{PP_r} = \frac{2OB}{OB} = 2$$

The relative angular velocity is(相对角速度为)

$$\omega_r = 2\omega_r = 2\omega_O$$

(2) Determine the instantaneous absolute angular velocity of the connecting rod AC. Point P (absolute velocity instantaneous center) is located outside the segment P_eP_r. Therefore, the directions of the transport angular velocity and relative angular velocity are opposite. And since point P is closer to P_r rather than P_e, then $\omega_r > \omega_e$. The absolute angular velocity of connecting rod AB is equal to(确定连杆 AC 的瞬时绝对角速度.点 P(绝对速度瞬心)位于区段 P_eP_r 的外部,因此牵连角速度与相对角速度的方向相反.并且由于点 P 更靠近 P_r 而不是 P_e,因此 $\omega_r > \omega_e$,故连杆 AB 的绝对角速度等于)

$$\omega = \omega_r - \omega_e = 2\omega_O - \omega_O = \omega_O$$

The relative angular velocity of the AC connecting rod is twice that of the OB crank, with opposite direction. At this point, the absolute angular velocity of the AC connecting rod is equal to the angular velocity of the OB crank, but in the opposite direction. ((AC 连杆的相对角速度是 OB 曲柄角速度的 2 倍,方向相反.此时,AC 连杆的绝对角速度等于 OB 曲柄的角速度,但是方向相反.)

7.5 Questions (思考题)

1. In the composite motion of the point, is the absolute acceleration of the moving point always equal to the vector sum of the transport acceleration and the relative acceleration? (在点的合成运动中,动点的绝对加速度总是等于牵连加速度与相对加速度的矢量和吗?)

2. The emergency return mechanism of the planer is shown in Fig. 7.23. One end A of the crank OA is connected to the slider with a hinge. When the crank OA rotates around the fixed axis O at a constant angle velocity ω, the slider slides on the rocker O_1B and drives the rocker O_1B swing around the fixed axis O_1. Let the crank length $OA = r$, the distance between the two axes $OO_1 = l$. When the crank is in the horizontal position, why should the crank OA not be used as the dynamic reference system when solve the angular velocity ω_1 of the rocker? If taking Point A on O_1B as the moving point, with crank OA as the moving reference system, whether the angular velocity and angular acceleration of O_1B can be found? (刨床的急回机构如图 7.23 所示.曲柄 OA 的一端 A 与滑块用铰链连接.当曲柄 OA 以匀角速度 ω 绕固定轴 O 转动时,滑块在摇杆 O_1B 上滑动,并带动摇杆 O_1B 绕固定轴 O_1 摆动.设曲柄长 $OA=r$,两轴间距离 $OO_1=l$.当曲柄在水平位置求摇杆的角速度 ω_1 时,为什么不宜以曲柄 OA 为动参考系?若以 O_1B 上的点 A 为动点,以曲柄 OA 为动参考系,是否可求出 O_1B 的角速度和角加速度?)

3. In Fig. 7.24, taking slider A as the moving point, and rod OC as the moving reference system, then we have(图 7.24 中取滑块 A 为动点,杆 OC 为动参考系,则有)

$$v_e = \omega \cdot OA, \quad v_a = v_e \cos\varphi$$

Is the above analysis right? What's wrong? (上面的分析是否正确?错在哪里?)

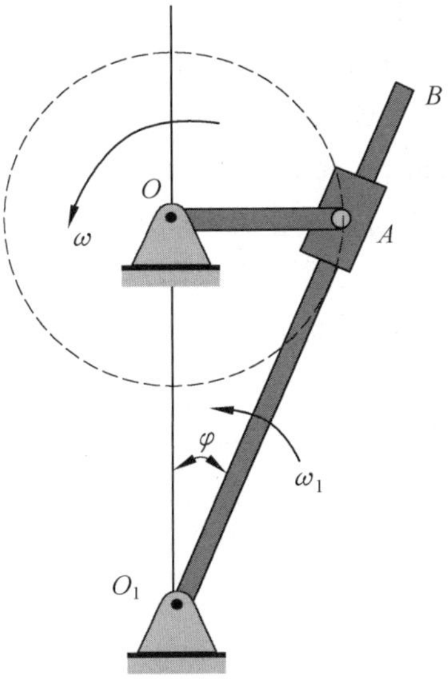

Fig. 7.23 Diagram of Question 2
(图 7.23 思考题 2 图)

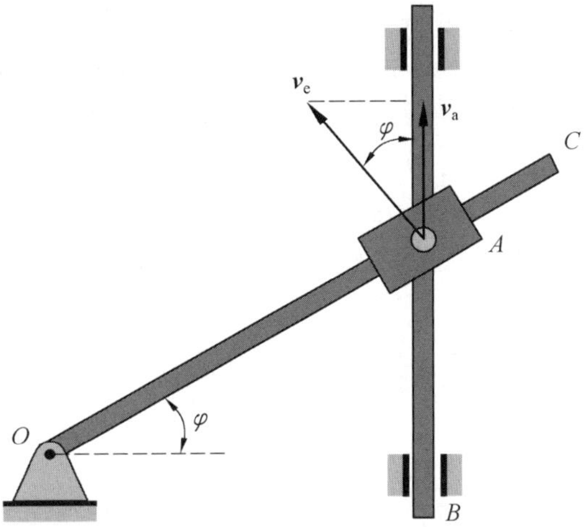

Fig. 7.24 Diagram of Question 3
(图 7.24 思考题 3 图)

4. When the transport motion is rotation, is the relative acceleration equal to the first derivative of the relative velocity to time? (当牵连运动为转动时,相对加速度就等于相对速度对时间的一阶导数吗?)

5. What is the formal difference between the projection theorem of the velocity composite theorem and the equilibrium equation in statistics? (速度合成定理及加速度合成定理的投影方程在形式上与静力学中的平衡方程有何不同?)

6. Using the concept of composite motion of points, prove that the acceleration formula at a point in polar coordinates is(试引用点的合成运动的概念,证明在极坐标中点的加速度公式为)

$$a_\rho = \ddot{\rho} - \rho\dot{\varphi}^2 \ , \ a_\varphi = \ddot{\varphi}\rho - 2\dot{\varphi}\dot{\rho}$$

where ρ and φ are the polar coordinates of points in polar coordinates; a_ρ and a_φ are the projection of the acceleration of the point along the polar diameter and in its vertical direction. (式中,ρ 和 φ 为极坐标下点的极坐标; a_ρ 和 a_φ 是点的加速度沿极径和其垂直方向的投影.)

7.6 Exercises(习题)

1. In Fig. 7.25, car A runs along a straight road with $v_1 = 40$ km/h, and car B drives along another fork road with $v_2 = 40\sqrt{2}$ km/h. Find the speed of car A observed in car B. (如图 7.25 所示,汽车 A 以 $v_1 = 40$ km/h 沿直线道路行驶,汽车 B 以 $v_2 = 40\sqrt{2}$ km/h 沿另一岔道行驶. 求在 B 车上观察到 A 车的速度.)

2. The banks of the river are parallel to each other, a ship is sailing at a constant speed from point A towards the opposite shore in a direction perpendicular to the shore, after 10 min, the ship arrives to the other side, the ship reached point C at the opposite side downstream 120 m, as

shown in Fig. 7.26. In order for the ship to depart from point A and reach point B on the opposite bank, it should sail in the opposite direction and maintain a certain angle with the line AB. In this case, the ship will arrive on the opposite bank after 12.5 min. Given the river width is l, try to find the relative velocity u of the ship to the water and the magnitude of the water flow velocity v. (河的两岸相互平行,一船由 A 点沿与岸垂直的方向向对岸匀速驶去,经过 10 min 后到达对岸,这时船到达对岸下游 120 m 处的 C 点,如图 7.26 所示. 为使船从 A 点出发能到达对岸的 B 点处,船应逆流并保持与 AB 线成某角度的方向航行,在此情况下,船经过 12.5 min 到达对岸. 求河宽 l、船相对水的相对速度 $\boldsymbol{u}$ 和水流速度 $\boldsymbol{v}$ 的大小.)

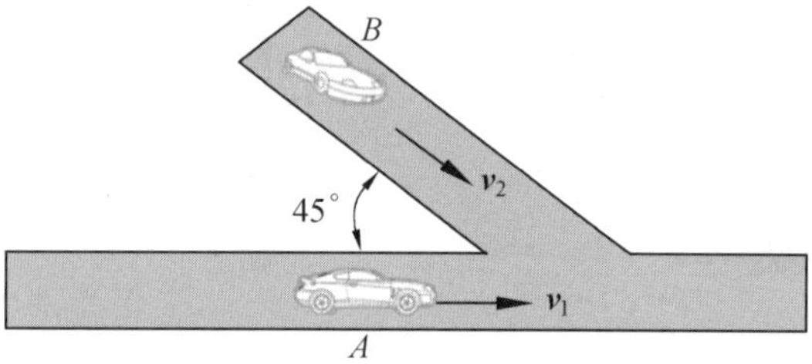

Fig. 7.25 Diagram of Exercise 1
(图 7.25 习题 1 图)

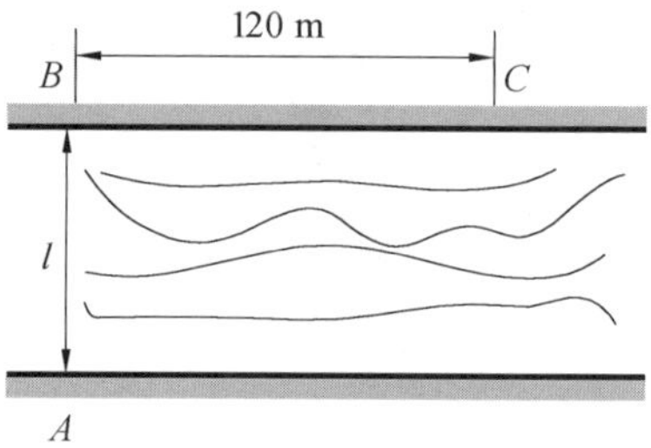

Fig. 7.26 Diagram of Exercise 2
(图 7.26 习题 2 图)

3. As shown in Fig. 7.27, light spot M undergoes harmonic vibration along the y-axis, and its equation of motion is(如图 7.27 所示,光点 M 沿 y 轴做简谐振动,其运动方程为)

$$x = 0 \text{ , } y = a\cos(kt + \beta)$$

If the point M is projected on the photo sensitive recording paper, while this paper moves to the left at an equal speed v_e, find the trajectory of the dot M on the recording paper. (若将点 M 投影到感光记录纸上,此纸以等速度 $\boldsymbol{v}_e$ 向左运动,求点 M 在记录纸上的轨迹.)

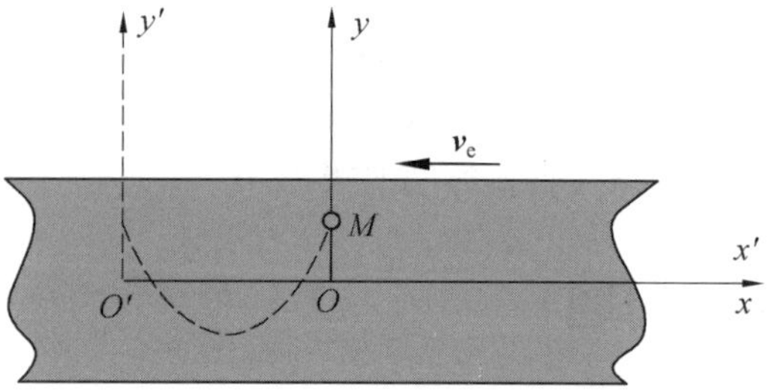

Fig. 7.27 Diagram of Exercise 3
(图 7.27 习题 3 图)

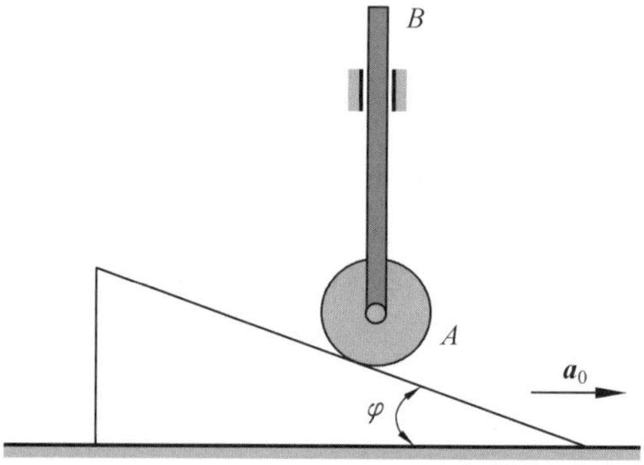

Fig. 7.28 Diagram of Exercise 4
(图 7.28 习题 4 图)

4. A smooth inclined surface at an angle of φ to the horizontal line moves in the horizontal direction, as shown in Fig. 7.28. The A end of the rod AB is hinged with a small wheel. When the small wheel rolls along the inclined plane, the AB rod can be driven to slide up and down along the guide rail in the vertical direction. If the inclined plane moves to the right with uniform acceleration a_0, try to find the acceleration of the AB bar. (与水平线成 φ 角的光滑斜面沿水平

方向运动,如图 7.28 所示. 杆 AB 的 A 端与一小轮相铰接,小轮沿斜面滚动时,可带动 AB 杆沿导轨在垂直方向上下滑动. 若斜面以匀加速度 a_0 向右运动,求 AB 杆的加速度.)

5. The length of rod OA is l, pushed by the push rod, and rotated around the point O in the diagram plane, as shown in Fig. 7.29. Suppose that the speed of the pushrod is $\boldsymbol{v}$, and its elbow height is a. Find the magnitude of the speed of the rod end A (expressed as A function of x). (杆 OA 长 l,由推杆推动而在图面内绕点 O 转动,如图 7.29 所示. 假定推杆的速度为 $\boldsymbol{v}$,其弯头高为 a,求杆端 A 的速度的大小(表示为 x 的函数).)

6. The speed of the lathe spindle $n = 30$ r/min, and the diameter of the workpiece d = 40 mm, as shown in Fig. 7. 30. If the transverse shifting speed of the turning blade is v = 10 mm/s, try to prove the relative motion trajectory of the workpiece is a spiral line, and find the pitch of the spiral line. (车床主轴的转速 $n = 30$ r/min,工件的直径 $d = 40$ mm,如图 7.30 所示. 若车刀横向走刀速度 $v = 10$ mm/s,证明车刀对工件的相对运动轨迹为螺旋线,并求出该螺旋线的螺距.)

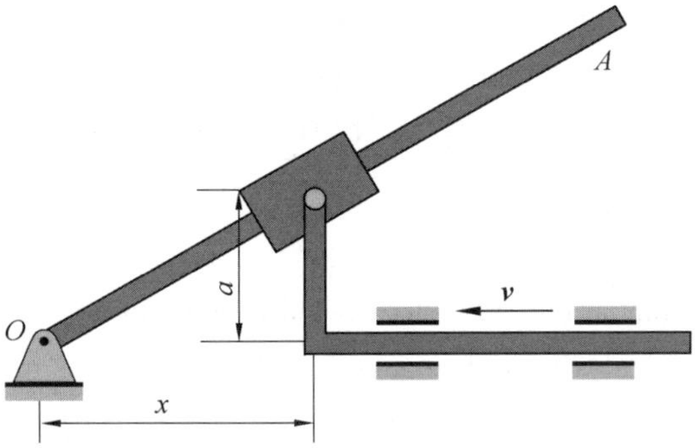

Fig. 7.29　Diagram of Exercise 5
(图 7.29　习题 5 图)

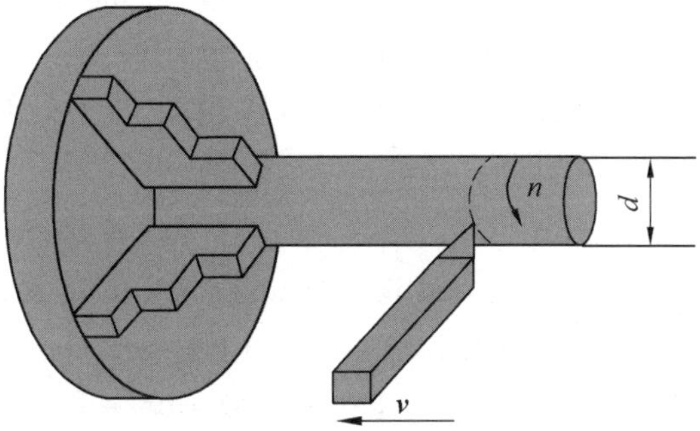

Fig. 7.30　Diagram of Exercise 6
(图 7.30　习题 6 图)

7. As shown in Fig. 7. 31, point M moves at v uniform speed along the disk diameter AB, and when starting, point M is in the center of the disk, and the diameter AB coincides with the x-axis. If the disk rotates around the O point at a uniform angular velocity ω, find the absolute trajectory of the point M. (如图 7.31 所示,点 M 沿圆盘直径 AB 以 v 匀速运动,当开始时,点 M 在圆盘中心,且直径 AB 与 x 轴重合. 若圆盘以均匀角速度 ω 绕 O 点转动,求点 M 的绝对轨迹.)

8. Cam with radius r and offset e, rotate around the O axis with uniform angular speed ω, and AB pole length is l, the A end is placed on the cam, and the B end is supported with a hinge. At the moment shown in Fig. 7.32, the AB rod is in the horizontal position. Try to find the angular speed of the AB rod. (半径为 r,偏心距为 e 的凸轮,以均匀角速度 ω 绕 O 轴转动,AB 杆长 l,A 端置于凸轮上,B 端用铰链支撑. 在图 7.32 所示瞬间,AB 杆处于水平位置. 试求 AB 杆的角速度.)

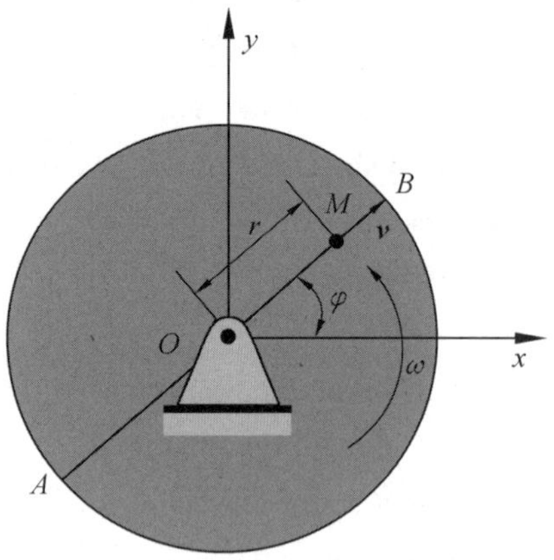

Fig. 7. 31 Diagram of Exercise 7
(图 7.31 习题 7 图)

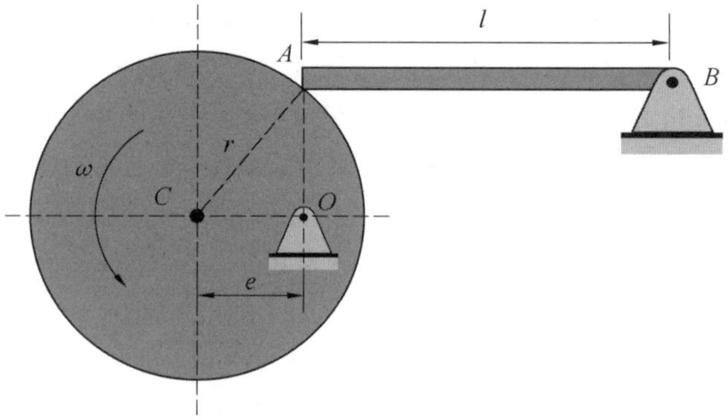

Fig. 7. 32 Diagram of Exercise 8
(图 7.32 习题 8 图)

9. In Fig. 7. 33, the small ring M moves along the rod OA, and the rod OA rotates around the O axis, so that the small ring has the following equation of motion in the Oxy plane, that is (图 7.33所示小环 M 沿杆 OA 运动,杆 OA 绕 O 轴转动, 从而使小环在 Oxy 平面内具有如下运动方程,即)

$$x = 10\sqrt{3}t, y = 10\sqrt{3}t^2$$

where t is measured in s; x and y are measured in mm. Find the velocity and acceleration of the small ring M relative to the rod OA, and the angular velocity and angular acceleration of the rod OA rotation at $t=1$ s. (式中,t 以 s 计;x 和 y 以 mm 计. 求 $t=1$ s 时,小环 M 相对于杆 OA 的速度和加速度,以及杆 OA 转动的角速度及角加速度.)

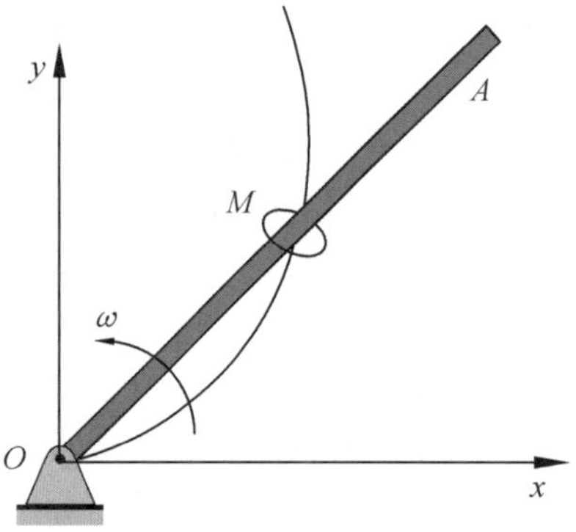

Fig. 7. 33 Diagram of Exercise 9
(图 7.33 习题 9 图)

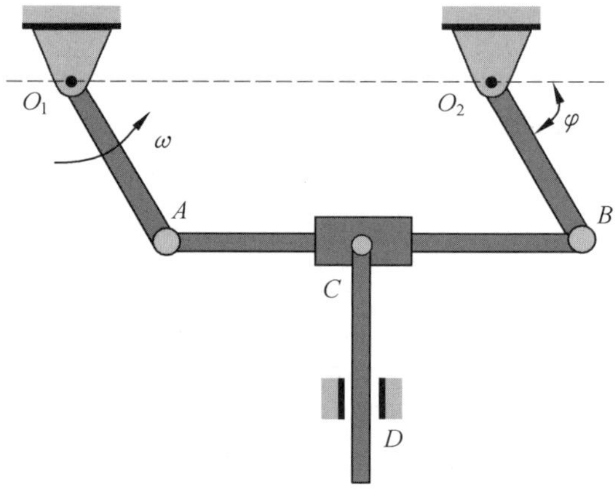

Fig. 7. 34 Diagram of Exercise 10
(图 7.34 习题 10 图)

10. As shown in Fig. 7. 34, an articulated quadrangle structure, $O_1A = O_2B = 100$ mm, and $O_1O_2 = AB$, rod O_1A turns around the axis O_1 at an equal angular velocity $\omega = 2$ rad/s. On the rod AB, there is a sleeve C, which is hinged to the rod CD. All parts of the mechanism are in the same vertical plane. Find the velocity and acceleration of the rod CD when $\varphi = 60°$. (图 7.34 所

示铰接四边形结构中，$O_1A=O_2B=100$ mm，又有 $O_1O_2=AB$，杆 O_1A 以等角速度 $\omega=2$ rad/s 绕轴 O_1 转动. 杆 AB 上有一套筒 C，此套筒与杆 CD 相铰接，机构的各部件都在同一铅垂面内. 求当 $\varphi=60°$时，杆 CD 的速度和加速度.)

11. As shown in Fig. 7.35, the length of the crank OA is 0.4 m and rotates counterclockwise around the O-axis at an equal angular velocity $\omega=0.5$ rad/s. The A end of the crank pushes the horizontal plate B, sliding rod C rises in the vertical direction. Find the velocity and acceleration of slide C when the angle between the crank and the horizontal line is $\theta=30°$. (如图 7.35 所示，曲柄 OA 长 0.4 m，以等角速度 $\omega=0.5$ rad/s 绕 O 轴逆时针转动. 曲柄的 A 端推动水平板 B，使滑杆 C 沿铅直方向上升. 求当曲柄与水平线间的夹角 $\theta=30°$时，滑杆 C 的速度和加速度.)

12. As shown in Fig. 7.36, a semi-circular cam with radius r rolls with a uniform velocity v_0 on the horizontal surface, the straight rod OA with a length of $\sqrt{2}r$ can rotate around the O axis. Find the velocity and acceleration of the illustrated instantaneous point A, and find the angular velocity and angular acceleration of the rod OA. (如图 7.36 所示，半径为 r 的半圆形凸轮以匀速 v_0在水平面上滚动，长为 $\sqrt{2}r$ 的直杆 OA 可绕 O 轴转动. 求图中瞬时点 A 的速度和加速度，并求杆 OA 的角速度和角加速度.)

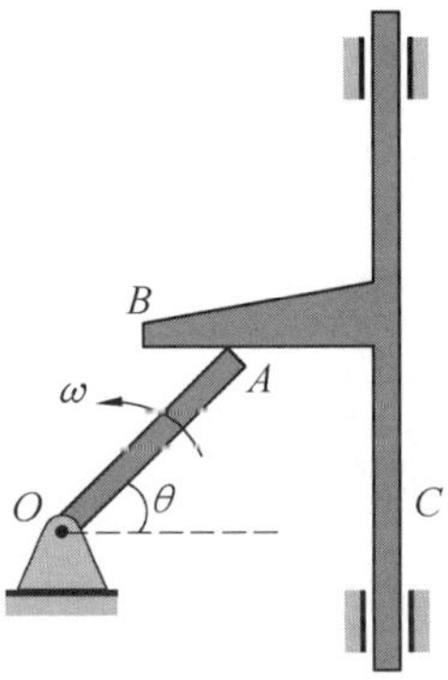

Fig. 7.35 Diagram of Exercise 11
(图 7.35 习题 11 图)

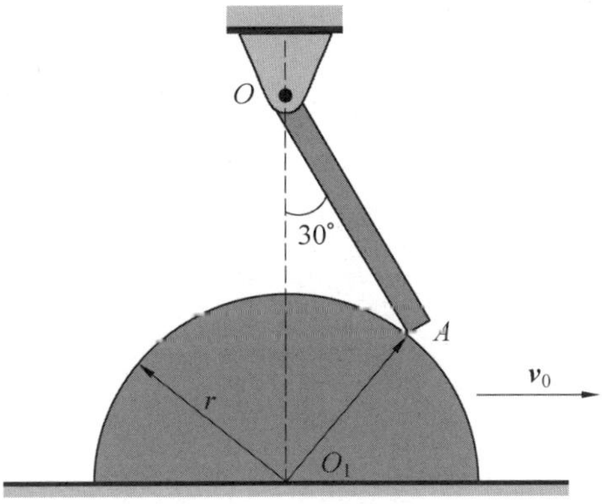

Fig. 7.36 Diagram of Exercise 12
(图 7.36 习题 12 图)

13. As shown in Fig. 7.37, the ring with radius r is filled with liquid. The liquid moves at a uniform relative speed v in the direction of the arrow in the ring. The ring shown in the figure rotates at an equal angular velocity ω around the O-axis. Find the absolute acceleration of the liquid at points 1 and 2 inside the ring. (如图 7.37 所示，半径为 r 的圆环内充满液体，液体按箭头方向以相对速度 v 在环内做匀速运动. 图中，环以等角速度 ω 绕 O 轴转动，求在圆环内点 1 和 2 处液体的绝对加速度的大小.)

14. As shown in Fig. 7.38, the discs rotates around the AB axis with an angular velocity $\omega=2t$(φ in rad, t in s). Point M moves away from the center to the outer edge with a pattern of $OM=40t^2$(OM in mm, t in s). The radius OM is a 60° angle with the AB axis. Find the magnitude of the absolute acceleration of point M at the time $t=1$ s. (如图 7.38 所示，圆盘绕

AB 轴转动,其角速度 $\omega=2t$ (φ 以 rad 计,t 以 s 计). 点 M 沿圆盘直径离开中心向外缘运动,其运动规律为 $OM=40t^2$(OM 以 mm 计,t 以 s 计),半径 OM 与 AB 轴间成 60°. 求当 $t=1$ s 时,点 M 的绝对加速度的大小.)

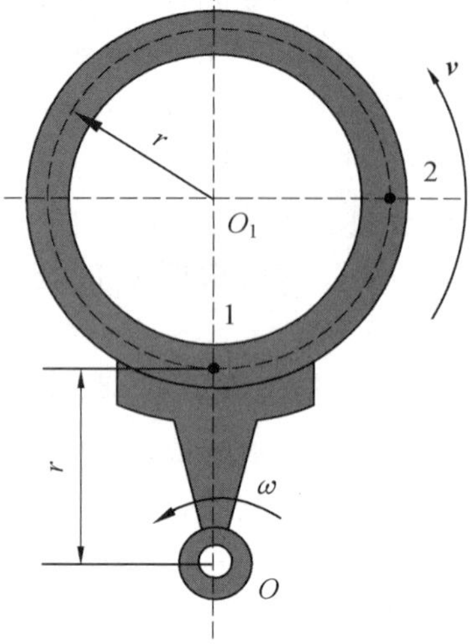

Fig. 7.37 Diagram of Exercise 13
(图 7.37 习题 13 图)

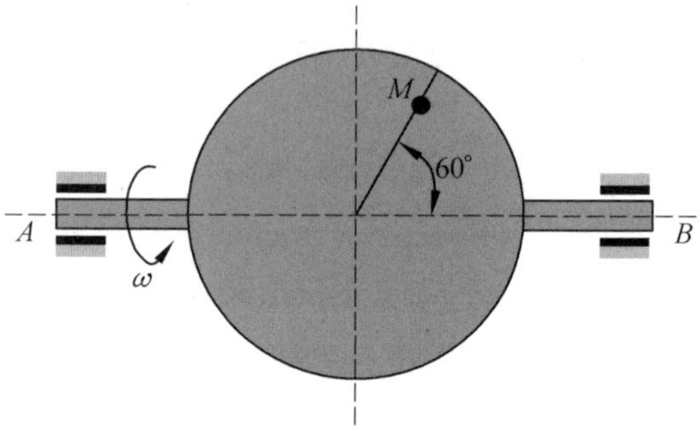

Fig. 7.38 Diagram of Exercise 14
(图 7.38 习题 14 图)

Chapter 8 Planar Motion of a Rigid Body (刚体的平面运动)

In Chapter 6, we discuss the translate and rotation of a rigid body, which are the two most common and simple forms of rigid body motion. Rigid bodies have more complex forms of motion, among which planar motion of rigid bodies is a common type of motion in construction machinery. It can be seen as the synthesis of translational and rotational motion of rigid bodies, as well as rotation around an axis of constant motion. This chapter introduces the decomposition of the plane motion of a rigid body, the angular velocity and angular acceleration of a rigid body in plane motion, and the method of solving for the velocity and acceleration of each point on a rigid body. (第6章讨论了刚体的平动和定轴转动,这是两种最常见、最简单的刚体运动. 刚体还有更复杂的运动形式. 其中,刚体的平面运动是工程机械中较为常见的一种运动,它可视为刚体的平动和转动的合成,也可视为绕不断运动的轴的转动. 本章介绍刚体平面运动的分解、平面运动刚体的角速度与角加速度、刚体上各点的速度及加速度的求解方法.)

8.1 Overview and Motion Decomposition of Rigid Body Plane Motion(刚体平面运动的概述和运动分解)

During motion, the distance between any point on a rigid body and a fixed plane is always constant, and this motion of a rigid body is called plane motion. In planar motion, all points on the body move in a plane parallel to a fixed plane. (在运动过程中,刚体上任意一点与某固定平面的距离始终保持不变,刚体的这种运动称为平面运动. 平面运动时,其上各点都在平行于某一固定平面的平面内运动.)

The motion of the connecting rod AB, slider B and wheel C in Fig. 8.1 is all planar motion. (图8.1中,连杆AB、滑块B及车轮C的运动都是平面运动.)

The line A_1A_2, which is perpendicular to the fixed plane in Fig. 8.2, is always translating, and its intersection with the plane figure S is A. According to the nature of the translational motion of a rigid body, the motion of the intersection A can represent the motion of the line A_1A_2. Numerous lines parallel to A_1A_2 form the rigid body, and each line has an intersection with the plane figure S. Thus, the motion of the plane figure S represents the motion of the rigid body in plane motion. That is, the motion of a rigid body in the plane can be reduced to the motion of a plane figure in its own plane. (图8.2中,与固定面垂直的直线A_1A_2始终做平动,其与平面图形S的交点为A,根据刚体平动的性质,交点A的运动可代表直线A_1A_2的运动. 无数与A_1A_2平行的直线组成了刚体,每一直线都与平面图形S有一交点,于是平面图形S的运动就代表

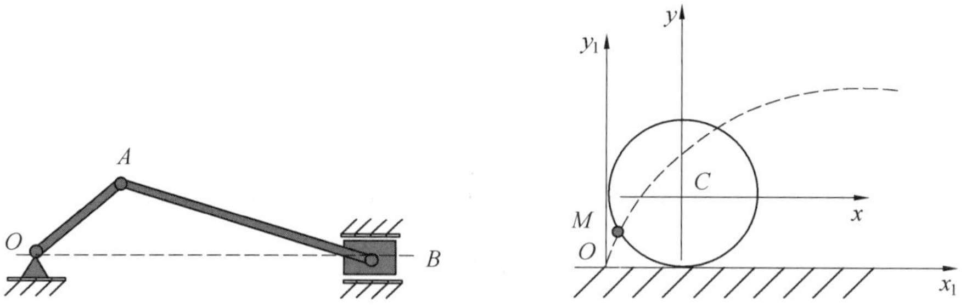

(a) Connecting rod AB in the crank-slider structure
(曲柄-滑块结构中的连杆AB)

(b) A wheel rolling on a flat track
(在平直的轨道上滚动的车轮)

Fig. 8. 1 Example of planar motion(图 8. 1 平面运动实例)

平面运动中刚体的运动,即刚体平面运动可简化为平面图形在其自身平面内的运动.)

Thus, the motion of all points on a rigid body in plane motion can be studied by simply determining the motion of points within a certain S-section parallel to the plane of motion. (因此,只需确定平行于运动平面某一 S 截面内各点的运动,就可以研究平面运动刚体上所有点的运动.)

So, in the following analysis, we have drawn only section S, and the points of this section are moving in the plane shown in Fig. 8. 3. (因此,在后续的分析中,仅画了截面 S,而这一截面的各点是在图 8. 3 所示平面上运动的.)

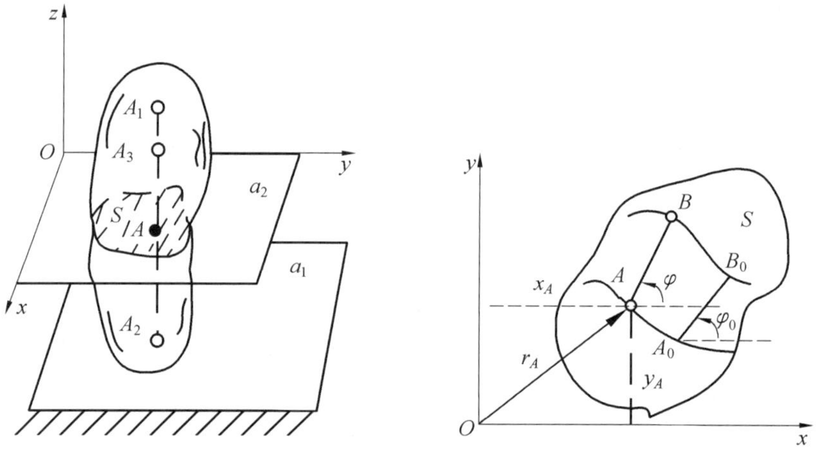

Fig. 8. 2 Simplification of rigid body plane motion
(图 8. 2 刚体平面运动的简化)

Fig. 8. 3 Motion of the cross-section S
(图 8. 3 截面 S 的运动)

The position of the section S and its points will be determined by the xy coordinate axis system. During the motion of a rigid body, the section S will rotate as it moves in the plane. Therefore, the position of the section will be determined using the coordinates x_A and y_A of a point A (the base point) and the angle φ between an arbitrarily drawn line segment AB and the x-axis. (截面 S 及其点的位置将通过 xy 坐标轴系确定. 在刚体运动时,S 截面在平面上移动的同时也会转动. 因此,将使用某点 A(基点)的坐标 x_A 和 y_A 以及任意绘制的线段 AB 与 x 轴之间的角度 φ 来确定截面的位置.)

The motion of any line segment AB on the plane figure in Fig 8. 3 can represent the motion of the plane figure, i. e. the plane motion of a rigid body. (图 8.3 中平面图形上任意线段 AB 的运动可以代表平面图形的运动,也就是刚体的平面运动.)

To determine the position of a section S in real time and its points at any given time, simply set the time function, that is(要实时确定截面 S 及其各点在任意时间的位置,只需设定时间函数即可,即)

$$x_A = f_1(t), y_A = f_2(t), \varphi = f_3(t) \tag{8.1}$$

These functions are called the equations of plane motion. Of course, if we choose another point as the base point, i. e. point B, the first two equations will change because point B moves differently and along a different trajectory. (这些函数称为平面运动方程. 当然,如果将另一个点选为基点,如 B 点,则前两个方程式将发生变化,因为 B 点的运动方式不同并且沿着不同的运动轨迹.)

If the point A in the plane figure is fixed, the motion of the plane figure is the fixed-axis rotation of the rigid body; if the orientation of the line AB remains the same, i. e. φ unchanged, the motion of the plane figure is the advection of the rigid body. Thus, the plane motion of the rigid body consists of two forms of simple motion of the rigid body: advection and fixed-axis rotation. (如果平面图形中的 A 点固定不动,则平面图形的运动为刚体的定轴转动;如果线段 AB 的方位不变,即 φ 不变,则平面图形的运动为刚体的平动. 因此,刚体的平面运动包含刚体简单运动的两种形式:平动和定轴转动.)

The planar motion of a rigid body can be divided into translation motion with the chosen base point and rotation with respect to the chosen base point. The rigid body in position Ⅰ in Fig. 8.4 can be moved to position Ⅱ in two ways. (刚体的平面运动可以分为随所选基点的平移和相对于所选基点的旋转. 在图 8.4 中,位置Ⅰ的刚体可以通过两种方式移动到位置Ⅱ.)

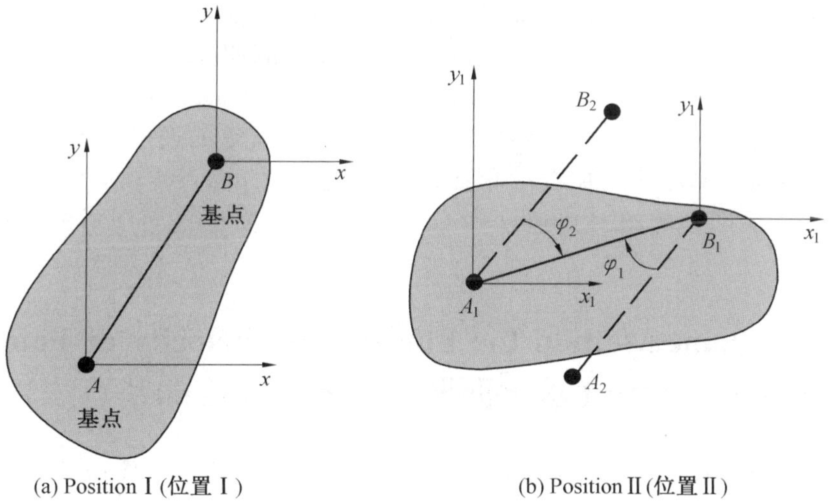

Fig. 8.4 Selection of different base points to study plane motion (图 8.4 选取不同基点研究平面运动)

(1) Moving the rigid body parallel to the original position so that the line AB parallel to the original position occupies position A_2B_1 in space, we rotate the rigid body around point B_1 at an angle φ_1. (平行移动刚体,使平行于原始位置的直线 AB 在空间中占据位置 A_2B_1,围绕点 B_1

以角度 φ_1 旋转刚体.）

（2）Move the rigid body parallel from position Ⅰ so that the line AB is aligned with the line A_1B_2 parallel to it. Afterwards, we will rotate the rigid body around point A_1 until point B_2 is turned to point B_1. Since $A_1B_2 /\!/ A_2B_1$, the angle $\varphi_1 = \varphi_2$.（从位置Ⅰ平移刚体，使得直线 AB 与平行于它的直线 A_1B_2 对齐. 然后，将围绕 A_1 点旋转刚体，直到 B_2 点转到 B_1 点. 由于 $A_1B_2 /\!/ A_2B_1$，因此角度 $\varphi_1 = \varphi_2$.）

Therefore, the angular velocity relationshipis are（因此，角速度关系为）

$$\omega_1 = \dot{\varphi}_1, \omega_2 = \dot{\varphi}_2$$

The angular acceleration relationships are（角加速度关系为）

$$\varepsilon_1 = \ddot{\varphi}_1, \varepsilon_2 = \ddot{\varphi}_2$$

Both angular velocity and angular acceleration are equal.（角速度和角加速度均是相等的.）

Thus, in order to occupy position Ⅱ, the rigid body can perform various translational motions（depending on the chosen base point）and the rotation is the same in both cases. Any planar motion of the rigid body can therefore be decomposed into a translational motion of the rigid body following the base point and a rotation with respect to the base point.（因此，为占据位置Ⅱ，刚体可以执行各种平移运动（取决于所选择的基点），并且在上述两种情况下的旋转是相同的. 因此，刚体的任何平面运动可以分解为刚体跟随基点的平移运动和相对于基点的转动.）

Conclusion：The trajectory and velocity of the translational part of the rigid body in plane motion are related to the selection of the base point, while the angle of rotation, the angular velocity and angular acceleration are independent of the selection of the base point. Therefore, the relative angular velocity and relative angular acceleration of a plane figure relative to the base point of rotation is also called the absolute angular velocity and absolute angular acceleration of a plane figure, which simply called as the angular velocity and angular acceleration of a plane figure.（**结论**：刚体平面运动中平动部分的轨迹及速度与基点的选取有关，而转动的角度、角速度和角加速度与基点的选取无关. 因此，平面图形相对于基点转动的相对角速度和相对角加速度又称平面图形的绝对角速度和绝对角加速度，简称平面图形的角速度和角加速度.）

8.2 Base Point Method for Finding the Velocity of Points in a Plane Figure（求平面图形内各点速度的基点法）

The velocity theorem of a point in a plane figure and its corollary: The velocity of any point on a plane figure is equal to the vector sum of the velocity following the motion of the base point and the linear velocity of the rotation of that point with respect to the base point.（平面图形内点的速度定理及其推论：平面图形上任意点的速度等于跟随基点运动的速度与该点相对于基点转动的线速度的矢量和.）

Take point O as the base point, whose velocity is known and equal to $\boldsymbol{v}_O$. It is possible to determine the velocity of any point belonging to this plane figure, i. e. point A (Fig. 8.5). From any fixed point O_1 to point O and point A in the plane, we plot the radius vector $\boldsymbol{\rho}_O$ and $\boldsymbol{\rho}_A$, and the radius vector $\boldsymbol{r}_{OA}$ from point O to point A. It rotates around the base point O with the angular velocity $\boldsymbol{\omega}$ of the plane figure, and the magnitude of this vector remains constant because the distance between point A and the base point O does not change. (取 O 点作为基点,其速度已知且等于 $\boldsymbol{v}_O$. 可以确定属于这个平面图形的任何点的速度,如点 A(图 8.5). 从平面内任意定点 O_1 至 O 点和 A 点绘制矢径 $\boldsymbol{\rho}_O$ 和 $\boldsymbol{\rho}_A$,以及从 O 点到 A 点的矢径 $\boldsymbol{r}_{OA}$. 它以平面图形 $\boldsymbol{\omega}$ 的角速度绕基点 O 旋转运动,并且该矢量的大小保持不变,因为点 A 与基点 O 之间的距离不会改变.)

Furthermore, as can be seen in Fig. 8.5, we have(另外,从图 8.5 中可以看出)

$$\boldsymbol{\rho}_A = \boldsymbol{\rho}_O + \boldsymbol{r}_{OA}$$

This determines the velocity of point A, that is(由此确定点 A 的速度,即)

$$\boldsymbol{v}_A = \mathrm{d}\boldsymbol{\rho}_A/\mathrm{d}t = \mathrm{d}\boldsymbol{\rho}_O/\mathrm{d}t + \mathrm{d}\boldsymbol{r}_{OA}/\mathrm{d}t \tag{8.2}$$

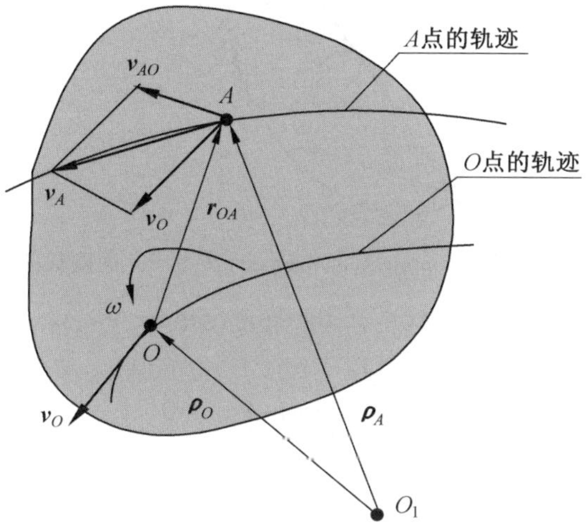

Fig. 8.5 Base point method (图 8.5 基点法)

The time derivative of the radius vector $\boldsymbol{\rho}_O$ is the velocity of the base point, and the time derivative of the radius vector $\boldsymbol{r}_{OA}$ is exactly the linear velocity of point A as it rotates around O, which we will denote as $\boldsymbol{v}_{AO}$. Thus, the theorem can be proved as(矢径 $\boldsymbol{\rho}_O$ 的时间导数是基点的速度,矢径 $\boldsymbol{r}_{OA}$ 的时间导数正是 A 点围绕 O 旋转时的线速度,将其表示为 $\boldsymbol{v}_{AO}$. 因此,可以证明该定理,有)

$$\boldsymbol{v}_A = \boldsymbol{v}_O + \boldsymbol{v}_{AO} \tag{8.3}$$

The velocity $\boldsymbol{v}_{AO}$ can be expressed as the vector product of the radius vector $\boldsymbol{r}_{OA}$ and the angular velocity vector of the plane figure, that is(速度 $\boldsymbol{v}_{AO}$ 可表示为矢径 $\boldsymbol{r}_{OA}$ 与平面图形的角速度向量的矢量乘积,即)

$$\boldsymbol{v}_{AO} = \boldsymbol{\omega} \times \boldsymbol{r}_{OA}$$

The velocity vector $\boldsymbol{v}_{AO}$ follows the steering of the rigid body perpendicular to the OA line segment (Fig. 8.5), and its magnitude is(速度矢量 $\boldsymbol{v}_{AO}$ 顺着刚体的转向垂直于 OA 线段(图 8.5),其大小为)

$$v_{AO} = \omega \times OA$$

Corollary 1: The points of a plane figure have equal velocity projections on the line passing through them. (**推论** 1:平面图形内各点在通过这些点的连线上的速度投影相等.)

As shown in Fig. 8.6, with point A as the base point, the velocity of point B is(如图 8.6 所示,以 A 点为基点,B 点的速度为)

$$\boldsymbol{v}_B = \boldsymbol{v}_A + \boldsymbol{v}_{BA} \tag{8.4}$$

of which(其中)

$$v_{BA} = \omega_{AB} \cdot BA$$

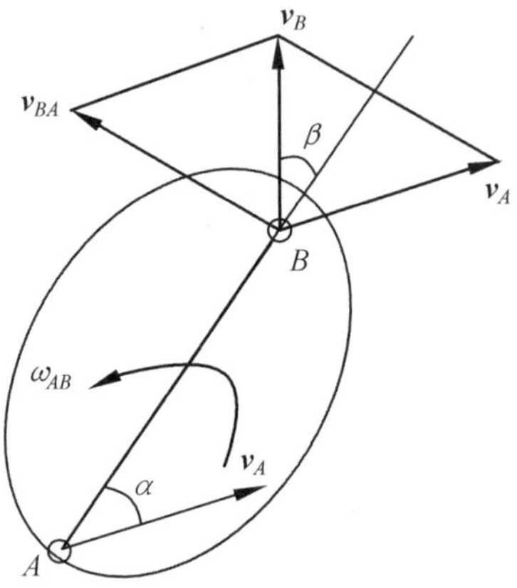

Fig. 8.6 Velocity projection theorem(图 8.6 速度投影定理)

Projecting Eq. (8.4) onto the axis of the line joining the two points of AB, we get(将式(8.4)向 AB 两点连线的轴投影,得到)

$$(\boldsymbol{v}_B)_{AB} = (\boldsymbol{v}_A)_{AB} + 0$$

That is(即)

$$(\boldsymbol{v}_B)_{AB} = (\boldsymbol{v}_A)_{AB}$$

That is, the projection of the velocity of any two points on a plane figure on the line joining those two points is equal. Corollary 1 is proved below. (即平面图形上任意两点的速度在这两点连线上的投影相等. 下面证明推论 1.)

Assume that the angular velocity ω of the plane figure (its magnitude and direction) and the velocity $\boldsymbol{v}_M$ of the point M in that figure are known at a given moment (Fig. 8.7). Take the base point M and determine the velocities of points B and C, which lie on the same line with the base point M, according to the velocity theorem for points of a plane figure, we have(假设在给定的时刻,平面图形的角速度 ω (其大小和方向)和该图中点 M 的速度 $\boldsymbol{v}_M$ 是已知的(图 8.7). 取基点 M,并根据平面图形点的速度定理确定与基点 M 位于同一直线上的点 B 和 C 的速度,有)

$$\boldsymbol{v}_B = \boldsymbol{v}_M + \boldsymbol{v}_{BM}, \ \boldsymbol{v}_C = \boldsymbol{v}_M + \boldsymbol{v}_{CM} \tag{8.5}$$

The velocity vector $\boldsymbol{v}_{BM}$ and $\boldsymbol{v}_{CM}$ is perpendicular to the MB segment and the velocity direction follows the rotation direction of the plane shape. Plot the x-axis through points M, C, B and project the velocity onto the x-axis, we have(速度矢量 $\boldsymbol{v}_{BM}$ 和 $\boldsymbol{v}_{CM}$ 垂直于 MB 段并且指向顺着平面图形的转向. 通过点 M、C、B 绘制 x 轴并将速度投射到 x 轴上,有)

$$v_{Bx} = v_{Mx} + v_{BMx}, \quad v_{Cx} = v_{Mx} + v_{CMx} \tag{8.6}$$

The projection of v_{BMx} and v_{CMx} on the x-axis is zero because the vector $\boldsymbol{v}_{BM}$ and $\boldsymbol{v}_{CM}$ is perpendicular to this axis. Therefore, we have(x 轴上的 v_{BMx} 和 v_{CMx} 投影为零，因为 $\boldsymbol{v}_{BM}$ 和 $\boldsymbol{v}_{CM}$ 矢量与此轴垂直. 因此，有)

$$v_{Bx} = v_{Mx} = v_{Cx}$$

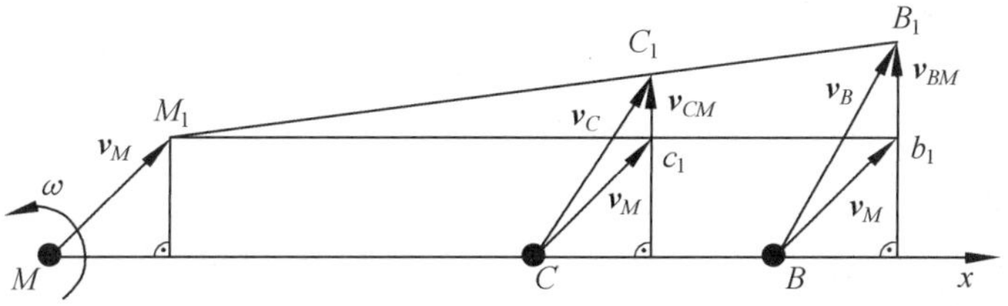

Fig. 8.7 Equal projection of the velocities of the points of a plane figure on the line passing through these points
(图 8.7 平面图形的各点在通过这些点的连线上的速度投影相等)

Corollary 2: The end of the velocity vector of all points located on the same line segment lies on a straight line and divides it into segments proportional to the distance between the corresponding points of the segment. (**推论** 2:位于一条线段上的点的速度矢量末端位于一条直线上,并将其分成与该段相应点之间距离成比例的几段.)

From Fig 8.7, it is clear that there is(由图 8.7,很明显有)

$$b_1B_1 = v_{BM} = MB \cdot \omega, \quad c_1C_1 = v_{CM} = MC \cdot \omega$$

Therefore(因此)

$$\frac{c_1C_1}{b_1B_1} = \frac{MC}{MB}$$

$MC = M_1C_1$ and $MB = M_1B_1$, opposite sides of a parallelogram are equal. This leads to($MC = M_1c_1$, $MB = M_1b_1$,平行四边形的对边相等,由此可得)

$$\frac{c_1C_1}{b_1B_1} = \frac{M_1c_1}{M_1b_1}$$

It follows that $M_1C_1B_1$ is a straight line. From the similarity of the triangles of $\triangle M_1c_1C_1$ and $\triangle M_1b_1B_1$, we have(由此可得, $M_1C_1B_1$是一条直线段. 由$\triangle M_1c_1C_1$和$\triangle M_1b_1B_1$的三角形相似性可得)

$$\frac{M_1C_1}{M_1B_1} = \frac{M_1c_1}{M_1b_1}$$

or(或者)

$$\frac{M_1C_1}{M_1B_1} = \frac{MC}{MB}, \quad \frac{M_1C_1}{C_1B_1} = \frac{MC}{CB}$$

These are the proofs we need.(以上便是所需要的证明.)

Example 1: $AB = l = 200$ mm and $v_A = 200$ mm/s are known in Fig. 8.8(a). Find: (1) The velocity v_B at the end of the rod B; (2) AB the angular velocity ω_{AB} of the rod. (**例题** 1:已知图 8.8(a)中 $AB = l = 200$ mm, $v_A = 200$ mm/s. 求: (1)杆端 B 的速度 v_B; (2) AB 杆角速度 ω_{AB}.)

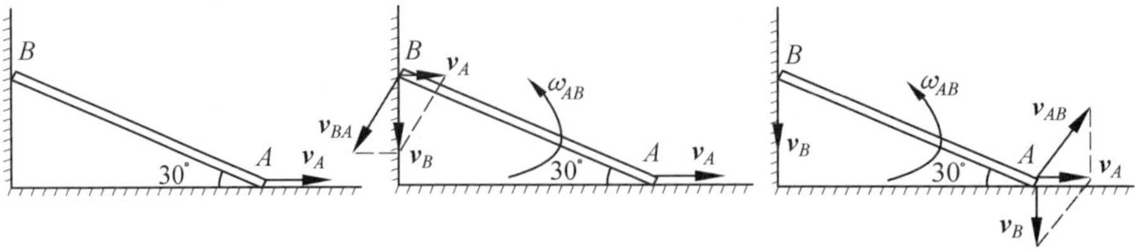

(a) Original diagram(原图)　(b) Take A as base point (以A为基点)　(c) Take B as base point (以B为基点)

Fig. 8. 8　Diagram of Example 1(图 8. 8　例题 1 图)

Solution 1:(解法 1:)

Take point A as the base point, analyze the velocity of point B, draw the velocity vector diagram (Fig. 8. 8(b)), and from the velocity synthesis theorem, we have(取 A 点为基点,分析 B 点的速度,画速度矢量图(图 8. 8(b)),由速度合成定理可得)

$$\boldsymbol{v}_B = \boldsymbol{v}_A + \boldsymbol{v}_{BA}$$

Solution is(得到)

$$v_B = v_A \cot 30° = 200\sqrt{3}\ \text{mm/s}\ ,\ v_{BA} = v_A / \sin 30° = 400\ \text{mm/s}$$

We obtain(则有)

$$\omega_{AB} = \frac{v_{BA}}{AB} = 2\ \text{rad/s}$$

Solution 2:(解法 2:)

Take point B as the base point, analyze the velocity of point A, and draw the velocity vector diagram (Fig. 8. 8(c)). By the velocity synthesis theorem, we have(取 B 点为基点,分析 A 点的速度,画速度矢量图(图 8. 8(c)). 由速度合成定理有)

$$\boldsymbol{v}_A = \boldsymbol{v}_B + \boldsymbol{v}_{AB}$$

We obtain(得到)

$$v_B = v_A \cot 30° = 200\sqrt{3}\ \text{mm/s}\ ,\ v_{AB} = v_A / \sin 30° = 400\ \text{mm/s}$$

$$\omega_{AB} = \frac{v_{AB}}{AB} = 2\ \text{rad/s}$$

Example 2: In the crank-slider mechanism of Fig. 8. 9(a), the crank $OA = r$, it rotates around the O axis with equal angular velocity ω_0 and the linkage $AB = l$. In the case, the linkage AB is perpendicular to the crank. Find: (1) the velocity of the slider v_B; (2) the angular velocity ω_{AB} of the linkage AB. (**例题** 2:图 8. 9(a)所示曲柄-滑块机构中,曲柄 $OA = r$,以等角速度 ω_0 绕 O 轴转动,连杆 $AB = l$. 在图示情形下连杆与曲柄垂直. 求:(1)滑块的速度 v_B;(2)连杆 AB 的角速度 ω_{AB}.)

Solution:(解:)

Take point A as the base point, analyze the velocity of point B, draw the velocity vector diagram (Fig. 8. 9(b)), and from the velocity synthesis theorem, we have(取 A 点为基点,分析 B 点的速度,画速度矢量图(图 8. 9(b)),由速度合成定理有)

$$\boldsymbol{v}_B = \boldsymbol{v}_A + \boldsymbol{v}_{BA}$$

where (其中)

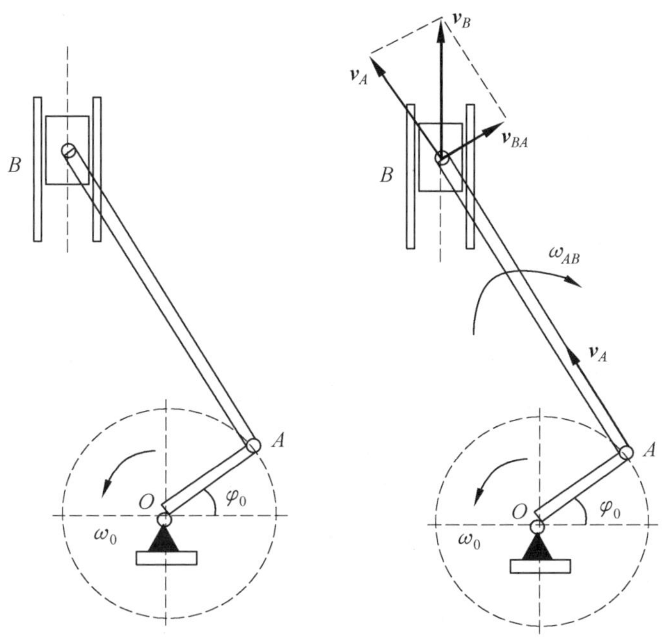

(a) Original diagram(原图)　(b) Velocity analysis diagram(速度分析图)

Fig. 8.9 Diagram of Example 2(图 8.9 例题 2 图)

$$v_A = r\omega_0$$

We obtain(得到)

$$v_B = \frac{v_A}{\cos\varphi_0} = \frac{r\omega_0}{\cos\varphi_0}$$

$$v_{BA} = v_A \tan\varphi_0 = r\omega \tan\varphi_0$$

$$\omega_{AB} = \frac{v_{BA}}{l} = \frac{r\omega_0}{l}\tan\varphi_0$$

Example 3: In Fig. 8.10(a), $OA = OO_1 = r$, $BC = 2r$, $O_1B \perp AB$, $\angle OAB = 45°$, ABC is a folding rod. Find the speed $\boldsymbol{v}_C$ of point C at this position. (**例题** 3:图 8.10(a)中,$OA = OO_1 = r$, $BC = 2r$, $O_1B \perp AB$, $\angle OAB = 45°$, ABC 是一根折杆.求此位置时 C 点的速度 $\boldsymbol{v}_C$.)

Solution:(解:)

(1) Perform motion analysis on the mechanism.(对机构进行运动分析.)

(2) Take A as the base point and study the velocity at point B. The velocity vector diagram is shown in Fig. 8.10(b). By the velocity synthesis theorem, we have(取 A 为基点,研究 B 点的速度,其速度矢量图如图 8.10(b).由速度合成定理,有)

$$\boldsymbol{v}_B = \boldsymbol{v}_A + \boldsymbol{v}_{BA}$$

where(其中)

$$v_A = r\omega_0$$

We obtain(得到)

$$v_B = v_{BA} = v_A \cos 45° = \frac{\sqrt{2}}{2} r\omega_0$$

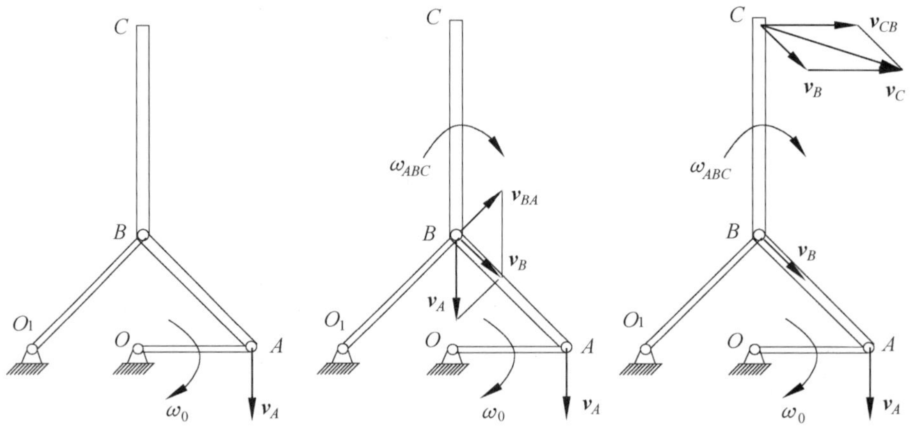

(a) Original diagram(原图)　(b) Velocity analysis of point B (B点速度分析)　(c) Velocity analysis of point C (C点速度分析)

Fig. 8. 10　Diagram of Example 3(图 8. 10　例题 3 图)

$$\omega_{ABC}=\frac{v_{BA}}{AB}=\frac{1}{2}\omega_0$$

(3) Then take B as the base point and study the velocity of point C. The velocity vector diagram is shown in Fig. 8. 10(c). By the velocity synthesis theorem, we have(再取 B 为基点,研究 C 点的速度,其速度矢量图如图 8. 10(c)所示. 由速度合成定理有)

$$\boldsymbol{v}_C=\boldsymbol{v}_B+\boldsymbol{v}_{CB}$$

where(其中)

$$v_{CB}=BC\cdot\omega_{ABC}=r\omega_0$$

We obtain(得到)

$$v_B=\frac{\sqrt{2}}{2}r\omega_0$$

$$v_C=\sqrt{v_B^2+v_{CB}^2+2v_Bv_{CB}\cos 45^\circ}=\frac{\sqrt{10}}{2}r\omega_0$$

Example 4: Use the velocity projection method to find the velocity $\boldsymbol{v}_B$ in Example 1 (Fig. 8. 11). $AB=l=200$ mm, $v_A=200$ mm/s. (**例题** 4:用速度投影法求例题 1 (图 8. 11) 中的速度 $\boldsymbol{v}_B$. $AB=l=200$ mm, $v_A=200$ mm/s.)

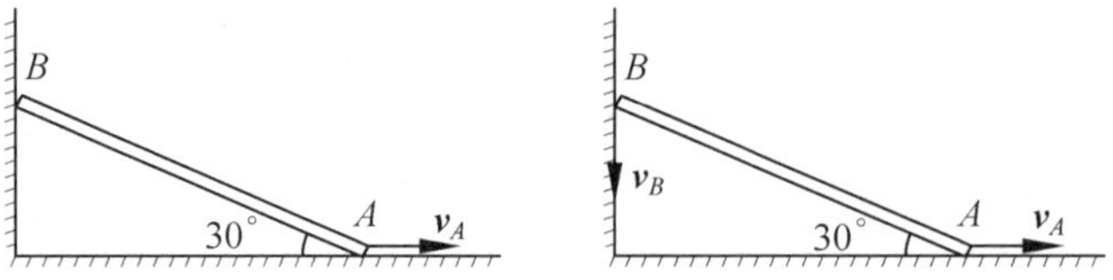

(a) Original diagram(原图)　(b) Velocity projection method (速度投影法)

Fig. 8. 11　Diagram of Example 4 (图 8. 11　例题 4 图)

Solution:(解:)

From the velocity projection theorem, we get(由速度投影定理得到)

$$v_B \cos 60° = v_A \cos 30°$$

that is(即)

$$v_B = v_A \cot 30° = 200\sqrt{3}\ \text{mm/s}$$

8.3 Instantaneous Centroid Method for Finding the Velocity of Points in a Plane Figure(求平面图形内各点速度的瞬心法)

1. Instantaneous velocity center(瞬时速度中心)

A point on a rigid body, which has zero velocity at a given moment, is called the instantaneous velocity center.(位于刚体上的点,在给定时刻的速度为零,这一点称为瞬时速度中心.)

2. Characteristics of the instantaneous velocity center(瞬时速度中心的特点)

The formula for solving the velocity of a point in a plane figure according to the base point method is(根据基点法求平面图形点的速度公式为)

$$\boldsymbol{v}_B = \boldsymbol{v}_A + \boldsymbol{v}_{BA}$$

In Fig. 8.12, if it makes $v_B = 0$, $\boldsymbol{v}_{BA}$ should be co-linear with the reverse $\boldsymbol{v}_A$ and $\boldsymbol{v}_{BA}$ is perpendicular to AB, then $\boldsymbol{v}_A$ should also be perpendicular to AB. Therefore, the point of $v = 0$ should be on the line perpendicular to the velocity of the base point.(图 8.12 中,若使 $v_B = 0$,则 $\boldsymbol{v}_{BA}$ 应该与 $\boldsymbol{v}_A$ 反向共线,而 $\boldsymbol{v}_{BA}$ 又与 AB 垂直,则 $\boldsymbol{v}_A$ 也应与 AB 垂直. 故 $v = 0$ 的点应在与基点速度垂直的直线上.)

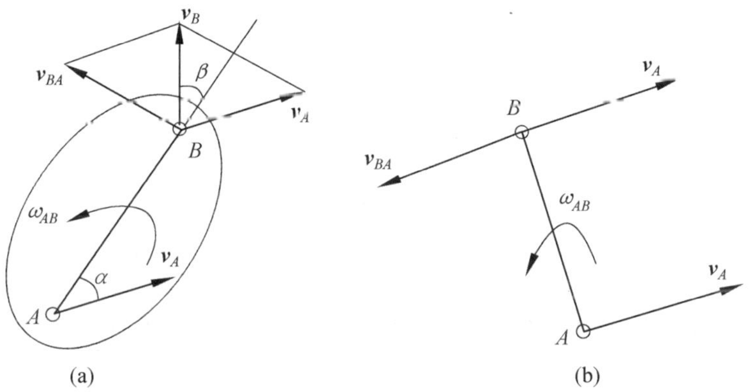

Fig. 8.12 Determination of the instantaneous velocity center position of a plane figure
(图 8.12 平面图形瞬时速度中心位置的确定)

Thus, when an object is moving in a plane, it is always possible to find such a point, whose velocity at a given moment is zero. This point is called the instantaneous velocity center.(因此,当物体在做平面运动时,总是可以找到这样的点,其在给定时刻的速度为零. 这一点称为速度瞬时中心.)

Characteristics of the instantaneous velocity center is as follows.(速度瞬时中心的特点如下.)

(1) Instantaneity. (瞬时性.)

Different transients have different instantaneous velocity centre. (不同的瞬时有不同的速度瞬心.)

(2) Uniqueness. (唯一性.)

There is only one instantaneous velocity cennter at a given instant. (某一瞬时只有一个速度瞬心.)

(3) The instantaneous rotation property. (瞬时转动特性.)

The motion of a plane figure at a given instant can be regarded as an instantaneous rotation around the instantaneous center of velocity at that instant. (平面图形在某一瞬时的运动都可以视为绕这一瞬时的速度瞬心做瞬时转动.)

In Fig. 8.13, if point C is the instantaneous velocity center, this point can be chosen as the base point for calculating the velocities of other points on the planar graph, the absolute velocities of which are also the relative velocities around the instantaneous center of velocity. (图 8.13 中，若 C 为速度瞬心，则可选该点为基点，计算平面图形上其他点的速度，这些点的绝对速度也就是绕速度瞬心的相对速度.)

Thus, if the center of instantaneous velocity is known, the velocity calculation of point is greatly simplified. We assume that the object simply rotates about the axis passing through the center of instantaneous velocity and obtain the velocity under this rotational condition. (因此，如果已知瞬时速度中心，则极大地简化了点速度的计算. 假设物体只是围绕通过瞬时速度中心的轴旋转，并在这种旋转条件下得到速度.)

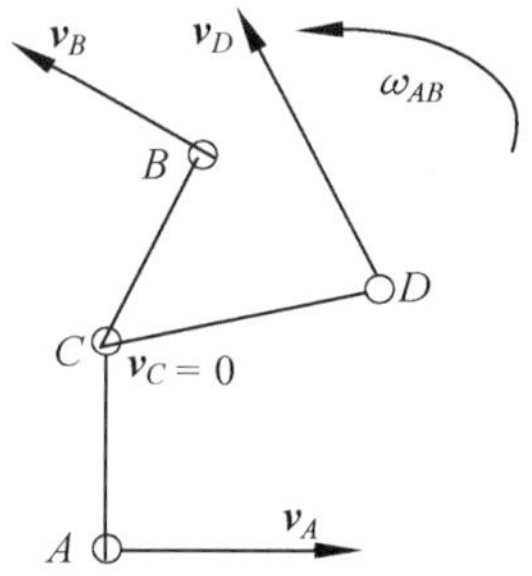

Fig. 8.13 Point C as the instantaneous velocity center

(图 8.13 速度瞬心 C 点)

3. The instantaneous velocity center in several special cases

(几种特殊情况下的速度瞬时中心位置)

Pure rolling: If a plane figure is known to be rolling without sliding along a fixed surface, the point of contact between the figure and the fixed surface is its velocity instantaneous center, as shown in Fig. 8.14. (纯滚动：已知平面图形沿一个固定表面做无滑动的滚动，则图形与固定面的接触点就是其速度瞬心，如图 8.14 所示.)

The instantaneous velocity center is located at the intersection of lines perpendicular to the velocity of each point in the rigid body (Fig. 8.15(a)). Different cases for determining the instantaneous velocity center (denoted by the letter P) are shown in Fig. 8.15(b)-(d). (瞬时

速度中心位于垂直于刚体各点速度的直线的交点上(图 8.15(a)). 确定速度瞬心(由字母 P 表示)的不同情况如图 8.15(b) ~ (d)所示.)

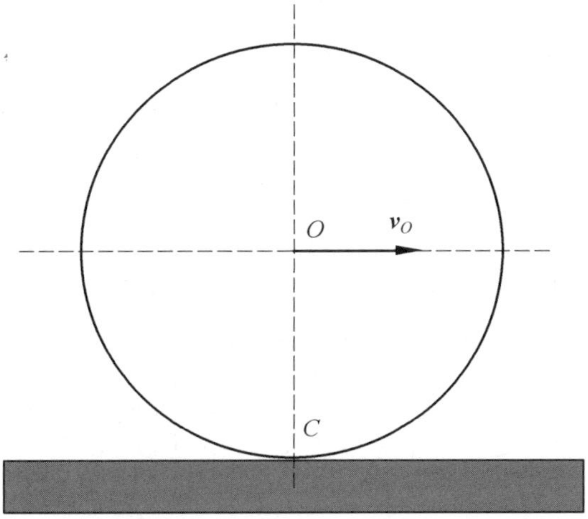

Fig. 8.14 Pure rolling(图 8.14 纯滚动)

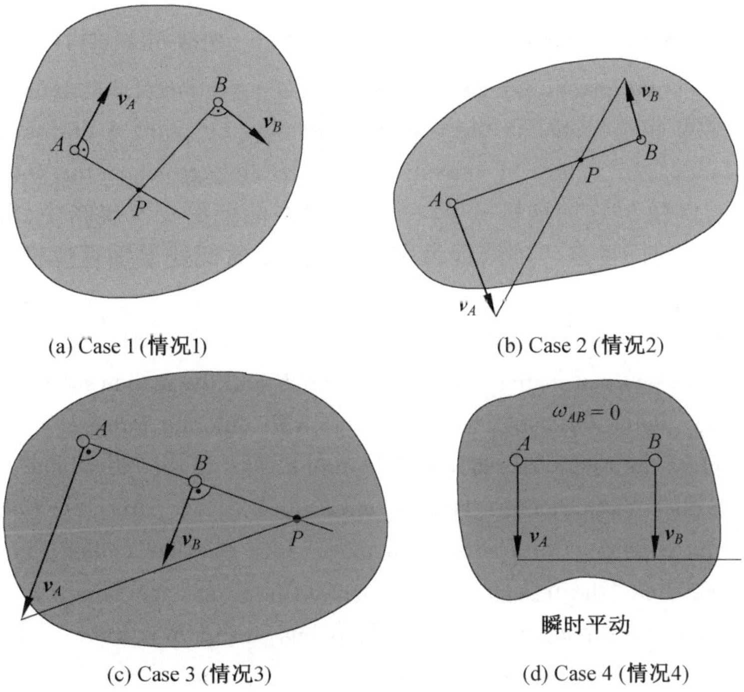

Fig. 8.15 Instantaneous velocity center for different cases(图 8.15 不同情况下的速度瞬心)

Note: In Fig 8.15(d), the velocity vectors of two points in a plane are known to be parallel to each other and in the same direction, both equal in magnitude and perpendicular to the line connecting the two points. At this point, the instantaneous velocity center is located at infinity, then $\omega = 0$, this situation is called instantaneous translate, at this instant, the distribution of velocity of the points in the graph is like as the case that plane figure does the transition motion, but the acceleration is different. (注意:图 8.15(d)中,已知平面图形上两点的速度矢量互相

平行、方向相同,二者大小相等且垂直于两点的连线.此时速度瞬心位于无穷远处,则 $\omega=0$,此种情形称为瞬时平动.该瞬时,图形上各点的速度分布如同图形做平动的情形一样,但加速度不同.)

4. Transformation of movement(运动的转换)

In machines, usually one motion is converted into another. For example, in a slider-crank mechanism (Fig. 8.16), the *OA* crank makes a fixed-axis rotation and this rotational motion is converted into a translational motion of the slider *B*. In solving practical problems, it is necessary to find the law of this motion or velocity.(在机器中,通常一个运动被转换成另一个运动.例如,在曲柄滑块机构中(图 8.16),*OA* 曲柄做定轴转动,该旋转运动被转换成滑块 *B* 的平动.在解决实际问题时,有必要找到该运动或速度的规律.)

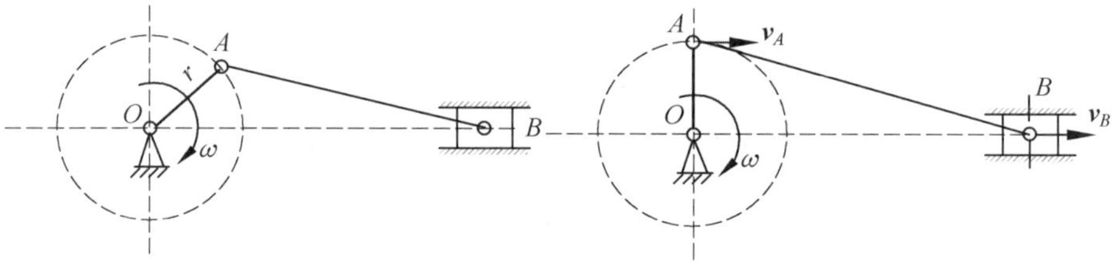

Fig. 8.16 Slider-crank mechanism(图 8.16 曲柄-滑块机构)

Example 5: In crank machinery (Fig. 8.16), in one rotation of the crank, the travel of the slider is equal to 400 mm. What is the distance travelled by point *A* during this travel time? Where is the instantaneous centre of velocity of the *AB* linkage when the *OA* crank is in the vertical position?(**例题 5**:在曲柄机械中(图 8.16),在曲柄的一次旋转中,滑块的行程等于 400 mm.在这个行程时间内点 *A* 的路程为多少?当 *OA* 曲柄处于垂直位置时,*AB* 连杆的速度瞬心在哪里?)

Solution:(解:)

(1) Consider the paths of motion for points *A* and *B* and the motion modes of the objects the two points belong to. Point *A* belongs to two objects with different forms of motion. On the one hand, point *A* involves the rotational motion of the crank *OA*; on the other hand, it belongs to the linkage *AB*, which does plane motion. Point *B* also connects two parts: the linkage *AB* and the slider *B*. Since point *B* belongs to a slider that moves in a translate motion, in which all points move along a straight line, the trajectory of the motion is always the same, a horizontal straight line. Therefore, when the direction of velocity of points *A* and *B* is known, and also when the crank *OA* is in a vertical position, the position of the instantaneous velocity center of the crank mechanism can be obtained. From Fig. 8.16, it can be seen that the instantaneous velocity center is located at infinity. Therefore, all points of the linkage *AB* have the same velocity.(考虑 *A*、*B* 点运动的路径及它们所属的物体有哪些运动方式. *A* 点属于两个运动形式不同的物体.一方面,点 *A* 涉及曲柄 *OA* 的旋转运动;另一方面,它属于连杆 *AB*,做平面运动. *B* 点也连接两部分:连杆 *AB* 和滑块 *B*.由于 *B* 点属于一个做平动的滑块,其中所有点都沿直线移动,运动轨迹始终不变,是一条水平直线,因此当知道点 *A* 和 *B* 的速度方向,同时当曲柄 *OA* 处于垂直位置时,可以得到曲柄机构的瞬时速度中心的位置.由图 8.16 可知,速度瞬心位于无穷远

处. 因此,连杆 AB 的所有点都具有相同的速度.)

(2) In one rotation of the crank, point A passes through with path $S=2\pi r$. Slider B in one rotation will cover a distance of $4r$. Therefore, if the distance travelled by point B is known, the radius of the crank can be found, we have(在曲柄的一次旋转中,点 A 通过路径 $S=2\pi r$. 一次旋转中的滑块 B 将覆盖 $4r$ 的路程. 因此,如果知道 B 点的路程,则可以找到曲柄的半径,有)

$$4r=400\ \text{mm},\ r=100\ \text{mm}$$

(3) Knowing the radius r of the crank, the distance travelled by point A in one round of the crank can be determined as(已知曲柄的半径 r,可以确定曲柄转一周中 A 点经过的路程为)

$$S=2\pi r=2\pi\times 100=628\ (\text{mm})$$

Example 6: As shown in Fig. 8.17, $AB=l=200$ mm; $v_A=200$ mm/s. Find: (**例题** 6: 如图 8.17所示, $AB=l=200$ mm; $v_A=200$ mm/s,求:)

(1) The velocity $\boldsymbol{v}_B$ at rod end B; (杆端 B 的速度 v_B;)

(2) Angular velocity $\boldsymbol{\omega}_{AB}$ of the rod AB. (AB 杆角速度 ω_{AB}.)

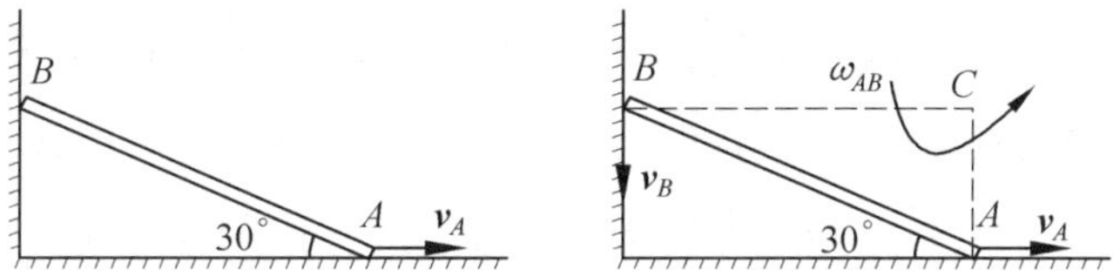

Fig. 8.17 Diagram of Example 6(图 8.17 例题 6 图)

Solution: (解:)

From its velocity distribution, it is clear that its instantaneous center is at point C, we have (由其速度分布可知其瞬心为 C 点,有)

$$\omega_{AB}=\frac{v_A}{AC}=2\ \text{rad/s}$$

$$v_B=\omega_{AB}\cdot BC=200\sqrt{3}\ \text{mm/s}$$

Example 7: The circular wheel of radius R in Fig. 8.18 makes a pure rolling on a straight track, the velocity of the center of the wheel is v_O. Find velocity at points A, B, C, D on the rim. (**例题** 7: 图 8.18 中半径为 R 的圆轮在直线轨道上做纯滚动,轮心速度为 v_O. 求轮缘上 A、B、C、D 四点的速度.)

Solution: (解:)

The circular wheel is in contact with the ground at point A. Since there is no relative sliding, the velocity at this instant is $v_A=0$. Point A is the instantaneous velocity center. Assume that the angular velocity at this instant is ω. (圆轮与地面接触点 A,由于没有相对滑动,因此在这一瞬时, A 点的速度 $v_A=0$. A 点即为速度瞬心. 假设这一瞬时的角速度为 ω.)

Obtained $\omega=\dfrac{v_O}{R}$ from $v_O=\omega R$. (由 $v_O=\omega R$ 得到 $\omega=\dfrac{v_O}{R}$.)

Because(因为)

$$\frac{v_B}{v_O}=\frac{BA}{OA}$$

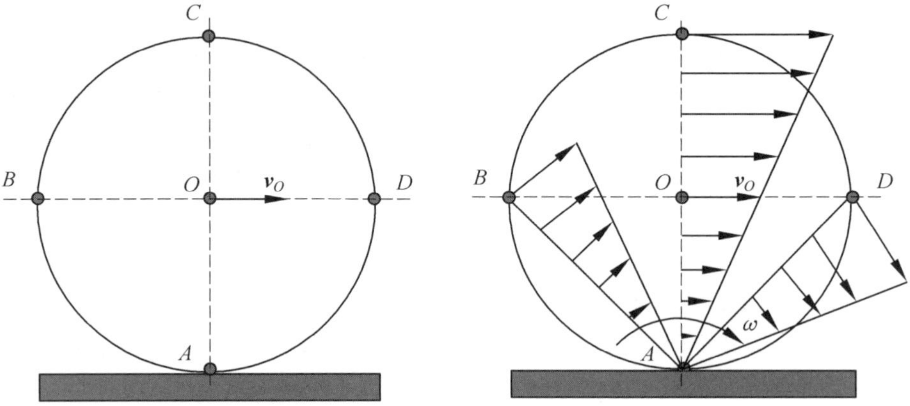

Fig. 8.18 Diagram of Example 7(图 8.18 例题 7 图)

So we get(所以得到)

$$v_B = \sqrt{2} v_O,\ v_C = 2v_O, v_D = \sqrt{2} v_0$$

8.4 The Acceleration of a Point Within an Object, Center of Instantaneous Acceleration (物体内点的加速度,瞬时加速度中心)

8.4.1 Find the acceleration of a point in a plane figure (求平面图形内点的加速度)

For example, the velocity of point A in Fig. 8.19 is(例如,图 8.19 中点 A 的速度为)

$$\boldsymbol{v}_A = \boldsymbol{v}_C + \boldsymbol{v}_{AC}$$

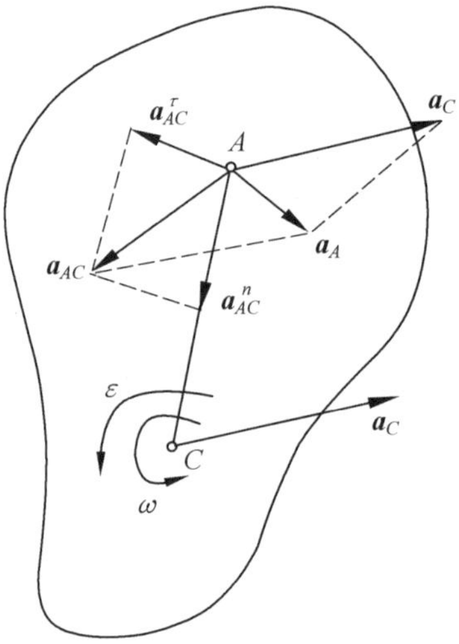

Fig. 8.19 Find the acceleration of a point in a plane figure(图 8.19 求平面图形内点的加速度)

Its acceleration is(它的加速度为)

$$\boldsymbol{a}_A = \frac{\mathrm{d}\boldsymbol{v}_A}{\mathrm{d}t} = \frac{\mathrm{d}\boldsymbol{v}_C}{\mathrm{d}t} + \frac{\mathrm{d}\boldsymbol{v}_{AC}}{\mathrm{d}t}$$

The first term of this sum is the base point acceleration, that is(该总和的第一项是基点加速度,即)

$$\frac{\mathrm{d}\boldsymbol{v}_C}{\mathrm{d}t} = \boldsymbol{a}_C$$

If the motion of the base point C is stopped, so that $v_C = 0$, retaining only the rotation of the object, the derivative is(如果基点 C 的运动停止,使 $v_C = 0$,只保留物体的旋转,则其导数为)

$$\frac{\mathrm{d}\boldsymbol{v}_{AC}}{\mathrm{d}t} = \boldsymbol{a}_A$$

That is, the acceleration at point A is equal to the acceleration $\boldsymbol{a}_{AC}$ of the object when it is assumed to be rotating around a stationary base point, and therefore(即点 A 的加速度等于假设物体绕着静止的基点旋转时的加速度 $\boldsymbol{a}_{AC}$,因此)

$$\boldsymbol{a}_A = \boldsymbol{a}_C + \boldsymbol{a}_{AC}$$

However, since in rotation the acceleration of point is the sum of the normal and tangential acceleration, that is(但是,因为在旋转中,点的加速度是法向加速度和切向加速度的总和,即)

$$\boldsymbol{a}_A = \boldsymbol{a}_{\mathrm{C}} + \boldsymbol{a}_{AC}^{\tau} + \boldsymbol{a}_{AC}^{n} \tag{8.7}$$

of which(其中)

$$a_{AC}^{n} = AC \cdot \omega^2$$

and the vector directionof $\boldsymbol{a}_{AC}^{n}$ points to the base point C (Fig. 8. 19), that is(而矢量方向指向基点 C(图 8. 19),即)

$$a_{AC}^{\tau} = AC \cdot \varepsilon$$

and the direction of its vector is perpendicular to AC, following the rotation direction of the angular acceleration ε. (并且其矢量的方向垂直于 AC,顺着角加速度 ε 的转动方向.)

Example 8: The disc rolls along a straight line without sliding. Its center C has velocity $\boldsymbol{v}_C$ and acceleration $\boldsymbol{a}_C$ (Fig. 8. 20). Find the acceleration at point A. (**例题** 8:圆盘沿直线滚动而不滑动. 它的中心 C 具有速度 $\boldsymbol{v}_C$ 和加速度 $\boldsymbol{a}_C$ (图 8. 20). 求点 A 加速度.)

Use the instantaneous velocity center to solve the angular velocity, we have(利用瞬时速度中心求得角速度,有)

$$\omega = \frac{v_C}{CP} = \frac{v_C}{R}$$

When the point C is moving along a straight line, i. e. $CP = R = \text{const}$, it can be found that the angular acceleration in this motion is the derivative of the angular velocity. We obtain(当点 C 沿直线运动时,即 $CP = R = \text{const}$,可以发现在这种运动中的角加速度是角速度的导数,得到)

$$\varepsilon = \frac{\mathrm{d}\omega}{\mathrm{d}t} = \frac{d}{\mathrm{d}t}\frac{v_C}{CP} = \frac{1}{R}\frac{\mathrm{d}v_C}{\mathrm{d}t} = \frac{a_C}{R}$$

If point C is the base point, then we have(若点 C 为基点,则有)

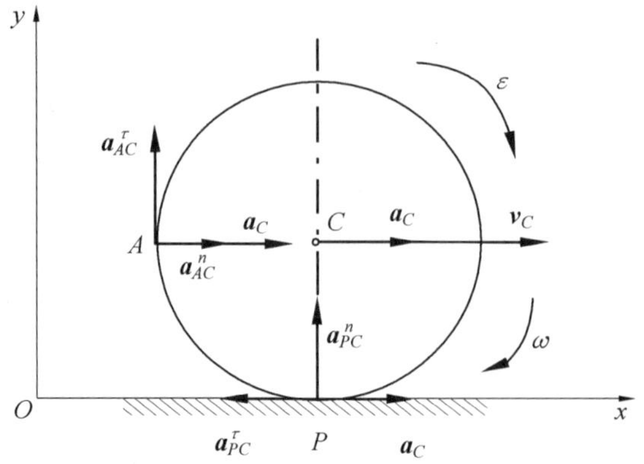

Fig. 8.20 Diagram of Example 8(图 8.20 例题 8 图)

$$\boldsymbol{a}_A = \boldsymbol{a}_C + \boldsymbol{a}_{AC}^{\tau} + \boldsymbol{a}_{AC}^{n} \tag{8.8}$$

$$a_{AC}^{n} = AC \cdot \omega^2 = \frac{v_C^2}{R}$$

$$a_{AC}^{\tau} = AC \cdot \varepsilon = a_C$$

By the projections of Eq. (8.8) on the x and y axes, we obtain the magnitude of the acceleration as (利用在 x 轴和 y 轴上的投影得到加速度的大小为)

$$a_{Ax} = a_C + a_{AC}^{n} = a_C + \frac{v_C^2}{R}, a_{Ay} = a_{AC}^{\tau} = a_C$$

$$a_A = \sqrt{a_{Ax}^2 + a_{Ay}^2} = \frac{1}{R}\sqrt{(v_C^2 + Ra_C)^2 + R^2 a_C^2}$$

The acceleration of the instantaneous velocity centre P is(瞬时速度中心 P 的加速度为)

$$\boldsymbol{a}_P = \boldsymbol{a}_C + \boldsymbol{a}_{PC}^{\tau} + \boldsymbol{a}_{PC}^{n}$$

of which(其中)

$$a_{PC}^{n} = R\omega^2 = \frac{v_C^2}{R}, a_{PC}^{\tau} = R\varepsilon = a_C$$

and because(且由于)

$$\boldsymbol{a}_{PC}^{\tau} = -\boldsymbol{a}_C$$

so the acceleration is(因此加速度为)

$$\boldsymbol{a}_P = \boldsymbol{a}_{PC}^{n}$$

and we have(且有)

$$a_P = \frac{v_C^2}{R} \neq 0$$

Therefore, the acceleration at the instantaneous velocity center is not equal to zero.(因此,瞬时速度中心的加速度不等于零.)

Example 9: Find the acceleration of point A (Fig. 8.21), set $\boldsymbol{v}_B = \boldsymbol{u} = \text{const}$, $\boldsymbol{a}_B = 0$, $AB = l$.(**例题 9**:求点 A 的加速度(图 8.21),设 $\boldsymbol{v}_B = \boldsymbol{u} = \text{const}$,$\boldsymbol{a}_B = 0$,$AB = l$.)

Solution:(解:)

$$\boldsymbol{a}_A = \boldsymbol{a}_B + \boldsymbol{a}_{AB}^{\tau} + \boldsymbol{a}_{AB}^{n} \tag{8.9}$$

where(其中)

$$a_{AB}^{n}=l\omega^{2}=\frac{u^{2}}{l\sin^{2}\alpha},a_{AB}^{\tau}=l\varepsilon$$

However, the direction of the vector $\boldsymbol{a}_{AB}^{\tau}$ is unknown and the angular acceleration ε is unknown.(但向量 $\boldsymbol{a}_{AB}^{\tau}$ 的方向未知,角加速度 ε 未知.)

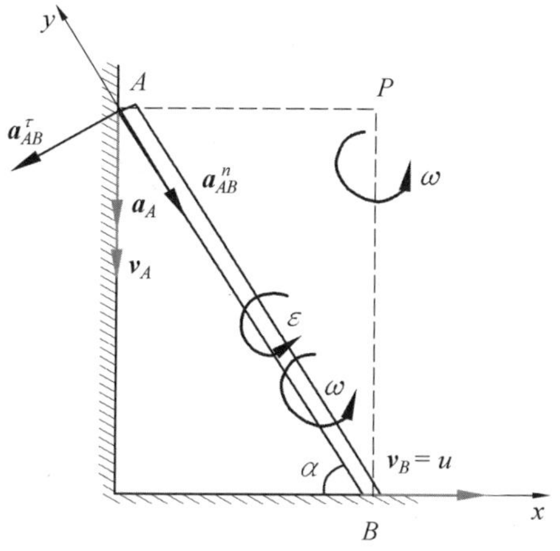

Fig. 8.21 Diagram of Example 9(图 8.21 例题 9 图)

Assume that the vector $\boldsymbol{a}_{AB}^{\tau}$ is directed to the left perpendicular to AB. The acceleration $\boldsymbol{a}_{A}$ follows the trajectory of the linear motion of point A, assuming a downward direction. The projections of the two sides of Eq. (8.9) on the x and y axes are equal, respectively, and we obtain(假设向量 $\boldsymbol{a}_{AB}^{\tau}$ 朝左垂直于 AB. 加速度 $\boldsymbol{a}_{A}$ 沿着点 A 直线运动的轨迹,假设方向向下. 式(8.9)两边分别在 x 轴和 y 轴上的投影相等,得到)

$$0=a_{AB}^{n}\cos\alpha-a_{AB}^{\tau}\sin\alpha \tag{8.10}$$

and (和)

$$-a_{A}\sin\alpha=-a_{AB}^{n} \tag{8.11}$$

From Eq. (8.11), we get the acceleration of point A as (由式(8.10)可得到点 A 的加速度为)

$$a_{A}=\frac{a}{\sin\alpha}=\frac{u^{2}}{l\sin^{3}\alpha}$$

a_{A} is positive indicates that the assumption direction of vector $\boldsymbol{a}_{A}$ is correct. (a_{A} 的值为正值,则表明矢量 $\boldsymbol{a}_{A}$ 假设的方向是正确的.)

From Eq. (8.10), the acceleration is(由式(8.9)可得加速度为)

$$a_{AB}^{\tau}=a_{AB}^{n}\frac{\cos\alpha}{\sin\alpha}=\frac{u^{2}}{l\sin^{3}\alpha}\cos\alpha$$

The angular acceleration is(角加速度为)

$$\varepsilon=\frac{a_{AB}^{\tau}}{l}=\frac{u^{2}}{l^{2}\sin^{3}\alpha}\cos\alpha$$

The assumption direction of $\boldsymbol{a}_{AB}^{\tau}$ and ε is also right. ($\boldsymbol{a}_{AB}^{\tau}$ 和 ε 假设的方向也是正确的.)

8.4.2 Instantaneous acceleration centre(瞬时加速度中心)

The plane parallel motion of the object is again analyzed (Fig. 8.22). The base acceleration $\boldsymbol{a}_C$, the angular velocity ω and the angular acceleration of the object ε are known.(再次分析物体的平面平行运动(图 8.22). 已知基点加速度 $\boldsymbol{a}_C$、角速度 ω 和物体的角加速度 ε.)

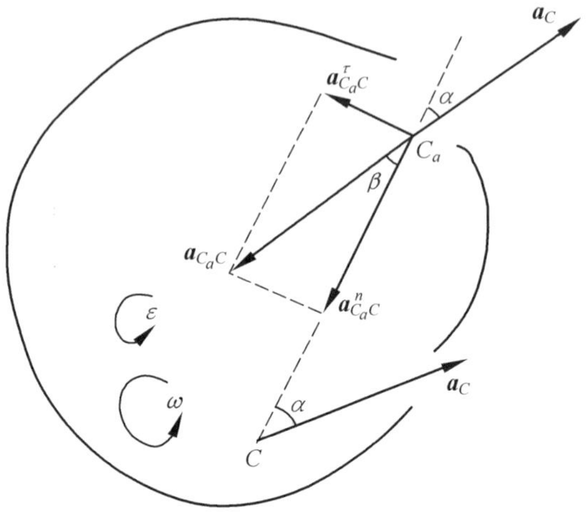

Fig. 8.22 Location of the instantaneous acceleration center in a planar graph
(图 8.22 平面图形内瞬时加速度中心的位置)

From the base point C, we draw a line at angle α such that(从基点 C 以角度 α 绘制一条直线,使得)

$$\tan\alpha = \frac{\varepsilon}{\omega^2} \tag{8.12}$$

Vector $\boldsymbol{a}_C$ relative to the point C_a follows the turn direction of ε.(矢量 $\boldsymbol{a}_C$ 相对点 C_a 顺着 ε 转的方向.)

On this line, we find the point C_a whose distance from point C is(在这条直线上找到点 C_a,其与 C 点的距离为)

$$CC_a = \frac{a_C}{\sqrt{\varepsilon^2 + \omega^4}}$$

Let's prove that the acceleration C_a of the point is zero, we have(下面证明点 C_a 的加速度为零,有)

$$\boldsymbol{a}_{C_a} = \boldsymbol{a}_C + \boldsymbol{a}_{C_aC} = \boldsymbol{a}_C + \boldsymbol{a}^n_{C_aC} + \boldsymbol{a}^{\tau}_{C_aC}$$

of which(其中)

$$a^n_{C_aC} = CC_a \cdot \omega^2,\ a^{\tau}_{C_aC} = CC_a \cdot \varepsilon$$

and(和)

$$a_{C_aC} = \sqrt{(a^n_{C_aC})^2 + (a^{\tau}_{C_aC})^2} = CC_a\sqrt{\varepsilon^2 + \omega^4} = a_C$$

The angle β between the vector $\boldsymbol{a}_{C_aC}$ and the line CC_a is obtained.(得到矢量 $\boldsymbol{a}_{C_aC}$ 与直线 CC_a 之间的角度 β.)

Its tangent is(它的正切为)

$$\tan\beta = \frac{a_{C_aC}^{\tau}}{a_{C_aC}^{n}} = \frac{\varepsilon}{\omega^2} = \tan\alpha$$

This means that $\beta=\alpha$ and the vector sum of $\boldsymbol{a}_C$ and the vectors $\boldsymbol{a}_{C_aC}$ is zero, i. e. the acceleration at that point C_a is zero.(即$\beta=\alpha$,且向量$\boldsymbol{a}_C$与$\boldsymbol{a}_{C_aC}$的矢量和为零,即该点C_a的加速度为零.)

Thus, in planar parallel motion, a point can be found on the object where its acceleration is zero at that moment. This point C_a is known as the center of instantaneous acceleration.(因此,在平面平行运动中,可以在物体上找到一个在该时刻其加速度为零的点.点C_a称为瞬时加速度中心.)

If one can manage to find this pointin the object, then the definition of the acceleration of the point on the object is greatly simplified. In fact, by specifying a point whose acceleration is zero, it will be easier to obtain by means of the acceleration synthesis formula,we have(如果能设法在物体上找到这个点,就大大简化了物体上点的加速度的定义.事实上,通过指定一个其加速度为零的点,通过加速度合成公式就将更容易得到,有)

$$\boldsymbol{a}_A = \boldsymbol{a}_{ACa} = \boldsymbol{a}_{AC_a}^{\tau} + \boldsymbol{a}_{AC_a}^{n} \tag{8.13}$$

That is, the acceleration of a point in an object is defined as a rotation about an axis that passes through the center of instantaneous acceleration perpendicular to the plane of motion. For example, in Example 9, where the center of instantaneous acceleration is at point B, the acceleration of point A would be calculated as a rotation about the axis $C_a$passing through point B.(也就是说,物体上点的加速度被定义为绕通过垂直于运动平面的瞬时加速度中心轴旋转.例如,在例题9中,瞬时加速度中心位于点B,A点的加速度将以绕通过点B的轴C_a旋转来计算.)

Since the vector of acceleration at a point and the angle α between the line connecting the point and the centre of instantaneous acceleration is determined only by the angular acceleration and angular velocity, as shown in Eq. (8.12), and is the same for all points, this angle α, which determines the direction of acceleration, is the same for all points.(由于点的加速度矢量和连接该点与瞬时加速中心的直线之间的角度α仅由角加速度和角速度确定,见式(8.12),对于所有点都相同,因此确定加速度方向的角度α对于所有点都是相同的.)

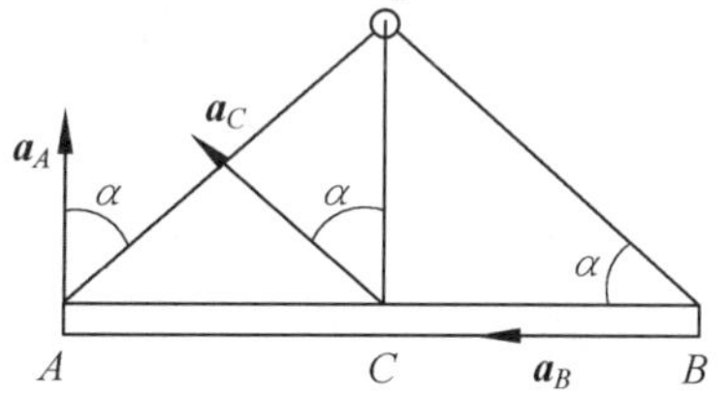

Fig. 8.23 Acceleration of points on rod AB(图8.23 杆AB上各点的加速度)

For example, the acceleration of the point on rod AB is shown in Fig. 8.23, $a_A = a_B$, the angle $\alpha=45°$.(例如,图8.23显示了杆AB上点的加速度,$a_A = a_B$,角度$\alpha=45°$.)

It should be noted that the center C_a of instantaneous acceleration and the center P of instantaneous velocity of an object are usually different points.(应该注意的是,瞬时加速中心C_a和物体的瞬时速度中心P通常是不同的点.)

8.5 Questions(思考题)

1. Is it possible for the velocity of points A and B on a plane figure to be in this direction as shown in Fig. 8.24? Why? (如图 8.24 所示,平面图形上点 A、B 的速度方向可能是这样的吗? 为什么?)

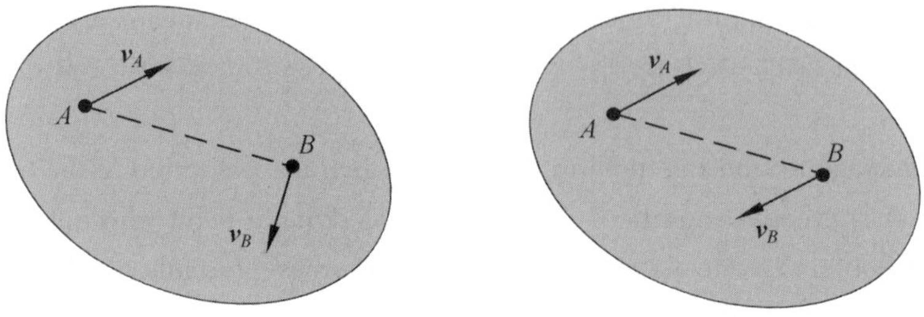

Fig. 8.24 Diagram of Question 1(图 8.24 思考题 1 图)

2. As shown in Fig. 8.25, the acceleration of rod O_1A is ω_1 and that plate ABC and rod O_1A are hinged. Whether the velocity distribution patterns for the points of O_1A and AC in the figure are correct? (如图 8.25 所示,杆 O_1A 的加速度为 ω_1,板 ABC 与杆 O_1A 铰链. 则图中 O_1A 和 AC 上各点的速度分布规律对不对?)

3. A wheel is rolling along a curved surface as shown in Fig 8.26. The velocity $\boldsymbol{v}_O$ and acceleration $\boldsymbol{a}_O$ of the wheel center O at a given instant are known. Whether the angular acceleration of the wheel is equal to $a_O\times\cos\beta/R$? How are the magnitude and direction of the acceleration of the velocity instant center C determined? (如图 8.26 所示,车轮沿曲面滚动. 已知轮心 O 在某一瞬时的速度 $\boldsymbol{v}_O$ 和加速度 $\boldsymbol{a}_O$. 则车轮的角加速度是否等于 $a_O\times\cos\beta/R$? 速度瞬心 C 的加速度大小和方向如何确定?)

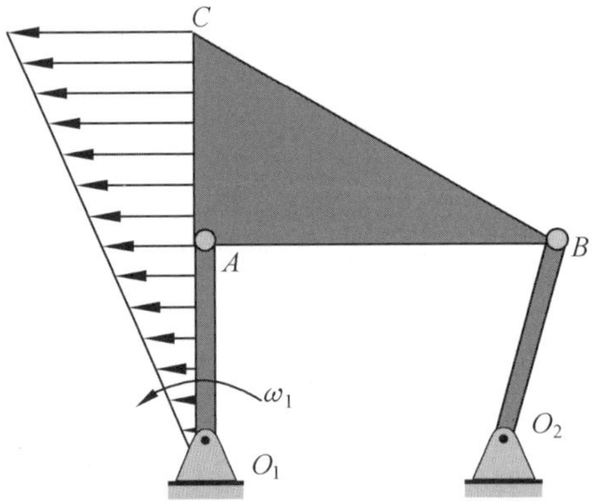

Fig. 8.25 Diagram of Question 2
(图 8.25 思考题 2 图)

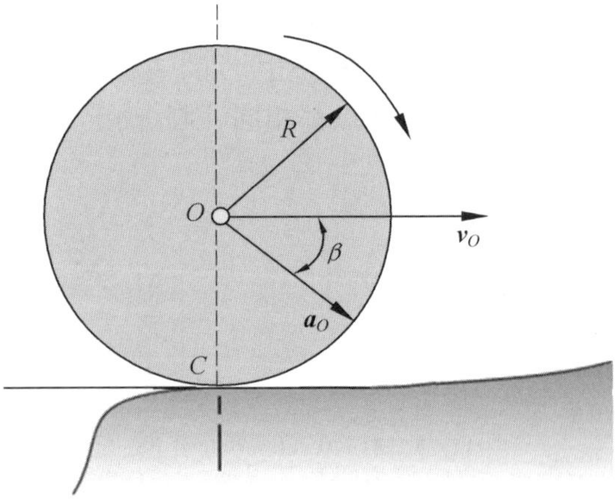

Fig. 8.26 Diagram of Question 3
(图 8.26 思考题 3 图)

4. A plane figure moving in its plane has two points in it with the same vector of acceleration at a given instant. Try to determine whether the following statement is correct:(平面图形在其平面内运动,某瞬时其上有两点的加速度矢量相同.试判断下述说法是否正确:)

(1)The velocities of all points on it must be equal at that instant;(其上各点速度在该瞬时一定都相等;)

(2)The acceleration at each point on it must be equal at that instant.(其上各点加速度在该瞬时一定都相等.)

5. As shown in Fig. 8.27, when a rigid body is in plane motion, the angular velocity of a certain instantaneous plane shape is ω and the angular acceleration is α. Try to determine the projection relationship of the acceleration of any two points A and B on the line connecting A and B of the plane figure.(如图8.27所示,刚体做平面运动,某瞬时平面图形的角速度为ω,角加速度为α.试确定其上任意两点A、B的加速度在A、B连线上的投影关系.)

6. As shown in Fig. 8.28, $O_1A \perp\!\!\!\!= O_2B$ is given. Whether ω_1 and ω_2, α_1 and α_2 are equal?(如图8.28所示,已知$O_1A \perp\!\!\!\!= O_2B$,则ω_1与ω_2、α_1与α_2是否相等?)

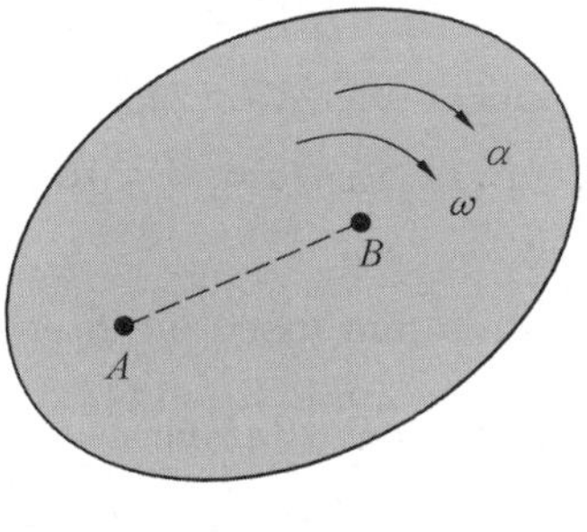

Fig. 8.27 Diagram of Question 5
(图8.27 思考题5图)

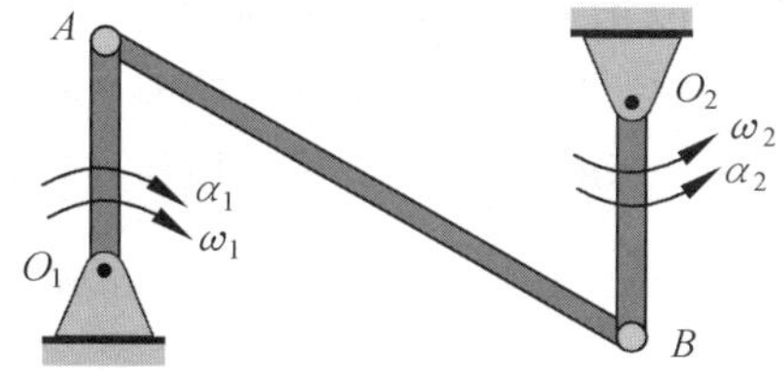

Fig. 8.28 Diagram of Question 6
(图8.28 思考题6图)

7. The rod AB for plane motion, the instantaneous A, B of speed the size and direction $\boldsymbol{v}_A$, $\boldsymbol{v}_B$ are known in the diagram, C, D are the $\boldsymbol{v}_A$, $\boldsymbol{v}_B$ vector end, as shown in Fig 8.29.(杆AB做平面运动,图示A、B两点的瞬时速度$\boldsymbol{v}_A$、$\boldsymbol{v}_B$的大小和方向均为已知,C、D两点分别是$\boldsymbol{v}_A$、$\boldsymbol{v}_B$的矢端,如图8.29所示.)

(1)Are the endpoints of the velocity vectors at each point on the rod AB all on the line CD?(杆AB上各点速度矢的端点是否都在直线CD上?)

(2)For any point E on the rod AB, let its velocity vector end be H, where is the point H?(对杆AB上任意一点E,设其速度矢端为H,那么点H在什么位置?)

(3)Let the rod AB be of infinite length and it intersects the extension of CD at the point P, try to determine whether the following statement is correct.(设杆AB为无限长,它与CD的延长线交于点P,试判断下述说法是否正确.)

①The instantaneous velocity of point P is zero.(点P的瞬时速度为零.)

②The instantaneous velocity of the point P must not be zero and its velocity vector end must be on the line AB.(点P的瞬时速度必须不为零,其速度矢端必须在直线AB上.)

③The instantaneous velocity of point P is not zero and its velocity vector end must be on the extension of CD.(点P的瞬时速度不为零,其速度矢端必在CD的延长线上.)

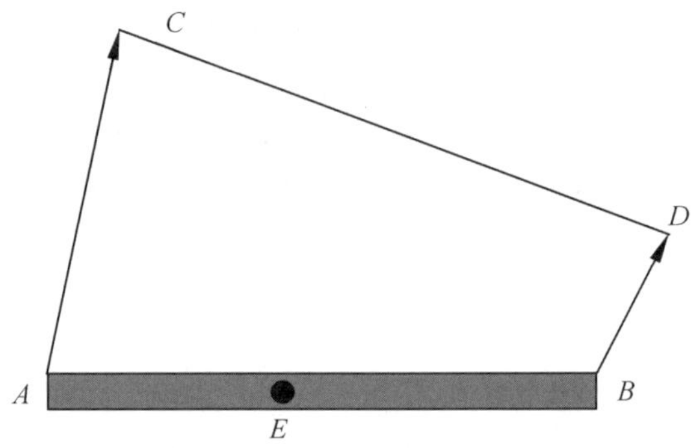

Fig. 8.29 Diagram of Question 7(图 8.29 思考题 7 图)

8.6 Exercises(习题)

1. An elliptical ruler AB is driven by a crank OC, which rotates with angular velocity ω_O about the O axis at constant speed, as shown in Fig. 8.30. If $OC=BC=AC=r$, and taking C as the base point, find the equation of plane motion of the elliptical ruler AB. (椭圆规尺 AB 由曲柄 OC 带动,曲柄以角速度 ω_0 绕 O 轴匀速转动,如图 8.30 所示. 若 $OC=BC=AC=r$,并取 C 为基点,求椭圆规尺 AB 的平面运动方程.)

2. As shown in Fig. 8.31, a cylinder A is wrapped with a thin rope, the B end of which is fixed to the ceiling. The cylinder falls from rest with a velocity $v=\frac{2}{3}\sqrt{3gh}$ on its axis, where g is a constant and h is the distance from the axis of the cylinder to its initial position. If the radius of the cylinder is r, find the equation of plane motion of the cylinder. (如图 8.31 所示,圆柱 A 缠以细绳,绳的 B 端固定在天花板上. 圆柱自静止落下,其轴心的速度为 $v=\frac{2}{3}\sqrt{3gh}$,其中 g 为常量,h 为圆柱轴心到初始位置的距离. 若圆柱半径为 r,求圆柱的平面运动方程.)

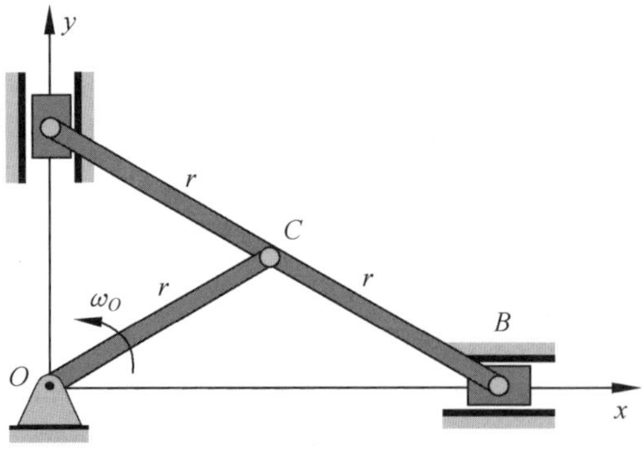

Fig. 8.30 Diagram of Exercise 1
(图 8.30 习题 1 图)

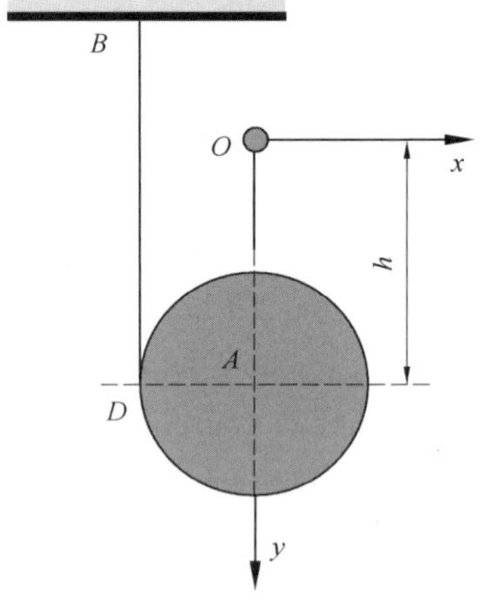

Fig. 8.31 Diagram of Exercise 2
(图 8.31 习题 2 图)

3. A gear of radius r is driven by the crank OA and rolls along a fixed gear of radius R, as shown in Fig. 8.32. If the crank OA rotates around the O axis with equal angular acceleration α, when the motion starts, the angular velocity $\omega_1 = 0$ and the angle of rotation $\varphi = 0$. Find the equation of plane motion of the moving gear with the center A as the base point. (半径为 r 的齿轮由曲柄 OA 带动,沿半径为 R 的固定齿轮滚动,如图 8.32 所示. 若曲柄 OA 以等角加速度 α 绕 O 轴转动,当运动开始时,角速度 $\omega_1 = 0$,转角 $\varphi = 0$. 求动齿轮以中心 A 为基点的平面运动方程.)

4. As shown in Fig. 8.33, in a screening mechanism, the oscillation of the sieve is driven by a crank linkage mechanism. The speed of the crank OA is known to be $n_{OA} = 40$ r/min and $OA = 0.3$ m. When the sieve BC moves to the point O on the same horizontal line, $\angle BAO = 90°$. Find the velocity of the sieve BC at this position. (如图 8.33 所示,在筛动机构中,筛子的摆动是由曲柄连杆机构带动的. 已知曲柄 OA 的转速 $n_{OA} = 40$ r/min, $OA = 0.3$ m. 当筛子 BC 运动到点 O 在同一水平线上时, $\angle BAO = 90°$. 求此瞬时筛子 BC 的速度.)

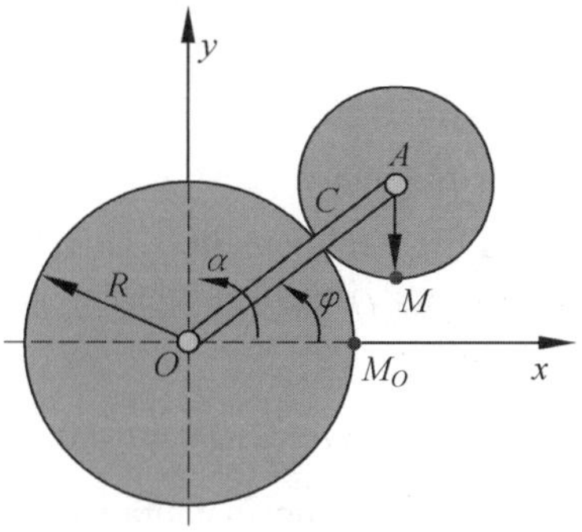

Fig. 8.32 Diagram of Exercise 3
(图 8.32 习题 3 图)

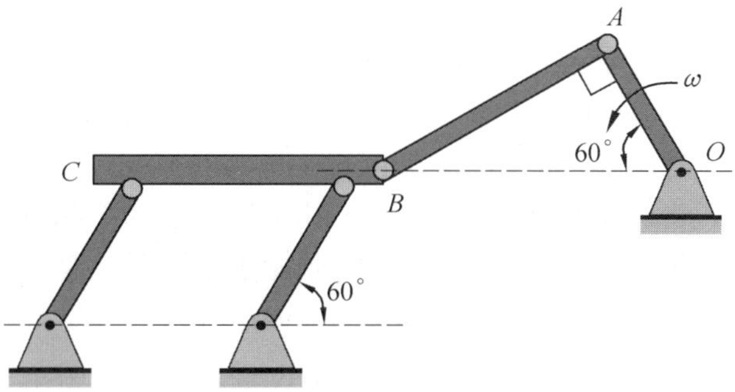

Fig. 8.33 Diagram of Exercise 4
(图 8.33 习题 4 图)

5. As shown in Fig. 8.34, the rollers of the rolling mechanism roll along the horizontal plane without sliding. Given that the length of the crank OA is $R = 10$ cm, it rotates at a uniform speed of $n = 30$ r/min. The length of connecting rod AB is 17.3 cm, and the rolling radius is 10 cm. Calculate the angular velocity and acceleration of the roller at the illustrated position. (如图 8.34 所示,滚压机构的滚子沿水平面滚动而不滑动,已知曲柄 OA 长 $r = 10$ cm,以匀转速 $n = 30$ r/min转动. 连杆 AB 长 $l = 17.3$ cm,滚动半径 $R = 10$ cm. 求在图示位置时滚子的角速度及角加速度.)

6. In the Watt planetary transmission mechanism, the bar O_1A rotates around the O_1 axis and drives the crank OB by means of the linkage AB, the crank OB is movably mounted on the O axis, as shown in Fig. 8.35. The O axis is equipped with gear Ⅰ, gear Ⅱ and connecting rod AB solidly connected in one piece. It is known that $r_1 = r_2 = 0.3\sqrt{3}$ m, $O_1A = 1.5$ m, and the angular velocity $\omega = 6$ rad/s of the rod O_1A. Find the angular velocity of crank OB and gear Ⅰ when $\gamma = 60°$ and $\beta = 90°$. (在瓦特行星传动机构中,杆 O_1A 绕 O_1 轴转动,并利用连杆 AB 带动曲柄 OB,曲柄 OB 活动地装置在 O 轴上,如图 8.35 所示. 在 O 轴上装有齿轮Ⅰ,齿轮Ⅱ与连杆 AB 固连于一体. 已知 $r_1 = r_2 = 0.3\sqrt{3}$ m, $O_1A = 1.5$ m,又有杆 O_1A 的角速度 $\omega = 6$ rad/s. 求当 $\gamma = 60°$ 且 $\beta = 90°$时,曲柄 OB 和齿轮Ⅰ的角速度.)

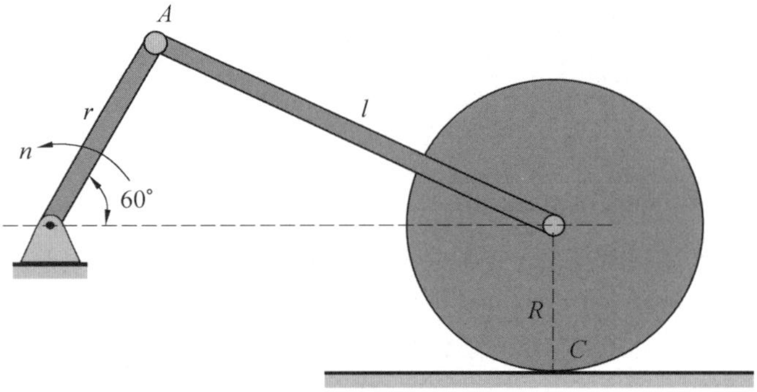

Fig. 8.34 Diagram of Exercise 5

(图 8.34 习题 5 图)

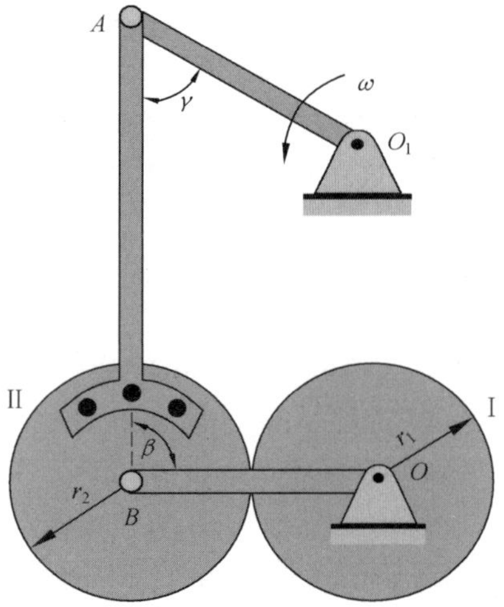

Fig. 8.35 Diagram of Exercise 6

(图 8.35 习题 6 图)

7. As shown in Fig. 8.36, gear Ⅰ rolls within gear Ⅱ with radii r and $R = 2r$. The crank OO_1 rotates about axis O with equal angular velocity ω_O and drives planetary gear Ⅰ. Find the acceleration of the center of instantaneous velocity C on this wheel Ⅰ at this moment. (如图 8.36 所示,齿轮Ⅰ在齿轮Ⅱ内滚动,其半径分别为 r 和 $R=2r$. 曲柄 OO_1 绕 O 轴以等角速度 ω_O 转动,并带动行星齿轮Ⅰ. 求该瞬时轮Ⅰ上瞬时速度中心 C 的加速度.)

8. The four-bar mechanism shown in Fig. 8.37 is known to have a crank $OA = r = 0.5$ m, rotating at a uniform angular velocity of $\omega_O = 4$ rad/s, $AB = 2r$, $BC = \sqrt{2}r$, the instantaneous OA shown in the figure is horizontal, AB is vertical, and $\varphi = 45°$. Try to calculate the acceleration of instantaneous point B and the angular velocity and acceleration of joystick BC. (四杆机构如图 8.37 所示,已知曲柄 $OA = r = 0.5$ m,以匀角速度 $\omega_O = 4$ rad/s 转动,$AB = 2r$,$BC = \sqrt{2}r$,图示瞬时 OA 水平,AB 铅直,$\varphi = 45°$. 试求该瞬时点 B 的加速度和摇杆 BC 的角速度和角加速度.)

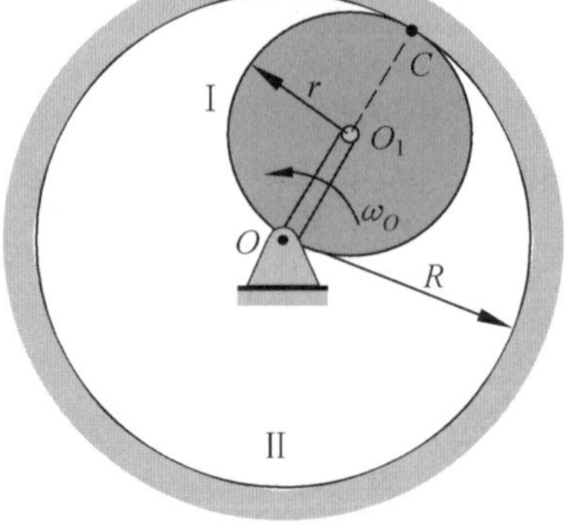

Fig. 8.36 Diagram of Exercise 7

(图 8.36 习题 7 图)

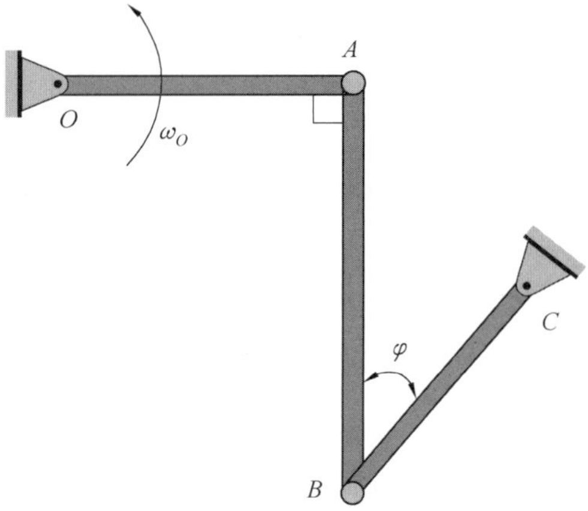

Fig. 8.37 Diagram of Exercise 8

(图 8.37 习题 8 图)

9. In a crank gear elliptical compass, gear A and crank O_1A are cemented together, and both gear C *and gear* A have radius r and mesh with each other, as shown in Fig. 8. 38. In the diagram, $AB = O_1O_2$, $O_1A = O_2B = 0.4$ m. O_1A rotates around O_1 with a constant angular velocity ω, $\omega = 0.2$ rad/s. M is a point on the wheel C, $CM = 0.1$ m. At the instant shown, CM is long vertical direction. Find the velocity and acceleration of the point M at this position. (在曲柄齿轮椭圆规中,齿轮 A 和曲柄 O_1A 固结为一体,齿轮 C 和齿轮 A 半径均为 r 并相互啮合,如图 8.38所示. 图中,$AB = O_1O_2$,$O_1A = O_2B = 0.4$ m. O_1A 以恒定的角速度 ω 绕 O_1 转动,$\omega = 0.2$ rad/s. M 为轮 C 上一点,$CM = 0.1$ m. 在图示瞬时,CM 铅垂. 求此时点 M 的速度和加速度.)

10. In the mechanism in Fig. 8. 39, the crank OA is of length r and rotates about the O axis with equal angular velocity ω_O, $AB = 6r$, $BC = 2\sqrt{3}\,r$. Find the velocity and acceleration of the slider C at the position shown. (在图 8.39 所示机构中, 曲柄 OA 长为 r, 绕 O 轴以等角速度 ω_O转动, $AB = 6r$,$BC = 2\sqrt{3}\,r$. 求图示位置时,滑块 C 的速度和加速度.)

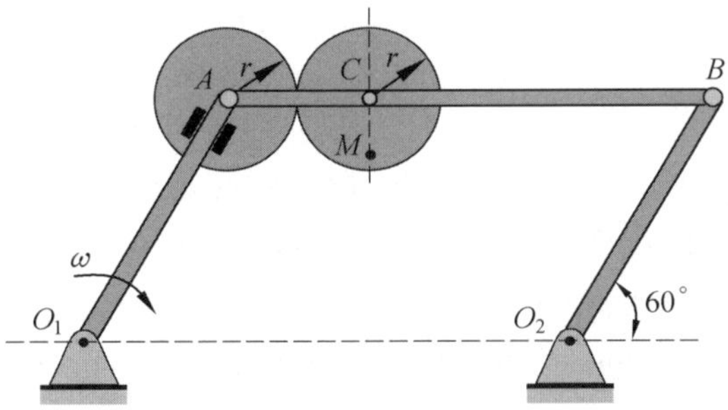

Fig. 8. 38 Diagram of Exercise 9
(图 8.38 习题 9 图)

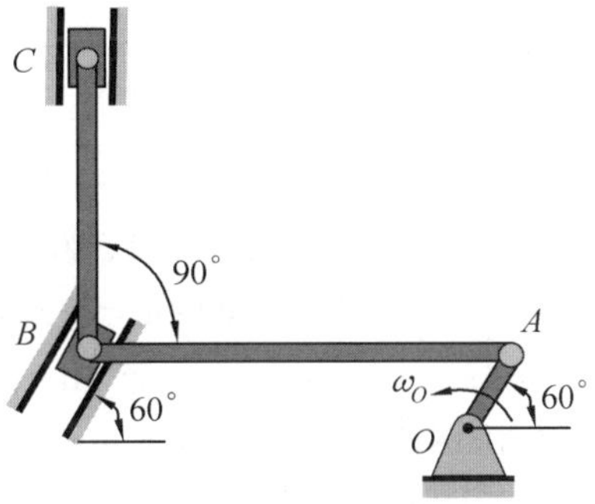

Fig. 8. 39 Diagram of Exercise 10
(图 8.39 习题 10 图)

11. The tower pulley 1 shown in Fig. 8. 40 has radii $r = 0.1$ m and $R = 0.2$ m, rotates about axis O with the law $\varphi = t^2 - 3t$ (where φ in rad and t in s) and is coiled by a non-extendable rope to dynamic pulley 2, which has a radius of $r_2 = 0.15$ m. Given that there is no relative sliding between the rope and each pulley. Find the angular velocity and angular acceleration of pulley 2 at $t = 1$ s, and find the velocity and acceleration of points C, D, E on the horizontal diameters at this instant. (图 8.40 所示塔轮 1 半径为 $r = 0.1$ m,$R = 0.2$ m,绕轴 O 转动的规律是 $\varphi = t^2 - 3t$ (式中 φ 以 rad 计,t 以 s 计),并通过不可伸长的绳子卷动动滑轮 2,滑轮 2 的半径为 $r_2 = 0.15$ m. 设绳子与各轮之间无相对滑动,求 $t = 1$ s 时轮 2 的角速度和角加速度,并求该瞬时水平直径上 C、D、E 各点的速度和加速度.)

12. The wheel with radius R rolls along the horizontal plane without sliding, as shown in Fig. 8. 41. There is a cylindrical part on the wheel with a radius of r. Wrap the wire around a cylinder, and the B-end of the wire moves horizontally with velocity v and acceleration a. Calculate the velocity and acceleration of the axis O of the wheel. (半径为 R 的轮子沿水平面滚动而不滑动,如图 8.41 所示. 在轮上有圆柱部分,其半径为 r. 将线绕于圆柱上,线的 B 端以速度 v 和加速度 a 沿水平方向运动. 求轮的轴心 O 的速度和加速度.)

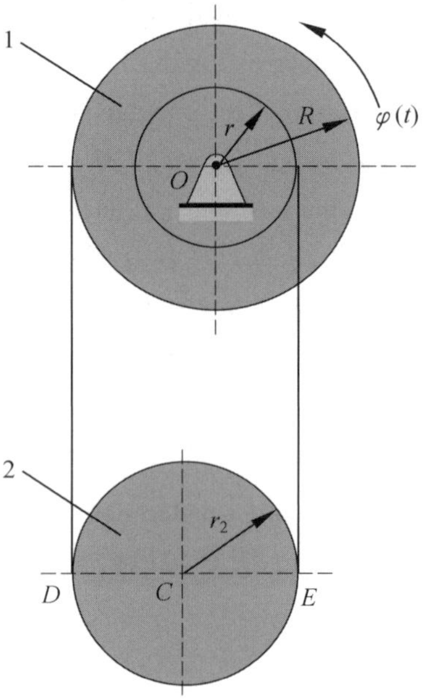

Fig. 8.40　Diagram of Exercise 11
(图 8.40　习题 11 图)

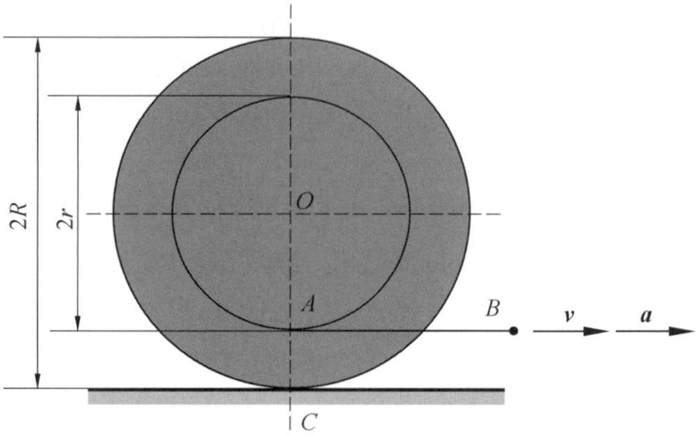

Fig. 8.41　Diagram of Exercise 12
(图 8.41　习题 12 图)

13. As shown in Fig. 8.42, the crank OA rotates around the axis O at a constant angular velocity $\omega = 0.5$ rad/s, and drives the wheel with radius r to perform non sliding rolling in the circular arc groove with radius R through the connecting rod AB. Assuming $OA = AB = R = 2r = 1$ m, calculate the velocity and acceleration of points B and C as shown in the diagram at the instant. (如图 8.42 所示,曲柄 OA 以恒定的角速度 $\omega = 2$ rad/s 绕轴 O 转动,并借助连杆 AB 驱动半径为 r 的轮子在半径为 R 的圆弧槽中做无滑动滚动. 设 $OA = AB = R = 2r = 1$ m,求图示瞬时点 B 和点 C 的速度和加速度.)

14. A light lever type steel pusher, with crank OA driving rod O_1B around the O_1 axis by means of connecting rod AB, and rod EC connected by a hinge to slider C, which slides along rod O_1B, the swing of the rocker drives rod EC to push the steel, as shown in Fig. 8.43. It is known that $OA = \sqrt{3}r$, $AB = r$, $O_1B = \frac{2}{3}l$ ($r = 0.2$ m, $l = 1$ m), $\omega_{OA} = \frac{1}{2}$ rad/s, $\alpha_{OA} = 0$. At the position shown, $BC = \frac{4}{3}l$. Find: (1) Absolute velocity of slider C and velocity relative to rocker O_1B; (2) Absolute acceleration of slider C and acceleration relative to rocker O_1B. (轻型杠杆式推钢机,曲柄 OA 借连杆 AB 带动连杆 O_1B 绕 O_1 轴摆动,杆 EC 以铰链与滑块 C 相连,滑块 C 可沿杆 O_1B 滑动,摇杆摆动时带动杆 EC 推动钢材,如图 8.43 所示. 已知 $OA = r$, $AB = \sqrt{3}r$, $O_1B = \frac{2}{3}l$ ($r = 0.2$ m, $l = 1$ m), $\omega_{OA} = \frac{1}{2}$ rad/s, $\alpha_{OA} = 0$. 在图示位置时, $BC = \frac{4}{3}l$. 求:(1)滑块 C 的绝对速度和相对于摇杆 O_1B 的速度;(2)滑块 C 的绝对加速度和相对于摇杆 O_1B 的加速度.)

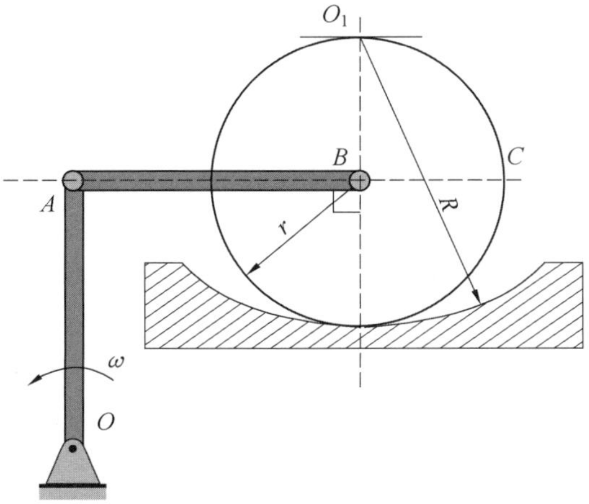

Fig. 8.42 Diagram of Exercise 13
(图 8.42 习题 13 图)

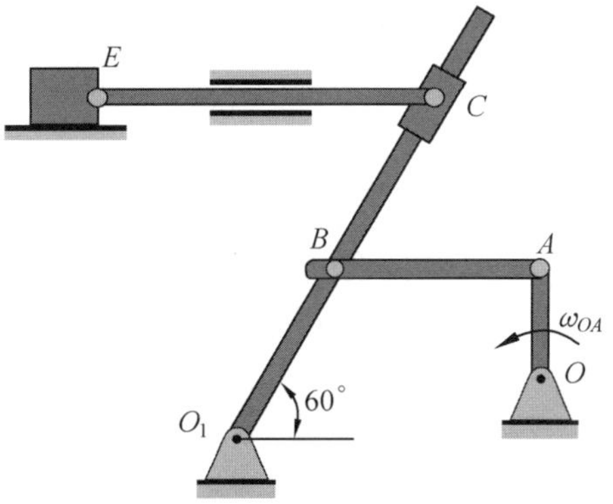

Fig. 8.43 Diagram of Exercise 14
(图 8.43 习题 14 图)

15. A rod OC is hinged to a wheel Ⅰ at the center of the wheel O and translates horizontally to the left at a uniform $\boldsymbol{v}$ as shown in Fig. 8.44. At the start, the distance between the point O and point A is l. Rod AB is rotatable about the fixed axis of axis A and keep in contact with wheel Ⅰ at point D. There is sufficient friction at the point of contact to keep it from slipping and the radius of wheel Ⅰ is r. Find the angular velocity ω_1 of wheel Ⅰ and the angular velocity of rod AB when $\theta=30°$. (杆 OC 与轮 Ⅰ 在轮心 O 处铰接并以均匀 $\boldsymbol{v}$ 水平向左平移,如图 8.44 所示. 起始时点 O 与点 A 相距 l,杆 AB 可绕 A 轴定轴转动,与轮 Ⅰ 在点 D 接触,接触处有足够大的摩擦使之不打滑,轮 Ⅰ 的半径为 r. 求当 $\theta=30°$时,轮 Ⅰ 的角速度 ω_1 和杆 AB 的角速度.)

16. As shown in Fig. 8.45, two identical large rings of radius $R=0.2$ m roll along the ground in opposite directions without sliding, with a constant velocity at the center of the rings, $v_A=0.1$ m/s and $v_B=0.4$ m/s. When $\angle MAB=30°$, find the velocity and acceleration of the small ring M attached to these two large rings with respect to each large ring, and the absolute velocity and absolute acceleration of the small ring M. (如图 8.45 所示,半径 $R=0.2$ m 的两个相同的大圆环沿地面向相反的方向无滑动地滚动,环心的速度为常数,$v_A=0.1$ m/s,$v_B=0.4$ m/s. 当$\angle MAB=30°$时,求套在这两个大圆环上的小圆环 M 相对于每个大圆环的速度和加速度,以及小圆环 M 的绝对速度和绝对加速度.)

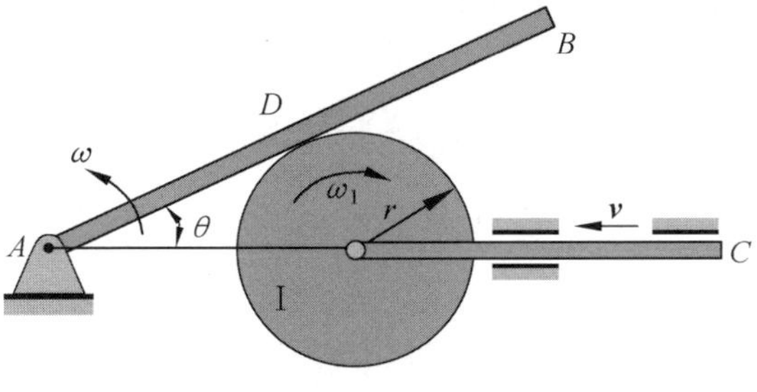

Fig. 8.44 Diagram of Exercise 15
(图 8.44 习题 15 图)

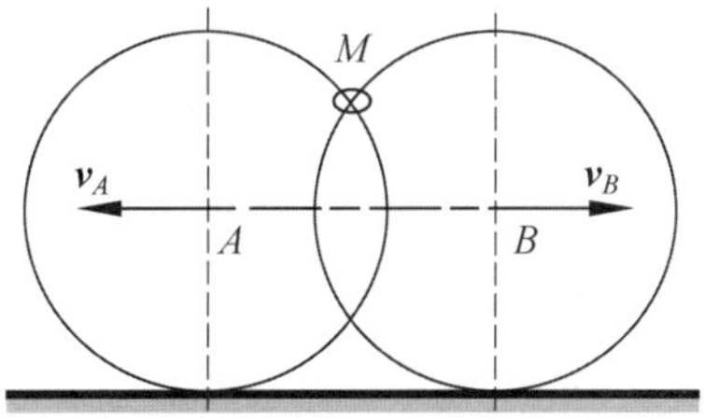

Fig. 8.45 Diagram of Exercise 16
(图 8.45 习题 16 图)

Part 3 Dynamics(第 3 篇　动力学)

Preface(引言)

Dynamics studies the relationship between the mechanical motion of an object and the forces acting upon it.(动力学研究物体的机械运动与作用力之间的关系.)

In statics, we analyze the forces acting on an object and study the equilibrium of the object under the action of the force system. In kinematics, we analyze the motion of an object only from a vector perspective without considering the forces acting upon it. Dynamics provides a comprehensive analysis of the mechanical motion of an object and studies the relationship between the forces acting upon the object and changes in its motion state, establishing the general laws of mechanical motion for objects.(在静力学中分析了作用于物体的力,研究了物体在力系作用下的平衡问题.在运动学中,仅从矢量方面分析了物体的运动,而不涉及作用力.动力学则对物体的机械运动进行全面的分析,研究作用于物体的力与物体运动状态变化之间的关系,建立物体机械运动的普遍规律.)

The formation and development of dynamics are closely related to the development of production, especially in today's rapidly developing modern industry and science and technology, which presents more complex issues for dynamics. For example, the calculation of the dynamics of high-speed operating machinery, the effects of wind loads and earthquakes on high-rise structures, space flight and rocket propulsion technology, and the dynamic characteristics of robots all require the application of dynamics theory.(动力学的形成和发展是与生产的发展密切联系的,特别是在现代工业和科学技术迅速发展的今天,对动力学提出了更加复杂的课题.例如,高速运转机械的动力计算、高层结构受风载及地震的影响、宇宙飞行及火箭推进技术以及机器人的动态特性等,都需要应用动力学的理论.)

The abstract models of objects in dynamics include particles and particle systems. A particle is an object with a certain mass whose shape and size can be ignored.(动力学中物体的抽象模型有质点和质点系.质点是具有一定质量而其形状和尺寸大小可以忽略不计的物体.)

For example, when studying the orbit of an artificial satellite, the shape and size of the satellite have little effect on the problem being studied. The satellite can be abstracted as a pointwhich mass concentrated at its center of mass. Similarly, when a rigid body undergoes translation, its shape and size can be ignored since the motion of all points within the rigid body is identical. It can also be abstracted as a point particle for analysis.(例如,在研究人造地球卫星的轨道时,卫星的形状和大小对所研究的问题几乎没有影响,可将卫星抽象为一个质量集中在质心的质点.刚体平移时,因刚体内各点的运动情况完全相同,所以也可以不考虑这个刚体的形状和大小,而将它抽象为一个质点来研究.)

If the shape and size of an object cannot be ignored in the problem being studied, the object should be abstracted as a system of particles. A system of particles refers to a system composed of several or an infinite number of interconnected particles.(如果物体的形状和大小在所研究的问题中不可忽略,则物体应抽象为质点系.质点系是由几个或无限个相互有联系的质点组成的系统.)

The solids, fluids, structures composed of several objects, and the solar system that we commonly encounter are all systems of particles. A rigid body is a special case of a system of particles where the distance between any two particles remains constant, also known as an invariant system of particles. For example, when studying the motion posture of artificial satellite, it should be abstracted as a mechanical model of a system of particles.(常见的固体、流体、由几个物体组成的机构,以及太阳系等都是质点系.刚体是质点系的一种特殊情形,其中任意两个质点间的距离保持不变,又称不变的质点系.例如,在研究人造卫星的运动姿态时,就要把它抽象为质点系的力学模型.)

Dynamics can be divided into particle dynamics and particle system dynamics, with the former serving as the foundation of the latter.(动力学可分为质点动力学和质点系动力学,前者是后者的基础.)

Chapter 9 The Basic Equations of Particle Dynamics (质点动力学的基本方程)

A particle is the simplest and most fundamental model for the motion of complex objects. In this chapter, the basic equations of particle dynamics are derived based on the fundamental laws of dynamics. These equations describe the relationship between the forces acting on a particle and its motion. By using calculus methods on the basic equations, we can solve dynamic problems of a single particle. (质点是构成复杂物体运动最简单最基本的模型. 本章根据动力学基本定律得出质点动力学的基本方程. 质点动力学基本方程描述质点受力与其运动变化之间的关系. 由基本方程运用微积分方法,可以求解一个质点的动力学问题.)

9.1 Dynamic Laws, the Equations of Motion of a Particle, D'Alembert's Principle (动力学定律,质点的运动方程,达朗伯尔原理)

Dynamics is a branch of mechanics that studies the laws governing the motion of particles under the action of forces. The laws developed by Newton form the basis of dynamics. (动力学是力学的一部分,研究质点在受力作用下的运动规律. 牛顿创建的定律是动力学的基础.)

1. The first law: Galileo's law of inertia(第一定律:伽利略惯性定律)

Any object will remain at rest or in uniform motion in a straight line unless compelled to change its state by the action of another object. (任何物体都保持静止或匀速直线运动状态,直到受到其他物体的作用力迫使它改变这种状态为止.)

2. The second law: the fundamental law of dynamics(第二定律:动力学基本定律)

The relationship between the acceleration $\boldsymbol{a}$, mass m and force $\boldsymbol{F}$ acting on particle M is determined (Fig. 9.1(a)): the acceleration of the particle is directly proportional to the applied force and in the same direction as the force. (确定质点 M 的加速度 $\boldsymbol{a}$、质量 m 和力 $\boldsymbol{F}$ 之间的关系(图 9.1(a)):质点的加速度与所受到的作用力成正比,且加速度的方向与作用力的方向相同.)

We can express this law using Euler's formula as (可以将此定律用欧拉公式表述为)

$$m\boldsymbol{a} = \boldsymbol{F} \tag{9.1}$$

In classical mechanics, mass m is a fixed value. Mass is a measure of inertia for a particle in translational motion. We can express the fundamental laws of dynamics in scalar form by projecting the vector in the equation onto the coordinate axes. (在经典力学中,质量 m 为固定值. 质量是质点在平移运动中的惯性度量. 可以将运动学的基本定律表述为标量形式,将等

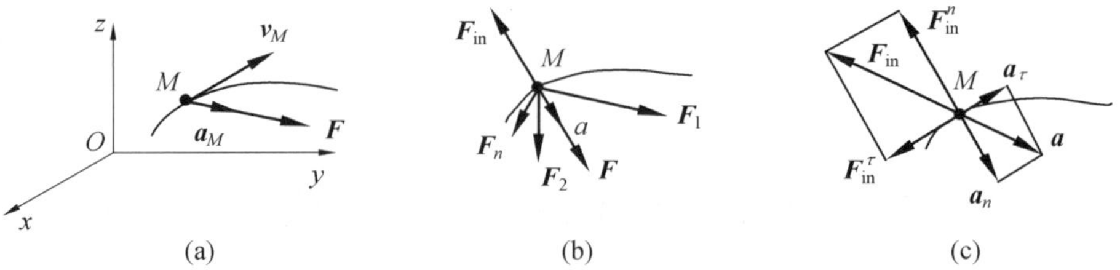

Fig. 9.1 Relationship between acceleration and force of a mass point
（图 9.1 质点的加速度和作用力的关系）

式中的矢量投影到坐标轴上得）

$$ma_x = F_x$$
$$ma_y = F_y$$
$$ma_z = F_z$$

3. The third law: law of action and reaction forces

（第三定律：作用力与反作用力定律）

The content of the third law is every actionforce has an equal and opposite reaction force.（第三定律的内容是任何作用力都有大小相等、方向相反的反作用力.）

This law states that when two objects interact with each other, regardless of their motion state, the forces they exert on each other are equal in magnitude, opposite in direction, and collinear.（该定律说明在两个物体相互作用时，无论它们处于什么运动状态，它们受到的力都在同一条直线上，大小相等，方向相反.）

The fundamental laws of mechanics can be expressed in scalar form or by projecting vectors onto Cartesian coordinates or natural coordinate axes. In the first case, we can derive equations of motion for a particle in a Cartesian coordinate system.（力学的基本定律可以表述为标量形式，或者将矢量投影到笛卡儿坐标上，或者投影到自然坐标轴上.在第一种情况下，可以得出质点在笛卡儿坐标系中的运动公式为）

$$m\ddot{x} = F_x\ ,\ m\ddot{y} = F_y\ ,\ m\ddot{z} = F_z$$

Among which(其中)

$$\ddot{x} = a_x\ ,\ \ddot{y} = a_y\ ,\ \ddot{z} = a_z$$

In the second scenario, we can derive the equation of motion in natural coordinates, that is（在第二种情况下，可以得出自然坐标下的运动公式，即）

$$ma_n = F_n,\ ma_\tau = F_\tau,\ ma_b = F_b$$

where(式中)

$$a_n = \frac{v^2}{r}\ ,\ a_\tau = \frac{d^2S}{dt^2}$$

where S is arc coordinate.（其中，S 是弧坐标.）

The projection of acceleration on the bi-normal axis is zero , $a_b = 0$, so $F_b = 0$.（加速度在副法线上的投影等于零，$a_b = 0$，所以 $F_b = 0$.）

Example 1: The motion equation of a particle M with mass m is（**例题** 1：质量为 m 的质点 M 的运动公式为）

$$x = r\cos kt,\ y = r\sin kt$$

Determine the resultant force acting on the particle and its trajectory of motion. (确定质点所受的合力,以及质点的运动轨迹.)

Solution:(解:)

(1) Determine the projection of acceleration on the coordinate axis. To do so, first determine the projection of velocity on the corresponding axis, we have (确定加速度在坐标轴上的投影. 为此,首先确定速度在相应轴上的投影,有)

$$v_x = \dot{x} = -kr\sin kt,\ v_y = \dot{y} = kr\cos kt$$

From this, we can conclude that(由此可以得出)

$$a_x = \dot{v}_x = -k^2 r\cos kt,\ a_y = \dot{v}_y = -k^2 r\sin kt$$

(2) Determine the projection of the resultant force. (确定合力的投影.)

Because (因为)

$$F_x = m\ddot{x} = ma_x,\ F_y = m\ddot{y} = ma_y$$

so(所以)

$$F_x = -mk^2 r\cos kt\ ,\ F_y = -mk^2 r\sin kt$$

(3) Determine the magnitude of the resultant force, we have (确定合力的大小,有)

$$F = \sqrt{F_x^2 + F_y^2} = mk^2 r$$

(4) Determine the direction of the resultant force, we have (确定合力的方向,有)

$$\cos\alpha = \frac{F_x}{F} = -\cos kt = -\frac{x}{r},\cos\beta = \frac{F_y}{F} = -\sin kt = -\frac{y}{r}$$

Clearly, the angle between the resultant force and the coordinate axis changes with time. (显然,合力与坐标轴的夹角随时间发生改变.)

(5) Determine the trajectory of the particle. To eliminate the variable t, we perform some exponentiation and derive the formula for the trajectory. We obtain the equation of a circle with radius r, that is(确定质点的运动轨迹. 为消除变量 t,进行乘方并推导轨迹公式. 可以得到半径为 r 的圆方程,即)

$$x^2 + y^2 = r^2$$

According to the given solution, we can draw the following conclusion: the particle moves along a circular path with a radius of r under the influence of force, and the force always points toward the center of this circle. (根据所给解答可以得出下列结论:质点在力的作用下沿半径为 r 的圆周运动,作用力始终指向该圆周的中心.)

4. D'Alembert's principle(达朗伯尔原理)

D'Alembert's principle is a general method that can transform the dynamic equation into the form of static equations. To achieve this, the concept of "inertial force" of a particle is introduced, where the magnitude of the inertial force is equal to the product of the particle's mass and acceleration, and its direction is opposite to that of the acceleration, we have(达朗伯尔原理是普遍方法,使用该方法可使动力方程具有静力学方程的形式. 为此,引入了"质点惯性力"的概念,质点惯性力的大小等于质点质量与其加速度的乘积,方向与加速度方向相反,有)

$$\boldsymbol{F}_{\text{in}} = -m\boldsymbol{a}$$

We can assume that particle M undergoes acceleration $\boldsymbol{a}$ under the action of a force system $\boldsymbol{F}_1, \boldsymbol{F}_2, \ldots, \boldsymbol{F}_n$ (as shown in Fig. 9.1(b)). The basic dynamic equation for this is(可以假设质点 M 在力系 $\boldsymbol{F}_1, \boldsymbol{F}_2, \cdots, \boldsymbol{F}_n$ 作用下以加速度 $\boldsymbol{a}$ 运动(图 9.1(b)),此时基本的动力学方程为)

$$m\boldsymbol{a} = \boldsymbol{F}_1 + \boldsymbol{F}_2 + \cdots + \boldsymbol{F}_n$$

Move the $m\boldsymbol{a}$ on the left side of the equation to the right side. Then we have(把等式左边的 $m\boldsymbol{a}$ 挪到右边,则有)

$$\boldsymbol{F}_1 + \boldsymbol{F}_2 + \cdots + \boldsymbol{F}_n - m\boldsymbol{a} = 0$$

Because of $-m\boldsymbol{a} = \boldsymbol{F}_{\text{in}}$, then (因为 $-m\boldsymbol{a} = \boldsymbol{F}_{\text{in}}$,所以)

$$\boldsymbol{F}_1 + \boldsymbol{F}_2 + \cdots + \boldsymbol{F}_n + \boldsymbol{F}_{\text{in}} = 0$$

The resulting equation reflects the principle of D'Alembert, which can be expressed as the vector sum of all external forces acting on a point and its inertial force is equal to zero. (所得的关系式反映了达朗伯原理,可以表述为所有作用于点上的外力与该点的惯性力的矢量和等于零.)

The principle of D'Alembert applies to both free particles and constrained particles, as constraints are removed from the particle and replaced with constraint reaction forces. We can consider that the particle is subject to active and constraint forces, resulting in motion and acceleration. (达朗伯原理既适用于自由质点,也适用于非自由质点,因为质点去掉了约束,并用约束反力替代了其作用. 可以认为质点在主动力和约束力的作用下发生运动,产生了加速度.)

Remember that the inertial force is hypothetical force acting on a particle. In fact, the inertial force does not act on the particle itself but on the object that causes its acceleration. This method has been widely used in calculating the intensity of dynamic loads. (记住,惯性力是假想作用于质点上的. 实际上惯性力并没有作用于质点上,而是作用于使其产生加速度的物体上. 在计算动载荷作用的强度时该方法得到了广泛应用.)

The inertial force can be decomposed into tangential $\boldsymbol{F}_{\text{in}}^{\tau}$ (tangential inertial force) and normal $\boldsymbol{F}_{\text{in}}^{n}$ (centrifugal inertial force) components (Fig. 9.1(c)), we have (惯性力可以分解为切向的 $\boldsymbol{F}_{\text{in}}^{\tau}$ (切向惯性力)和法向的 $\boldsymbol{F}_{\text{in}}^{n}$ (离心惯性力)分量(图 9.1(c)),有)

$$F_{\text{in}}^{\tau} = ma^{\tau}, \quad F_{\text{in}}^{n} = m\frac{v^2}{\rho}$$

where ρ is the curvature radius of the trajectory. (式中,ρ 为轨迹的曲率半径.)

If an object rotates with angular velocity ω and angular acceleration ε, and the trajectory of a point on it is circular (with radius r), then the tangential and centrifugal components of the inertial force can be expressed as (如果物体以角速度 ω 和角加速度 ε 旋转,其上的点轨迹为圆(圆周半径为 r),则惯性力的切向和离心分量可表述为)

$$F_{\text{in}}^{\tau} = mr\varepsilon, \quad F_{\text{in}}^{n} = mr\omega^2$$

9.2 The Force Acting on a System of Particles (作用于质点系上的力)

A collection of particles that interact with each other is called a system of particles, which is sometimes also referred to as an object or a particle system. There are free systems of particles (such as the solar system) and constrained systems of particles (whose motion is limited by constraints). Examples of constrained systems of particles can be found in any mechanical or machine system. All forces acting on a non-free system of particles can be divided into known forces (active forces) and constraint forces (passive forces). (相互作用的质点的集合称为质点系统. 质点系统有时又称物系或质点系. 存在自由质点系(如太阳系)和非自由质点系(其运动受到约束限制). 非自由质点系的范例可以是任何机械或机床. 所有作用于非自由点系上的力都可分为已知力(主动力)和约束力(被动力).)

It is also possible to separate the forces acting on any particle system into external forces and internal forces using another method. Let $\boldsymbol{F}^{\mathrm{E}}$ represent external forces and $\boldsymbol{F}^{\mathrm{I}}$ represent internal forces. (也可以按照另外一种方法,将作用于任何质点系统点上的力分为外力和内力. 用$\boldsymbol{F}^{\mathrm{E}}$代表外力, $\boldsymbol{F}^{\mathrm{I}}$ 代表内力.)

A particle that does not belong to the system acts on particles within the system is called an external force. (不属于该系统的质点对该系统内质点的作用力称为外力.)

The force of interaction between particles in a particle system is called internal force. An example of internal force can be the elastic force between particles in an elastic body particle system. (该质点系统质点间相互作用的力称为内力. 内力的例子可以是质点系统弹性体质点间相互作用的弹力.)

According to the system of forces under study, the same force can be both an external force and an internal force. For example, the reactive force of a bearing constraint is an external force for the shaft, but for the entire machine or device, this reactive force can be considered an internal force. (根据研究的力系,同一种力既可以是外力,也可以是内力. 例如,轴承的约束反力对于轴来说是外力. 而对整个机器及装置,这一约束反力就可以视为内力.)

In this way, any force can be an external or internal force. At the same time, it can also be a known force or a constraint reaction force. The motion of a system's particle depends on both external and internal force. (这样,任何力都可以是外力或内力. 同时,它也可以是已知力或约束反力. 系统质点的运动既取决于外力,也取决于内力.)

According to Newton's third law, for every action force, there is an equal and opposite reaction. Based on this principle, the following conclusions can be drawn. (按照牛顿第三定律,每个力都有另一个大小相等、方向相反的反作用力. 在这一基础上可以得出以下结论.)

(1) The total vector sum of all internal forces in the system is equal to zero, we have (系统所有内力的总矢量和等于零,有)

$$\boldsymbol{R}^{\mathrm{I}} = \sum_{i=1}^{k} \boldsymbol{F}_i^{\mathrm{I}} = 0$$

Therefore, the algebraic sum of their projections on the coordinate axis is also equal to zero, we have(因此,它们在坐标轴上投影的代数和也等于零,有)

$$\sum_{i=1}^{k} F_{ix}^{\mathrm{I}} = 0,\ \sum_{i=1}^{k} F_{iy}^{\mathrm{I}} = 0,\ \sum_{i=1}^{k} F_{iz}^{\mathrm{I}} = 0$$

(2) The sum of all internal forces in a system with respect to any center and coordinate axis is equal to zero, we have (系统所有内力相对于任何中心和坐标轴的主矩等于零,有)

$$\boldsymbol{M}_O^{\mathrm{I}} = \sum_{i=1}^{k} \boldsymbol{M}_{iO}^{\mathrm{I}} = 0$$

or(或者)

$$\sum_{i=1}^{k} M_x(\boldsymbol{F}_i^{\mathrm{I}}) = 0,\ \sum_{i=1}^{k} M_y(\boldsymbol{F}_i^{\mathrm{I}}) = 0,\ \sum_{i=1}^{k} M_z(\boldsymbol{F}_i^{\mathrm{I}}) = 0$$

These equations have the form of equilibrium equations for a system of arbitrary forces in space, but they also include unbalanced internal forces because they act at different points within the system and can cause relative displacement of these points. (这些等式具有空间任意力系平衡方程的形式,但是其中还包括不平衡的内力,因为它们作用于体系的不同点上,并且可以引起这些点发生相对位移.)

9.3 Law of Motion of the Mass Center of a Particle System (质点系统质心运动定律)

Assuming a particle system of mass m composed of k particles (Fig. 9.2). It is well-known that if the masses m_i and coordinates of each point are given, then the position of the center of mass of the system can be determined as(假设质量为 m 的质点系统由 k 个质点组成(图9.2). 众所周知,如果给出点的质量 m_i 和其坐标,那么可以找到该系统的质心位置,即)

$$x_C = \frac{\sum_{i=1}^{k} m_i x_i}{m},\ y_C = \frac{\sum_{i=1}^{k} m_i y_i}{m},\ z_C = \frac{\sum_{i=1}^{k} m_i z_i}{m}$$

or(或者)

$$m x_C = \sum_{i=1}^{k} m_i x_i,\ m y_C = \sum_{i=1}^{k} m_i y_i,\ m z_C = \sum_{i=1}^{k} m_i z_i$$

Taking the second derivative of these equations, we can obtain (对这些等式求两次导数,可以得出)

$$m \ddot{x}_C = \sum_{i=1}^{k} m_i \ddot{x}_i,\ m \ddot{y}_C = \sum_{i=1}^{k} m_i \ddot{y}_i,\ m \ddot{z}_C = \sum_{i=1}^{k} m_i \ddot{z}_i$$

According to the main laws of dynamics, the right-hand side of the equation obtained is the algebraic sum of the external force $\boldsymbol{F}_i^{\mathrm{E}}$ and the internal force $\boldsymbol{F}_i^{\mathrm{I}}$ acting on the particle projected onto the corresponding coordinate axes. Therefore, the final equation can be transformed into(按照动力学的主要定律,所获等式的右边部分是作用于质点的外力 $\boldsymbol{F}_i^{\mathrm{E}}$ 和内力 $\boldsymbol{F}_i^{\mathrm{I}}$ 在相应坐标轴投影的代数和. 因此,最后的等式可以转化为)

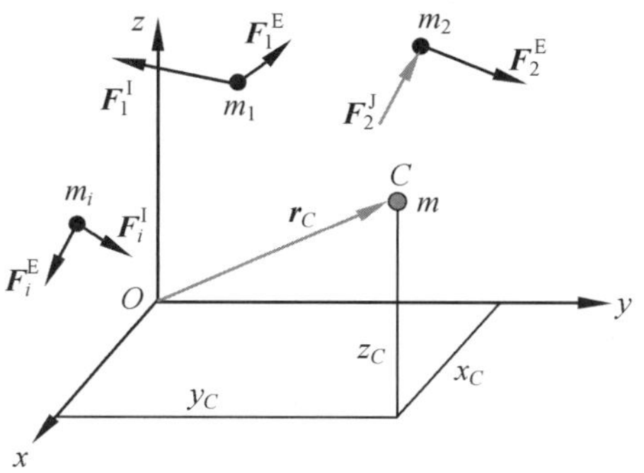

Fig. 9.2 A system of particle consisting of k particles(图 9.2 由 k 个质点组成的质点系)

$$m\ddot{x}_C = \sum_{i=1}^{k} F_{ix}^{E} + \sum_{i=1}^{k} F_{ix}^{I}$$

$$m\ddot{y}_C = \sum_{i=1}^{k} F_{iy}^{E} + \sum_{i=1}^{k} F_{iy}^{I}$$

$$m\ddot{z}_C = \sum_{i=1}^{k} F_{iz}^{E} + \sum_{i=1}^{k} F_{iz}^{I}$$

Considering that the main vector of internal force is equal to zero ($\boldsymbol{R}^{I} = 0$), it can be concluded that(考虑到内力的主矢量等于零($\boldsymbol{R}^{I} = 0$),可以得出)

$$m\ddot{x}_C = \sum_{i=1}^{k} F_{ix}^{E}$$

$$m\ddot{y}_C = \sum_{i-1}^{k} F_{iy}^{E}$$

$$m\ddot{z}_C = \sum_{i=1}^{k} F_{iz}^{E}$$

These equations can reflect the laws of motion of the system's center of mass, expressed as follows.(这些等式可以反映系统质心运动的规律,表述如下.)

The motion of the mass center of an object system is the same as the motion of aparticle with the same mass subject to the same external forces.(物体系统质心的运动与受到相同外力作用的同质量质点的运动相同.)

It can be concluded that internal forces do not affect the motion of the center of mass of a system.(由此可以得出,内力不会影响物体系统质心的运动.)

Example 2: Determine the displacement of a floating crane capable of lifting a 2 000 kg weight when the crane's boom is rotated to a vertical position (Fig. 9.3). The weight of the crane is 20 t and the length of the AB boom is 8 m. Water resistance can be ignored.(**例题** 2:在起重臂转动至垂直位置时确定提升质量 2 000 kg 重物的浮游起重机的位移(图 9.3).起重机质量为 20 t,AB 起重臂长度 8 m,可以忽略水的阻力.)

Solution:(解:)

(1) Select a coordinate system (Fig. 9.3(a)).(选择坐标系(图 9.3(a)).)

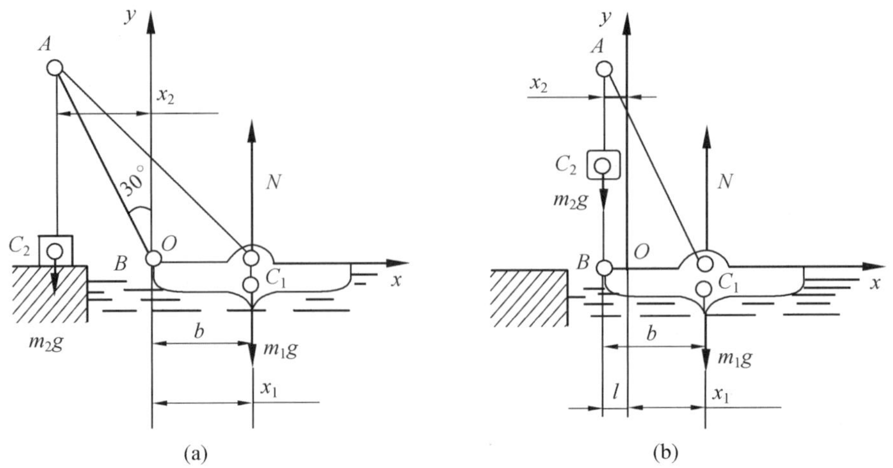

Fig. 9.3 Diagram of Example 2 (图 9.3 例题 2 图)

(2) List all external forces acting on the object system. The gravitational force $m_1\boldsymbol{g}$ (known force) and the force $\boldsymbol{N}$ (passive force) act on the floating crane, only one external force acts on the heavy object, which is its own weight $m_2\boldsymbol{g}$. (列出所有作用于该物体系统上的外力. 重力 $m_1\boldsymbol{g}$ (已知力)和力 $\boldsymbol{N}$ (被动力)作用于浮游起重机,只有一个外力作用于重物,其重力为 $m_2\boldsymbol{g}$.)

(3) The equation of motion for the mass center of a particle system can be expressed as(质点系统的质心运动方程可表述为)

$$m\ddot{x}_C = \sum_{i=1}^{2} F_{ix}^{\mathrm{E}},\ m\ddot{y}_C = \sum_{i=1}^{2} F_{iy}^{\mathrm{E}}$$

or(或者)

$$m\ddot{x}_C = 0,\ m\ddot{y}_C = -m_1g - m_2g + N$$

(4) Study the first equation, because we are concerned with the motion of the center of gravity along the horizontal direction. Due to $m\ddot{x}_C = 0$, the velocity of the gravity center along the Ox axis is $v_{x_C} = \text{const}$. This indicates that the velocity of the gravity center in this direction is constant at all times, which means that $v_{x_C\text{begin}} = v_{x_C\text{end}}$ is truly valid. (研究第一个等式,关心的是重心沿水平方向的运动. 由于 $m\ddot{x}_C = 0$,因此重心沿 Ox 轴的速度 $v_{x_C} = \text{const}$. 这说明重心在该方向上的速度在任何时候都是不变的,也就是说 $v_{x_C\text{begin}} = v_{x_C\text{end}}$ 真实成立.)

At the beginning, the system is in a stationary state, so $v_{x_C\text{begin}} = v_{x_C\text{end}} = 0$. Because $v_{x_C} = \dfrac{\mathrm{d}x_C}{\mathrm{d}t}$, so $x_C = \text{const}$. (在初始时,系统处于静止,因此 $v_{x_C\text{begin}} = v_{x_C\text{end}} = 0$. 因为 $v_{x_C} = \dfrac{\mathrm{d}x_C}{\mathrm{d}t}$,所以 $x_C = \text{const}$.)

This implies that the analysis of the equation of motion for the center of mass along the Ox axis shows that the initial and final coordinates of the center of mass coincide, we have(这样,对质心沿 Ox 轴运动方程的分析表明,质心的初始和最终坐标是重合的,有)

$$x_{C\text{begin}} = x_{C\text{end}}$$

(5) We can write formulas to determine the initial and final positions of the center of mass of a particle system which are(可以写出用于确定质点系统重心的初始和最终位置的公式,有)

$$\begin{cases} x_{Cbegin} = \dfrac{m_1 \cdot x_{1begin} + m_2 \cdot x_{2begin}}{m_1 + m_2} \\ x_{Cend} = \dfrac{m_1 \cdot x_{1end} + m_2 \cdot x_{2end}}{m_1 + m_2} \end{cases}$$

(6) We can write the initial and final coordinates of the system's center of mass in the chosen coordinate system (Fig. 9.3(a) and(b)),and we have (可以按照选择的坐标系写出系统质心的初始和最终坐标(参加图 9.3(a)和(b),然后有)

$$\begin{cases} x_{1begin} = b\ ,\ x_{2begin} = -AB\sin 30^\circ = -4\ \text{m} \\ x_{1end} = b - l\ ,\ x_{2end} = -l \end{cases}$$

(7) We can determine the displacement l of the floating crane. By setting $x_{Cbegin} = x_{Cend}$, we can obtain(可以确定浮游起重器的位移 l,使 $x_{Cbegin} = x_{Cend}$,可以得出)

$$m_1 \cdot x_{1begin} + m_2 \cdot x_{2begin} = m_1 \cdot x_{1end} + m_2 \cdot x_{2end}$$

or(或者)

$$m_1 b + m_2(-4) = m_1(b - l) + m_2(-l), m_1 b - m_2 \cdot 4 = m_1 b - m_1 l - m_2 l$$

$$-2\ 000 \cdot 4 = -2\ 000l - 2\ 000l, l = (4 \cdot 2\ 000)/(2\ 000 + 2\ 000) = 0.36\ (\text{m})$$

Then, we obtain l=0.36 m. The answer is l=0.36 m. (这样,得到 l=0.36 m,则答案为 l=0.36 m.)

9.4 Questions(思考题)

1. Can the force acting on a free particle M moving along the curve AB appear as shown in Fig. 9.4? (已知自由质点 M 沿曲线 AB 运动,质点上所受的力能否出现图 9.4 所示的情况?)

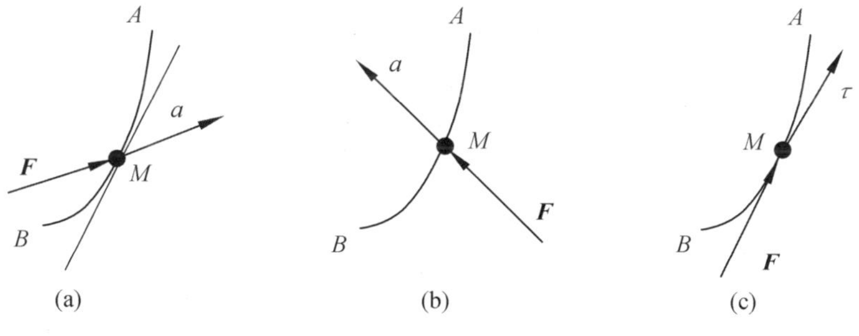

Fig. 9.4 Diagram of Exercise 1(图 9.4 思考题 1 图)

2. A small car moves along the positive x-axis under the action of force $\boldsymbol{F}$, with an initial velocity of $v_0>0$. If the direction of $\boldsymbol{F}$ is the same as that of the positive x-axis, and its magnitude decreases with time, then the velocity of the car will also gradually decrease with time. Is this statement correct? (一小车在力 $\boldsymbol{F}$ 作用下沿 x 轴正向运动,其初速度 $v_0>0$. 如果 $\boldsymbol{F}$ 的方向与 x 轴正方向一致,大小随时间减小,则小车的速度是否也随时间逐渐减小?)

3. Does the velocity of a particle necessarily increase as the force acting upon it increases? (质点受到的力越大,其运动的速度是否一定也越大?)

4. If two small balls with the same mass are thrown simultaneously with the same initial velocity at the same location and in the same coordinate system, ignoring air resistance, will their velocities be the same when they hit the ground? (同一地点、同一坐标系内,以相同大小的初速度斜抛两质量相同的小球,若不计空气阻力,则它们落地时速度的大小是否相同?)

5. Compared to a stone thrown horizontally with a certain initial velocity, which stone has a greater acceleration when falling freely without considering air resistance? Why? (若不计空气阻力,自由下落的石块与以一定初速度水平抛出的石块相比,哪一个下落的加速度较大? 为什么?)

6. Assuming air resistance is proportional to the square of the projectile's velocity, please write the differential equations for the motion of an object in three scenarios: vertical upward motion, vertical downward motion and inclined projectile motion. (设空气阻力与抛射体的速度平方成正比,试写出物体铅垂向上、铅垂向下及倾斜抛射三种情况下的运动微分方程.)

9.5 Exercises(习题)

1. A pendulum has a rope length of l and a mass of m. The pendulum is released from position OA, which is 30° from the vertical, without any initial velocity, as shown in Fig. 9.5. When the pendulum reaches the vertical position, the midpoint of the rope is blocked by a wooden peg C, and only the lower half continues to swing. Find the speed of the pendulum and the tension in the rope when the rope rises to an angle of φ with the vertical. (单摆的摆绳长为 l,摆锤质量为 m,单摆由偏离铅垂线30°的位置 OA 无初速度地释放,如图9.5所示. 当摆到铅垂位置时,绳的中点被木钉 C 挡住,只有下半段继续摆动. 求当摆绳升到与铅垂线成 φ 角时,摆锤的速度和摆绳的拉力.)

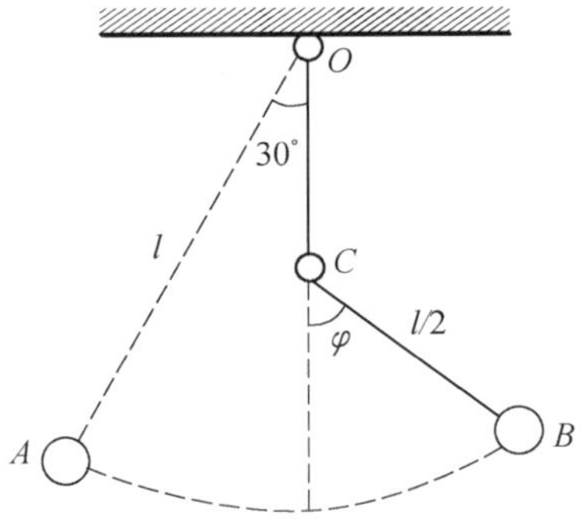

Fig. 9.5 Diagram of Exercise 1

(图9.5 习题1图)

2. A steel wire rope with a length of $l = 5$ m is suspended on the trolley of a bridge crane, carrying a weight of $m = 10$ t, as shown in Fig. 9.6. When the trolley moves to the right at a constant speed of $v_0 = 1$ m/s, the steel wire rope remains vertically oriented. When the trolley

brakes suddenly, the weight swings around the suspension point O due to inertia. Calculate the instantaneous tension $\boldsymbol{F}$ in the steel wire rope when the swinging begins and the maximum swing angle φ. (在桥式起重机的小车上用长度为 $l=5$ m 的钢丝绳悬吊着质量 $m=10$ t 的重物，如图 9.6 所示. 小车以匀速 $v_0=1$ m/s 向右运动时，钢丝绳保持铅直方向. 当小车紧急刹车时，重物因惯性而绕悬挂点 O 摆动. 试求刚开始摆动的瞬时钢丝绳的拉力 $\boldsymbol{F}$ 及最大摆角 φ.)

3. A balloon with a total mass of m, as shown in Fig. 9.7, rises with a constant acceleration $\boldsymbol{a}$. What is the increase in mass of the balloon needed for it to descend with the same acceleration? (图 9.7 所示气球的总质量为 m,以匀加速度 $\boldsymbol{a}$ 上升. 则气球的质量增加多少，方能使它以相同的加速度下降?)

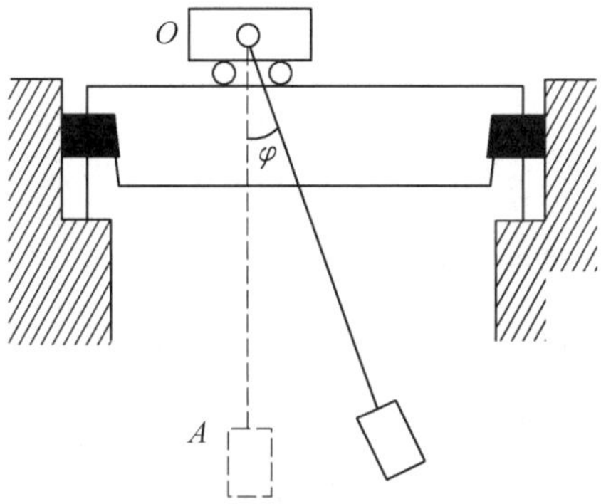

Fig. 9.6 Diagram of Exercise 2
(图 9.6 习题 2 图)

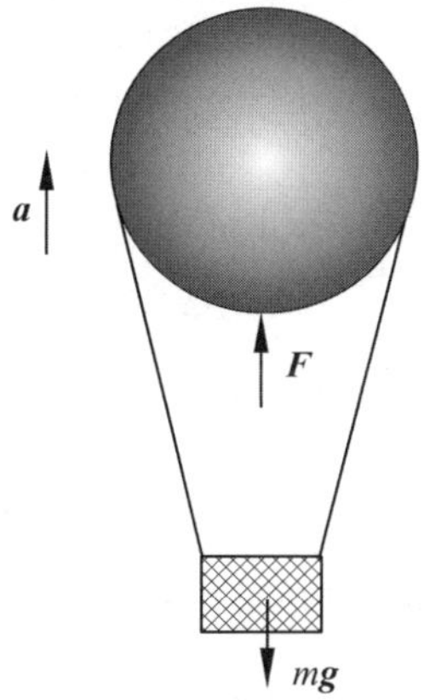

Fig. 9.7 Diagram of Exercise 3
(图 9.7 习题 3 图)

4. The masses of block A and B are $m_1=20$ kg and $m_2=40$ kg respectively, and are connected to each other by springs, as shown in Fig. 9.8. Block A moves harmonically along the plumb line with $y=H\cos\frac{2\pi}{T}t$, in which the amplitude $H=10$ mm and the period $T=0.25$ s. The mass of the spring is neglected. Find the maximum and minimum values of pressure on CD at the support surface. (物块 A、B 的质量分别为 $m_1=20$ kg 和 $m_2=40$ kg,并以弹簧互相连接,如图 9.8 所示. 物块 A 沿铅垂线以 $y=H\cos\frac{2\pi}{T}t$ 做简谐运动,式中振幅 $H=10$ mm,周期 $T=0.25$ s. 弹簧的质量不计. 求支撑面 CD 所受压力的最大值和最小值.)

5. As shown in Fig. 9.9, two objects have masses of m_1 and m_2, respectively, connected by a rope passing over a pulley with radius r. If the initial height difference between the two objects is h and $m_1>m_2$, neglecting the mass of the pulley, find the time it takes for the two objects to reach the same height after being released from rest. (如图 9.9 所示,两物体质量分别为 m_1 和 m_2,用绳跨过一滑轮,滑轮的半径为 r. 若在开始时,两物体的高度差为 h,且 $m_1>m_2$,不计滑轮质量,求由静止释放后,两物体达到相同的高度时所需的时间.)

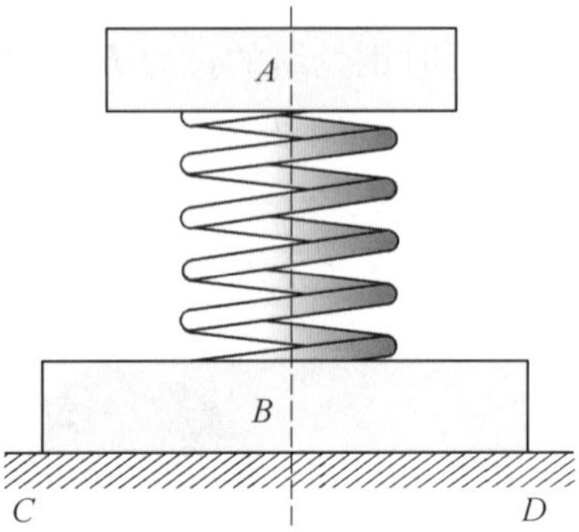

Fig. 9.8 Diagram of Exercise 4
(图 9.8 习题 4 图)

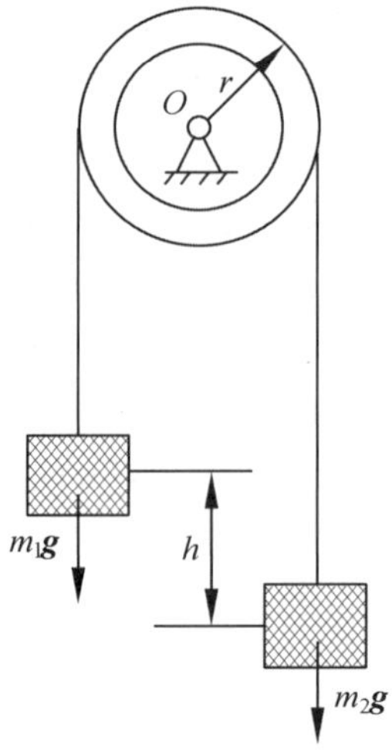

Fig. 9.9 Diagram of Exercise 5
(图 9.9 习题 5 图)

6. The mass of a ship is m, and it travels in a straight line with the rule $s = \frac{m}{b}v_0(1 - e^{\frac{-b}{m}t})$ before stopping, where v_0 is the initial velocity and b is a constant. Find the drag force on the ship and express it as a function of velocity v. (一船的质量为 m,在停止前以 $s = \frac{m}{b}v_0(1 - e^{\frac{-b}{m}t})$ 规律沿直线行驶,式中 v_0 为初速度,b 为常数. 求船所受的阻力并将其表示成速度 v 的函数.)

7. The length of a simple pendulum is l, and the mass of the pendulum is m. When $t=0$, the angle between the swing rope and the plumb line is $\varphi=\varphi_0$. And it swings according to the angular velocity law of $\dot{\varphi} = \pm\sqrt{2\frac{g}{l}(\cos\varphi - \cos\varphi_0)}$. Find the tension on the rope when the pendulum passes the lowest position. (单摆的摆长为 l,摆锤的质量为 m. $t=0$ 时,摆绳与铅垂线的夹角 $\varphi=\varphi_0$,且按 $\dot{\varphi} = \pm\sqrt{2\frac{g}{l}(\cos\varphi - \cos\varphi_0)}$ 的角速度规律摆动. 求当摆锤经过最低位置时,绳所受的拉力.)

8. The mass of the wheel is m and it moves uniformly along the horizontal road surface, as shown in Fig. 9. 10. There is a pit in the road whose shape is determined by equation $y = \frac{\delta}{2}\left(1 - \cos\frac{2\pi}{l}x\right)$. Assuming the road surface and wheels are both rigid bodies and the carriage applies pressure $\boldsymbol{F}$ to the wheels through springs. When the car passes over a pothole, what are the maximum and minimum constraint forces that the road surface exerts on the wheels? (车轮的质量为 m,沿水平路面做匀速运动,如图 9. 10 所示. 路面上有一凹坑,其形状由方程 $y = \frac{\delta}{2}\left(1 - \cos\frac{2\pi}{l}x\right)$ 确定. 路面和车轮均看成刚体,车厢通过弹簧给车轮以压力 $\boldsymbol{F}$. 当车子经过凹坑时,路面给车轮的最大和最小约束力是多少?)

9. A wheel with a radius of R is rotating about axis O with an angular velocity of ω. It drives a sliding block along a vertical track, as shown in Fig. 9. 11. A mass m is placed on top of the sliding block A, and the eccentric distance of the wheel is e. At the beginning, OC lies along the horizontal line. Find: (1) the maximum pressure that the mass A exerts on the sliding block; (2) the maximum angular velocity ω_{max} that prevents the mass A from leaving the sliding block. (半径为 R 的偏心轮绕轴 O 以匀角速度 ω 转动,推动导板沿铅直轨道运动,如图 9.11 所示. 导板顶部放有一质量为 m 的物块 A,设偏心距 $OC=e$,开始时 OC 沿水平线. 求:(1)物块 A 对导板的最大压力;(2)使物块 A 不离开导板的最大角速度 ω_{max}.)

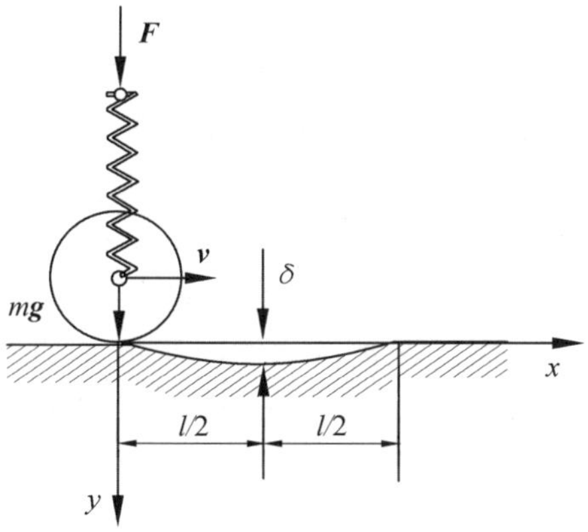

Fig. 9. 10 Diagram of Exercise 8
(图 9.10 习题 8 图)

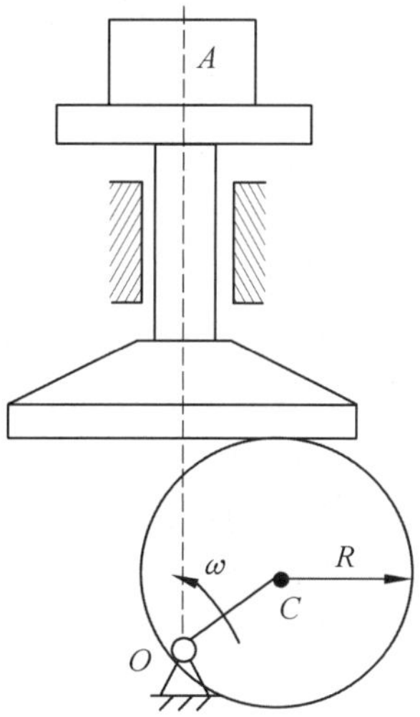

Fig. 9. 11 Diagram of Exercise 9
(图 9.11 习题 9 图)

10. The motion equation of a small ball with a mass of 0.001 kg falling under the influence of gravity and subject to air resistance is given as $x = 4.9t - 2.45 + 0.01e^{-2t}$, where x is in m and t is in s, with the Ox axis pointing vertically downwards. Find the relationship between the air resistance force F and the velocity of the ball. (质量为 0.001 kg 的小球在重力作用下降落,并受到空气阻力,其运动方程为 $x = 4.9t - 2.45 + 0.01e^{-2t}$,其中 x 以 m 计,t 以 s 计,Ox 轴铅垂向下. 求小球所受阻力 F 与速度的关系.)

11. As shown in Fig. 9. 12, a car turns a curved path with a radius of R, while the angle of inclination of the lane towards the center of the circle is θ, and the static friction coefficient between the tire and the road surface is f_s. Given that $R = 20$ m, $\tan\theta = 2f_s = 0.5$, find the maximum speed of the car when passing through the curve. (如图 9.12 所示,一汽车转过半径为 R 的圆弯,而车道向圆心方向的倾斜角为 θ,车胎与道面间的静摩擦因数为 f_s. 已知 $R=$ 20 m, $\tan\theta=2f_s=0.5$,求汽车经过弯道时的最大速度.)

12. A small car moves up a slope at an angle of θ with the horizontal plane with a constant acceleration of $\boldsymbol{a}$. A block of mass m is placed on the flat top of the car and moves with the car, as shown in Fig. 9. 13. What is the minimum coefficient of static friction (f_s) between the block and the car required to prevent slipping? (一小车以等加速度 $\boldsymbol{a}$ 沿与水平面夹角为 θ 的斜面向

上运动,在小车的平顶上放一质量为 m 的物块,随车一同运动,如图 9.13 所示. 则物块与小车间的静摩擦因数 f_s 最小应为多少?)

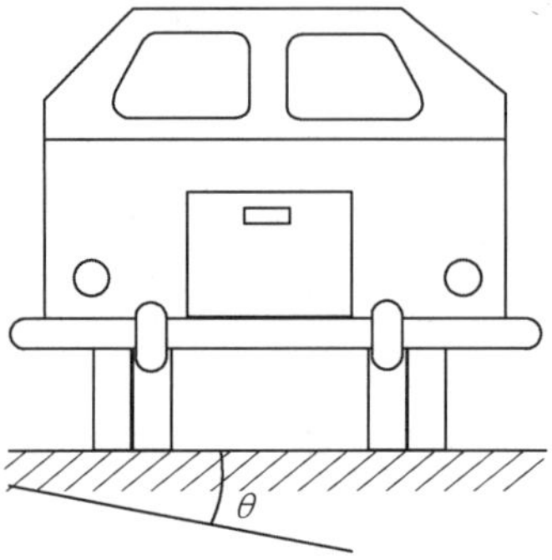

Fig. 9.12　Diagram of Exercise 11
(图 9.12　习题 11 图)

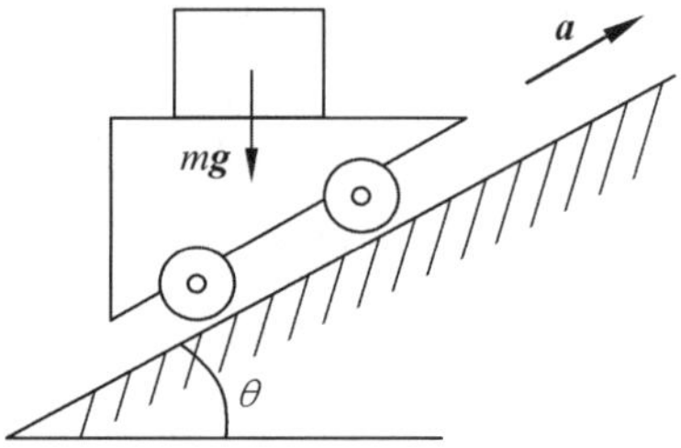

Fig. 9.13　Diagram of Exercise 12
(图 9.13　习题 12 图)

13. The tube A, with a mass of m, is pulled up vertically along the pole by a rope that winds around the pulley B located at a distance l from the pole, as shown in Fig. 9.14. As the drum rotates, the velocity of each point on its edge is denoted by v. Neglecting friction, determine the relationship between the tension in the rope and the distance x. (套管 A 的质量为 m,因受绳子牵引而沿铅垂直杆向上滑动,绳子的另一端绕过离杆距离为 l 的滑车 B 缠在鼓轮上,如图 9.14 所示. 当鼓轮转动时,其边缘上各点的速度大小为 v. 若不计摩擦,求绳子的拉力与距离 x 之间的关系.)

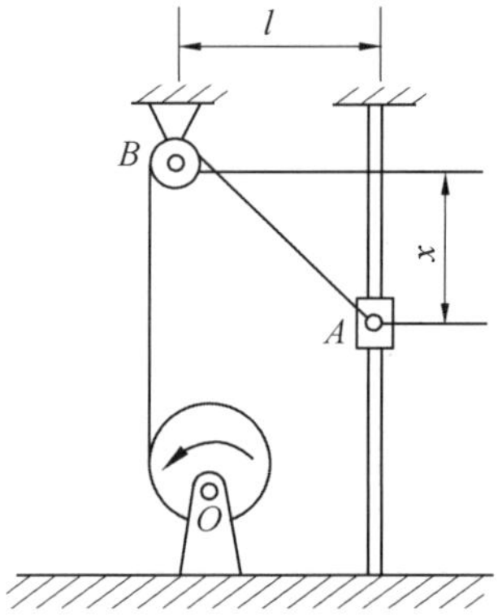

Fig. 9.14　Diagram of Exercise 13
(图 9.14　习题 13 图)

14. An object has a mass of $m = 10$ kg and moves in a straight line along a frictionless horizontal surface under the influence of a varying force $F = 100(1-t)$, where t is measured in s and F is measured in N. The initial velocity of the object is $v_0 = 200$ mm/s. Initially, the direction of the force is the same as the direction of motion. What is the time it takes for the object to come to a stop? How far does it travel before stopping? (一物体的质量为 $m = 10$ kg,在变力 $F = 100(1-t)$ 作用下沿光滑水平面作直线运动,其中 t 以 s 计,F 以 N 计. 设物体的初速为 $v_0 = 200$ mm/s. 初始时,力的方向与速度方向相同,则经过多少时间物体停止运动? 停止前走了多少路程?)

15. A projectile is launched vertically upwards from the ground, and it is desired that it never returns. What initial velocity should be given to the projectile? Only assuming Earth's gravity, which varies inversely with the square of the distance between the object and the center of the Earth, with a known radius of 6 370 km and a gravitational acceleration of $g = 9.8\ \text{m/s}^2$ at the surface of the Earth.（在地面上铅垂向上射出一物体，欲使它一去不复返，则应给物体多大初速度？设只考虑地球引力，此力与物体到底薪的距离的平方成反比，已知地球半径等于6 370 km，在地球表面的重力加速度 $g = 9.8\ \text{m/s}^2$.）

16. A particle of mass m moves along a chord on a circle, as shown in Fig. 9.15. The particle is subjected to a force directed towards the center O of the circle, with the magnitude of the force inversely proportional to the distance between the particle and O, with proportionality constant k. At the beginning, the particle is at position M_0 with zero initial velocity. If the radius of the circle is R and the perpendicular distance from O to the chord is r, find the velocity of the particle when it passes through the midpoint O_1 of the chord.（质量为 m 的质点沿圆上的弦运动，如图 9.15 所示. 此质点受一指向圆心 O 的吸力作用，吸力大小与质点到 O 的距离成反比，比例常数为 k. 开始时，质点处于 M_0位置，初速度为零. 若圆的半径为 R，O 点到弦的垂直距离为 r，求质点经过弦中点 O_1时的速度.）

17. As shown in Fig. 9.16, a particle of mass m is subjected to a force $F = kr$ directed towards the origin O, where the magnitude of the force is proportional to the distance between the particle and point O. At the initial moment, the coordinates of the particle are $x = x_0$ and $y = 0$, and the components of its velocity are $v_x = 0$ and $v_y = v_0$. Find the trajectory of the particle.（如图 9.16所示，质点的质量为 m，受指向原点 O 的力 $F = kr$ 作用，力与质点到 O 点的距离成正比. 在初瞬时，质点坐标为 $x = x_0$，$y = 0$，而速度的分量为 $v_x = 0$，$v_y = v_0$. 求质点的轨迹.）

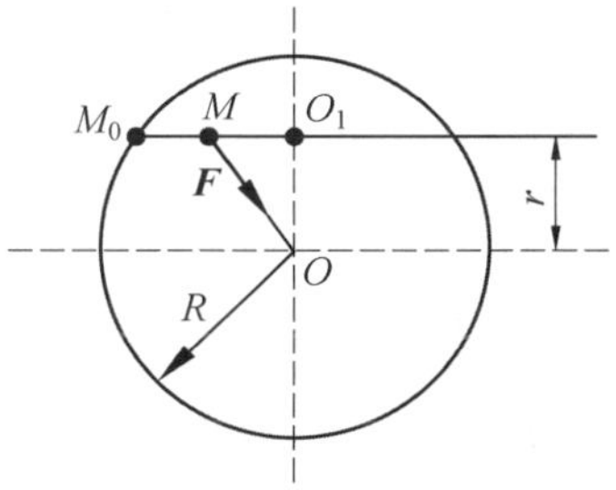

Fig. 9.15 Diagram of Exercise 16
（图 9.15 习题 16 图）

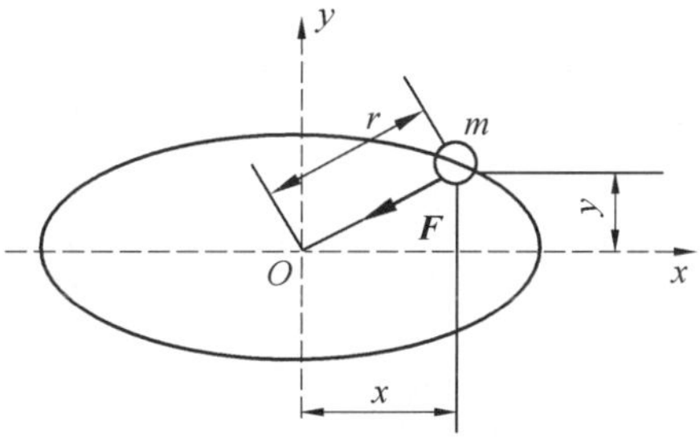

Fig. 9.16 Diagram of Exercise 17
（图 9.16 习题 17 图）

18. A particle of mass m is launched at an initial speed of $\boldsymbol{v}_0$ and angle φ with respect to the horizontal in a medium, where it experiences both gravitational and resistive forces. Assuming the resistive force is proportional to velocity, i. e. $F = kmgv$, where k is a known constant, find the equation of motion and trajectory of the particle.（质量为 m 的质点在介质中以初速度 $\boldsymbol{v}_0$与水平面成仰角 φ 抛出，在重力和介质阻力的作用下运动. 设阻力可视为与速度的一次方成正比，即 $F = kmgv$，k 为已知常数. 试求该质点的运动方程和轨迹方程.）

19. A paratrooper with a parachute falls from high altitude with zero initial velocity. The

mass of the paratrooper is 65 kg, and the air resistance is $F = \frac{1}{2}c\rho Av^2$, where c is the dimensionless drag coefficient, A is the maximum cross-sectional area in the vertical direction, ρ is the air density. For a fully open spherical parachute, $c=0.96$, $A=50\ \mathrm{m}^2$, $\rho=1.225\ \mathrm{kg/m}^3$ under standard conditions. Find the terminal velocity and the time required to achieve 95% of the terminal velocity. (伞兵带降落伞从高空无初速度落下,伞兵的质量为 65 kg,所受空气阻力 $F = \frac{1}{2}c\rho Av^2$,其中 c 为量纲一的阻力系数,A 为垂直运动方向的最大截面积,ρ 为空气密度. 对完全张开的球面降落伞而言,$c=0.96$,$A=50\ \mathrm{m}^2$,在标准状态下,$\rho=1.225\ \mathrm{kg/m}^3$. 求伞兵下降的极限速度和达到 95% 的极限速度值时所需的时间.)

20. A non-moving submarine with a mass of m receives a small sinking force $\boldsymbol{F}$ and sinks to the bottom. When the sinking force is not large, the water resistance can be considered proportional to the sinking velocity and equal to kAv. Here, k is a constant of proportionality, A is the horizontal projected area of the submarine, and v is the sinking velocity. If $v=0$ at $t=0$, calculate the sinking velocity and the distance H that the submarine sinks in time t. (质量为 m 的不前进的潜水艇受到较小的沉力 $\boldsymbol{F}$ 向水底下沉,在沉力不大时,水的阻力可视为与下沉速度的一次方成正比,并等于 kAv. 其中 k 为比例常数,A 为潜水艇的水平投影面积,v 为下沉速度. 如果在 $t=0$ 时,$v=0$,求下沉速度及在时间 t 内潜水艇下沉的路程 H.)

Chapter 10 Momentum Theorem · Angular Momentum Theorem · Kinetic Energy Theorem (动量定理·动量矩定理·动能定理)

This chapter introduces three theorems of dynamics: the momentum theorem, the angular momentum theorem and the kinetic energy theorem. These three theorems reveal the relation between the motion change and the action of the particle and the particle system from different aspects, and can be used to solve the dynamics problem of the particle system. The momentum theorem, the angular momentum theorem and the kinetic energy theorem are called as the general theorem of dynamics.(本章介绍动力学的三大定理:动量定理、动量矩定理和动能定理.这三个定理从不同的侧面揭示了质点和质点系总体的运动变化和作用量之间的关系,用以求解质点系动力学问题.动量定理、动量矩定理和动能定理称为动力学普遍定理.)

10.1 The Work and Power of the Force(力的功和功率)

1. Work of force(力的功)

(1) The work of constant force.(常力的功.)

The work of the force is calculated by the magnitude and direction of the force, as shown in Fig. 10.1(a). Let's say that point M moves to point M_1. The force $\boldsymbol{F}$ is in the direction of angle α. In this case, the work done by the force is equal to the product of the projection of the force in the displacement direction and the displacement $\boldsymbol{U}$.(通过力的大小和方向计算力的功,如图 10.1(a)所示.假设点 M 移动到点 M_1.力 $\boldsymbol{F}$ 的方向是角度 α.此时,力做的功等于力 $\boldsymbol{F}$ 在位移方向的投影 $\boldsymbol{F}\cos\alpha$ 与位移 $\boldsymbol{U}$ 的乘积.)

$$A = FU\cos\alpha = FU\cos\langle \boldsymbol{F}, \boldsymbol{U}\rangle$$

In vector algebra, we know the product of two vectors, that is(在矢量代数中,已知两个向量的积,即)

$$\boldsymbol{F} \cdot \boldsymbol{U} = FU\cos\langle \boldsymbol{F}, \boldsymbol{U}\rangle$$

Therefore, the magnitude and direction of the work done by a force on a linear displacement are determined by the dot product of the force vector of that point and its displacement vector, that is(因此,力在直线位移上做功的大小和方向由该点的力矢量和其位移矢量的点积来确定,即)

$$A = \boldsymbol{F} \cdot \boldsymbol{U}$$

Special case of determine the work done by costant force are as follows.(确定常力做功的特例如下.)

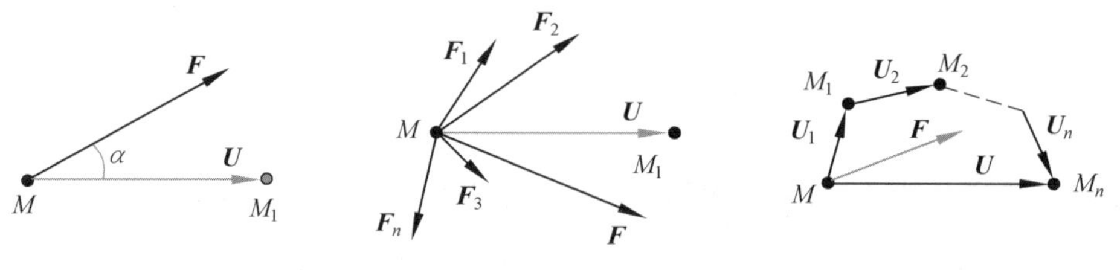

Fig. 10.1 The work of force(图 10.1 常力的功)

① The force $\boldsymbol{F}$ is in the same direction as the displacement vector $\boldsymbol{U}$, we have (力 $\boldsymbol{F}$ 与位移矢量 $\boldsymbol{U}$ 的方向一致,有)

$$A = F \cdot U$$

② The force F is perpendicular to the displacement vector $\boldsymbol{U}$, we have(力 $\boldsymbol{F}$ 垂直于位移矢量 $\boldsymbol{U}$,有)

$$A=0$$

③ The force is in the opposite direction of the displacement vector $\boldsymbol{U}$, we have(力 $\boldsymbol{F}$ 与位移矢量 $\boldsymbol{U}$ 的方向相反,有)

$$A = -F \cdot U$$

Theorem 1: The work done by the resultant force on a given displacement is equal to the algebraic sum of work done by the components on the same displacement. (**定理** 1:合力在一定位移上做的功等于各分力在相同位移上做功的代数和.)

Suppose that forces $\boldsymbol{F}_1, \boldsymbol{F}_2, \ldots, \boldsymbol{F}_n$ (Fig. 10.1(b)) with definite direction and magnitude act on the point M, and their resultant force is $\boldsymbol{F}$, namely $\boldsymbol{F} = \boldsymbol{F}_1 + \boldsymbol{F}_2 + \ldots + \boldsymbol{F}_n$. If the displacement of a point is $\boldsymbol{U}$, then the work done by this force $\boldsymbol{F}$corresponding to this displacement is (假设方向与大小确定的力 $\boldsymbol{F}_1, \boldsymbol{F}_2, \cdots, \boldsymbol{F}_n$ 作用于点 M(Fig. 10.1(b)),这些力的合力是 $\boldsymbol{F}$, $\boldsymbol{F} = \boldsymbol{F}_1 + \boldsymbol{F}_2 + \cdots + \boldsymbol{F}_n$. 如果一个点的位移为 $\boldsymbol{U}$,则力 F 在该位移上所做的功将等于)

$$A = \boldsymbol{F} \cdot \boldsymbol{U} = (\boldsymbol{F}_1 + \boldsymbol{F}_2 + \cdots + \boldsymbol{F}_n) \cdot \boldsymbol{U} = \boldsymbol{F}_1 \cdot \boldsymbol{U} + \boldsymbol{F}_2 \cdot \boldsymbol{U} + \cdots + \boldsymbol{F}_n \cdot \boldsymbol{U}$$

The total work is equal to the algebraic sum of the work done by the forces at displacement $\boldsymbol{U}$, so we have(得到的总功等于各个力在位移 $\boldsymbol{U}$ 做功的代数和,所以有)

$$A = A_1 + A_2 + \cdots + A_n$$

Theorem 2: The work of a force is equal to the algebraic sum of the work done by that force on each displacement. (**定理** 2: 力的功等于该力在各个位移上做功的代数和.)

Suppose that a constant force $\boldsymbol{F}$ obtains a continuous displacement $\boldsymbol{U}_1, \boldsymbol{U}_2, \ldots, \boldsymbol{U}_n$ (Fig. 10.1(c)), the composite displacement of point M is (假设一常力 $\boldsymbol{F}$ 获得连续位移 $\boldsymbol{U}_1$, $\boldsymbol{U}_2, \cdots, \boldsymbol{U}_n$ (图 10.1(c)),点 M 的合成位移为)

$$\boldsymbol{U} = \boldsymbol{U}_1 + \boldsymbol{U}_2 + \cdots + \boldsymbol{U}_n$$

Define the work done by force $\boldsymbol{F}$ in displacement $\boldsymbol{U}$ as (定义力 $\boldsymbol{F}$ 在位移做的功为)

$$A = \boldsymbol{F} \cdot \boldsymbol{U} = \boldsymbol{F} \cdot (U_1 + U_2 + \cdots + U_n) = \boldsymbol{F} \cdot \boldsymbol{U}_1 + \boldsymbol{F} \cdot \boldsymbol{U}_2 + \cdots + \boldsymbol{F} \cdot \boldsymbol{U}_n$$

The resulting sum is the sum of the work done by the force $\boldsymbol{F}$ on each displacement. Thus, we get (得到的总和是力 $\boldsymbol{F}$ 在各个位移上做功的总和. 因此,得到)

$$A = A_1 + A_2 + \cdots + A_n$$

Note that work in SI systems is measured in J(1 J=1 N · m). (请注意,SI 系统功的测量单位是 J(1 J=1 N · m).)

The work of gravity is independent of the trajectory of motion, but is determined only by the vertical distance between the initial and final points of motion (height difference H). If the point moves from the top to the bottom, the work done by gravity is positive, that is(重力的功与运动轨迹无关,而是仅由初始和最终运动点之间的垂直距离(高度差 H)确定. 如果点从顶部移动到底部,则重力做功为正,即)

$$A = mgH$$

If the point moves upward, the work of gravity is negative, that is(如果该点向上移动,则重力做功为负,即)

$$A = -mgH$$

Therefore, an important conclusion is drawn: the work done by gravity along the closed path is zero. (由此得出一个重要结论:重力沿着闭合路径所做的功为零.)

Example 1: A glider weighing 1 200 kg descends from point A to point B (Fig. 10.2), ignoring the air resistance, calculate the work done by gravity. (**例题** 1:重 1 200 kg 的滑翔机从 A 点降到 B 点(图 10.2),忽略空气阻力,计算重力所做的功.)

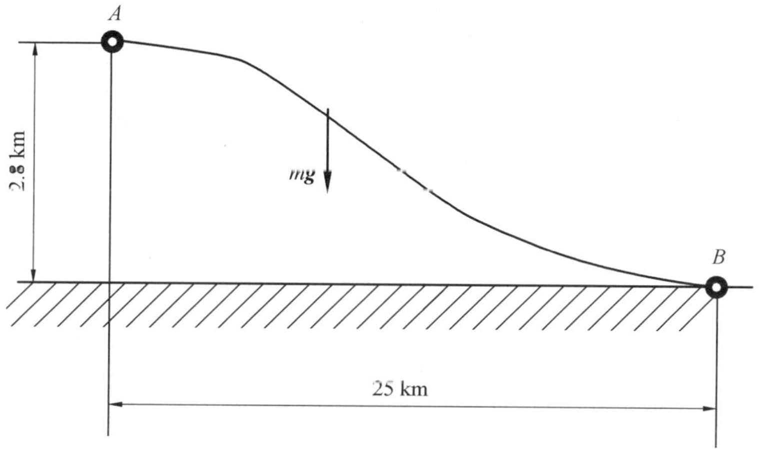

Fig. 10.2 Diagram of Example 1 (图 10.2 例题 1 图)

Solution: (解:)

We treat the glider as a point, only affected by gravity. The work doneby gravity for this point moving from top to bottom is calculated as(将滑翔机作为一点,仅受重力影响. 从顶部到底部移动时,其重力做功计算为)

$$A = mgH = 1\ 200 \times 9.8 \times 2\ 800 = 32\ 928\ 000(\text{N} \cdot \text{m}) = 32.93\ (\text{MN} \cdot \text{m})$$

(2) Elementary work. (元功.)

Suppose a point moves along a curved path from M_1 to M_2 under a force $\boldsymbol{F}$ of variable direction and magnitude (Fig. 10.3). Take a small displacement ΔS_i on the trajectory, and at this displacement you can assume that the force $\boldsymbol{F}_i$ remains constant and calculate the element work on the microsegment, we have(假设一个点在方向和大小可变的力 $\boldsymbol{F}$ 作用下沿着从 M_1 到 M_2 的曲线路径移动(图 10.3). 在轨迹上取微小位移 ΔS_i ,在这个位移上可以假设力 $\boldsymbol{F}_i$ 保持不变,计算微段上的元功,有)

$$\delta A = F_i \Delta S_i \cos \alpha_i$$

where α_i is the angle between the tangent line of the trajectory at the given point and the force $\boldsymbol{F}_i$. (式中, α_i 是给定点处轨迹的切线与力 $\boldsymbol{F}_i$ 之间的角度.)

In fact, this is a formula for determining the work done by a constant force on a microdisplacement. The sum of the work in moving a point from M_1 to M_2 is determined by the element work, that is(事实上,这是确定常力在微段位移上做功的一个公式. 将点从 M_1 移动到 M_2 时的功的总和由元功决定,即)

$$A = \sum \delta A$$

It should be noted that $\delta A \neq \mathrm{d}A$, because in general, micro work is not a differential of a function. (应该注意 $\delta A \neq \mathrm{d}A$,因为在一般情况下,微量功不是函数的微分.)

Under the limit condition that the number of intervalsn increases without limit and ΔSi decreases without limit, we obtain the expression for the work done for the point moving from M_1 to M_2, that is(在区间数 n 无限制地增加并且 ΔS_i 无限减小的极限条件下,得到点从 M_1 移动到 M_2 做功的表达式,即)

$$A_{1,2} = \lim_{\substack{n \to \infty \\ \Delta S \to 0}} \sum F_i \Delta S_i \cos \alpha_i$$

This limit is called the curvilinear integral of the first kind along the arc length $\widehat{M_1 M_2}$, expressed as(这种极限称为沿弧长 $\widehat{M_1 M_2}$ 的第一类曲线积分,表示为)

$$A_{1,2} = \int_{\widehat{M_1 M_2}} F \cos \alpha \mathrm{d}S$$

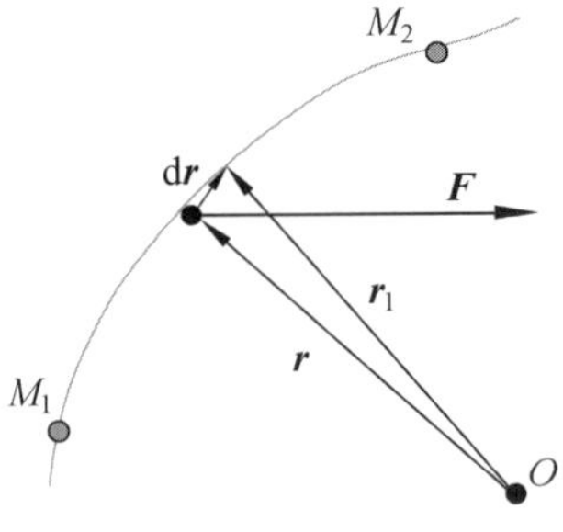

Fig. 10.3 Force does work along a curved path (图 10.3 力沿曲线路径做功)

At the same time, the element work of a tiny displacement can be expressed as the scalar product of two vectors (force vector $\boldsymbol{F}$ and displacement vector $\mathrm{d}\boldsymbol{r}$) (Fig. 10.3), that is (同时,微小位移的元功可以表示为两个矢量(力向量 $\boldsymbol{F}$ 和位移向量 $\mathrm{d}\boldsymbol{r}$)的数量积(图 10.3),即)

$$\delta A = \boldsymbol{F} \cdot \mathrm{d}\boldsymbol{r}$$

Elemental work can be calculated by projecting these vectors, we have(可以通过这些矢量的投影来计算元功,有)

$$\delta A = F_x \mathrm{d}x + F_y \mathrm{d}y + F_z \mathrm{d}z$$

(3) The work done by the force on the total path. (力在总路径上做的功.)

A force acting on a particle is replaced by an equal effect $\boldsymbol{F}$ varying in direction and magnitude. Since the work of the elements can be expressed by their projection on the coordinate axis as $\delta A = F_x \mathrm{d}x + F_y \mathrm{d}y + F_z \mathrm{d}z$, the work on the final displacement of the point from position M_1 to M_2 is determined by the curve integral along the arc $\widehat{M_1M_2}$, that is(用方向和大小可变的等效力 $\boldsymbol{F}$ 代替作用在质点上的力. 由于元功可以用它们在坐标轴上的投影来表示 $\delta A = F_x \mathrm{d}x + F_y \mathrm{d}y + F_z \mathrm{d}z$,因此确定了点从位置 M_1 到 M_2 的最终位移上的功,沿弧线 $\widehat{M_1M_2}$ 的曲线积分为)

$$A_{1,2} = \int_{\widehat{M_1M_2}} (F_x \mathrm{d}x + F_y \mathrm{d}y + F_z \mathrm{d}z)$$

or(或者)

$$A_{1,2} = \int_{t_1}^{t_2} (F_x \dot{x} + F_y \dot{y} + F_z \dot{z}) \mathrm{d}t$$

In the above formula for the relation of work done by a force on a certain path, $F_x \dot{x} \mathrm{d}t$ is the work done by the component forces. Therefore, the work done by the combined force at a given displacement is equal to the sum of work done by the component forces at the same displacement, we have (上述关于力在某一路径上做功的关系式, $F_x \dot{x} \mathrm{d}t$ 是分力做的功. 因此,在给定位移上合力做的功等于各分力在同一位移上做功之和,有)

$$A = A_1 + A_1 + \cdots + A_n$$

(4) Work on a rigid body rotating on a fixed axis. (定轴转动刚体上作用力的功.)

A rigid body is a system of forces in which the distance between points remains constant, assuming that the rigid body is rotating about a fixed axis, an external force is acting on the rigid body, $\boldsymbol{F}_1^{\mathrm{E}}$, $\boldsymbol{F}_2^{\mathrm{E}}$,. . . , $\boldsymbol{F}_n^{\mathrm{E}}$ (Fig. 10.4), and a restraining reaction is generated at the supports A and B (their projections are shown in Fig. 10.4). It is necessary to determine how a force causes a rigid body to rotate. In addition to external forces, there are internal forces and moments, but for an absolutely rigid body, the work of internal forces is zero. Calculate the elemental work of a single force $\boldsymbol{F}_i^{\mathrm{E}}$ on a tiny displacement $\mathrm{d}S_i$ at its point of operation. The locus of a point D_i is a circle with radius $r_i = D_iO$. The rigid body rotates purely, its angle is $\mathrm{d}\varphi$, and the arc coordinate of the D_i point is $\mathrm{d}S_i = r_i \mathrm{d}\varphi$. The calculated element work is first decomposed into three component forces along the path of D_i according to three coordinate axes. Because the work of the sum of forces $\boldsymbol{F}_{in}^{\mathrm{E}}$, $\boldsymbol{F}_{ib}^{\mathrm{E}}$ perpendicular to the velocity vector of the point D_i is zero, the elemental work of the force is determined only by its tangential component, that is (刚体是点之间距离保持不变的力系,假设刚体绕定轴转动,刚体上作用有外力 $\boldsymbol{F}_1^{\mathrm{E}}$, $\boldsymbol{F}_2^{\mathrm{E}}$,…, $\boldsymbol{F}_n^{\mathrm{E}}$ (图 10.4),在支承 A 和 B 处产生约束反力(它们的投影如图 10.4 所示). 有必要确定力是如何使刚体旋转的. 除外力外,还有内力和力矩,但对于绝对刚体,内力的功为零. 计算单独的力 $\boldsymbol{F}_i^{\mathrm{E}}$ 对其作用点微小位移 $\mathrm{d}S_i$ 的元功. 点 D_i 的轨迹是半径为 $r_i = D_iO$ 的圆. 刚体纯转动,它的转角是 $\mathrm{d}\varphi$,而 D_i 点的弧坐标是 $\mathrm{d}S_i = r_i \mathrm{d}\varphi$. 计算 $\boldsymbol{F}_i^{\mathrm{E}}$ 的元功,先把它沿着 D_i 的轨迹按三个坐标轴分解为三个分力. 因为与点 D_i 的速度矢量垂直的力 $\boldsymbol{F}_{in}^{\mathrm{E}}$ 和 $\boldsymbol{F}_{ib}^{\mathrm{E}}$ 的功为零,所以力 $\boldsymbol{F}_i^{\mathrm{E}}$ 的元功仅由其切向

分量 $F_{i\tau}^{E}$ 确定,即)

$$\delta A_i^E = F_{i\tau}^E \mathrm{d}S_i = F_{i\tau}^E r_i \mathrm{d}\varphi = M_{iz}^E \mathrm{d}\varphi$$

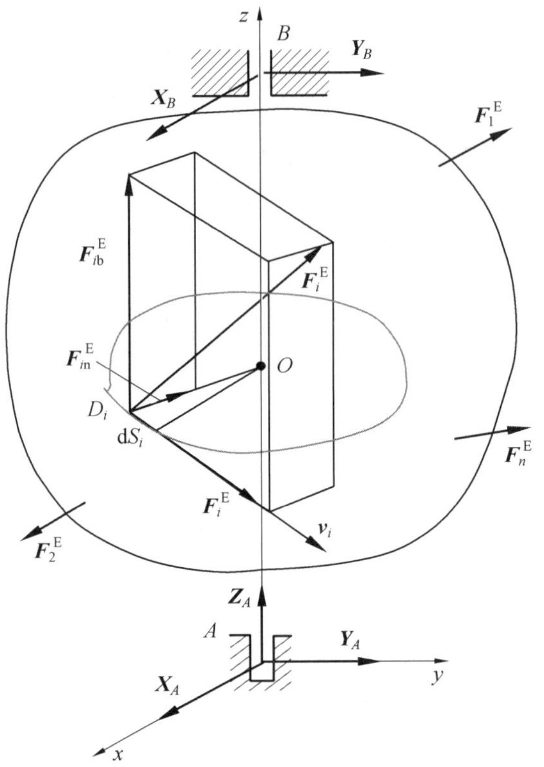

Fig. 10.4 Work done by force on a fixed axis rotating rigid body (图 10.4 定轴转动刚体上力做功)

The elemental work applicable to all external forces of a rigid body is(适用于刚体所有外力的元功为)

$$\delta A = \sum \delta A_i^E = \sum M_{iz}^E \mathrm{d}\varphi = \mathrm{d}\varphi \sum M_{iz}^E = M_z^E \mathrm{d}\varphi$$

where $M_z^E = \sum M_{iz}^E$ is the principal moment of the external force with respect to the rotation axis Oz. It should be noted here that the constrained reactions do not produce moments with respect to the Oz axis as they pass through that axis. So, we have (其中, $M_z^E = \sum M_{iz}^E$ 是外力相对于转轴 Oz 的主矩. 这里应该注意约束反力不会产生相对于 Oz 轴的力矩,因为它们通过该轴. 因此,有)

$$\delta A = \sum \delta A_i^E = M_z^E \mathrm{d}\varphi$$

The elemental work of the force acting on a fixed rotating rigid body is equal to the product of the external force on the principal moment of the rotating axis and the increment of the rotation angle. (作用在定轴转动刚体上力的元功等于外力对转轴主矩与转角增量的乘积.)

If the angle changes from φ_1 to φ_2, then the sum of the work done by the force on the given path will be (如果转角从 φ_1 变为 φ_2,那么力在所给路径上做功的总和将是)

$$\sum A_i = \int_{\varphi_1}^{\varphi_2} M_z^E \mathrm{d}\varphi$$

If the principal torque of the external force with respect to axis Oz is constant, then (如果外力相对于轴 Oz 的主力矩是恒定的,那么)

$$\sum A_i = M_z^{\mathrm{E}} \int_{\varphi_1}^{\varphi_2} \mathrm{d}\varphi = M_z^{\mathrm{E}}(\varphi_2 - \varphi_1)$$

In this case, the sum of the work done on the final angular displacement is equal to the product of the principal moment of the external force and the change in the rotationangle of the rigid body with respect to the axis of rotation. (在这种情况下,最终角位移上做功的总和等于外力的主矩与刚体相对于转轴转角变化的乘积.)

2. Power(功率)

The same work can be done at different times, so power is introduced. The unit of measurement for power in the SI system is W(1 W=1 J/s). (同样的功可以在不同的时间内完成,因此引入功率. 在国际单位制系统,功率测量单位是 W(1 W=1 J/s).)

If the force does the same work in the same amount of time, power can be defined as the ratio of work to time. When a point undergoes a uniform linear motion displacement of $\boldsymbol{U}=\boldsymbol{v}t$, power can be expressed by force and motion velocity, we have(如果力在相同时间内做相同的功,则可以将功率定义为功与时间的比率. 当点发生 $\boldsymbol{U}=\boldsymbol{v}t$ 的匀速直线运动位移时,可以通过力和运动速度来表示功率,有)

$$N = Fv\cos\alpha$$

If the object is rotating at a constant speed and no angular velocity is given (instead, therotational speed n is given), we can use(若物体匀速转动,没有给出角速度(而给出转速 n),可以使用)

$$N = M_{\mathrm{kp}}\omega = M_{\mathrm{kp}} \frac{\pi n}{30}$$

where M_{kp} is the torque acting on a rigid body; n is the speed, the unit is $\mathrm{r \cdot min^{-1}}$. (式中, M_{kp} 为作用在刚体上的转矩;n 为转速,单位为 $\mathrm{r \cdot min^{-1}}$.)

Consider the general case, when the force does work at a variable speed, the formula for calculating the work from a fixed point M_1 to the current position M_2 is (考虑一般情况,当力非匀速做功时,从某个固定点 M_1 到当前位置 M_2 的功的计算式为)

$$A = \int_{M_1}^{M_2} (F_x \mathrm{d}x + F_y \mathrm{d}y + F_z \mathrm{d}z)$$

or(或者)

$$A = \int_{t_1}^{t_2} (F_x \dot{x} + F_y \dot{y} + F_z \dot{z})\mathrm{d}t$$

Power N defines the rate at which work is done by force $\boldsymbol{F}$, we have (功率 N 定义为力 $\boldsymbol{F}$ 做功的速度,有)

$$N = \lim_{\Delta t \to 0} \frac{\Delta A}{\Delta t} = \frac{\mathrm{d}A}{\mathrm{d}t}$$

where A is considered to be a function of time t. In this case, the total differential of work $\mathrm{d}A = (F_x \dot{x} + F_y \dot{y} + F_z \dot{z})\mathrm{d}t$, expressed as a function of time t, is equal to $\mathrm{d}A(t) = \delta A$, or, as mentioned earlier, $\mathrm{d}A = \boldsymbol{F} \cdot \mathrm{d}\boldsymbol{r}$. Then we have (式中,$A$ 被认为是时间 t 的函数. 在这种情况

下,功的全微分 $dA = (F_x\dot{x} + F_y\dot{y} + F_z\dot{z})dt$,表示为时间 t 的函数,微功等于 $dA(t)= \delta A$,或者如前所述, $dA = \boldsymbol{F} \cdot d\boldsymbol{r}$. 此时,有)

$$dA(t) = F_x dx + F_y dy + F_z dz$$

Thus we obtain(这样可以得到)

$$N = \frac{dA}{dt} = \boldsymbol{F} \cdot \frac{d\boldsymbol{r}}{dt} = \boldsymbol{F} \cdot \boldsymbol{v} = F_x\dot{x} + F_y\dot{y} + F_z\dot{z}$$

That is, power N is the scalar product of the velocity at the point of force application and the force $\boldsymbol{F}$. (即功率 N 表示在力的作用点的速度与力 $\boldsymbol{F}$ 的标量乘积.)

The efficiency of useful work: In order to accomplish useful work, more work needs to be done than expected, because part of the work is used to overcome resistance (friction in gears and bearings, and resistance to the movement of particles in air and other environments). The efficiency of any device or machine should be evaluated in terms of the efficiency η of the available work. (有用功的效率:为完成有用功,需要做的功比预期的要多些,因为其中一部分功用于克服阻力(齿轮和轴承中的摩擦力和质点在空气及其他环境中运动的阻力). 任何装置或机器的工作效率应根据有用功的效率 η 加以评估.)

The useful work efficiency (that is, efficiency) of a machine is the ratio of useful power to total power, we have(机器的有用功效率(即效率)是指有用功率与全部功率的比值,有)

$$\eta = \frac{A_{\text{use}}}{A_{\text{input}}} < 1$$

where A_{use} is useful power; A_{input} is input power of full power. (式中, A_{use} 为有用功率; A_{input} 为输入功率或全部功率.)

10.2 The Moment of Inertia of a Rigid Body (刚体的惯性矩)

1. The definition of moment of inertia (惯性矩的定义)

When a rigid body is in translation, the measure of inertia is its mass, and when a rigid body is rotating, it is the moment of inertia. The moment of inertia can be considered in terms of planes, axes and poles. (在刚体平动时,惯性的度量是其质量,而在刚体转动时是惯性矩. 可以从平面、轴和极点的角度考虑惯性矩.)

The moment of inertia J of an object with respect to the plane, axis or pole is the sum of the object's mass multiplied by the square of their distance from the plane, axis or pole(Fig. 10.5), we have (物体相对于平面、轴或极点的惯性矩 J 是物体的质量与它们到平面、轴或极点距离的平方的乘积之和(图 10.5),有)

$$J = \int r^2 dm = \sum r_i^2 m_i$$

According to this definition, the moment of inertia with respect to the plane is(根据这一定义,平面的惯性矩表示为)

$$J_{yOz} = \sum m_i x_i^2,\ J_{xOy} = \sum m_i z_i^2,\ J_{zOx} = \sum m_i y_i^2$$

Relative to the axes, we have(相对于坐标轴有)

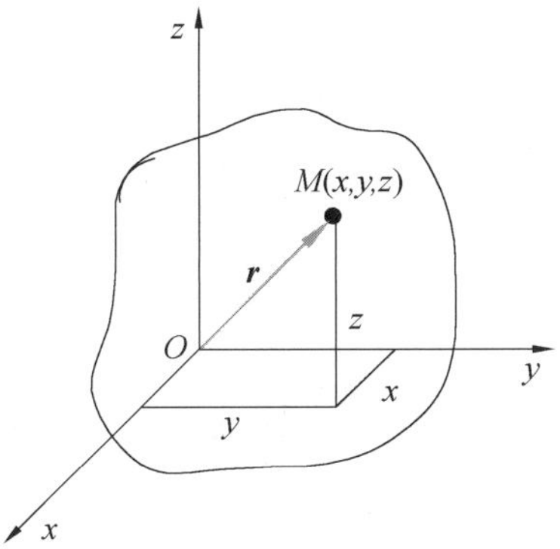

Fig. 10.5 The moment of inertia (图 10.5 惯性矩)

$$J_x = \sum m_i(y_i^2 + z_i^2), J_y = \sum m_i(z_i^2 + x_i^2), J_z = \sum m_i(x_i^2 + y_i^2)$$

Relative to pole O, we have (相对于极点 O 有)

$$J_O = \sum m_i(x_i^2 + y_i^2 + z_i^2)$$

The relationship between the moments of inertia is(惯性矩之间的关系为)

$$J_O = \sum m_i(x_i^2 + y_i^2 + z_i^2) = J_{xOy} + J_{yOz} + J_{zOx}$$

$$J_x + J_y + J_z = 2\sum m_i(x_i^2 + y_i^2 + z_i^2) = 2J_O$$

2. Parallel axis theorem of moment of inertia (惯性矩的平行轴定理)

The moment of inertia of a rigid body with respect to any axis is equal to the moment of inertia of a rigid body with respect to an axis passing through the center of mass and parallel to that axis plus the product of the mass of the rigid body and the square of the distance between the two axes. (刚体对于任何轴的惯性矩等于刚体对于通过质心并与该轴平行的轴的转动惯量加上刚体的质量与这两轴间距离平方的乘积.)

To prove this theorem, we will draw three axes perpendicular to each other through the center of mass of a rigid body (Fig. 10.6(a)). It is necessary to obtain the moment of inertia of the rigid body with respect to the axis parallel to the C axis of the center of mass, and the two axes are d apart. The moment of inertia of any point A_i with respect to axes Oz_1 and Cz is expressed as (为证明该定理,将通过通过刚体的质心 C 绘制三个相互垂直且的轴(图 10.6(a)). 需要得到刚体相对于轴的惯性矩,该轴平行于质心 C 轴,两轴相距为 d. 任意点 A_i 关于轴 Oz_1 和 Cz 的惯性矩表示为)

$$J_{iz_1} = m_i h_i^2 \ , \ J_{izC} = m_i r_i^2$$

As shown in Fig. 10.6(b), we can get (由图 10.6(b)可以得到)

$$r_i^2 = x_i^2 + y_i^2, \ h_i^2 = (y_i - d)^2 + x_i^2 = r_i^2 - 2y_i d + d^2$$

Define the moment of inertia of the object with respect to axis Oz_1, we have (定义物体相对于轴 Oz_1 的惯性矩,有)

$$J_{z_1} = \sum m_i r_i^2 - 2\sum m_i y_i d + \sum m_i d^2$$

or(或者)

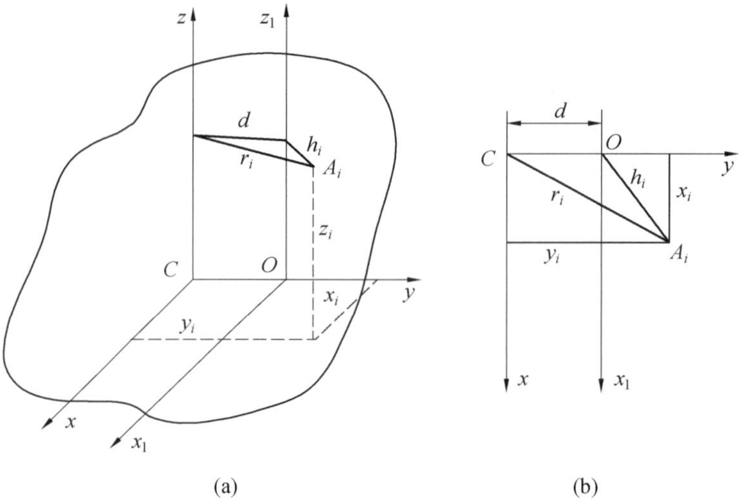

Fig. 10.6 Parallel axis theorem of moment of inertia (图 10.6 惯性矩的平行轴定理)

$$J_{z_1} = J_{zC} - 2d\sum m_i y_i d + d^2 \sum m_i$$

Since $\sum m_i = m$ (the mass of the object) and $\sum m_i y_i = m y_C$, given $y_C = 0$, so we get (因为 $\sum m_i = m$ (物体的质量)和 $\sum m_i y_i = m y_C$,已知 $y_C = 0$,故可以得到)

$$J_{z_1} = J_{zC} + md^2$$

The above is the proof. (以上就是证明.)

Example 2:If the radius of the circular wheel is R, its thickness is h, and its density is ρ, try to calculate the moment of inertia of the circular wheel with respect to the center of mass. (**例题** 2:若已知圆轮半径为 R,其厚度为 h,密度为 ρ,试计算圆轮相对于质心的惯性矩.)

Since the circular wheel is a rotating body of small thickness and its mass is evenly distributed on the circumference, the moment of inertia of the circular wheel with respect to the center of mass can be calculated by selecting the microelement mass $m_i = \rho h \mathrm{d}S$ on the circle (Fig. 10.7), we have (由于圆轮是一个小厚度的旋转体,其质量均匀分布在圆周上,因此可以通过选择圆上的微元质量 $m_i = \rho h \mathrm{d}S$ (图 10.7)来计算圆轮相对于质心的惯性矩,有)

$$J_{zC} = \int_0^{2\pi R} \rho h \mathrm{d}S R^2 = \rho h R^2 2\pi R = mR^2$$

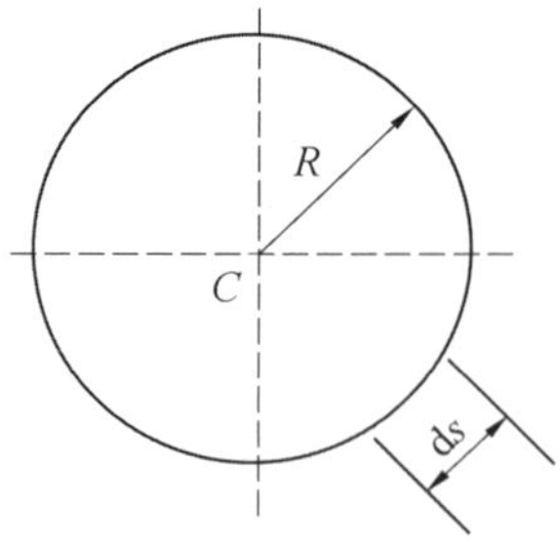

Fig. 10.7 Figure of Example 2 (图 10.7 例题 2 图)

The moment of inertia respect to the center of mass of the circular wheel is the product of the mass of the object and the square of the radius.(圆轮相对于质心的惯性矩为该物体的质量与半径平方的乘积.)

10.3 Momentum Theorem for Particles and Systems of Particles (质点和质点系的动量定理)

10.3.1 Impulse (冲量)

If the force $\boldsymbol{F}$ is constant in magnitude and direction over the t_2-t_1 interval, then the change in momentum of the particle, the impulse is (如果力 $\boldsymbol{F}$ 在 t_2-t_1 时间间隔内大小和方向上是恒定的,那么质点动量的变化即冲量为)

$$\boldsymbol{S}=\boldsymbol{F}(t_2-t_1)$$

The direction of the impulse is the same as the direction of the force, and its magnitude is (冲量的方向与力的方向相同,它的大小是)

$$S=F(t_2-t_1)$$

The impulse of force is a physical quantity that expresses the cumulative effect of an action on a particle over a period of time and is the cause of changing the mechanical motion of a particle. (力的冲量表述了对质点作用一段时间的积累效应的物理量,是改变质点机械运动状态的原因).

The impulse of the variable force $\boldsymbol{F}=\boldsymbol{F}(t)$, which changes in direction and magnitude, is expressed by (大小和方向发生变化的变力 $\boldsymbol{F}=\boldsymbol{F}(t)$,其冲量可表示为)

$$\boldsymbol{S}=\lim_{t_k\to 0}\sum \boldsymbol{S}_k=\lim_{t_k\to 0}\sum \boldsymbol{F}\cdot t_k$$

or(或者)

$$\boldsymbol{S}=\int_{t_1}^{t_2}\boldsymbol{F}\mathrm{d}t$$

The projection of this vector on the axis is equal to(该矢量在坐标轴上的投影等于)

$$S_x=\int_{t_1}^{t_2}F_x\mathrm{d}t\ ,\ S_y=\int_{t_1}^{t_2}F_y\mathrm{d}t\ ,\ S_z=\int_{t_1}^{t_2}F_z\mathrm{d}t$$

The magnitude of the impulse is(冲量大小为)

$$S=\sqrt{S_x^2+S_y^2+S_z^2}$$

Its direction is determined by the direction cosine, that is(它的方向由方向余弦决定,即)

$$\cos(\boldsymbol{S},\mathbf{i})=S_x/S,\ \cos(\boldsymbol{S},\mathbf{j})=S_y/S,\ \cos(\boldsymbol{S},\mathbf{k})=S_z/S$$

If there are multiple forces at a point, then $\boldsymbol{F}$ should be understood as the resultant force, F_x , F_y, F_z are the projection on the coordinate axis, and the impulse would be the impulse of the resultant force. (如果一个点上有多个作用力,那么 $\boldsymbol{F}$ 应该被理解为合力, F_x , F_y, F_z 为力在坐标轴上的投影,而冲量将是合力的冲量.)

10.3.2 Momentum theorem for a particle (质点的动量定理)

The momentum of a particle is a vector in the same direction as its velocity, and the magnitude is equal to the product of its mass m and its velocity v. The momentum of a particle is a measure of its mechanical motion. (质点的动量是与速度方向相同的矢量,其大小等于质量 m 与其速度 v 的乘积. 质点的动量是其机械运动的度量.)

The concept of momentum was introduced to mechanics by Descartes and is based on Newtonian mechanics. (动量的概念是由笛卡儿引入力学的,其以牛顿力学为基础).

Let the force on a particle be $\boldsymbol{F}$. We introduce a fundamental equation of dynamics, that is (设一质点上的作用力为 $\boldsymbol{F}$. 引入动力学的一个基本方程,即)

$$m\boldsymbol{a} = \boldsymbol{F}$$

Replace $\boldsymbol{a} = \frac{\mathrm{d}\boldsymbol{v}}{\mathrm{d}t}$ to transform this equation into(替换 $\boldsymbol{a} = \frac{\mathrm{d}\boldsymbol{v}}{\mathrm{d}t}$,将此等式转换为)

$$\frac{\mathrm{d}(m\boldsymbol{v})}{\mathrm{d}t} = \boldsymbol{F}$$

The resulting relation expresses the differential form of the particle momentum theorem. The theorem is formulated as follows: The time derivative of the momentum vector of a particle is equal to the vector sum of the forces applied to that point. (所得的关系式表示质点动量定理的微分形式. 将此定理表述如下:质点动量矢量的时间导数等于施加在该点的力的矢量和.)

Establish the relationship between the change in momentum acting on a particle and the impulse. To do this, we integrate both sides of the equation $\mathrm{d}(m\boldsymbol{v}) = \boldsymbol{F}\mathrm{d}t$, we have (建立作用在质点上动量的变化量和冲量之间的关系. 为此,对等式 $\mathrm{d}(m\boldsymbol{v}) = \boldsymbol{F}\mathrm{d}t$ 的两边进行积分,有)

$$\int_{v_1}^{v_2} \mathrm{d}(m\boldsymbol{v}) = \int_0^t \boldsymbol{F}\mathrm{d}t$$

The right part of the above equation is the integral of impulse $\boldsymbol{S}$, it equal to force $\boldsymbol{F}$over time interval $t_2 - t_1$, so we get (上式右边的部分是冲量 $\boldsymbol{S}$,其等于力 $\boldsymbol{F}$ 在时间间隔 $t_2 - t_1$ 内的积分,因此得到)

$$m\boldsymbol{v}_2 - m\boldsymbol{v}_1 = \boldsymbol{S}$$

or(或者)

$$m\boldsymbol{v}_2 = \boldsymbol{S} + m\boldsymbol{v}_1$$

That is, a vector $m\boldsymbol{v}_2$ is the diagonal of a parallelogram drawn $m\boldsymbol{v}_1$and $\boldsymbol{S}$ as adjacent sides (Fig. 10.8). (也就是说,矢量 $m\boldsymbol{v}_2$ 是以 $m\boldsymbol{v}_1$ 和 $\boldsymbol{S}$ 为邻边画的平行四边形的对角线(图 10.8).)

If there is not one force acting on the particle, but several force, then $\boldsymbol{S} = \sum \boldsymbol{S}_i$, and the change in momentum of the particle will be written as(如果在质点上作用的不是一个力,而是几个,那么 $\boldsymbol{S} = \sum \boldsymbol{S}_i$,此时,质点的动量变化将写成)

$$m\boldsymbol{v}_2 - m\boldsymbol{v}_1 = \sum \boldsymbol{S}_i$$

This equation is the integral form of the particle momentum theorem, that is, the change in momentum in a given time is equal to the sum of the impulses in the same time. (这一方程是表示质点动量定理的积分形式,即一定时间内动量的变化等于同一时间内冲量的总和.)

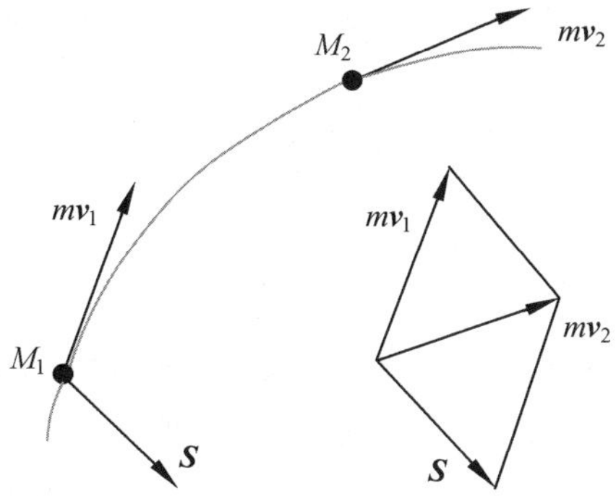

Fig. 10.8 Schematic diagram of the relationship between momentum and impulse
(图 10.8 动量与冲量关系示意图)

10.3.3 Momentum theorem for particle systems (质点系的动量定理)

The momentum of a particle system is a vector, equal to the vector sum of all the momenta of the particles in that particle system. If the momentum of a particle is $\boldsymbol{K}_i = m_i \boldsymbol{v}_i$, then the momentum of the entire system of particles is (质点系的动量是矢量,其等于这个质点系所有质点动量的矢量和. 如果质点的动量为 $\boldsymbol{K}_i = m_i \boldsymbol{v}_i$,则整个质点系的动量为)

$$\boldsymbol{K} = \sum m_i \boldsymbol{v}_i$$

Transform and we get(变换后得到)

$$\boldsymbol{K} = \sum m_i \boldsymbol{v}_i = \sum (m_i \mathrm{d}\boldsymbol{r}_i / \mathrm{d}t) = \mathrm{d}(\sum m_i \boldsymbol{r}_i)/\mathrm{d}t$$

Because(因为)

$$\sum m_i \boldsymbol{r}_i = m\boldsymbol{r}_C$$

So(则有)

$$\boldsymbol{K} = m\mathrm{d}\boldsymbol{r}_C/\mathrm{d}t$$

or(或者)

$$\boldsymbol{K} = m\boldsymbol{v}_C$$

That is to say, the momentum of a system of particles is equal to the mass of the system m times the velocity of its center of mass and in the same direction as the velocity. (也就是说,质点系的动量等于该系统的质量 m 乘它质心的运动速度并且与速度方向相同.)

Project the vector $\boldsymbol{K} = m\boldsymbol{v}_C$ onto the axis, and we obtain (把矢量 $\boldsymbol{K} = m\boldsymbol{v}_C$ 投影在坐标轴上,得到)

$$K_x = \sum m_i v_{ix} = mv_{Cx},\ K_y = \sum m_i v_{iy} = mv_{Cy},\ K_z = \sum m_i v_{iz} = mv_{Cz}$$

Take the derivative of the projection of momentum, we have(对动量的投影进行求导,有)

$$\mathrm{d}K_x/\mathrm{d}t = m\mathrm{d}v_{Cx}/\mathrm{d}t = m\ddot{x}_C$$

$$\mathrm{d}K_y/\mathrm{d}t = m\mathrm{d}v_{Cy}/\mathrm{d}t = m\ddot{y}_C$$

$$\mathrm{d}K_z/\mathrm{d}t = m\mathrm{d}v_{Cz}/\mathrm{d}t = m\ddot{z}_C$$

According to the momentum theorem of particles system, we have(依据质点系的动量定理,有)

$$m\ddot{x}_c = \sum F_{ix}^{E}, m\ddot{y}_c = \sum F_{iy}^{E}, m\ddot{z}_c = \sum F_{iz}^{E}$$

Therefore(因此)

$$dK_x/dt = \sum F_{ix}^{E}, dK_y/dt = \sum F_{iy}^{E}, dK_z/dt = \sum F_{iz}^{E}$$

Thus, we obtain the momentum theorem of a system of particles expressed in differential form.(因此,得到了用微分形式表达的质点系动量定理.)

The time derivative of the momentum of a system of particles projected on each axis is equal to the algebraic sum of the external forces acting on the system projected on the same axis.(质点系的动量在每一个轴上投影的时间导数等于作用于该系统的外力在同一轴上投影的代数和.)

The principal vector $\boldsymbol{R}^{E} = \sum \boldsymbol{F}_i^{E}$ of the external force, written in vector form about the momentum theorem of the system of particles, we have(外力的主矢量 $\boldsymbol{R}^{E} = \sum \boldsymbol{F}_i^{E}$,用矢量形式写下关于质点系动量定理,有)

$$\frac{d\boldsymbol{K}}{dt} = \boldsymbol{R}^{E}$$

It can be expressed as follows: The derivative of the momentum of a system of particles with respect to time is equal to the vector sum of all the external forces acting on the system of particles.(可以用以下方法表达:质点系的动量对于时间的导数等于作用在该质点系上所有外力的矢量和.)

It follows from this theorem that the change in momentum of a system of particles is only caused by external forces.(由这个定理可以得到质点系动量的变化仅仅是由外力引起的.)

Corollary: If the vector sum of the external forces acting on a system of particles at all times is identical to zero, then the momentum of the system of particles remains constant, we have(**推论**:如果在所有时间内作用于质点系外力的矢量和恒等于零,那么质点系的动量保持不变,有)

$$\boldsymbol{R}^{E} = 0, d\boldsymbol{K}/dt = 0, \boldsymbol{K} = \text{const}$$

This is the law of conservation of momentum for a system of particles.(这就是质点系的动量守恒定律.)

In order to find the relationship between the change of momentum of a particle system and the impulse of the external force acting on the system, we use the momentum theorem of the particle, just like the external force acting on every point of the system, the internal force is the same. So the change in momentum of the particle system is equal to(为找到质点系动量的变化量和作用于该系统外力的冲量的关系,利用质点的动量定理,就像外力 $\boldsymbol{F}_i^{E}$ 作用于系统的每一个点一样,内力 $\boldsymbol{F}_i^{I}$ 同样如此.这样,质点系统动量变化量等于)

$$(m_i\boldsymbol{v}_i)_2 - (m_i\boldsymbol{v}_i)_1 = \boldsymbol{S}_i^{E} + \boldsymbol{S}_i^{I}$$

where $\boldsymbol{S}_i^{E}$ and $\boldsymbol{S}_i^{I}$ is the impulse of the external and internal forces, and acts on the particle at the time interval of t_2-t_1. Sum the left and right sides of k equations, and we get (式中, $\boldsymbol{S}_i^{\mathrm{E}}$ 和 $\boldsymbol{S}_i^{\mathrm{I}}$ 为外力和内力的冲量,并且是在 t_2-t_1 的时间间隔内作用于质点. 对 k 个等式的左边和右边分别求和,得到)

$$\sum(m_i\boldsymbol{v}_i)_2-\sum(m_i\boldsymbol{v}_i)_1=\sum\boldsymbol{S}_i^{\mathrm{E}}+\sum\boldsymbol{S}_i^{\mathrm{I}}$$

Because the sum of the internal force vectors is equal to zero, that is, the sum of the impulses of the internal force is equal to zero, that is $\sum\boldsymbol{S}_i^{\mathrm{I}}=0$, so we have (因为内力矢量之和等于零,即内力的冲量之和等于零,也就是说 $\sum\boldsymbol{S}_i^{\mathrm{I}}=0$,所以有)

$$\boldsymbol{K}_2-\boldsymbol{K}_1=\sum\boldsymbol{S}_i^{\mathrm{E}}$$

The obtained equations represent the momentum theorem of the system of particles in integral form: The change in momentum of a particle system over a certain period of time is equal to the vector sum of the impulse of the external force applied to the system over the same period of time. (得到的方程表征了积分形式的质点系动量定理:质点系在一定时间内动量的变化量等于在相同时间内施加给系统的外力冲量的矢量和.)

Example 3: A disk of mass m and radius R is rotated around a fixed axis at angular velocity ω (Fig. 10.9) to determine the momentum of the disk. (**例题** 3:质量为 m 和半径为 R 的圆盘以角速度 ω 绕固定轴(图 10.9)旋转,请确定圆盘动量.)

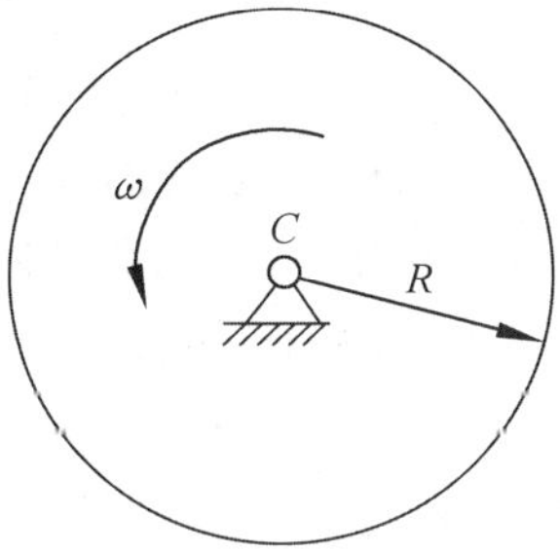

Fig. 10.9 Diagram of Example 3(图 10.9 例题 3 图)

Solution: (**解**:)

Point C is at the instantaneous center of velocity and is at the same time its center of mass, the velocity of the center of mass is zero, and therefore the momentum of the disk $\boldsymbol{K}=m\boldsymbol{v}_C$ will also be zero. (点 C 是速度瞬心,并且同时是其质心,质心的速度为零. 因此,盘 $\boldsymbol{K}=m\boldsymbol{v}_C$ 的动量也将为零.)

The momentum of a turntable rotating around the axis of mass is zero. (围绕质心轴旋转的转盘的动量为零.)

Example 4: A rod of weight $\boldsymbol{G}_1$ and length l rotates uniformly around a fixed axis O (Fig. 10.10) at an angular velocity $\boldsymbol{\omega}$. At the end of the rod is a rigid sphere with weight $\boldsymbol{G}_2$. If $G_1=4G_2=4G$, try to calculate the momentum of the system of particles. (**例题** 4:重力为 G_1、长度为 l 的杆绕固定轴 O(图 10.10)以角速度 $\boldsymbol{\omega}$ 匀速转动,在杆的末端是刚球,其重力为 G_2. 如果 $G_1=4G_2=4G$,试计算质点系的动量.)

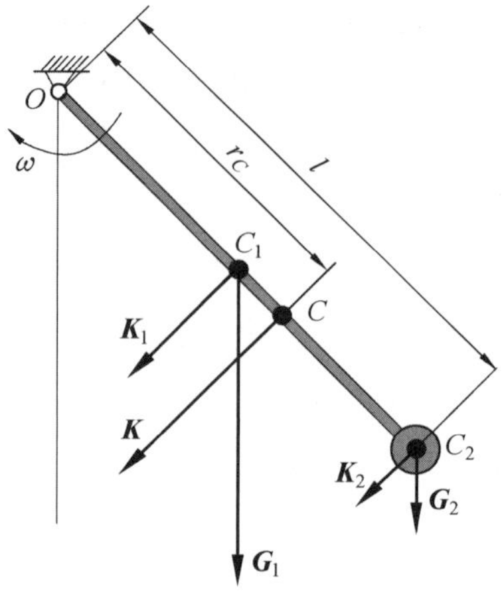

Fig. 10. 10 Diagram of Example 4 (图 10. 10 例题 4 图)

Solution: (解:)

There are two possible solutions to this problem: (1) use a relational expression $\boldsymbol{K}=\sum \boldsymbol{K}_i$; (2) Use formulas $\boldsymbol{K}=m\boldsymbol{v}_C$. (该问题有两种可能的解决方案:(1)使用关系式 $\boldsymbol{K}=\sum \boldsymbol{K}_i$;(2)使用公式 $\boldsymbol{K}=m\boldsymbol{v}_C$.)

Option 1(方案 1)

(1)Determine the momentum of the rod,we have (确定杆的动量,有)

$$\boldsymbol{K}_1=m_1\boldsymbol{v}_1, K=(4G/g)\omega(l/2)=2(G/g)\omega l$$

(2)Determine the momentum of the ball. (确定球的动量.)

Taking a ball as a particle, we calculate its momentum, that is(以球为质点,计算其动量,即)

$$\boldsymbol{K}_2=m_1\boldsymbol{v}_2, K_2=(G/g)v_2$$

(3)Calculate the momentum of the entire system of particles, we have (计算整个质点系的动量,有)

$$\boldsymbol{K}=\boldsymbol{K}_1+\boldsymbol{K}_2$$

Since the vector $\boldsymbol{K}_1$ and $\boldsymbol{K}_2$ are parallel, then we have (由于 $\boldsymbol{K}_1$ 和 $\boldsymbol{K}_2$ 向量是平行的,因此有)

$$K=K_1+K_2=2(G/g)\omega l+(G/g)\omega l=3(G/g)\omega l$$

Option 2 (方案 2)

(1) Determine the location of the system's center of mass, we have (确定系统质心的位置,有)

$$r_C=\frac{\sum m_i r_i}{\sum m_i}=\frac{G_1(l/2)+G_2 l}{G_1+G_2}=\frac{4G(l/2)+Gl}{4G+G}=0.6l$$

(2) Calculate the momentum of a system of particles, we have (计算质点系的动量,有)

$$\boldsymbol{K}=m\boldsymbol{v}_C, K=\frac{G_1+G_2}{g}\omega\cdot 0.6l=3(G/g)\omega l$$

10.4 Theorem of the Moment of Momentum(动量矩定理)

10.4.1 Theorem of moment of particle momentum (质点动量矩定理)

As shown in Fig. 10.11, assume that the motion of the point is caused by a force $\boldsymbol{F}$. Connect any selected center point O with the vector $\boldsymbol{r}$ of that point. The formula for determining the torque of the force $\boldsymbol{F}$ to the point O is (如图 10.11 所示,假设点的运动是由力 $\boldsymbol{F}$ 引起的.连接任意选定的中心点 O 与该点的矢径 $\boldsymbol{r}$.确定力 $\boldsymbol{F}$ 对点 O 的力矩的关系式为)

$$\boldsymbol{M}_O = \boldsymbol{r} \times \boldsymbol{F}$$

Calculate the moment of this point with respect to the same central point O, we have(计算这个点的动量对同一中心点 O 的矩,有)

$$\boldsymbol{L}_O = \boldsymbol{r} \times m\boldsymbol{v}$$

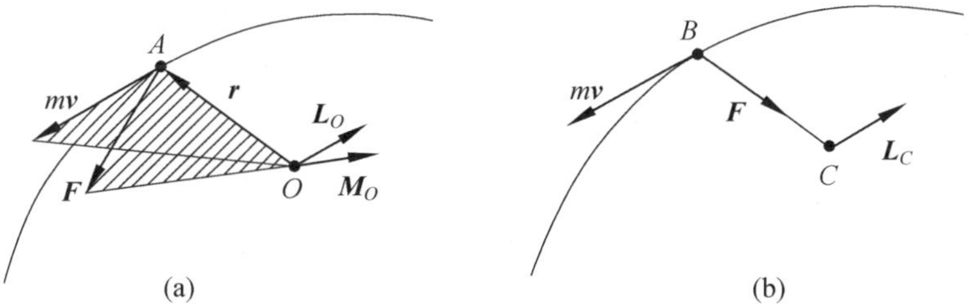

Fig. 10.11 Moment of particle momentum (图 10.11 质量动量矩)

To determine the relationship between vectors $\boldsymbol{M}_O$ and $\boldsymbol{L}_O$, we take the time derivative with respect to the moment of momentum, that is (为确定矢量 $\boldsymbol{M}_O$ 与 $\boldsymbol{L}_O$ 之间的关系,对动量矩求时间导数,即)

$$\frac{\mathrm{d}}{\mathrm{d}t}\boldsymbol{L}_O = \frac{\mathrm{d}}{\mathrm{d}t}(\boldsymbol{r} \times m\boldsymbol{v}) = \frac{\mathrm{d}\boldsymbol{r}}{\mathrm{d}t} \times m\boldsymbol{v} + \boldsymbol{r} \times \frac{\mathrm{d}}{\mathrm{d}t}m\boldsymbol{v} = \boldsymbol{v} \times m\boldsymbol{v} + \boldsymbol{r} \times m\boldsymbol{a} = 0 + \boldsymbol{r} \times \boldsymbol{F} = \boldsymbol{M}_O$$

If there are several forces acting on the particle, $\boldsymbol{M}_O$ should be regarded as the moment of its resultant force, we have (如果有几个力作用在质点上,则应将 $\boldsymbol{M}_O$ 视为其合力的力矩,有)

$$\frac{\mathrm{d}}{\mathrm{d}t}\boldsymbol{L}_O = \sum \boldsymbol{M}_{iO}$$

Therefore, the derivative of the moment of momentum of a particle about a fixed point with respect to time is equal to the vector sum of the moments of the forces acting on the particle about the same point, which is the theorem of moment of particle momentum. (因此,质点关于某一固定点的动量矩对于时间的导数等于作用在质点上的各力对于同一点之矩的矢量和,这就是关于质点的动量矩定理.)

The final relation can be expressed as a projection on a coordinate axis, that is(最后的关系式可用在坐标轴上的投影表示,即)

$$\mathrm{d}L_x/\mathrm{d}t = \sum M_{ix},\ \mathrm{d}L_y/\mathrm{d}t = \sum M_{iy},\ \mathrm{d}L_z/\mathrm{d}t = \sum M_{iz}$$

That is, the derivative of the moment of a particle momentum about a fixed axis with respect to time is equal to the moment of the force acting on a particle about the same axis, which is the theorem of the moment of momentum of a particle in projective form. (即质点的动量对于某一固定轴之矩对于时间的导数等于作用在质点上的力对于同轴之矩,这就是投影形式的质点动量矩定理).

The inference of theorem is as follows. (定理的推论如下.)

(1) If the line of action of the resultant force acting on a particle always passes through a center, then the moment of momentum of the particle with respect to that center remains constant. (如果作用在质点上合力的作用线总是通过某个中心,那么质点相对于该中心的动量矩保持不变.)

In this case the force always follows the vector direction of point B (Fig. 10.11(b)), so the vector product $\boldsymbol{r}\times\boldsymbol{F}$ is equal to zero, that is, the moment of momentum of the force with respect to point C remains constant. It can be expressed as $\boldsymbol{L}_C$ = const. (在这种情况下,力 $\boldsymbol{F}$ 总是沿着点 B 的矢径方向(图 10.11(b)),因此向量积 $\boldsymbol{r}\times\boldsymbol{F}$ 等于零,即力 $\boldsymbol{F}$ 相对于 C 点的动量矩保持不变,可表示为 $\boldsymbol{L}_C$ = const.)

(2) If the resultant force on a particle is zero with respect to a certain axis, then the moment of momentum of the particle with respect to that axis remains constant. (如果质点所受合力对某个轴的力矩为零,那么质点相对于该轴的动量矩保持不变.)

For example, when $\sum M_{iy}=0$, $\mathrm{d}L_y/\mathrm{d}t=0$, L_y = const (例如,当 $\sum M_{iy}=0$ 时, $\mathrm{d}L_y/\mathrm{d}t=0$, L_y = const.)

10.4.2 Theorem of moment of momentum for systems of particles (质点系的动量矩定理)

The moment of momentum of a system of particles with respect to a fixed center is equal to the vector sum of the moments of all external forces acting on the system of particles with respect to this center point. (质点系对于某一固定中心的动量矩,等于作用于质点系上的所有外力对于这一中心点之矩的矢量和.)

The moment of momentum of a system of particles is also called the principal moment of a system of particles. For example, for some central point B it can be calculated like(质点系的动量矩又称质点系的动量主矩. 例如,对某个中心点 B,可以计算为)

$$\boldsymbol{L}_B=\sum\boldsymbol{L}_{iB}=\sum(\boldsymbol{r}\times m_i\boldsymbol{v}_i)$$

where $\boldsymbol{r}$ is the vector diameter of the ith particle departing from point O; $m_i\boldsymbol{v}_i$ is the momentum of the particle. (式中, $\boldsymbol{r}$ 是第 i 个质点自 O 点出发的矢径; $m_i\boldsymbol{v}_i$ 为该质点的动量.)

The algebraic sum of the moments of momentum of all particles in a system relative to an axis is equal to the moment of momentum of the system relative to that axis, for example, with respect to the Oz axis, we have (质点系所有质点相对于轴的动量矩的代数和等于质点系对该轴的动量矩,例如,相对于 Oz 轴,有)

$$L_z=\sum L_{iz}$$

Consider a system of particles consisting of k particles. The particle system moves under the action of external $\boldsymbol{F}_i^{\mathrm{E}}$ and internal forces $\boldsymbol{F}_i^{\mathrm{I}}$. According to the theorem of moment of momentum of the particle, the moment of momentum of the particle system with respect to the fixed point O is (考虑到 k 个质点组成的质点系. 质点系在外力 $\boldsymbol{F}_i^{\mathrm{E}}$ 和内力 $\boldsymbol{F}_i^{\mathrm{I}}$ 的作用下运动,根据质点的动量矩定理,质点系对定点 O 的动量矩为)

$$\frac{\mathrm{d}}{\mathrm{d}t}\boldsymbol{L}_{iO} = \boldsymbol{M}_{iO}^{\mathrm{E}} + \boldsymbol{M}_{iO}^{\mathrm{I}}$$

Sum up k such equations, we have (对 k 个这样的方程求和,有)

$$\sum \frac{\mathrm{d}}{\mathrm{d}t}\boldsymbol{L}_{iO} = \sum \boldsymbol{M}_{iO}^{\mathrm{E}} + \sum \boldsymbol{M}_{iO}^{\mathrm{I}}$$

As mentioned before, the vector sum of the moment of the internal force acting on the system of particles with respect to a fixed point is always zero, that is (如前所述,作用在质点系上的内力对某固定点之矩的矢量和恒为零,也就是说)

$$\sum \boldsymbol{M}_{iO}^{\mathrm{I}} = 0$$

At this time, we have(此时)

$$\sum \frac{\mathrm{d}}{\mathrm{d}t}\boldsymbol{L}_{iO} = \sum \boldsymbol{M}_{iO}^{\mathrm{E}}$$

or (或者)

$$\frac{\mathrm{d}}{\mathrm{d}t}\sum \boldsymbol{L}_{iO} = \sum \boldsymbol{M}_{iO}^{\mathrm{E}}$$

By definition, use $\boldsymbol{L}_O$ instead of the moment of momentum $\sum \boldsymbol{L}_{iO}$ of a system of particles, we get (根据定义,用 $\boldsymbol{L}_O$ 代替质点系的动量矩 $\sum \boldsymbol{L}_{iO}$,得到)

$$\frac{\mathrm{d}}{\mathrm{d}t}\boldsymbol{L}_O = \sum \boldsymbol{M}_{iO}^{\mathrm{E}} = \boldsymbol{M}_O^{\mathrm{E}}$$

This relation is the theorem of moment of momentum of a system of particles: The derivative of the moment of momentum vector of a system of particles about the fixed point O with respect to time is equal to the vector sum of the moment of the external force acting on the particle system with respect to point O. (这个关系式即质点系动量矩定理:质点系对固定点 O 的动量矩矢量对时间的导数等于作用于质点系上的外力对 O 点之矩的矢量和.)

The projection of a vector expression onto three rectangular coordinates are(矢量式在三个直角坐标系上的投影为)

$$\mathrm{d}L_x/\mathrm{d}t = M_x^{\mathrm{E}}, \mathrm{d}L_y/\mathrm{d}t = M_y^{\mathrm{E}}, \mathrm{d}L_z/\mathrm{d}t = M_z^{\mathrm{E}}$$

In the formula, L_x、L_y、L_z are the projection of the kinetic moment L_O of the system of particles on the axes x, y, z respectively. M_x^{E}, M_y^{E}, M_z^{E} is the principal moment of the external forces of the system with respect to the same axes. (式中, L_x、L_y、L_z为质点系的动量矩 L_O分别在轴 x、y、z 上的投影; M_x^{E}、M_y^{E}、M_z^{E} 为系统外力对相同轴的主矩.)

The inference of theorem is as follows. (定理的推论如下.)

(1) If the principal moment of an external force acting on a system of particles with respect to a fixed point is always zero, then the moment of momentum of the system of particles respect to that point remains unchanged, we have (若作用在质点系上的外力对某固定点的主矩恒为零,则质点系对该点的动量矩保持不变,有)

$$\frac{d}{dt}\boldsymbol{L}_O = 0, \boldsymbol{L}_O = \text{const}$$

It is called the conservation law of the moment of momentum of the particles system. (其称为质点系的动量矩守恒定律.)

(2) If the algebraic sum of the moment of the external force acting on a system of particles with respect to a fixed axis is always zero, then the moment of momentum of the system with respect to that axis remains constant. (若作用在质点系上的外力对某固定轴之矩的代数和恒为零,则质点系对该轴的动量矩保持不变.)

For example, $M_z = 0$, then $dL_z/dt = 0$, so $L_z = \text{const}$. (例如,$M_z = 0$,然后 $dL_z/dt = 0$,因此 $L_z = \text{const}$.)

10.5 Kinetic Energy Theorem for a Particle (质点的动能定理)

We know from physics that the kinetic energy of a particlem is equal to half the product of its mass times its velocity squared, we have (由物理学可知,质点 m 的动能等于该点质量乘其速度平方的一半,有)

$$T = mv^2/2$$

Consider the motion of point M under the action of a force system $\boldsymbol{F}_1, \boldsymbol{F}_2, \ldots, \boldsymbol{F}_n$ (Fig. 10.12). Select a positive reference direction to write the basic equations of particle dynamics, we have (考虑点 M 在力系 $\boldsymbol{F}_1, \boldsymbol{F}_2, \cdots, \boldsymbol{F}_n$ 的作用下运动(图 10.12). 选择一个正的参考方向写出质点动力学的基本方程,有)

$$m\boldsymbol{a} = \boldsymbol{F}$$

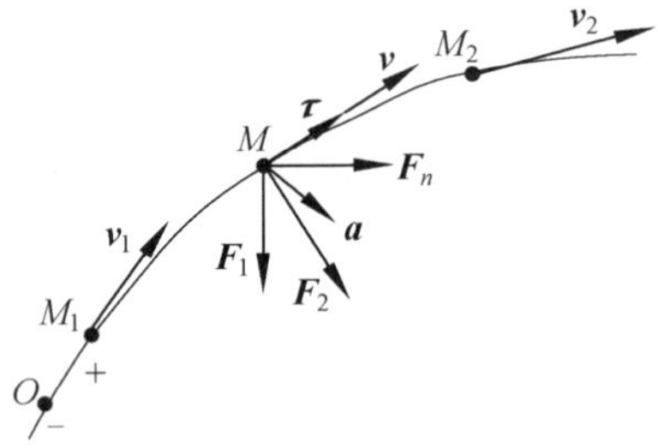

Fig. 10.12 A particle is subjected to the action of a force system (图 10.12 质点受到力系作用)

Here, the force is the resultant force of a conjunctive force system $\boldsymbol{F}_1, \boldsymbol{F}_2, \ldots, \boldsymbol{F}_n$, project it onto the τ axis, we have (这里,力 $\boldsymbol{F}$ 是汇交力系 $\boldsymbol{F}_1, \boldsymbol{F}_2, \cdots, \boldsymbol{F}_n$ 的合力,把它投影到 τ 轴上,有)

$$ma_\tau = F_\tau$$

For the natural coordinate method of point trajectories, see Section 5.1. (关于点运动轨迹的自然坐标方法,请参阅 5.1 节.)

Whereas $a_\tau = \frac{dv}{dt} = \frac{dv}{dS} \cdot \frac{dS}{dt} = \frac{dv}{dS} \cdot v$, we substitute the obtained values of tangential acceleration into the equation of motion along the tangential direction, we have (鉴于 $a_\tau = \frac{dv}{dt} =$

$\frac{dv}{dS} \cdot \frac{dS}{dt} = \frac{dv}{dS} \cdot v$,将获得的切向加速度值代入切向运动方程,有)

$$mv\mathrm{d}v/\mathrm{d}S = F_\tau$$

or(或者)

$$mv\mathrm{d}v = F_\tau \mathrm{d}S$$

or we have(或者,有)

$$\mathrm{d}(mv^2/2) = F\mathrm{d}S\cos(\boldsymbol{F}_i, \boldsymbol{\tau})$$

The left side of this formula is the differential of the kinetic energy of the particle, and the right side is the elemental work that causes the displacement dS (the resultant force is only the work done by the tangential component), so we have (该公式的左侧是质点动能的微分,右侧是发生移 dS 的元功(合力只有切线分量做功),所以有)

$$\mathrm{d}(mv^2/2) = \delta A$$

Because the displacement of the points of force is the same, the kinetic energy can therefore be calculated in different ways, we have (由于 $F_\tau = \sum F_{i\tau}$,所有力作用点的位移相同,则 $\delta A = \sum \delta A_i$, 因此可以用不同的方式计算动能,有)

$$\mathrm{d}(mv^2/2) = \sum \delta A_i$$

The change in kinetic energy at a point is equal to the sum of the elementary work done by the forces at that point. (点的动能变化等于在该点上各力所做元功的总和.)

When a point moves from the positionof M_1 to M_2, the velocity of that point will change from $\boldsymbol{v}_1$ to $\boldsymbol{v}_2$. In this case, the kinetic energy will change, that is (当一个点从 M_1 移动到 M_2 的位置时,该点的速度将从 $\boldsymbol{v}_1$ 变化到 $\boldsymbol{v}_2$. 在这种情况下,动能会发生变化,即)

$$m\int_{v_1}^{v_2} v\mathrm{d}v = \sum \int_{M_1}^{M_2} F_i \mathrm{d}S\cos(\boldsymbol{F}_i, \boldsymbol{\tau})$$

Hence(因此)

$$mv_2^2/2 - mv_1^2/2 = \sum \delta A_i$$

The resulting equation is a theorem of the kinetic energy of a particle: The change in the kinetic energy of a particle at any displacement is equal to the algebraic sum of the work forces acting on the particle at the same displacement. (所得方程是质点动能定理:质点在任一位移中动能的变化等于同一位移上作用在该质点上所有力做功的代数和).

If the sum of the work done is positive, then $v_2 > v_1$, the kinetic energy increases. If the sum of the work done is negative, then the kinetic energy decreases. (如果做功的总和为正,则 $v_2 > v_1$,动能增加. 如果做功的总和为负,则动能减小.)

10.6 Differential Equations for Parallel Movement of Rigid Bodies (刚体平行移动的微分方程)

When a rigid body moves in parallel, all its particles move in the same way as the center of mass, so the differential equation for the center of mass motion describes the parallel movement of the rigid body, we have(刚体在平行移动时,其所有质点都以质心相同的方式运动,所以质心运动的微分方程描述了刚体的平行移动,有)

$$m\ddot{x}_C = \sum_{i=1}^{k} F_{ix}^{\mathrm{E}},\ m\ddot{y}_C = \sum_{i=1}^{k} F_{iy}^{\mathrm{E}},\ m\ddot{z}_C = \sum_{i=1}^{k} F_{iz}^{\mathrm{E}}$$

where m is the mass of the rigid body; $\ddot{x}_C$, $\ddot{y}_C$, $\ddot{z}_C$ are the projection of the rigid body's acceleration on the coordinate axis; $F_{ix}^{\mathrm{E}}, F_{iy}^{\mathrm{E}}, F_{iz}^{\mathrm{E}}$ are the projection of the external force of a rigid body onto the corresponding coordinate axis. (式中,m 是刚体的质量;$\ddot{x}_C$ 、$\ddot{y}_C$ 、$\ddot{z}_C$ 是刚体加速度在坐标轴上的投影;F_{ix}^{E}、F_{iy}^{E}、F_{iz}^{E} 是刚体外力在相应坐标轴上的投影.)

10.7 Differential Equations of Rigid Body Rotating Around a Fixed Axis(刚体定轴转动的微分方程)

A rigid body under external force $\boldsymbol{F}_i^{\mathrm{E}}$ (Fig. 10. 13) rotates about a fixed axis at angular velocity ω. The moment of momentum around the axis A_z is equal to the sum of the moment of momentum of the particle relative to the same axis, that is (刚体在外力 $\boldsymbol{F}_i^{\mathrm{E}}$ 作用下(图 10.13)以角速度 ω 绕定轴转动. 其绕轴 A_z的动量矩等于质点相对于同一轴的动量矩的总和,即)

$$L_z = \sum m_i v_i r_i = \sum m_i r_i \omega r_i = \sum m_i \omega r_i^2 == \omega \sum m_i r_i^2 = \omega J_z$$

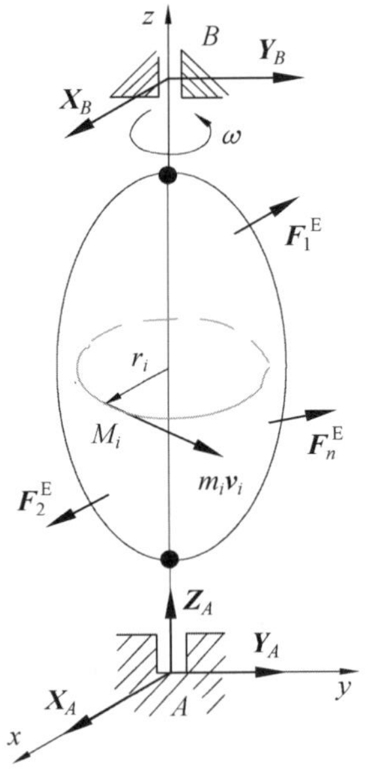

Fig. 10. 13 A righd body rotates about a fixed axis under external force
(图 10. 13 在外力作用下绕定轴转动的刚体)

Therefore, we obtain that the moment of momentum of the rigid body with respect to the axis of rotation is equal to the product of the moment of inertia of the rigid body with respect to the axis of rotation and its angular velocity, that is(因此,可以得到刚体对转轴的动量矩等于刚体对于转轴的惯性矩与其角速度的乘积,即)

$$L_z = J_z\omega$$

According to the moment of momentum theorem of the system of particles with respect to the axis, we write it in differential form with respect to time on the axis Az, we have (根据质点系对轴的动量矩定理,将它写成在轴 Az 上关于时间的微分形式,有)

$$\mathrm{d}L_z/\mathrm{d}t = \sum M_{iz}^{\mathrm{E}}$$

or(或者)

$$\mathrm{d}(J_z\omega)/\mathrm{d}t = \sum M_{iz}^{\mathrm{E}}$$

Hence(因此)

$$J_z\mathrm{d}\omega/\mathrm{d}t = \sum M_{iz}^{\mathrm{E}}$$

or(或者)

$$J_z\varepsilon = \sum M_{iz}^{\mathrm{E}}$$

Considering that angular acceleration $\ddot{\varphi}$ is the second derivative of rotation angle φ with respect to time, the formula can be written as(考虑到角加速度 $\ddot{\varphi}$ 是转角 φ 对于时间的二阶导数,公式写为)

$$J_z\ddot{\varphi} = \sum M_{iz}^{\mathrm{E}}$$

Therefore, we obtain the differential equation for fixed axis rotation of a rigid body. It should be noted that the right part of the equation is the main moment of the given force $\boldsymbol{F}_i^{\mathrm{E}}$ around the rotating axis Az, and the moment of the binding force around the rotating axis is zero, so for the rotating axis Az, we have (因此,获得了刚体定轴转动的微分方程. 应该注意的是它右边的部分是给定外力 $\boldsymbol{F}_i^{\mathrm{E}}$ 对转轴 Az 的主矩,而转轴处的约束力对转轴的矩是零,因此对转轴 Az,有)

$$J_z\ddot{\varphi} = \sum M_z^{\mathrm{E}}$$

If the combined torque of the external force is equal to zero, the moment of momentum around the rotation axis of the rigid body remains unchanged, that is (如果外力的合力矩等于零,刚体转动对转轴的动量矩保持不变,即)

$$J_z\omega = \mathrm{const}$$

In this case, if the system's moment of inertia remains constant, the rigid body rotates at a constant angular velocity; if the moment of inertia changes, the angular velocity changes. Thus we have(在这种情况下,如果系统的惯性矩保持不变,该刚体以恒定的角速度转动;如果惯性矩改变,角速度也会改变. 因此,有)

$$J_{z1}\omega_1 = J_{z2}\omega_2$$

10.8 Differential Equations of Plane Motion of Rigid Bodies (刚体的平面运动微分方程)

From the motion theorem of the mass center and the theorem of moment of momentum with respect to the center of mass, we get(由质心运动定理和相对于质心的动量矩定理,得到)

$$m\boldsymbol{a}_C = \sum \boldsymbol{F}_i^{\mathrm{E}}$$

$$\frac{\mathrm{d}}{\mathrm{d}t}J_C\omega = J_C\varepsilon = \sum M_C(\boldsymbol{F}_i^{\mathrm{E}})$$

we obtain(得到)

$$m\boldsymbol{a}_C = \sum \boldsymbol{F}_i^{\mathrm{E}}$$

$$J_C\frac{\mathrm{d}^2\varphi}{\mathrm{d}t^2} = \sum M_C(\boldsymbol{F}_i^{\mathrm{E}})$$

The above equation is the differential equation of the plane motion of a rigid body. (上式即为刚体平面运动的微分方程.)

Expressed in projection form as(用投影形式表示为)

$$m\ddot{x}_C = \sum_i F_x$$

$$m\ddot{y}_C = \sum_i F_y$$

$$J_C\ddot{\varphi} = \sum_i M_C(\boldsymbol{F}_i^{\mathrm{E}})$$

Example 5: As shown in Fig. 10.14, the radius of the homogeneous circular wheel is R, the mass is m, and the moment of inertia of the circular wheel against the rotating shaft is J_O. The wheel rotates around a fixed axis O driven by a weight P, the weight of which is known to be W. Find the acceleration a when the weight falls. (**例题** 5:如图 10.14 所示,均质圆轮半径为 R,质量为 m,圆轮对转轴的转动惯量为 J_O. 圆轮在重物 P 带动下绕固定轴 O 转动,已知重物重力为 W. 求重物下落的加速度 a.)

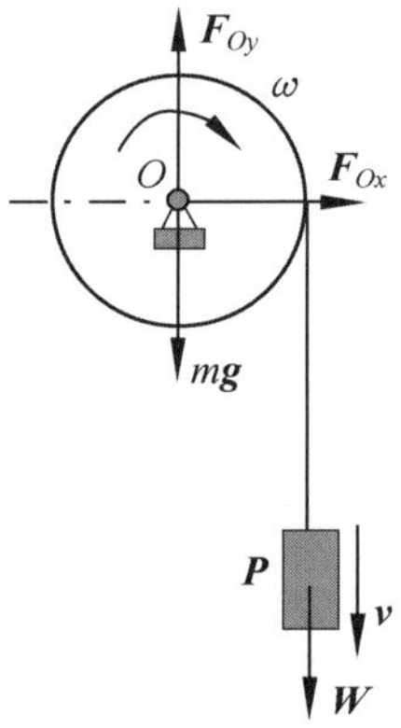

Fig. 10.14 Diagram of Example 5(图 10.14 例题 5 图)

Solution:(解:)

Take the system as the object of study, we have(取系统为研究对象,有)

$$L_O = J_O\omega + \frac{W}{g}vR$$

$$L_O = \left(\frac{J_O}{R} + \frac{W}{g}R\right)v\ , \ \omega = \frac{v}{R}$$

$$M_O^E = WR$$

$$\frac{dL_O}{dt} = M_O^E$$

$$\left(\frac{J_O}{R} + \frac{W}{g}R\right)\frac{dv}{dt} = WR$$

The acceleration of the weight is equal to(重物的加速度为)

$$a = \frac{WR^2}{J_O + \frac{W}{g}R^2}$$

Example 6: As shown in Fig. 10. 15, it is known that the mass of the ball is m, the pendulum length is a, and the moment of inertia of rotation is J_O. Find the period of tiny oscillation. (**例题 6**:如图 10. 15 所示,已知小球质量为 m,摆长为 a,转动转动惯性矩为 J_O. 求微小摆动的周期.)

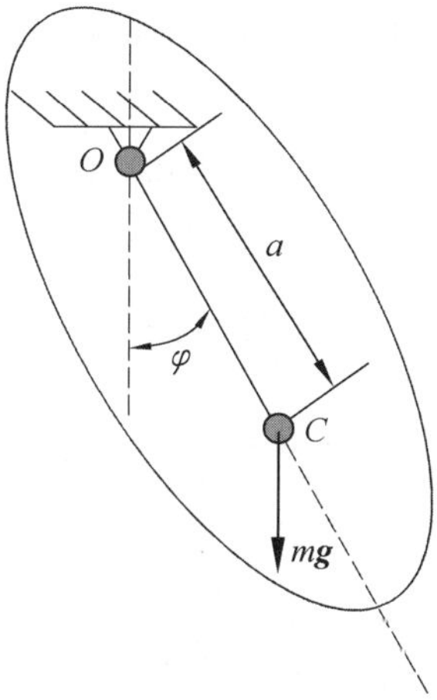

Fig. 10. 15 Diagram of Example 6(图 10. 15 例题 6 图)

Solution:(**解**:)

Take the pendulum as the research object and list the differential equation of fixed axis rotation, that is(取摆为研究对象,列出定轴转动的微分方程,即)

$$J_O\frac{d^2\varphi}{dt^2} = -mga\sin\varphi$$

Pendulum makes a small swing, there are(摆作微小摆动,有)

$$\sin\varphi \approx \varphi$$

$$\frac{d^2\varphi}{dt^2} + \frac{mga}{J_O}\varphi = 0$$

The general solution of this equation is(此方程的通解为)

$$\varphi = \varphi_0\sin\left(\sqrt{\frac{mga}{J_O}}t + \theta\right)$$

The period is(周期为)

$$T = 2\pi\sqrt{\frac{J_O}{mga}}$$

Example 7: As shown in Fig. 10.16, it is known that the flywheel rotation moment of inertia is J_O, angular velocity is ω_0, the positive pressure exerted by the brake shoe during braking is F_N, and the friction coefficient is f. Find the time t required for braking. (**例题**7:如图 10.16 所示,已知飞轮转动惯性矩为 J_O,角速度为 ω_0,在制动时受到闸瓦的正压力为 F_N,摩擦系数为 f. 求制动所需的时间 t.)

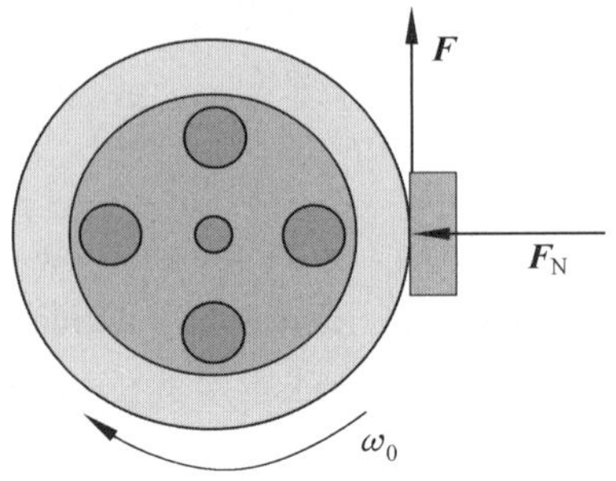

Fig. 10.16 Daigram of Example 7 (图 10.16 例题 7 图)

Solution:(**解**:)

Take the flywheel as the research object and list the dynamic equation of axis rotation, we have(取飞轮为研究对象,列定轴转动动力学方程,有)

$$J_O\varepsilon = J_O\frac{d\omega}{dt} = FR = fF_NR$$

$$\int_{-\omega_0}^{0} J_O d\omega = \int_0^t fF_NR dt$$

We obtain(解得)

$$t = \frac{J_O\omega_0}{fF_NR}$$

Example 8: As shown in Fig. 10.17, it is known that the mass of the homogeneous circular wheel is m and the radius is R, the angular acceleration of a fixed axis rotation is ε. Find the dynamic constraint reaction force at point O. (**例题** 8:如图 10.17 所示,已知均质圆轮质量为 m,半径为 R,定轴转动的角加速度为 ε. 求 O 处动约束反力.)

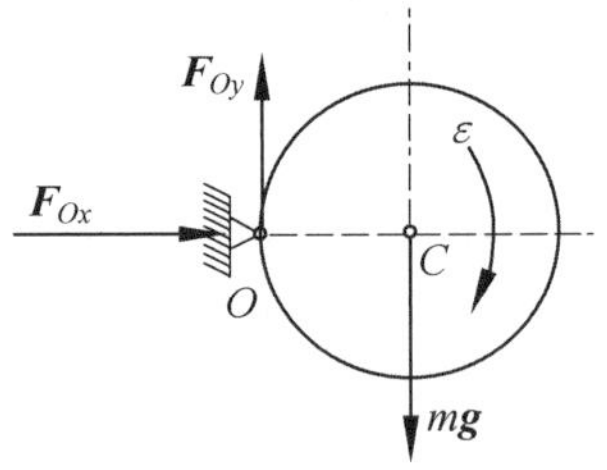

Fig. 10.17 Diagram of Example 8(图 10.17 例题 8 图)

Solution:(**解**:)

Take the circular wheel as the research object and rotate the differential equation by fixed axis, we have(取圆轮为研究对象,由定轴转动的微分方程,有)

$$J_O\varepsilon = mgR$$

where(其中)

$$J_O = \frac{1}{2}mR^2 + mR^2 = \frac{3}{2}mR^2$$

We obtain(解得)

$$\varepsilon = \frac{2g}{3R}$$

By the motion theorem of the center of mass, we have(由质心运动定理,有)

$$\begin{cases} \sum F_x^{\mathrm{E}} = F_{Ox} = ma_{Cx} \\ \sum F_y^{\mathrm{E}} = F_{Oy} - mg = ma_{Cy} \end{cases}$$

where(其中)

$$a_{Cx} = 0,\ a_{Cy} = -R\varepsilon = -\frac{2}{3}g$$

We obtain(解得)

$$F_{Ox} = 0, F_{Oy} = \frac{1}{3}mg$$

Example 9: As shown in Fig. 10. 18, it is known that the mass of the homogeneous circular wheel is m, the radius is R, and it only rolls without slipping on the plane under the action of the active torque M, and the angular acce is ε. Find the acceleration of the wheel center and the condition that the moment of the couple M satisfies in pure rolling. (**例题** 9:如图 10. 18 所示,已知均质圆轮质量为 m,半径为 R,受到主动力矩 M 的作用在平面上只滚不滑,角加速度为 ε. 求轮心的加速度 a_C 及纯滚动时力偶矩 M 满足的条件.)

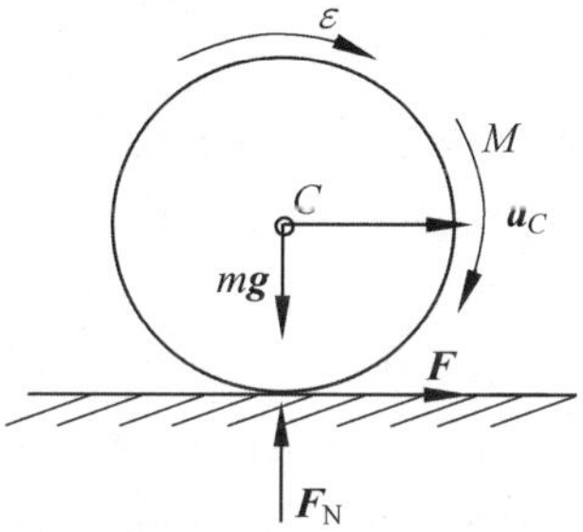

Fig. 10. 18 Diagram of Example 9(图 10. 18 例题 9 图)

Solution:(解:)

Take the circular wheel as the object of study, and the differential equation of plane motion is(取圆轮为研究对象,平面运动微分方程为)

$$\sum F_x^{\mathrm{E}} = ma_{Cx} = ma_C = F$$

$$\sum F_y^{\mathrm{E}} = F_{\mathrm{N}} - mg = ma_{Cy} = 0$$

$$J_C\varepsilon = M - FR$$

Because the round wheel only rolls and does not slip, then (因为圆轮只滚不滑,所以)

$$a_C = R\varepsilon$$

We obtain(得到)

$$F = ma_C, F_N = mg, \ a_C = \frac{MR}{J_C + mR^2}$$

Because the static friction force satisfied(因为静滑动摩擦力满足)

$$F \leqslant f_s F_N = f_s mg$$

Then we obtain(则可得到)

$$M \leqslant f_s g \frac{J_C + mR^2}{R}$$

Example 10: For sudden release of restraints, as shown in Fig. 10. 19(a), the system maintaining balance is suddenly cut off the rope, as shown in Fig. 10. 19(b). Find the restraint reaction force of hinge O at this time.(**例题** 10:关于突然解除约束问题,如图 10.19(a)所示,保持平衡状态的系统被突然剪断绳索,如图 10.19(b)所示.求此时铰链 O 的约束反力.)

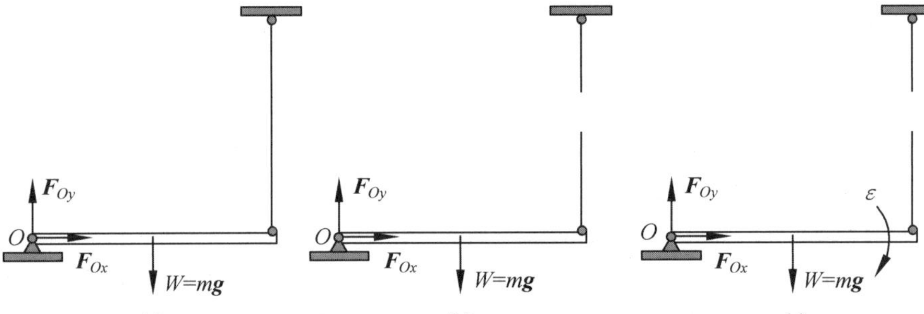

Fig. 10. 19 Diagram of Example 10(图 10.19 例题 10 图)

Solution:(解:)

Before lifting constraints, we have(解除约束前,有)

$$F_{Ox} = 0, \ F_{Oy} = mg/2$$

When the constraint is suddenly lifted, the bar OA will rotate around the O-axis, which is no longer a statics problem. In this case, $\omega = 0, \varepsilon \neq 0$. We need to solve the ε before determine the restraint reaction force. Apply the differential equation of rotation around thefixed axis, we have (突然解除约束瞬时,杆 OA 将绕 O 轴转动,不再是静力学问题. 这时,$\omega = 0, \varepsilon \neq 0$. 需要先求出 ε,再确定约束力. 应用定轴转动微分方程,有)

$$J_O \varepsilon = M_O(F^E)$$

We obtain(得到)

$$\frac{1}{3} ml^2 \varepsilon = mg \times \frac{l}{2}$$

And we obtain(然后得到)

$$\varepsilon = \frac{3g}{2l}$$

Apply the theorem of the centroid motion, we have (应用质心运动定理,得到)

$$m \times \frac{l}{2} \omega^2 = 0 = F_{Ox}$$

$$m \times \frac{l}{2} \varepsilon = mg - F_{Oy}$$

We obtain(得到)

$$F_{Ox}=0, F_{Oy}=mg-m\times\frac{l}{2}\varepsilon=\frac{mg}{4}$$

The characteristics of the problem about the sudden release constraints are as follows. (突然解除约束问题的特点如下.)

The degree of freedom of the system generally increases. At the moments before and after releasing constraints, the velocity and angular velocity are continuous, and the acceleration and angular acceleration will change abruptly. (系统的自由度一般会增加.解除约束的前、后瞬时,速度与角速度连续,加速度与角加速度将发生突变.)

Example 11: As shown in Fig. 10.20, given that the weight of the delivery truck is $\boldsymbol{G}$ when filled fully with goods, the friction resistance when running on the platform is 0.2 times of the weight of the truck, the compression deformation of the spring when arriving at the unloading place is set to be δ_m, and the empty weight after unloading is $\boldsymbol{G}_0$. Find G/G_0. (**例题** 11:如图 10.20 所示,已知送料车装满货物时的重力为 $\boldsymbol{G}$,在平台上运行时的摩擦阻力为车重的 0.2,设到达卸货地点时弹簧的压缩变形量为 δ_m,卸载后空车重力为 $\boldsymbol{G}_0$,求 G/G_0.)

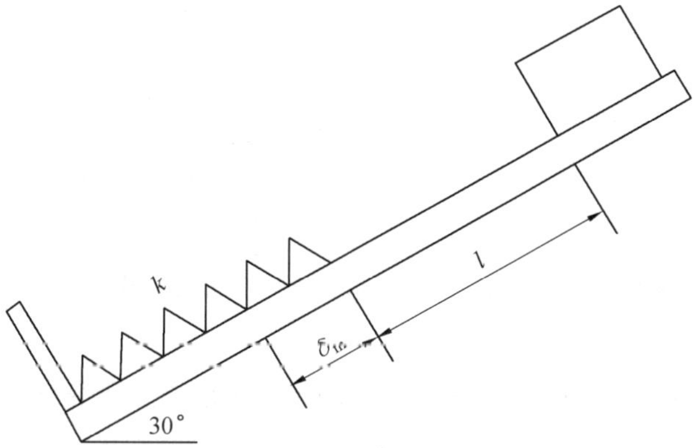

Fig. 10.20 Diagram of Example 11(图 10.20 例题 11 图)

Solution:(解:)

Take the truck as the research object and set the maximum deformation of the spring is δ_m. (取货车为研究对象,设弹簧的最大变形为 δ_m.)

(1) When the car slides down to the maximum compression of the spring, the work done by the external force is (车下滑到弹簧压缩量为最大时,外力做的功为)

$$W_{12}=G(l+\delta_m)\sin 30^\circ-0.2G(l+\delta_m)-\frac{k}{2}\delta_m^2$$

According to the kinetic energy theorem, we get(根据动能定理,可得)

$$0-0=G(l+\delta_m)\sin 30^\circ-0.2G(l+\delta_m)-\frac{k}{2}\delta_m^2$$

(2) After unloading, the truck is bounced back to its original position, according to the kinetic energy theorem, we have(车卸料后又弹回原位置,由动能定理得)

$$0-0=-G_0(l+\delta_m)\sin 30^\circ-0.2G_0(l+\delta_m)+\frac{k}{2}\delta_m^2$$

We obtoin(解得)

$$\frac{G}{G_0}=\frac{\sin 30^\circ + 0.2}{\sin 30^\circ - 0.2}=\frac{7}{3}$$

Example 12:As shown in Fig. 10. 21, the radius of the homogeneous circular wheel is R, the mass is m, and the moment of inertia of the circular wheel against the rotating shaft is J_O. The wheel rotates around a fixed axis O driven by a weight P, the weight of the object is W. Find the acceleration of the object when it falls down. (**例题** 12:如图 10.21 所示,均质圆轮半径为 R,质量为 m,圆轮对转轴的转动惯量为 J_O. 圆轮在重物 P 带动下绕固定轴 O 转动,已知重物重力为 **W**. 求重物下落的加速度.)

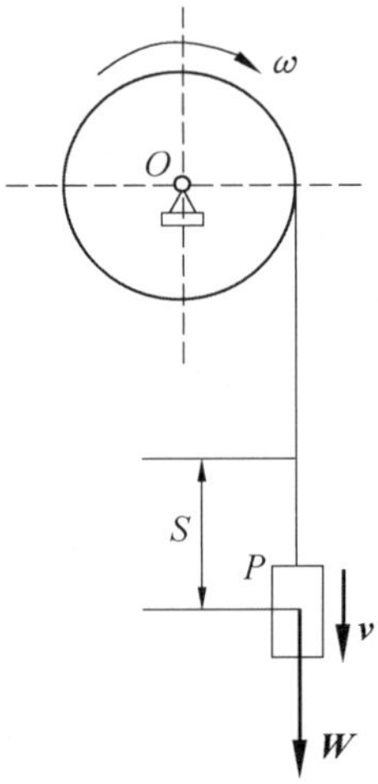

Fig. 10. 21 Diagram of Example 12(图 10.21 例题 12 图)

Solution:(解:)

Take the system as the object of study, we have (取系统为研究对象,得到)

$$T_1=0,T_2=\frac{1}{2}\frac{W}{g}v^2+\frac{1}{2}J_O\omega^2,\omega=\frac{v}{R}$$

The work done by the positive force is(主动力的功为)

$$W_{12}=WS$$

From the kinetic energy theorem, we have(由动能定理得)

$$\frac{1}{2}\frac{W}{g}v^2+\frac{1}{2}J_O\omega^2-0=WS$$

Take the derivative of the above formula with respect to time, and note that(将上式对时间求导,并注意到)

$$\frac{\mathrm{d}v}{\mathrm{d}t}=a,\frac{\mathrm{d}S}{\mathrm{d}t}=v$$

We obtain(解得)

$$a=\frac{WR^2}{J_O+\frac{W}{g}R^2}$$

Example 13: As shown in Fig. 10. 22, the mass of the homogeneous disk is m, the radius is

R, the coefficient of static friction between the disk and the bevel is f, and the inclination of the bevel is φ. Find the acceleration of the disk center when it roll purely. (**例题** 13:如图 10.22 所示,均质圆盘质量为 m,半径为 R,圆盘与斜面静摩擦系数为 f,斜面倾角为 φ.求纯滚时盘心的加速度.)

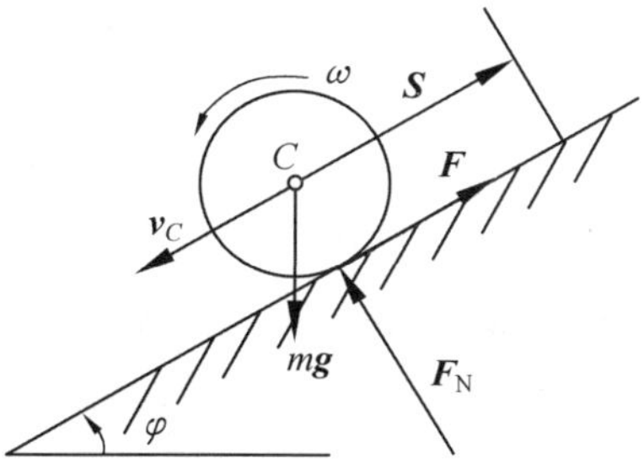

Fig. 10.22 Diagram of Example 13(图 10.22 例题 13 图)

Solution:(解:)

Take the system as the object of study, we have(取系统为研究对象,有)

$$T_1 = 0, T_2 = \frac{1}{2}mv^2 + \frac{1}{2}J_C\omega^2, \omega = \frac{v_C}{R}$$

We obtain(得到)

$$T_2 = \frac{3}{4}mv_C^2$$

The work done by the positive force is(主动力的功为)

$$W_{12} = mgS\sin\varphi$$

Because the disk makes roll purely, so the work done friction force is zero.(因为圆盘做纯滚动,所以摩擦力做功为零).

By the kinetic energy theorem, we get(由动能定理得)

$$\frac{3}{4}mv_C^2 - 0 = mgS\sin\varphi$$

We obtain(解得)

$$a = \frac{2}{3}g\sin\varphi$$

Example 14: As shown in Fig. 10.23, it is known that the moment of inertia of shaft Ⅰ and shaft Ⅱ are J_1 and J_2 respectively, the radius of meshing gear is R_1 and R_2 respectively, the transmission ratio $i_{12} = R_2/R_1$, and the torque acting on the two axes are M_1 and M_2 respectively. Find the angular acceleration of axis Ⅰ. (**例题** 14:如图 10.23 所示,已知轴Ⅰ、轴Ⅱ转动惯量分别为 J_1、J_2,啮合齿轮半径分别为 R_1、R_2,传动比 $i_{12} = R_2/R_1$,作用在两轴上的力矩分别为 M_1、M_2.求轴Ⅰ的角加速度.)

Solution:(解:)

Take the system as the object of study, we have(取系统为研究对象,有)

$$T_1 = 0, T_2 = \frac{1}{2}J_1\omega_1^2 + \frac{1}{2}J_2\omega_2^2$$

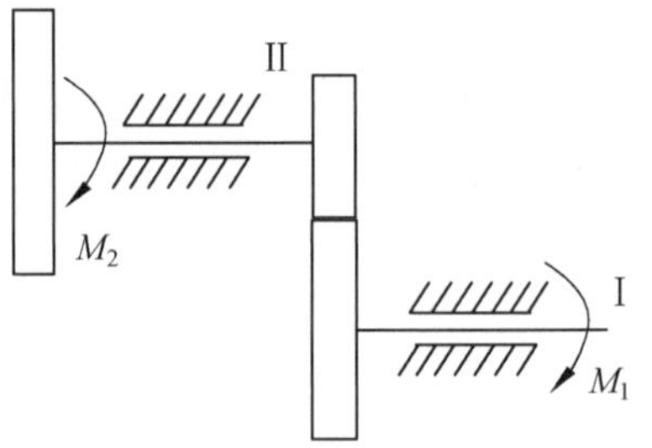

Fig. 10.23　Diagram of Exercise of 14(图 10.23　例题 14 图)

Let the angular velocity of the two gears be ω_1, ω_2 respectively, and the angles φ_1 and φ_2 respectively. It can be seen from the kinematics that(设两齿轮的角速度分别为 ω_1、ω_2,转角分别为 φ_1、φ_2,由运动学可知)

$$\frac{\omega_1}{\omega_2}=i_{12}=\frac{R_2}{R_1}=\frac{\varphi_1}{\varphi_2}$$

We obtain(得到)

$$T_2=\frac{1}{2}\left(J_1+\frac{J_2}{i_{12}^2}\right)\omega_1^2$$

The work done by the positive force is(主动力的功为)

$$W_{12}=M_1\varphi_1-M_2\varphi_2=\left(M_1-\frac{M_2}{i_{12}}\right)\varphi_1$$

From the kinetic energy theorem, we have(由动能定理得)

$$\frac{1}{2}\left(J_1+\frac{J_2}{i_{12}^2}\right)\omega_1^2-0=\left(M_1-\frac{M_2}{i_{12}}\right)\varphi_1$$

Take the derivative of the above formula with respect to time, and notice(将上式对时间求导,并注意)

$$\frac{d\omega_1}{dt}=\varepsilon_1,\frac{d\varphi_1}{dt}=\omega_1$$

We obtain(解得)

$$\varepsilon_1=\frac{M_1-\dfrac{M_2}{i_{12}}}{J_1+\dfrac{J_2}{i_{12}^2}}$$

10.9　Question(思考题)

1. If two objects A and B with the same weight are located at the same time interval, and object A moves s horizontally and B moves s vertically, is the impulse of gravity of the two objects the same during this time interval? (有两个相同质量的物体 A 与 B,设在相同时间间隔内,使 A 水平移动 s,使 B 垂直移动 s,则在此时间间隔内这两物体的重力的冲量是否相同?)

2. The more particles in a system of particles, is the greater its momentum? Let the mass of

each object be m. Find the momentum of each homogeneous body as shown in Fig. 10. 24. (质点系中质点越多,其动量是否越大? 设各物体质量均为 m. 求图 10. 24 所示各均质物体的动量.)

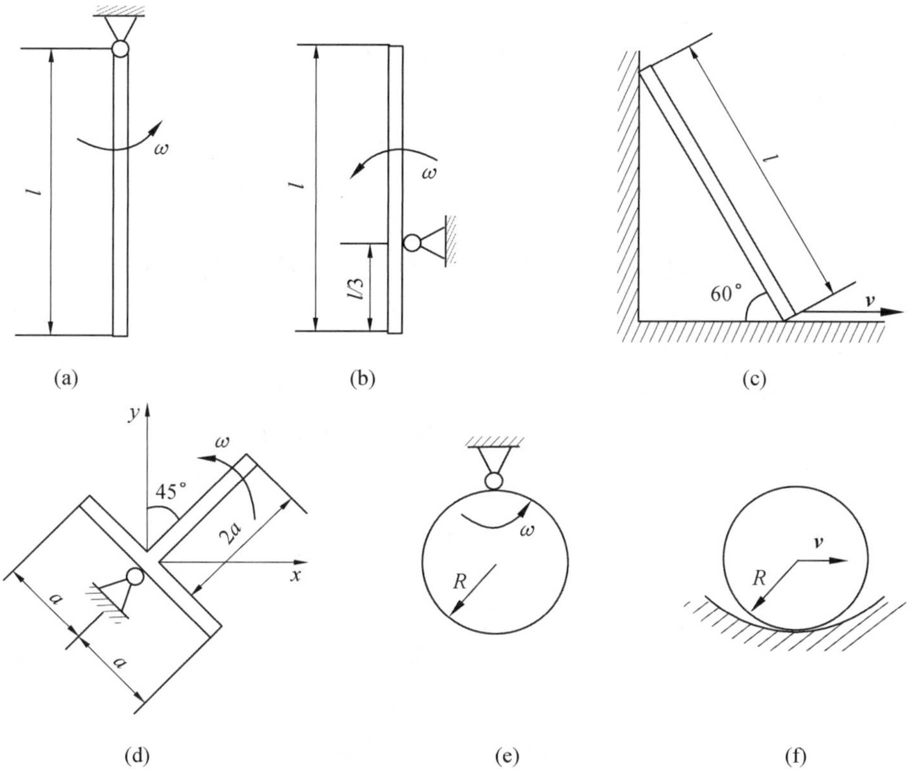

Fig. 10. 24 Diagram of Question 2(图 10. 24 思考题 2 图)

3. There are three identical homogeneous rods suspended in the air with the center of mass on the same horizontal line. One is horizontal, one is vertical, and another is tilted. If the three bars are released freely at the same time, is the motion law of center of mass for the three bars the same? (有三根相同的均质杆悬空放置,质心皆在同一水平线上. 其中一杆水平,一杆垂直,另一杆倾斜. 若同时自由释放此三杆,则三杆质心的运动规律是否相同?)

4. As shown in Fig. 10. 25, two identical homogeneous disks are placed on a smooth surface. Apply a horizontal force $\boldsymbol{F}$ and $\boldsymbol{F}'$ with the same magnitude and direction to each of the two disks at different positions respcetively, make the disks move from rest. Which disc has a faster center of mass movement? Why? (如图 10. 25 所示,在光滑面上放两个相同的均质圆盘,在两圆盘

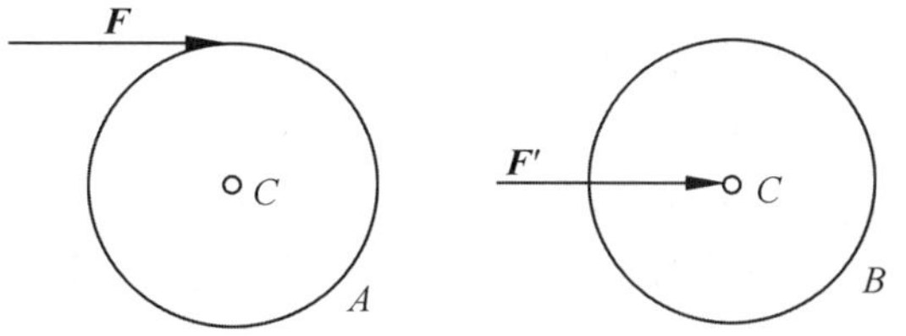

Fig. 10. 25 Diagram of Question 4(图 10. 25 思考题 4 图)

的不同位置上,各作用一大小和方向相同的水平力 $\boldsymbol{F}$ 和 $\boldsymbol{F}'$,使圆盘由静止开始运动. 试问哪个圆盘的质心运动得快? 为什么?)

5. There is a balloon above the ground, and a person is hanging on a ladder below the balloon. When the person climbs up the ladder, does the balloon move? (在地面的上空停着一个气球,气球下面吊一软梯,并挂着一人. 当此人沿着软梯往上爬时,气球是否运动?)

6. Homogeneous rods AC and BC, of the same length and mass m_1 and m_2 respectively, are hinged at point C and initially remain stationary in the plumb vertical, as shown in Fig. 10.26. If the ground is absolutely smooth, the two poles will fall apart to the ground when they are released. When m_1 and m_2 are equal or unequal, is the motion path of point C the same? (均质杆 AC 和 BC,长度相同,质量分别为 m_1 和 m_2,两杆在点 C 铰接,初始时维持在铅垂面内不动,如图 10.26 所示. 设地面绝对光滑,两杆被释放后将分开倒向地面. 则 m_1 和 m_2 相等或不相等时,C 点的运动轨迹是否相同?)

7. Is the acceleration of the center of a rigid body under the action of a force system the same regardless of the point of action of each force? (刚体受一力系作用,无论各力作用点如何,此刚体质心的加速度都一样吗?)

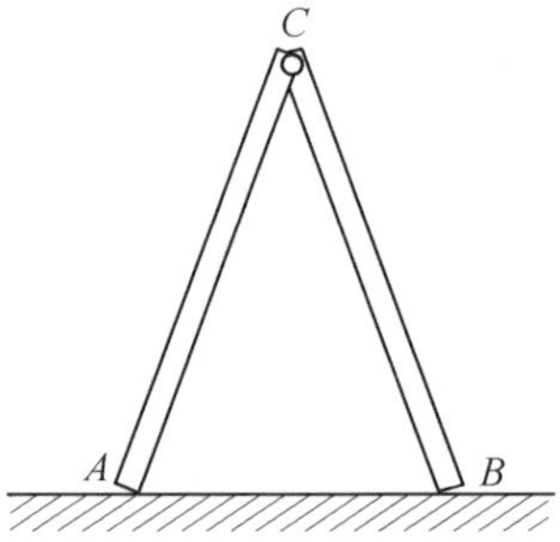

Fig. 10.26　Diagram of Question 6(图 10.26　思考题 6 图)

8. A rigid body rotates around a fixed axis, when the angular velocity is large, is the resultant external torque necessarily large? When the angular velocity is zero, is the resultant external torque necessarily zero? Is the turn of the angular velocity necessarily the same as that of the resultant external torque? (刚体绕定轴转动,当角速度很大时,所受的合外力矩是否一定很大? 当角速度为零时,合外力矩是否一定为零? 角速度的转向是否一定与合外力矩的转向相同?)

9. What is the radius of rotation? Is it equal to the vertical distance from the center of mass of a rigid body to the axis of rotation? (什么是回转半径? 它是否等于刚体质心到转轴的垂直距离?)

10. The link with mass m in Fig. 10.27 swings around point O at the angular velocity ω. The distance between the center of mass C and the fulcrum O is $OC = l$. Is it correct to use the following formula to calculate the momentum moment of link to point O? Why? (图 10.27 中质量为 m 的连杆以角速度 ω 绕 O 点摆动. 连杆质心 C 到支点 O 的距离为 $OC = l$. 用下式计算连杆对 O 点的动量矩是否正确? 为什么?)

$$L_O = \text{momentum(动量)} \times \text{distance(距离)} = (mv_C)l = m(l\omega)l = ml^2\omega$$

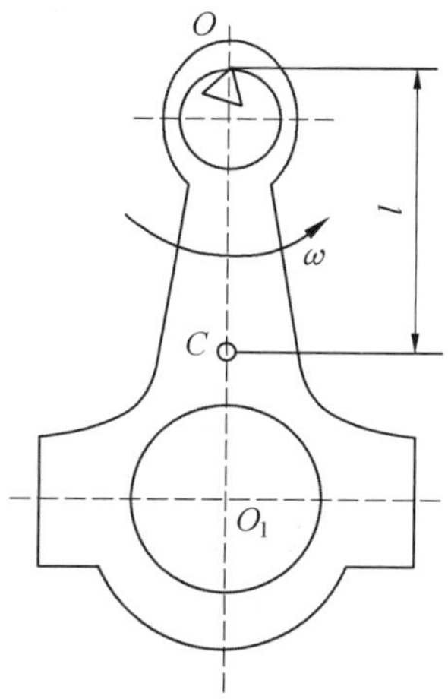

Fig. 10.27 Diagram of Question 10(图 10.27 思考题第 10 题图)

11. As shown in Fig. 10.28, the length of the rod is l, the mass is m, and the moment of inertia around the z axis is $J_z = \frac{7}{48}ml^2$. Now, the moment of inertia of the z' axis is obtained through the parallel axis theorem, and the calculation formula is $J_z' = J_z + m\left(\frac{l}{2}\right)^2 = \frac{19}{48}ml^2$. Is it correct?(如图 10.28 所示,一直杆长为 l,质量为 m,绕 z 轴的转动惯量为 $J_z = \frac{7}{48}ml^2$. 现通过平行轴定理求 z' 轴的转动惯量,列出算式 $J_z' = J_z + m\left(\frac{l}{2}\right)^2 = \frac{19}{48}ml^2$,其是否正确?)

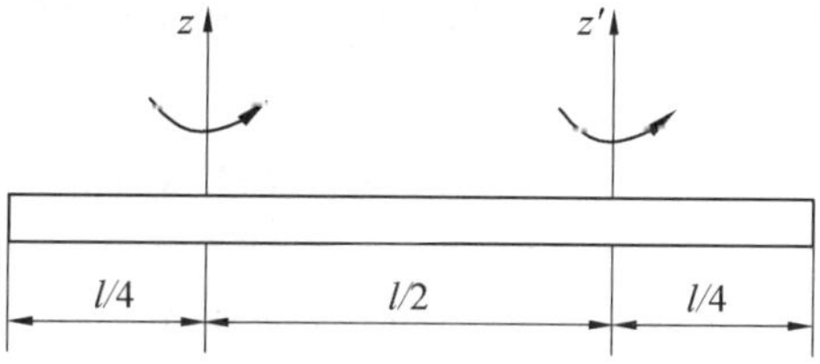

Fig. 10.28 Diagram of Question 11(图 10.28 思考题 11 图)

12. Try to find the moment of momentum of each object with respect to the axis of rotation shown in Fig. 10.29.(试求图 10.29 中所示各物体对转轴的动量矩.)

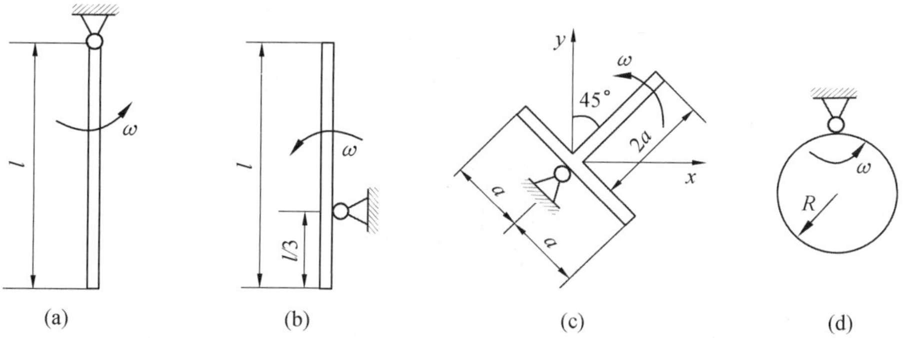

Fig. 10.29 Diagram of Question 12(图 10.29 思考题 12 图)

13. As shown in Fig. 10.30, J_1 and J_2 in the transmission system are the moment of inertia of wheel Ⅰ and wheel Ⅱ, and the angular acceleration of wheel Ⅰ is $\alpha_1 = \dfrac{M_1}{J_1 + J_2}$, is it right?(如图 10.30 所示,传动系统中 J_1、J_2 为轮Ⅰ、轮Ⅱ的转动惯量,轮Ⅰ的角加速度 $\alpha_1 = \dfrac{M_1}{J_1 + J_2}$,是否正确?)

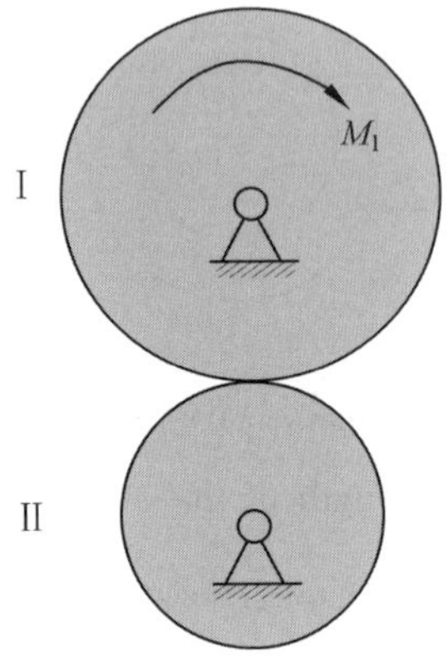

Fig. 10.30 Diagram of Question 13(图 10.30 思考题 13 图)

14. As shown in Fig. 10.31, both the mass of the homogeneous rod and the homogeneous disc are m, the rod length is $2R$, and the radius of the disc is R. They are hinged at point A, and the system is placed on a smooth horizontal surface with initial rest. Now loaded with a couple of moment M, which of the following statements is true?(如图 10.31 所示,均质杆、均质圆盘质量均为 m,杆长为 $2R$,圆盘半径为 R,二者铰接于点 A,系统放在光滑水平面上,初始静止. 现受一矩为 M 的力偶作用,则下列哪些说法正确?)

(1) If M acts on the rod, the rod rotates around A and the disc does not move.(若 M 作用于杆上,则杆绕 A 点转动,杆不动.)

(2) If M acts on a rod, the disc is translated.(若 M 作用于杆上,则盘平移.)

(3) If M acts on the disc, then the rod rotates around A point, the disc does not move.(若 M 作用于盘上,则杆绕 A 转动,盘不动.)

(4) The system moves the same no matter which object M acts on.(无论 M 作用于哪个物体上,系统运动都一样.)

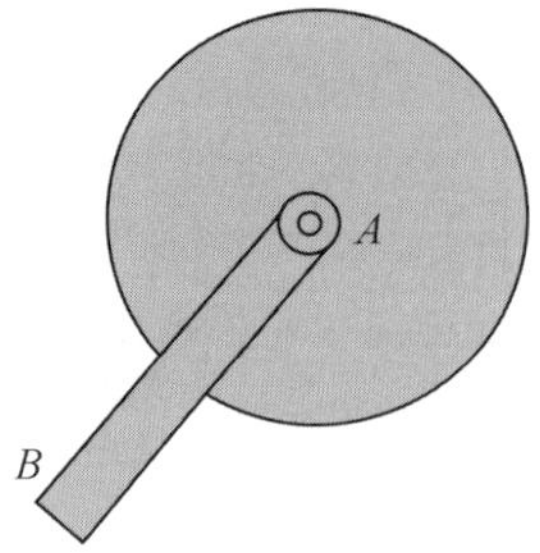

Fig. 10.31 Diagram of Question 14(图 10.31 思考题 14 图)

15. Fig. 10. 32 shows two identical homogeneous wheels, one end of the rope in Fig. 10. 32(a) is hung with a weight, and the end of the rope in Fig. 10. 32(b) is subjected to a tension $\boldsymbol{F}$, and $F=P$. Is the angular acceleration of the two wheels the same? Is the tension on the rope the same? Why?(图 10. 32 所示两个完全相同的均质轮,图 10. 32(a)轮中绳的一端挂一重物,重力为 $\boldsymbol{P}$,图 10. 32(b)中绳的一端受拉力 $\boldsymbol{F}$,且 $F=P$. 则两轮的角加速度是否相同? 绳子受的拉力是否相同? 为什么?)

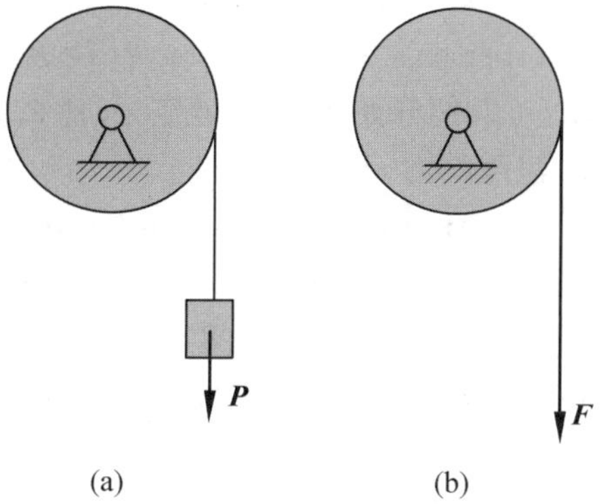

Fig. 10. 32 Diagram of Question 15(图 10. 32 思考题 15 图)

16. In general, when we use the kinetic energy theorem, do we need to take into account the work done by the restrained reaction of the system? Can we find the constraint reaction of the system?(一般来说,用动能定理时,是否要考虑系统约束反力做的功? 能否求出系统的约束反力?)

17. For non conservative systems, is it necessarily impossible to use the law of conservation of mechanical energy? Why?(对于非保守系统,是否一定不能用机械能守恒定律? 为什么?)

18. The calculation formula $W_{12} = \int_{M_1}^{M_2}(F_x\mathrm{d}x + F_y\mathrm{d}y + F_z\mathrm{d}z)$, can it be understood as the projection formula of calculated work? Does this formula hold if the x, y, z axes are not perpendicular?(力做功的计算公式 $W_{12} = \int_{M_1}^{M_2}(F_x\mathrm{d}x + F_y\mathrm{d}y + F_z\mathrm{d}z)$,能否理解为计算功的投影式? 如果 x、y、z 轴不垂直,该式成立吗?)

19. Three balls with the same mass are thrown from a certain height at the same speed, but in different directions, regardless of air resistance. When the three balls hit the same plane, do the three balls have the same velocity? Is the work of gravity the same for all three balls? Do all three balls hit the ground at the same time?(从某一高度同时抛出三个质量相同的小球,速度大小相等,但抛出方向各不相同,不计空气阻力. 这三个小球落到同一水平面时,三个小球的速度大小是否相同? 三个小球重力的功是否相同? 三个小球落地的时间是否相同?)

20. Block A of mass m slides from the vertex of height h along a smooth surface of three different shapes: flat, concave and convex, as shown in Fig. 10. 33. Is the speed at which block A slides to the bottom the same in the three situations shown in the diagram? Why?(质量为 m 的物块 A,从高为 h 的顶点由静止沿平、凹、凸三种不同形状的光滑面下滑,如图 10. 33 所示. 在图示三种情况下,物块 A 滑到底部时的速度是否相同? 为什么?)

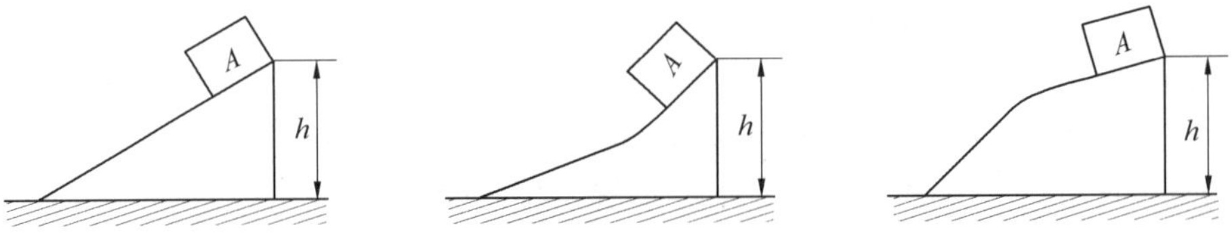

Fig. 10.33 Diagram of Question 20(图 10.33 思考题 20 图)

21. When people begin to walk or run, what force causes people to make accelerating motion? What force increases a person's kinetic energy? Does the force that produces the acceleration necessarily do work?(人们开始走动或起跑时,什么力可以使人做加速运动?什么力使人的动能增加?产生加速度的力一定做功吗?)

22. A stretches the spring by 0.03 m from the original length, and B continues to stretch the spring by 0.02 m after A. Who did the most work between A and B?(A 将弹簧由原长拉伸 0.03 m,B 继 A 之后再将弹簧继续拉伸 0.02 m. A、B 二人谁做的功多?)

23. Three particles with the same mass, thrown out from point A at the same time, have the same initial velocity magnitude v_0 but with different directions, as shown in Fig. 10.34. If the air resistance is not considered, when the three particles fall to the horizontal plane $H-H$, are the velocities of the three particles equal? Are the work done by gravity equal in three particles? Are the impulses of gravity in three particles equal?(三个质量相同的质点,同时由点 A 抛出,初速度 v_0 大小相同,而方向各不相同,如图 10.34 所示. 不计空气阻力,这三个质点落到水平面 $H-H$ 时,三质点的速度大小是否相等?三质点重力的功是否相等?三质点重力的冲量是否相等?)

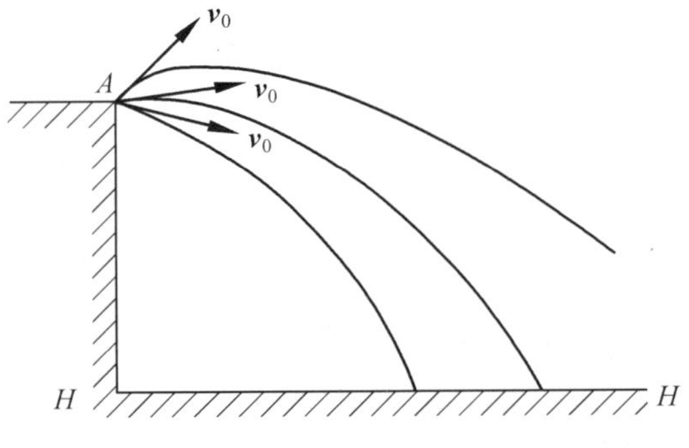

Fig. 10.34 Diagram of Question 23(图 10.34 思考题 23 图)

24. Two people A and B with the same weight, along the rope around the pulley, climb up at the same time from a standstill, regardless of the weight of the rope, as shown in Fig. 10.35. If A tries harder than B to climb: (1) Who reaches the top first? (2) Who has the bigger kinetic energy? (3) Who did the more work? (4) How to apply the kinetic energy theorem to A and B respectively?(相同质量的 A、B 两人,沿绕过滑轮的细绳,由静止起同时向上爬升,细绳质量不计,如图 10.35 所示. 若 A 比 B 更努力上爬:(1)谁先到达上端?(2)谁的动能大?(3)谁做的功多?(4)如何对 A、B 两人分别应用动能定理?)

25. Conservation of momentum means conservation of velocity, conservation of velocity means conservation of kinetic energy, so when momentum is conserved, kinetic energy must also

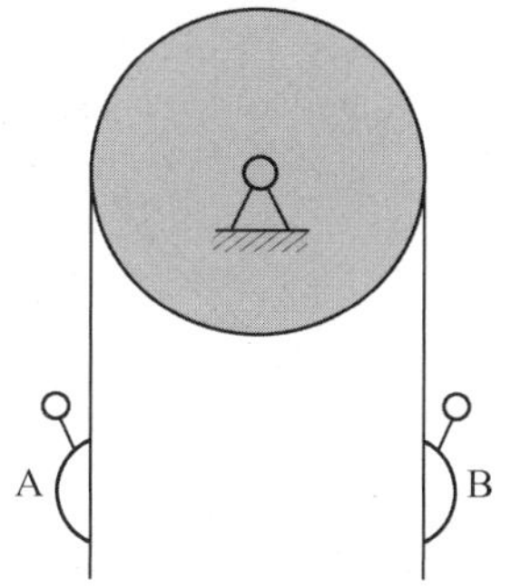

Fig. 10.35 Diagram of Question 24 (图 10.35 思考题 24 图)

be conserved. Is that right? Why? (动量守恒就意味着速度守恒,速度守恒就意味着动能守恒,因此动量守恒时动能必守恒. 上述说法对吗? 为什么?)

10.10 Exercises(习题)

Part 1—Exercise of Momentum theorem (第 1 部分——动量定理习题)

1. A force F is applied to a particle moving along a straight trajectory. The action line of this force always coincides with the trajectory, and the magnitude and direction of the force change with time, as shown in Fig. 10.36. What is the impulse of this force in the first 3 seconds and the first 4 seconds? (一沿直线轨迹运动的质点上作用一力 F,此力的作用线始终与轨迹重合,力的大小和方向随时间变化,如图 10.36 所示. 此力在最初 3 s 和最初 4 s 的时间间隔内的冲量分别是多少?)

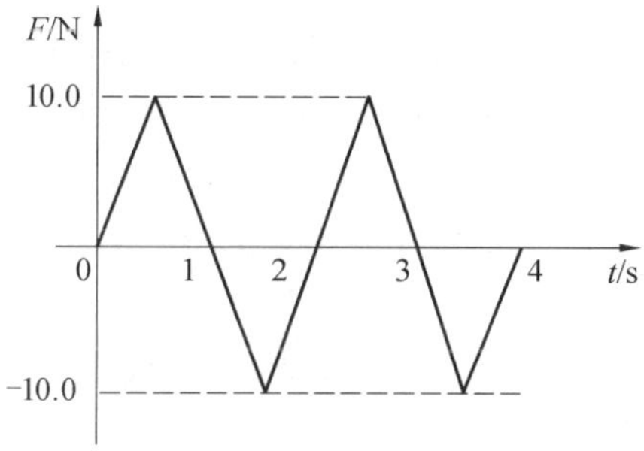

Fig. 10.36 Diagram of Exercise 1 (图 10.36 习题 1 图)

2. Block A and B are connected with a light rod. The mass of A and B is $m_A = 12$ kg and $m_B = 10$ kg respectively, and are placed on the vertical wall and horizontal floor respectively, as shown in Fig. 10.37. A horizontal force $F = 250$ N is applied to the block A to cause it to move to the right from rest. Suppose that after 1 s, block A moves 1.0 m, and velocity $v_A = 4.15$ m/s, friction is ignored, calculate the impulse acting on the wall and ground. (物块 A 和 B 用一轻杆连接,A、B 质量分别为 $m_A = 12$ kg,$m_B = 10$ kg,分别放在铅直墙面和水平地板上,如图 10.37 所示. 在物块 A 上作用一水平力 $F = 250$ N,使它从静止开始向右运动. 假设经过 1 s 后,物块 A 移动了 1.0 m,速度 $v_A = 4.15$ m/s,摩擦忽略不计,求作用在墙面和地面的冲量.)

3. The parachutist, weighing 70 kg, jumps out of a helicopter and drops 100 m before

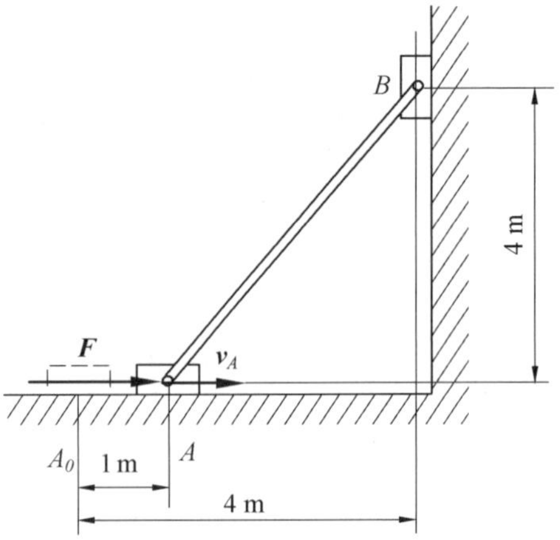

Fig. 10.37 Diagram of Exercise 2 (图 10.37 习题 2 图)

opening his parachute. Assume that before opening the umbrella, air resistance is not considered. After the umbrella is opened, the air resistance is unchanged. After 5 s, the skydiver's speed is reduced to 4.3 m/s. Find the tensile force of the rope holding the person to the umbrella. (跳伞者的质量为 70 kg,从停留在高空中的直升机中跳出,落下 100 m 后将降落伞打开. 假设打开伞以前,空气阻力不计. 伞张开后,所受的空气阻力不变. 经 5 s 后,跳伞者的速度减至 4.3 m/s. 求将人系于伞上的绳索受到的拉力.)

4. The machine gun fires 600 bullets per minute, with a bullet mass of $m = 12.5$ g, at a horizontal speed of $v_0 = 800$ m/s into the vertical target. Find the average pressure on the target. (机枪每分钟射出 600 发子弹,子弹质量 $m = 12.5$ g,以 $v_0 = 800$ m/s 的水平速度射入铅直的靶挡内. 求靶挡所受的平均压力.)

5. Someone with a mass of 70 kg jumps from a platform 0.6 m high, ignoring its initial speed. The supporting force of the ground facing the sole of the human foot increases linearly to the maximum value in the first 0.075 s, and decreases linearly to its weight in the second 0.075 s. Try to find the maximum force on the soles of the feet when the person lands. (某人的质量为 70 kg,从高 0.6 m 的站台上跳下,忽略其初速度. 设地面对人脚底的支撑力,在前 0.075 s内按线性变化增至最大值,在后 0.075 s 内又按线性变化递减至其体重. 试求此人着地时,脚底受到的最大约束力.)

6. As shown in Fig. 10.38, the mass of block D is $m = 5$ kg. At the beginning, the tape supporting block D moves to the right at a constant speed of $v = 1.6$m/s, and moves to the left at a speed of $v_{Dr} = 0.6$ m/s relative to the tape. If the dynamic friction factor between the block and the tape is $f = 0.3$, how long will the speed of the block relative to the tape be reduced by half (the direction is still to the right)? (如图 10.38 所示,物块 D 质量为 $m = 5$ kg,在开始时支承物块 D 的胶带以匀速 $v = 1.6$ m/s 向右运动,以相对于胶带的速度 $v_{Dr} = 0.6$ m/s 向左运动. 设物块与胶带间的动滑动摩擦因数 $f = 0.3$,试问经过多长时间,物块相对胶带的速度将减少一半(方向仍向右)?)

7. As shown in Fig. 10.39, box A, whose mass is $m = 50$ kg, is blocked by block B and parked on the inclined plane. Friction factor f_s between bevel and box is 0.3. Suppose that there

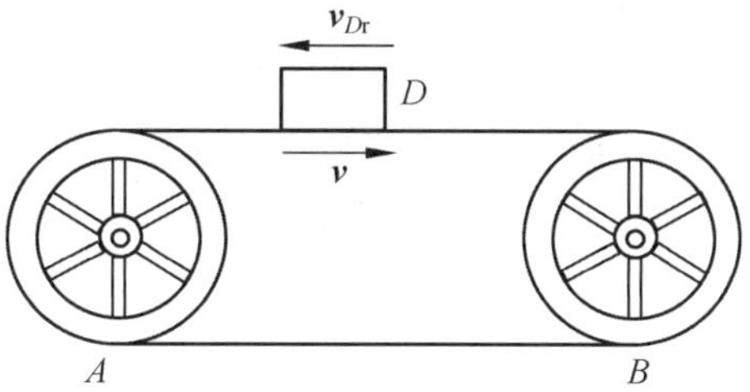

Fig. 10.38 Diagram of Exercise 6 (图 10.38 习题 6 图)

is a force $\boldsymbol{F}$ acting on the box parallel to the inclined plane, the magnitude of which is $F = 300t$ (N), where t is measured by s. How long does it take for this force to achieve an upward velocity of magnitude $v=2$ m/s in box A (Tip: Find the time it takes to overcome friction and start the upward motion)? (如图 10.39 所示,质量为 $m=50$ kg 的箱子 A 受挡块 B 的阻挡,停放在斜面上. 斜面与箱子间的摩擦因数 $f_s=0.3$. 假设有一平行于斜面的力 $\boldsymbol{F}$ 作用于箱子上,其大小为 $F=300t$ (N),式中 t 以 s 计. 过多长时间,该力才能使箱子 A 获得一个大小为 $v=2$ m/s、方向向上的速度(提示:先求克服摩擦力而开始向上运动所需的时间)?)

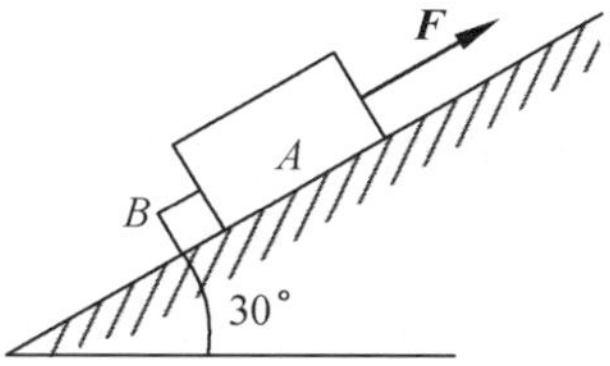

Fig. 10.39 Diagram of Exercise 7 (图 10.39 习题 7 图)

8. A platform car with a mass of $m_1=630$ kg moves along a smooth horizontal track. There is a person standing on the car with a mass $m_2=70$ kg. At the beginning, the car and the person are moving to the right at a common speed of v_0. If the person jump out to the left at the speed $v_r=2$ m/s relative to the platform car, ask how much speed the platform car increases? (质量 $m_1=630$ kg 的平台车,沿光滑的水平轨道运动. 车上站有一人,质量 $m_2=70$ kg. 开始时,车与人以共同速度 v_0 向右方运动. 若人相对平台车以速度 $v_r=2$ m/s 向左方跳出,则平台车增加的速度为多少?)

9. On a barge weighing 100 kN, a box weighing 5 kN is pulled by a winch. Assume that at the beginning, the boat and the box are in a static state, and the resistance of the water is ignored. Find: (1) The moving speed of the barge when the moving speed of the box relative to the barge is 3 m/s; (2) The horizontal distance the barge moves when the box is pulled over 10 m on the barge. (在重 100 kN 的一艘驳船上,用绞车拉动一重 5 kN 的箱子. 设开始时,船与箱子均处于静止状态,水的阻力忽略不计. 求:(1)当箱子在船上移动的速度为 3 m/s 时,驳船移动的速度;(2)当箱子在船上拉过 10 m 时,驳船移动的水平距离.)

10. As shown in Fig. 10.40, A bullet with a mass of $m_1=30$ g enters into a block A with a mass of $m_A=4.5$ kg at a velocity of $v_0=500$ m/s. The dynamic friction factor f between block A and trolley $BC=0.5$. Given the mass of the trolley is $m=3.5$ kg, and it can move freely on a smooth horizontal ground. Find: (1) Final velocity of trolley and block A; (2) The final position

of block A in the trolley from end B. (如图 10.40 所示,一颗质量 $m_1=30$ g 的子弹,以 $v_0=500$ m/s的速度射入质量 $m_A=4.5$ kg 的物块 A 中. 物块 A 与小车 BC 之间的动摩擦因数 $f=0.5$. 已知小车的质量 $m=3.5$ kg,可以在光滑的水平地面上自由运动. 求:(1)小车与物块 A 的末速度;(2)物块 A 在车上距离 B 端的最终位置.)

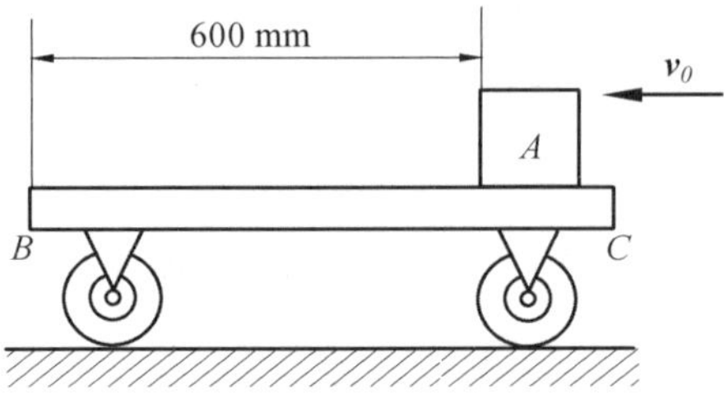

Fig. 10.40 Diagram of Exercise 10 (图 10.40 习题 10 图)

11. In the crank-chute mechanism shown in Fig. 10.41, a homogeneous crank of length l rotates around the O-axis at a uniform angular velocity ω, and the initial state φ is equal to 0°. The mass of the crank is m_1, the mass of the slider A is m_2, and the mass of the guide rod BD is m_3. Point G is its center of mass, and $BG=0.5l$. Find: (1) The motion equation of the mass center G of the mechanism; (2) The maximum horizontal force acting on the O-axis. (在图 10.41所示曲柄滑槽机构中,长为 l 的均质曲柄以匀角速度 ω 绕 O 轴转动,初始状态 φ 的角度为0°. 已知曲柄的质量为 m_1,滑块 A 的质量为 m_2,导杆 BD 的质量为 m_3. 点 G 为其质心,且 $BG=0.5l$. 求:(1)机构质心 G 的运动方程;(2)作用在 O 轴的最大水平力.)

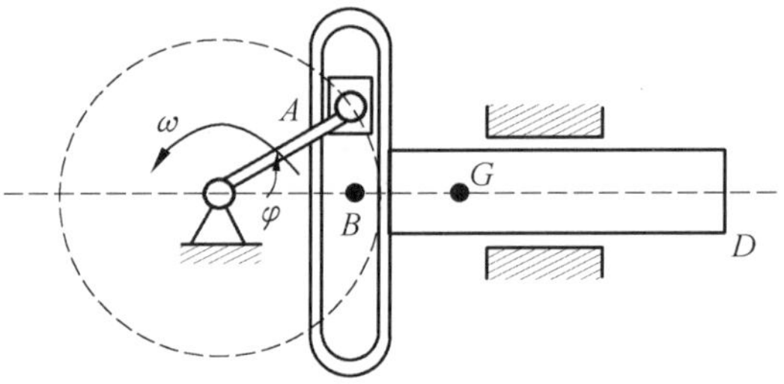

Fig. 10.41 Diagram of Exercise 11 (图 10.41 习题 11 图)

12. The three blocks, whose mass is $m_1=20$ kg, $m_2=15$ kg, $m_3=10$ kg respectively, are connected by a rope, bypassed two fixed pulleys M and N, and placed on the truncated cone $ABED$ with mass $m_4=100$ kg, as shown in Fig. 10.42. When block m_1 falls, block m_2 moves to the right along BE, while block m_3 rises along the slope. If all friction and the mass of the rope are ignored, find the displacement of the truncated cone relative to the ground when the weight m_1 drops by 1 m. (三个物块的质量分别为 $m_1=20$ kg、$m_2=15$ kg、$m_3=10$ kg,由一根绳连接,并绕过两个定滑轮 M 与 N,放在质量 $m_4=100$ kg 的截头锥 $ABED$ 上,如图 10.42 所示. 当物块 m_1下降时,物块 m_2沿着 BE 向右移动,而物块 m_3则沿斜面上升. 若略去一切摩擦和绳子的质量,求当重物 m_1下降 1 m 时,截头锥相对地面的位移.)

13. A homogeneous triprism A is placed on the smooth horizontal plane, and a homogeneous triprism B is placed on the inclined plane, as shown in Fig. 10.43. The cross sections of both triprisms are right triangles, and A has three times the mass of B. Find the distance that triprism

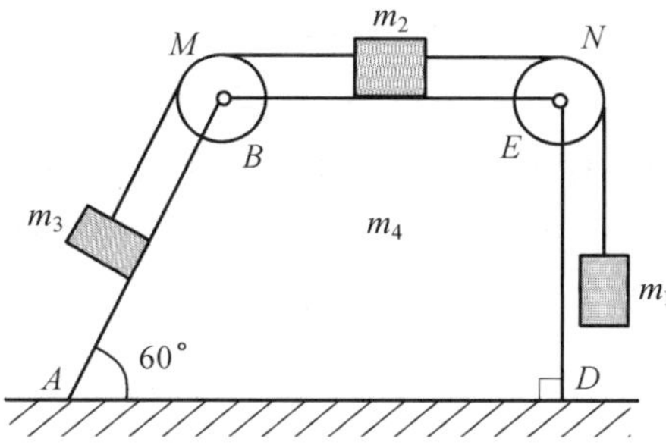

Fig. 10.42 Diagram of Exercise 12 (图 10.42 习题 12 图)

A moves when column B slides down along column A and just touches the horizontal plane. (光滑水平面上放一均质三棱柱 A,在其斜面上又放一均质三棱柱 B,如图 10.43 所示. 两个三棱柱的横截面均为直角三角形,A 的质量为 B 的 3 倍. 求当柱 B 沿柱 A 滑下刚接触水平面时,三棱柱 A 所移动的距离.)

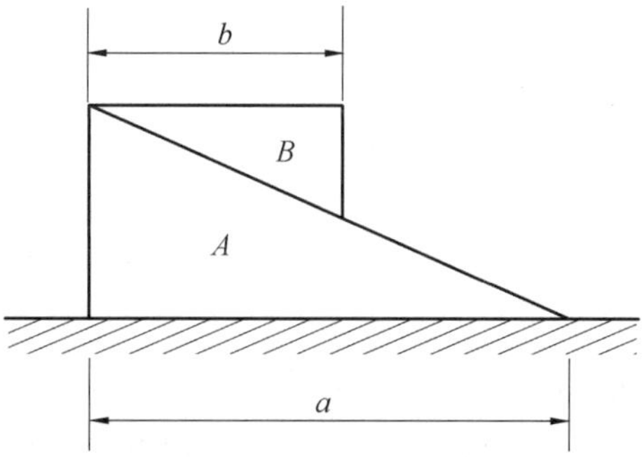

Fig. 10.43 Diagram of Exercise 13 (图 10.43 习题 13 图)

14. A particle with mass m_1 slides down a smooth right-angled triangular prism with an angle of θ and mass of m_2, and the triangular prism can slide freely on a smooth horizontal plane. Find: (1) The acceleration $\ddot{x}_1$ of the particle in the horizontal direction (2) Acceleration $\ddot{x}_2$ of the triangular prism; (3) The reaction force F_1 of the triprism on the particle; (4) Reaction force F_2 of horizontal plane on the triangular prism. (质量为 m_1 的质点,沿斜角为 θ、质量为 m_2 的光滑直角三棱柱滑下,三棱柱又可在光滑水平面上自由滑动,试求:(1)质点沿水平方向的加速度 $\ddot{x}_1$;(2)三棱柱的加速度 $\ddot{x}_2$;(3)三棱柱对质点的反作用力 F_1;(4)水平面对三棱柱的反作用力 F_2.)

15. As shown in Fig. 10.44, the motor weighing $\boldsymbol{P}$ is equipped with an eccentric wheel weighing $\boldsymbol{P}_1$ on the rotating shaft, and the eccentricity is e. The motor rotates at a uniform angular speed of ω. (1) The shell of the motor is fixed on the basis with a screw, find the maximum horizontal shear force acting on the screw; (2) If the motor is not fixed by the screw, at what speed will the motor jump off the ground? (如图 10.44 所示,重力为 $\boldsymbol{P}$ 的电动机,在转动轴上装一重力为 $\boldsymbol{P}_1$ 的偏心轮,偏心距为 e. 电动机以匀角速度 ω 转动. (1)设电动机的外壳用螺杆固定在基础上,求作用在螺杆上的最大水平剪力;(2)若不用螺杆固定,则转速多大时,电动机会跳离地面?)

16. As shown in Fig. 10.45, the homogenizing rod OA rotates in the lead plane around the horizontal axis O, the length of OA is $2l$ and the weight is $\boldsymbol{P}$. When turning to an angle φ to the horizontal line, the angular velocity and angular acceleration are ω and α, respectively. Find the binding force of O terminal at this time. (如图 10.45 所示,均质杆 OA 绕水平轴 O 在铅直面内转动,OA 长为 $2l$,重力为 $\boldsymbol{P}$. 当转到与水平线成 φ 角时,角速度和角加速度分别为 ω 和 α. 求此时 O 端的约束力.)

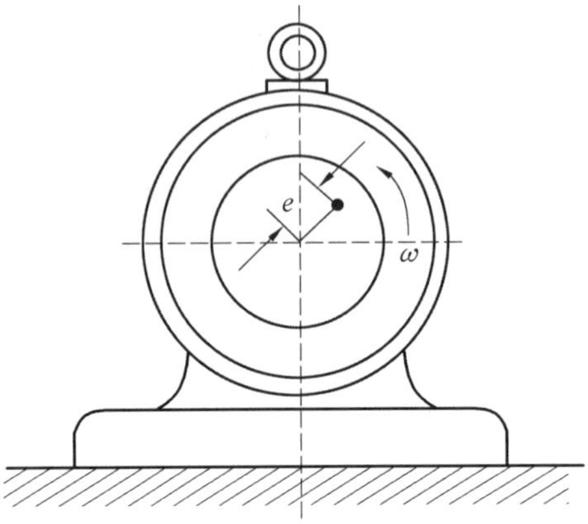

Fig. 10.44 Diagram of Exercise 15

(图 10.44 习题 15 图)

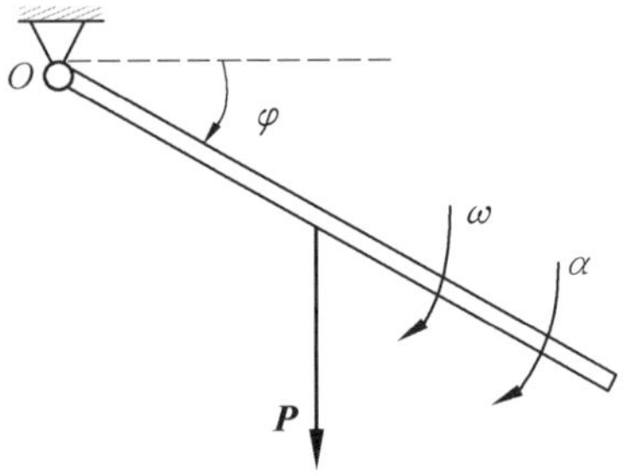

Fig. 10.45 Diagram of Exercise 16

(图 10.45 习题 16 图)

17. A small boat weighing 210 kg was docked on the shore. One person stands at the bow of the ship, with a mass of 70 kg. Estimate that this person can go ashore if he took a step 0.8 m. However, when he took a 0.8 m step, because his center of gravity moved forward by 0.4 m, causing the hull to move back, he stepped into the water and had to wade ashore. Assume that his original estimate of the distance from the shore is correct. Try to find the distance of the person's feet from the shore as he steps through the water. The resistance to a ship is negligible. (岸边停靠一质量为 210 kg 的小船. 设船头站立一人,质量为 70 kg. 此人估计其跨出一步 0.8 m 即可上岸. 不过,当他跨出 0.8 m 的一步时,由于其重心相应地向前移动了 0.4 m,因此船体后移,他却踏入水中,只好涉水上岸. 假设他原来估计离岸边的距离是正确的,试求此人跨步着水时,脚离岸边的距离. 船所受的阻力可以忽略.)

Part Two—— Exercises of the theorem of moment of momentum

(第二部分——动量矩定理习题)

1. A particle with mass m moves in the plane Oxy, and its equation of motion is $x=a\cos pt$, $y=b\sin 2pt$, where a, b and p are constants. Find the moment of momentum of the mass point with respect to the origin O. (质量为 m 的质点在平面 Oxy 内运动,其运动方程为 $x=a\cos pt$, $y=b\sin 2pt$,其中 a、b、p 为常量. 求质点对原点 O 的动量矩.)

2. As shown in Fig. 10.46, a small ball M with mass m attached to one end of the MOA, which passes through a plumb tubular. At first, the ball moves in a circular motion around the tube axis along the radius $MC=R$ and slowly pulls down the OA line segment. (1) If at the

beginning, the speed of the ball is 120 r/min, what is the rotating speed when it moves to $M_1C_1 = 0.5R$? (2) If $m = 1$ kg, $OM = 300$ mm, $R = 150$ mm, and $M_1C_1 = 100$ mm, find the downward pull distance of line segment OA. (图 10.46 所示质量为 m 的小球 M 系于 MOA 的一端,此线穿过一铅垂小管. 开始时,小球绕管轴沿半径 $MC = R$ 做圆周运动,缓慢地向下拉动 OA 线段. 若开始时,小球的转速为 120 r/min,求当其运动到 $M_1C_1 = 0.5R$ 时,转速为多少?

3. A ball of mass $m = 1$ kg rotates around a vertical axis at an angular velocity of $\omega_0 = 15$ rad/s. The ball is attached to a straight axis with two non-extendable strings with $l = 0.6$ m, $\theta = 30°$, as shown in Fig. 10.47. If the slider A moves 0.15 m upward, find the angular velocity ω of the ball. (质量 $m = 1$ kg 的小球以 $\omega_0 = 15$ rad/s 的角速度绕铅直轴转动. 该球用两根长 $l = 0.6$ m不可伸长的绳连接在铅直的轴上,$\theta = 30°$,如图 10.47 所示. 若滑块 A 向上移动 0.15 m,求此球的角速度 ω.)

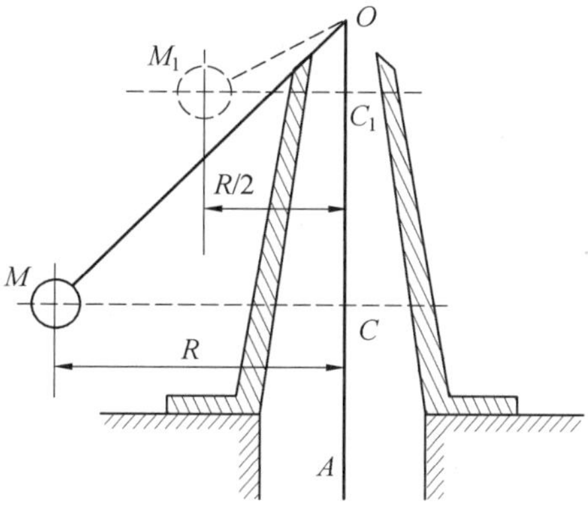

Fig. 10.46 Diagram of Exercise 2
(图 10.46 习题 2 图)

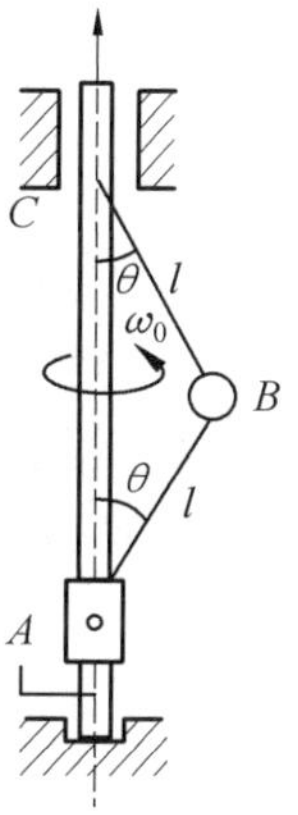

Fig. 10.47 Diagram of Exercise 3
(图 10.47 习题 3 图)

4. A homogeneous disk of mass m_1 and radius R is placed inside a vertical plane and rotated around its vertical diameter AB at an initial angular velocity ω_0. Another particle of mass m_2 starts at position A and moves along the edge of the disc. Find the angular velocity of the disc when the distance of the particle to the rotation axis AB is the maximum. (质量为 m_1,半径为 R 的均质圆盘,置于铅直面内,绕其铅垂直径 AB 以初角速度 ω_0转动. 另有一质量为 m_2的质点从位置 A 开始沿圆盘的边缘运动. 求质点运动到离转轴 AB 的距离为最大时圆盘的角速度.)

5. A small ball A of mass m is attached to a rod AB of length l and placed in a container with liquid, as shown in Fig. 10.48. The rod rotates around the vertical axis O_1O_2 at the initial velocity ω_0, and the resistance of the liquid is proportional to the product $m\omega$ of the mass of the ball and the angular velocity, that is, $F = km\omega$, where k is the constant of proportionality. How long does it take for the angular velocity to decrease to half of the initial angular velocity? (质量为 m 的小球 A,连接在长为 l 的杆 AB 上,并被放在盛有液体的容器内,如图 10.48 所示. 杆以初速度 ω_0 绕铅直轴 O_1O_2转动,液体的阻力与小球质量和角速度的乘积 $m\omega$ 成正比,即 $F = km\omega$,其中 k 是比例常数. 经过多少时间,角速度减为初角速度的一半?)

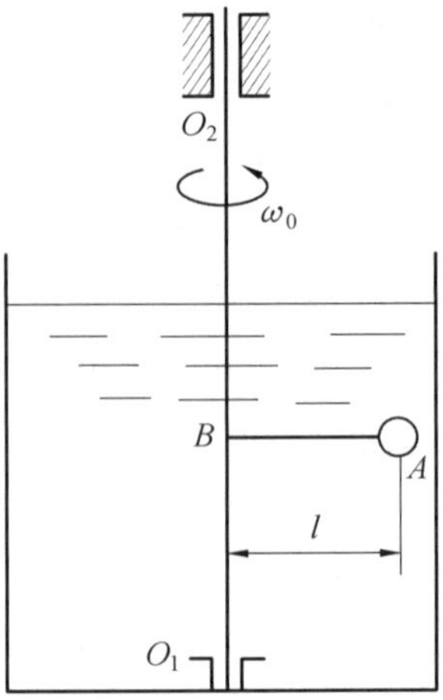

Fig. 10.48　Diagram of Exercise 5(图 10.48　习题 5 图)

6. As shown in Fig. 10.49, A frame supports a 5 kg ball A, which is initially at rest. A moment $M=3t$ is applied to the CD axis, and a horizontal force $F=10$ N is applied vertically to the transverse arm AB, where M is measured in N · m and t is measured in s. If the mass of the frame is negligible, find the velocity of ball A at $t=4$ s. (如图 10.49 所示,一构架支撑着一个 5 kg的小球 A,初始处于静止状态. 一力矩 $M=3t$ 作用于 CD 轴上,一水平力 $F=10$ N 垂直作用于横臂 AB 上,其中 M 以 N · m 计,t 以 s 计. 假如构架的质量可以忽略,求 $t=4$ s 时小球 A 的速度.)

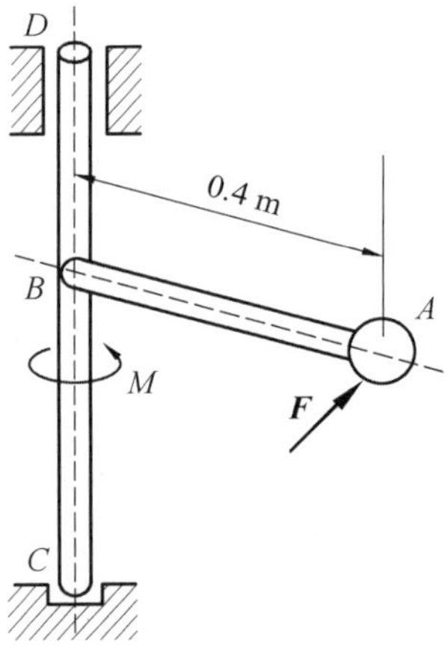

Fig. 10.49　Diagram of Exercise 6(图 10.49　习题 6 图)

7. As shown in Fig. 10.50, the homogeneous thin rod OA has a mass of m and a length of l, and rotates at a uniform angular velocity around a fixed axis Oz. Let the angle between the rod and Oz axis be θ, find the moment of momentum of the rod with respect to x, y, z axis and O point at the moment the rod just moves to Oyz plane. (图 10.50 所示均质细杆 OA 的质量为 m,长为 l,绕定轴 Oz 以匀角速度转动. 设杆与 Oz 轴夹角为 θ,求当杆运动到 Oyz 平面内的瞬时,杆对 x、y、z 轴及 O 点的动量矩.)

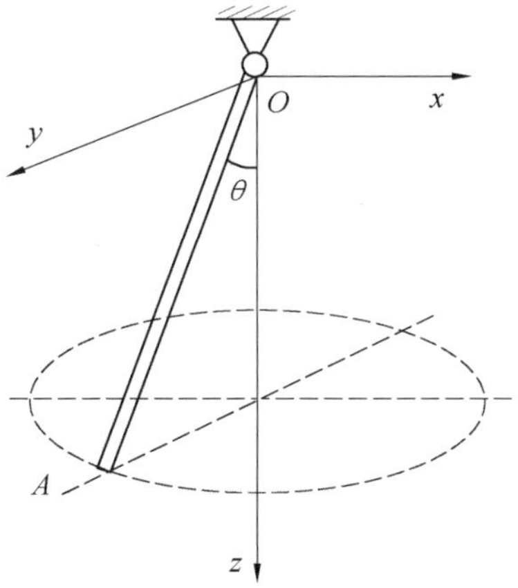

Fig. 10.50 Diagram of Exercise 7(图 10.50 习题 7 图)

8. As shown in Fig. 10.51, the right angle ruler ADB can be rotated around its plumb edge AD, and there is an object E with mass m on the side of BD. At the beginning, the system rotates around the AD axis at the angular velocity ω, and the object E is a from point D. Set the moment of inertia of the ruler to the AD axis as J, find the relationship between the angular velocity ω of the ruler and the distance $x=ED$. (图 10.51 所示直角曲尺 ADB 可绕其铅垂边 AD 旋转,在 BD 边上有一质量为 m 的物体 E. 开始时,系统以角速度 ω 绕 AD 轴转动,物体 E 距 D 点为 a. 设曲尺对 AD 轴的转动惯量为 J,求曲尺转动的角速度 ω 与距离 $x = ED$ 之间的关系.)

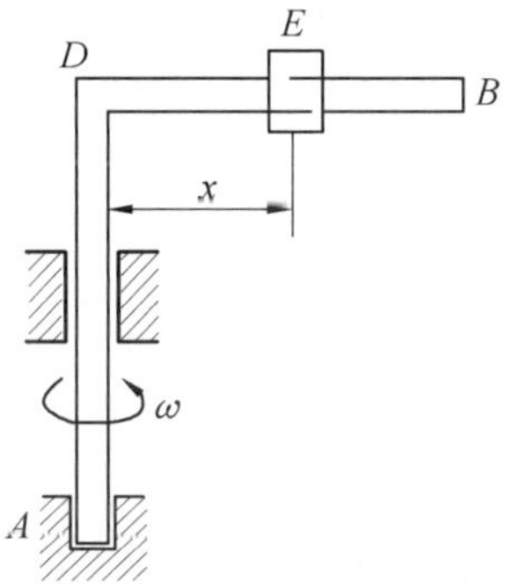

Fig. 10.51 Diagram of Exercise 8(图 10.51 习题 8 图)

9. A homogeneous disc of radius R and mass m_1 can be rotated about a plumb axis passing through its center without considering friction. A person of mass m walks by $s = 0.5at^2$ from point B along a circle of radius r from the O-axis, as shown in Fig. 10.52. At the beginning, the disc and the person is at rest, find the angular velocity and angular acceleration of the disc. (一半径为 R,质量为 m_1 的均质圆盘,可绕通过其中心的铅垂轴转动,摩擦不计. 另一质量为 m 的人由 B 点按规律 $s = 0.5at^2$ 沿距 O 轴半径为 r 的圆周行走,如图 10.52 所示. 开始时,圆盘和人均静止,求圆盘的角速度和角加速度.)

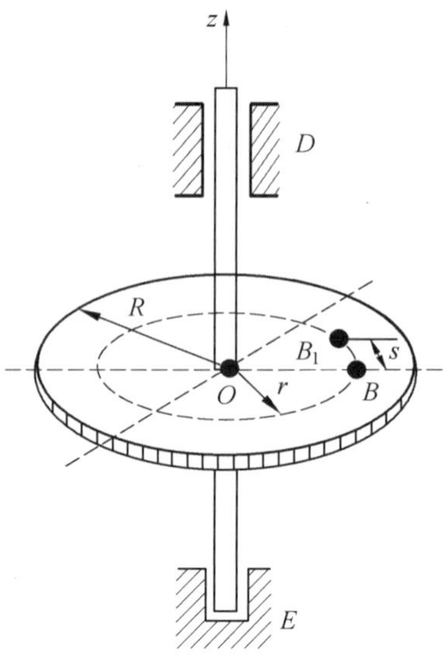

Fig. 10.52 Diagram of Exercise 9(图 10.52 习题 9 图)

10. As shown in Fig. 10.53, a homogeneous rectangular sheet of mass m has side lengths a and b. Find the moment of inertia J_x and J_y of the plate with respect to each edge, and its moment of inertia $J_{z_{C'}}$ with respect to the axis through centroid $z_{C'}$ perpendicular to the plate surface. (如图 10.53 所示,质量为 m 的均质矩形薄板,其边长分别为 a 和 b. 求薄板对于每条边的惯性矩 J_x 和 J_y,以及它对于与板面垂直的质心轴 $z_{C'}$ 的惯性矩 $J_{z_{C'}}$.)

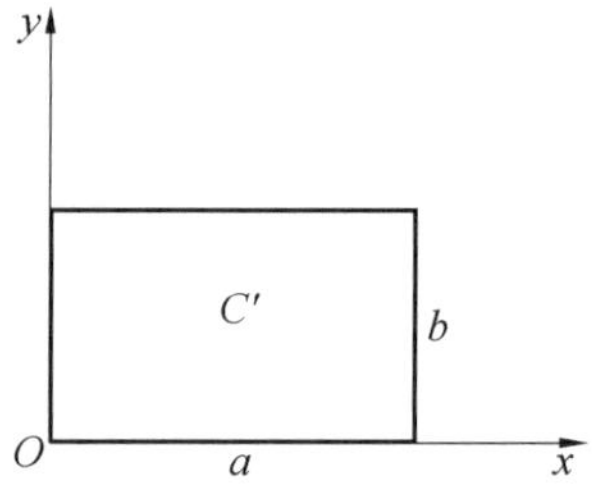

Fig. 10.53 Diagram of Exercise 10(图 10.53 习题 10 图)

11. As shown in Fig. 10.54, a rope crosses over a trolley, regardless of the mass of the trolley, a person of mass m holds the point A of the rope, and the point B of the rope is tied to an object with the same mass m. If a person climbs along the rope at a relative speed v_r, how will the object move and its speed (regardless of the mass of the rope and bearing friction)? (图 10.54 所示一绳跨过一滑车,滑车质量不计,质量为 m 的人抓住绳的 A 点,而绳的 B 点系有一等质量的物体. 若人沿绳以相对速度 v_r 上爬,则此物体将怎样运动,它的速度如何(不计绳的质量及轴承摩擦)?)

12. Two weights, A and B, of mass m_1 and m_2 respectively, are tied to two ropes respectively, which in turn are wound around the drum of radius r_1 and r_2 respectively, as shown in Fig. 10.55. The mass of the drum and rope and the friction of the shaft are ignored. Find the

angular acceleration of the drum. (质量分别为 m_1 和 m_2 的两重物 A 和 B,各系在两条绳上,这两条绳又分别绕在半径为 r_1 和 r_2 的鼓轮上,如图 10.55 所示. 鼓轮和绳的质量及轴的摩擦均略去不计. 求鼓轮的角加速度.)

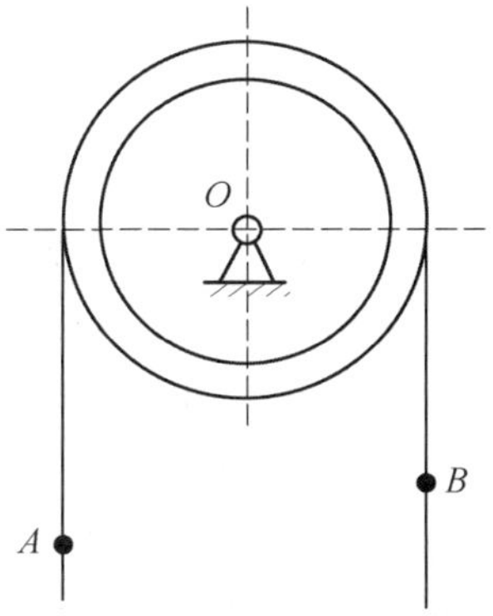

Fig. 10.54 Diagram of Exercise 11
(图 10.54 习题 11 图)

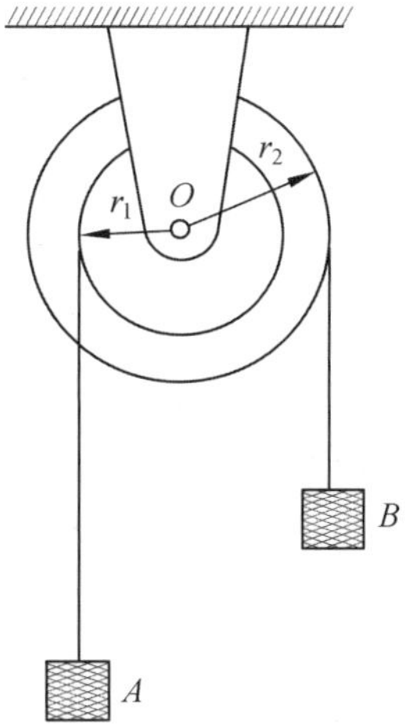

Fig. 10.55 Diagram of Exercise 12
(图 10.55 习题 12 图)

13. As shown in Fig. 10.56, in order to obtain the moment of inertia of flywheel A with radius $R=500$ mm through its centroid axis, a thin rope is wrapped around the flywheel, and a heavy hammer with mass $m_1=8$ kg is tied to the end of the rope. The heavy hammer falls from the height $h=2$ m, and the falling time $t_1=16$ s is measured. In order to eliminate the influence of bearing friction, a heavy hammer of $m_2=4$ kg was used for the second test, and the time of falling from the same height was $t_2=25$ s. Assuming that the friction moment is a constant and independent of the mass of the heavy hammer, find the moment of inertia of the flywheel with respect to its centroid axis. (如图 10.56 所示,为求半径 $R=500$ mm 的飞轮 A 对于通过其质心轴的惯性矩,在飞轮上绕一细绳,绳的末端系一质量为 $m_1=8$ kg 的重锤,重锤自高度 $h=2$ m 处落下,测得落下时间 $t_1=16$ s. 为消去轴承摩擦的影响,再用质量为 $m_2=4$ kg 的重锤做第二次试验,此重锤自同一高度处落下的时间为 $t_2=25$ s. 假定摩擦力矩是一常数,且与重锤的质量无关,求飞轮通过其质心轴的惯性矩.)

14. As shown in Fig. 10.57, the flywheel rotates around a fixed axis under the action of moment $M_0\cos\omega t$. There are two objects with the same mass m along the spoke of the flywheel, each of which moves periodically. What condition must the distance r satisfy in order for the flywheel to rotate uniformly at the angular velocity ω? (图 10.57 所示飞轮在力矩 $M_0\cos\omega t$ 作用下绕定轴转动,沿飞轮辐条有质量为 m 的两等质量物体,各做周期性运动. 则距离 r 应满足什么条件,才能使飞轮以角速度 ω 匀速转动?)

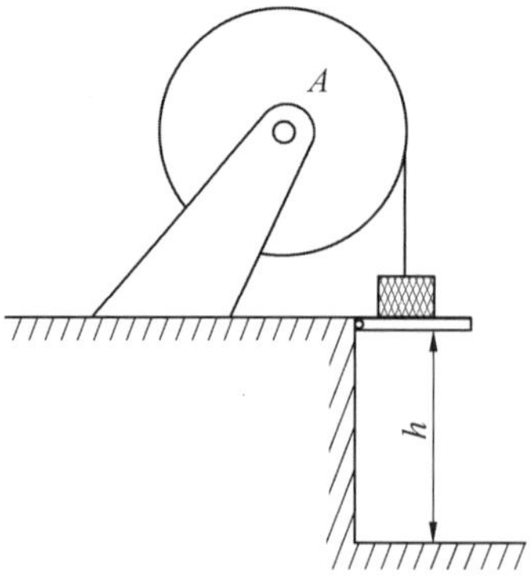

Fig. 10.56 Diagram of Exercise 13
(图 10.56 习题 13 图)

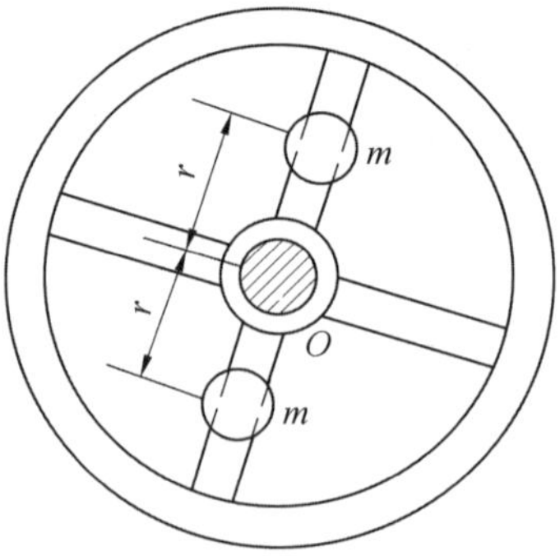

Fig. 10.57 Diagram of Exercise 14
(图 10.57 习题 14 图)

15. As shown in Fig. 10.58, to obtain an upward acceleration of 400 mm/s^2 for the object B with a mass of 8 kg, please calculate the force $\boldsymbol{F}_{TA}$ acting at the end of A. Assume that the mass of the homogeneous disc is 20 kg, the radius is 150 mm, and the rope has no relative slide on the surface of the disc. Calculate the tension of the rope on the object B and to explain that its tension is different from that of the rope at the end of A. (如图 10.58 所示,为使质量为 8 kg 的物体 B 获得 400 mm/s^2 向上的加速度,求作用于绳索 A 端的力 $\boldsymbol{F}_{TA}$. 假设均质圆盘质量为 20 kg,半径为 150 mm,绳索在圆盘表面上无相对滑动. 计算物体 B 上所受的绳的张力,说明此张力不同于 A 端绳的拉力.)

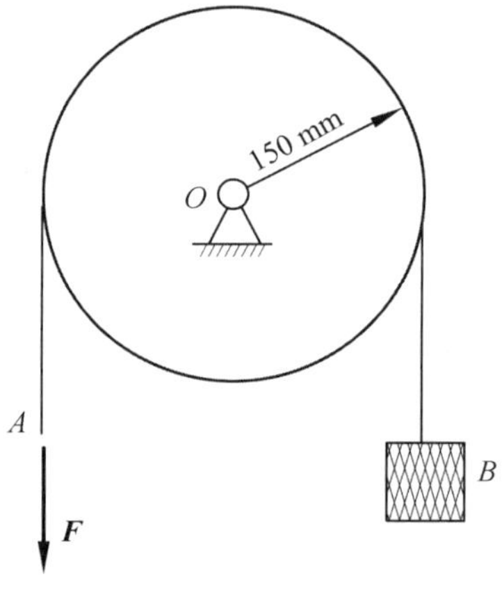

Fig. 10.58 Diagram of Exercise 15(图 10.58 习题 15 图)

16. There are two different objects, one is a homogeneous rod with a mass of m and a length of l; the other is a ball of mass m, consolidated on the rod end of a light rod of length l with negligible self-weight, as shown in Fig. 10.59. Both are hinged on a fixed horizontal plane and released at the same slightly inclined position. Which rod gets to the horizontal position first? Why? (有两个不同物体,一为均质细杆,其质量为 m,长为 l;另一为质量 m 的小球,固结于可忽略自重的长为 l 的轻杆的杆端,如图 10.59 所示. 二者均铰接于固定水平面上,并在同一微小倾斜位置释放. 哪一个先到达水平位置? 为什么?)

17. The pendulum of the clock shown in Fig. 10.60 consists of a rod and a disc. Rod length

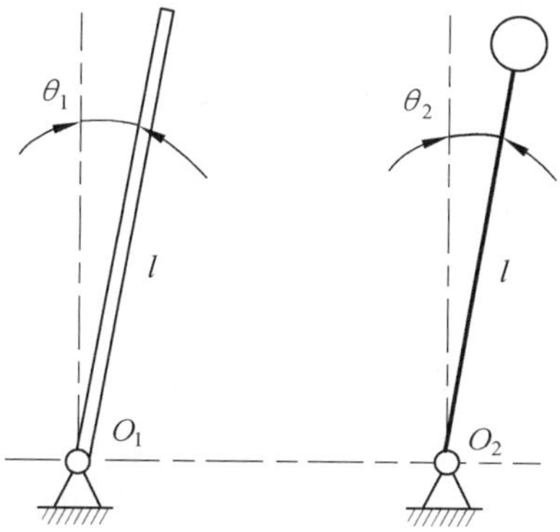

Fig. 10.59 Diagram of Exercise 16(图 10.59 习题 16 图)

l =1 m, mass m_1 =4 kg; the radius of the disc is R=0.2 m and the mass m_2 =6 kg. If the rod and the disc are considered homogeneous, find the moment of inertia of the pendulum with respect to the O-axis.(图 10.60 所示钟表的摆由杆和圆盘组成. 杆长 l=1 m,质量 m_1 =4 kg;圆盘的半径 R=0.2 m,质量 m_2= 6 kg. 若杆和圆盘视为均质,求摆对于 O 轴的惯性矩.)

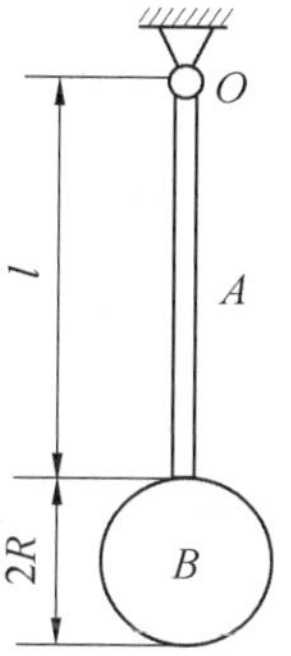

Fig. 10.60 Diagram of Exercise 17(图 10.60 习题 17 图)

18. In the two systems shown in Fig. 10.61, the OA bar is hinged at the O end, and the OA bar is in a horizontal position at point B due to the action of a straight spring. The spring stiffness coefficient is k, and a, l are known in the Fig. 10.61. In Fig. 10.61(a), the mass of the OA rod is m, regardless of the mass of the ball A. The OA in Fig. 10.61(b) is a homogeneous thin rod with a mass of m. If the rod makes a slight swing in the vertical plane, find the free vibration period of the above two systems.(在图 10.61 所示的两系统中,OA 杆在 O 端铰接,在 B 点因铅直弹簧的作用而使 OA 杆处于水平位置. 弹簧刚度系数为 k,图中 a、l 已知. 图 10.61(a)中

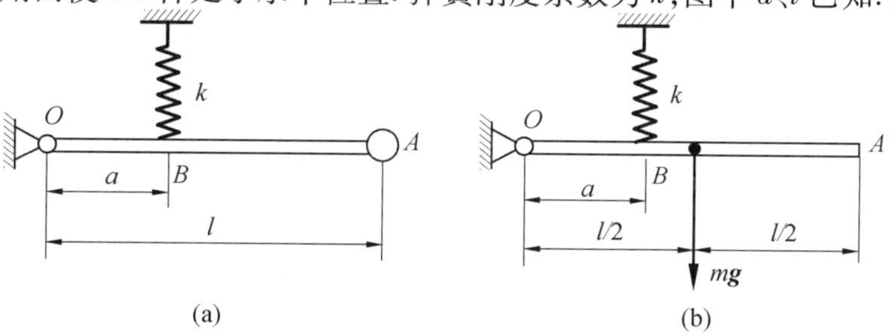

Fig. 10.61 Diagram of Exercise 18(图 10.61 习题 18 图)

OA 杆质量不计,小球 A 的质量为 m. 图 10.61(b)中的 OA 为均质细杆,其质量为 m. 若杆在铅垂面内做微小摆动,求上述两系统自由振动的周期.)

19. As shown in Fig. 10.62, the mass of the homogeneous roller is m, the radius is R, and the radius of rotation to its centroid axis C is ρ. The roller rests on a horizontal surface and is subjected to a horizontal tension $\boldsymbol{F}$. Assume the height of the action line of the tension be h, the roller only rolls and does not slide, and the rolling friction is ignored. Please solve the static friction force $\boldsymbol{F}_s$ and analyze the relationship between the magnitude, direction of $\boldsymbol{F}_s$ and the height h. (图 10.62 所示均质滚子的质量为 m,半径为 R,对其质心 C 轴的回转半径为 ρ. 滚子静止在水平面上,且受一水平拉力 $\boldsymbol{F}$ 作用. 设拉力 $\boldsymbol{F}$ 的作用线的高度为 h, 滚子只滚动不滑动,滚动摩阻忽略不计. 求静滑动摩擦力 $\boldsymbol{F}_s$,并分析 $\boldsymbol{F}_s$ 的大小和方向与高度 h 的关系.)

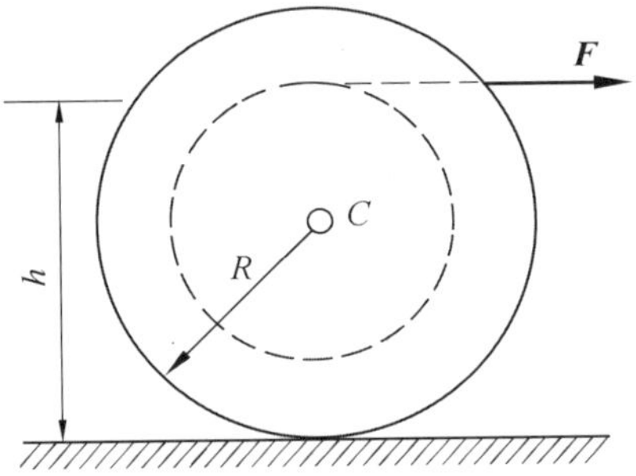

Fig. 10.62　Diagram of Exercise 19(图 10.62　习题 19 图)

20. As shown in Fig. 10.63, homogeneous cylinders A and B have the same mass of m and radius of r. A rope is wrapped on the cylinder A rotating around a fixed axis O, and the other end of the rope is wrapped on the cylinder B, the segment of the straight rope is in vertical position, and friction is not considered. Find: (1) The acceleration of the mass center B when cylinder B falls; (2) If A counterclockwise couple with moment M is applied to the cylinder A, under what conditions will the acceleration of its mass center of cylinder B be upward? (如图 10.63 所示,均质圆柱体 A 和 B,两圆柱质量均为 m,半径均为 r,一绳缠在绕固定轴 O 转动的圆柱 A 上,

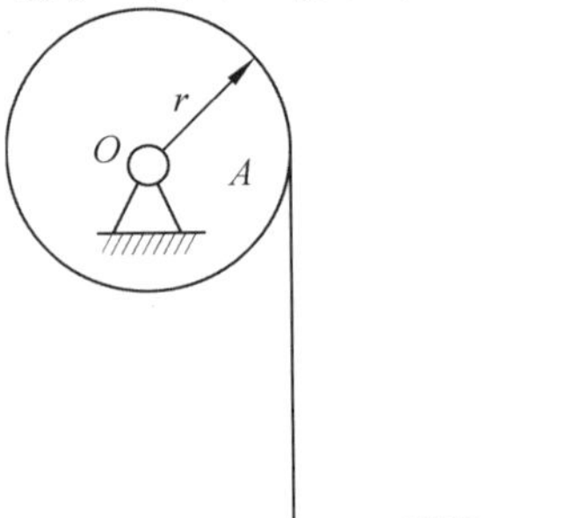

Fig. 10.63　Diagram of Exercise 20(图 10.63　习题 20 图)

绳的另一端绕在圆柱 B 上，直线绳段铅垂，摩擦不计. 求：(1)圆柱体 B 下落时质心的加速度；(2)若在圆柱体 A 上作用一逆时针转向、矩为 M 的力偶，则在什么条件下圆柱体 B 的质心加速度将向上？)

Part three——kinetic energy theorem part of the exercise

(第三部分——动能定理部分习题)

1. The natural length of a spring is $l_0=100$ mm, and the stiffness coefficient $k=0.5$ N/mm. One end of the spring is fixed on the O point of the circular axis with radius $R=100$ mm, and the other end is pulled from point B to point A as shown in Fig. 10.64, where OA and BC are perpendicular, OA is the cross section diameter of circular axis. Find the work done by the spring force. (一弹簧自然长度 $l_0=100$ mm，刚度系数 $k=0.5$ N/mm，一端固定在半径 $R=100$ mm 的圆轴的 O 点上，另一端由图 10.64 所示的 B 点拉至 A 点. 其中 OA 与 BC 垂直，OA 为圆轴截面直径. 求弹簧力所做的功.)

2. As shown in Fig. 10.65, particle M moves along the trajectory $\frac{x^2}{25}+\frac{y^2}{9}=1$. Find the work done by a force $\boldsymbol{F}=-5x\mathbf{i}-5y\mathbf{j}$ on the path from $M_0(5,0)$ to $M_1(0,3)$ (unit of $\boldsymbol{F}$ is N, unit of distance is m). (图 10.65 所示质点 M 沿轨迹 $\frac{x^2}{25}+\frac{y^2}{9}=1$ 运动. 求其上某一作用力 $\boldsymbol{F}=-5x\mathbf{i}-5y\mathbf{j}$ (F 以 N 计，距离以 m 计) 在由 $M_0(5,0)$ 至 $M_1(0,3)$ 的路程上所做的功.)

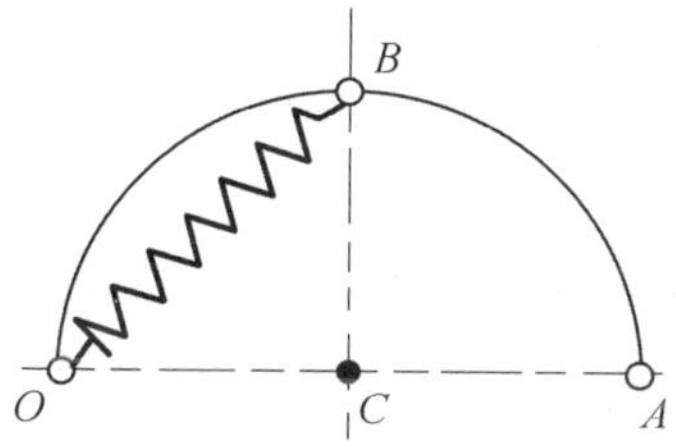

Fig. 10.64 Diagram of Exercise 1
(图 10.64 习题 1 图)

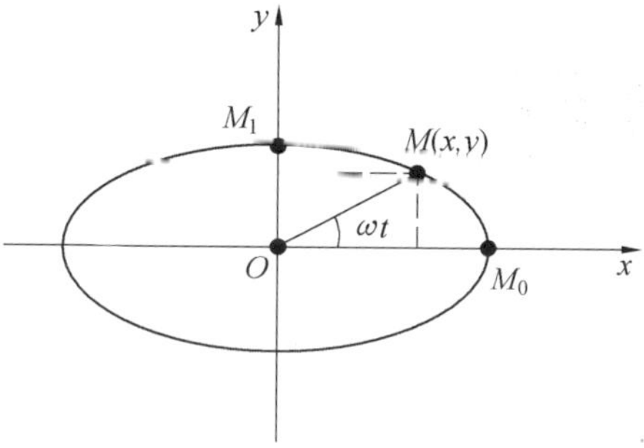

Fig. 10.65 Diagram of Exercise 2
(图 10.65 习题 2 图)

3. As shown in Fig. 10.66, the two springs are connected by strips of cloth, the initial tension of the spring is 600 N, and both the stiffness coefficient of the springs are $k=2$ N/mm. An object M with a mass of 40 kg falls freely from high h, and after the object falls on the strip, the its maximum distance of subsidence is 1 m. The mass of the spring and the strip are omitted, find the height h. (如图 10.66 所示，两根弹簧用布条连在一起，弹簧的拉力最初为 600 N，弹簧的刚度系数均为 $k=2$ N/mm. 质量为 40 kg 的物体 M 从高 h 处自由落下，重物落在布条上以后，下沉的最大距离为 1 m. 弹簧与布条的质量均略去，求高度 h.)

4. As shown in Fig. 10.67, the slide block with mass of 10 kg can slide along the vertical

guide bar CD, The block is initially static at position A and now pulled by a rope. Given that the tension of the rope is $F=400$ N, and the friction can be omitted everywhere. Find the velocity of the block arriving at position B. (如图 10.67 所示,质量为 10 kg 的滑块可沿铅直导杆 CD 滑动,最初静置于 A 处,现在用绳拉动. 已知绳的拉力 $F=400$ N,各处的摩擦均可略去. 求物块到达 B 处时的速度.)

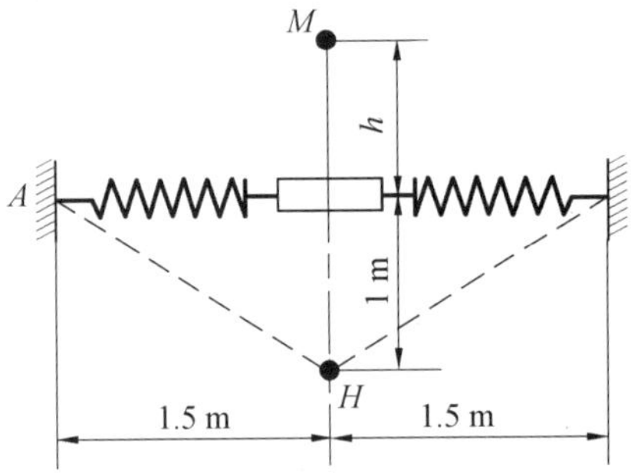

Fig. 10.66 Diagram of Exercise 3

(图 10.66 习题 3 图)

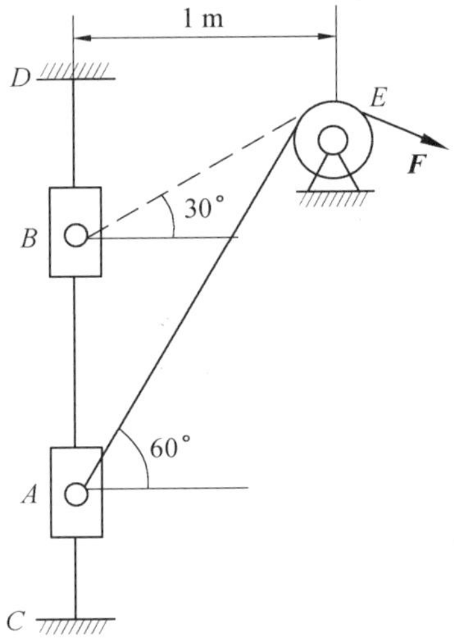

Fig. 10.67 Diagram of Exercise 4

(图 10.67 习题 4 图)

5. Two kinds of energy-absorbing shock and shock prevention devices used in wharfs can withstand loads with little deformation, and their force-deformation curves are shown in Fig. 10.68. When there is no planimeter, the force-deformation curve equation for the second device can be assumed to be $F=31.4x^{1.6}$, where x is measured in mm and F is measured in N. If a ship weighing 100 kN hits the anti-shock device at a speed of 1.8 km/h, find the maximum deformation of each device when the ship is stationary. (用于码头的两种能量吸震防冲装置可以承受变形不大的载荷,它们的力-变形曲线如图 10.68 所示. 当没有面积仪时,第二种装置的力-变形曲线方程可假设为 $F=31.4x^{1.6}$,其中 x 以 mm 计,F 以 N 计. 如果一艘重 100 kN 的轮船以 1.8 km/h 的速度撞击防冲装置,求当轮船静止下来时,每种装置的最大变形量.)

6. A homogeneous cylinder of radius r and mass m rolls without sliding inside a fixed cylinder, as shown in Fig. 10.69. The radius of the fixed cylinder is R, express the kinetic energy of the cylinder as a function of angle φ. (半径为 r,质量为 m 的均质圆柱体在固定的圆柱面内滚动而不滑动,如图 10.69 所示. 固定圆柱面的半径为 R,请将圆柱体的动能表达为 φ 角的函数.)

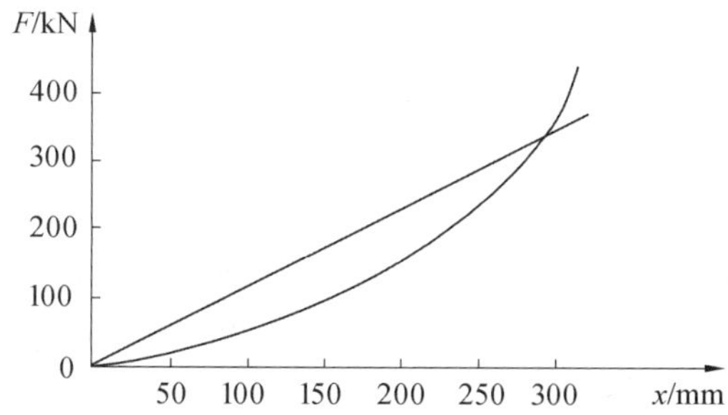

Fig. 10.68 Diagram of Exercise 5
(图 10.68 习题 5 图)

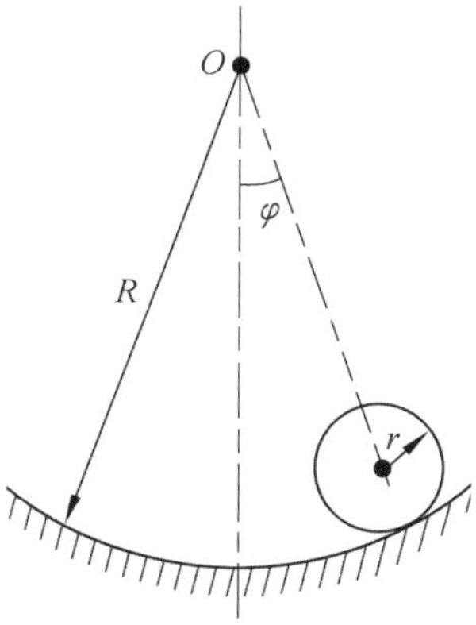

Fig. 10.69 Diagram of Exercise 6
(图 10.69 习题 6 图)

7. The planetary gear train located in the horizontal plane, as shown in Fig. 10.70, is driven by the connecting rod OA, which rotates at the angular velocity ω. The connecting rod is connected to the axes of three identical gears, and the gear Ⅰ is fixed. The mass of each gear is m_1, the radius is r, and the mass of the connecting rod is m_2. Assuming that the gear and connecting rod are homogeneous object. (1) Calculate the kinetic energy of the planetary gear mechanism; (2) What is the work done by the force acting on gear Ⅲ? (如图 10.70 所示,位于水平面内的行星轮系由连杆 OA 带动,连杆以角速度 ω 转动.此连杆与三个相同的齿轮的轴相连,齿轮Ⅰ是固定的.每个齿轮的质量为 m_1,半径为 r,连杆的质量为 m_2.设齿轮及连杆皆为均质物体.(1)计算此行星轮系机构的动能;(2)作用于齿轮Ⅲ之力的功等于多少?)

8. Two blocks are suspended from the pulley block as shown in Fig. 10.71, where the mass of block A is $m_A = 30$ kg and the mass of B is $m_B = 10$ kg. The radius of fixed pulley O_1 is

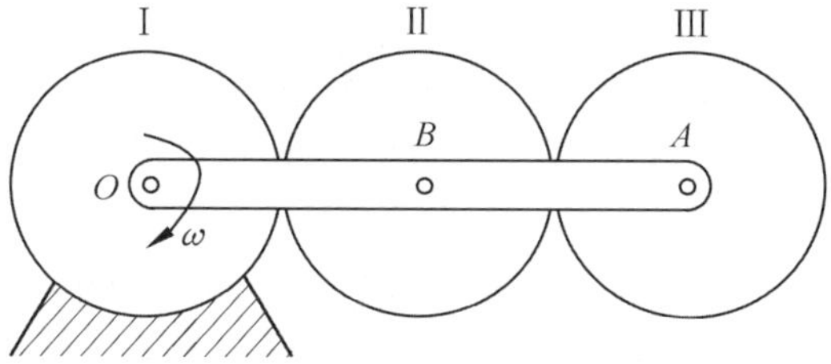

Fig. 10.70 Diagram of Exercise 7
(图 10.70 习题 7 图)

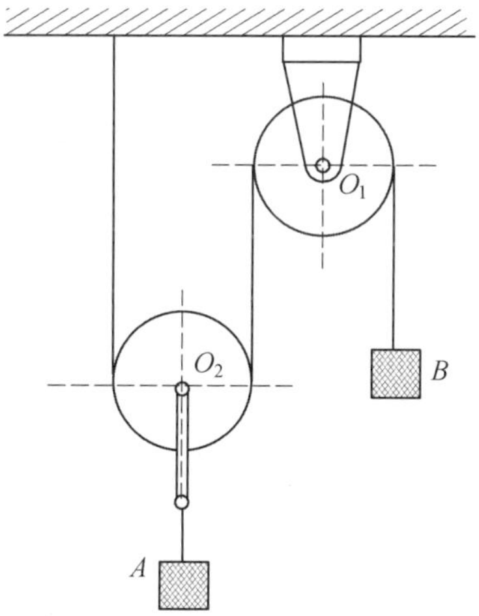

Fig. 10.71 Diagram of Exercise 8
(图 10.71 习题 8 图)

$r_1 = 0.1$ m, the mass $m_1 = 3$ kg. The radius of the pulley O_2 is $r_2 = 0.1$ m, and the mass $m_2 = 4$ kg. The two pulleys are assumed to be homogeneous discs, and the rope weight and friction are ignored. Find the velocity of weight A falling distance $h = 0.5$ m from rest. (在图 10.71 所示滑轮组中悬挂两个物块,其中 A 的质量 $m_A = 30$ kg,B 的质量 $m_B = 10$ kg. 定滑轮 O_1 的半径 $r_1 = 0.1$ m,质量 $m_1 = 3$ kg. 动滑轮 O_2 的半径 $r_2 = 0.1$ m,质量 $m_2 = 4$ kg. 设两滑轮均视为均质圆盘,绳重和摩擦都略去不计. 求重物 A 由静止下降距离 $h = 0.5$ m 时的速度.)

9. In Fig. 10.72, there are three balls A, B, C with negligible size, the same mass of 1 kg. Ball A and B can slide freely on the smooth horizontal bar, and ball C is connected with A and B with two thin wires 1 m long, and the thin wires cannot be extended. If the three balls are released at rest when they are in an equilateral triangle position, at what speed the spheres A and B will collide? (图 10.72 中,有 A、B、C 三个可以忽略大小的球,质量均为 1 kg. 球 A、B 可在光滑水平杆上自由滑动,球 C 则用两根长 1 m 的细线与 A、B 连接,细线不可伸长. 若三球在成等边三角形的位置时静止释放,则球 A、B 将以多大的速度碰撞?)

10. As shown in Fig. 10.73 the moment of inertia of shaft Ⅰ and shaft Ⅱ together with the belt wheels and gears installed on them are $J_1 = 5\ \text{kg}\cdot\text{m}^2$ and $J_2 = 4\ \text{kg}\cdot\text{m}^2$ respectively, the transmission ratio $\frac{\omega_2}{\omega_1} = \frac{3}{2}$, and the system starts to move from rest under the action of the torque $M = 50\ \text{N}\cdot\text{m}$ on shaft Ⅰ. After how many revolutions, the rotational speed of axis Ⅱ can reach $n_2 = 120$ r/min? (图 10.73 所示轴Ⅰ和轴Ⅱ连同安装在其上的带轮和齿轮的转动惯量分别为 $J_1 = 5\ \text{kg}\cdot\text{m}^2$ 和 $J_2 = 4\ \text{kg}\cdot\text{m}^2$,传动比 $\frac{\omega_2}{\omega_1} = \frac{3}{2}$,在轴Ⅰ上的转矩 $M = 50\ \text{N}\cdot\text{m}$ 作用下,系统由静止开始运动. 则轴Ⅱ经过多少转后,它的转速可达 $n_2 = 120$ r/min?)

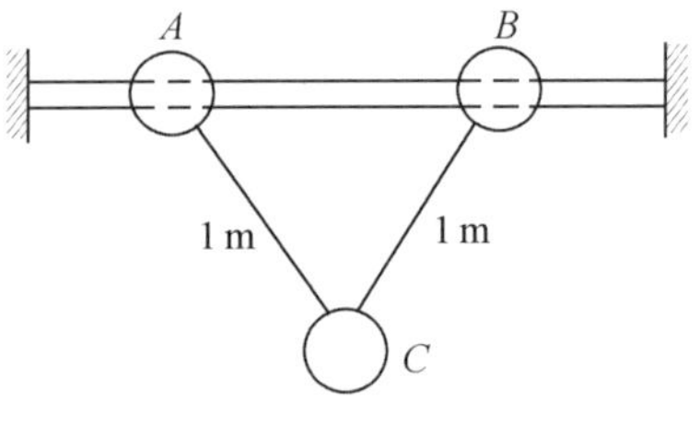

Fig. 10.72 Diagram of Exercise 9
(图 10.72 习题 9 图)

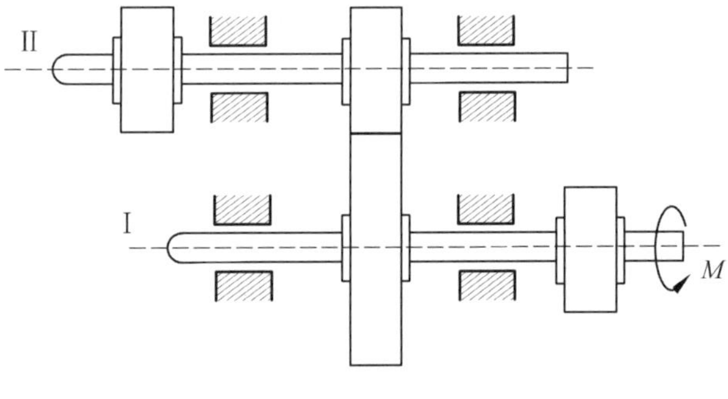

Fig. 10.73 Diagram of Exercise 10
(图 10.73 习题 10 图)

11. As shown in Fig. 10.74, when the punch presses the workpiece, the average working resistance $F = 520$ kN, working stroke $s = 10$ mm, and for the punch its moment of inertia $J = 39.2\ \text{kg}\cdot\text{m}^2$, and speed $n_0 = 415$ r/min. Assuming that all the energy required for stamping workpiece is supplied by the flywheel, calculate the speed of the flywheel after the stamping is completed. (如图 10.74 所示,冲床冲压工件时冲头所受的平均工作阻力 $F = 520$ kN,工作行程 $s = 10$ mm,飞轮的转动惯量 $J = 39.2\ \text{kg}\cdot\text{m}^2$,转速 $n_0 = 415$ r/min. 假定冲压工件所需的全部能量都由飞轮供给. 计算冲压结束后飞轮的转速.)

12. The homogeneous disc A rotates about the O-axis, and A has a mass of m_A and a radius of r. A string is wound around the disc and a block B of mass m_B is hung at the string end, as shown in Fig. 10.75. At the point D of $OD=e$ from the center of the disc, a spring is consolidated in the horizontal direction to restrain it. The stiffness coefficient of the spring is k. The system is in equilibrium at the shown position, regardless of the mass of rope and spring. Try to obtain the differential equation of the system with small motion. (均质圆盘 A 绕 O 轴转动,A 的质量为 m_A,半径为 r. 一细绳绕在盘上,其端部挂一质量为 m_B 的物块 B,如图 10.75 所示. 今在距盘心上部 $OD=e$ 的 D 点处,沿水平方向固结一弹簧加以约束,弹簧的刚度系数为 k. 该系统在图示位置处于平衡状态,绳及弹簧质量均不计. 试求该系统做微小运动的微分方程.)

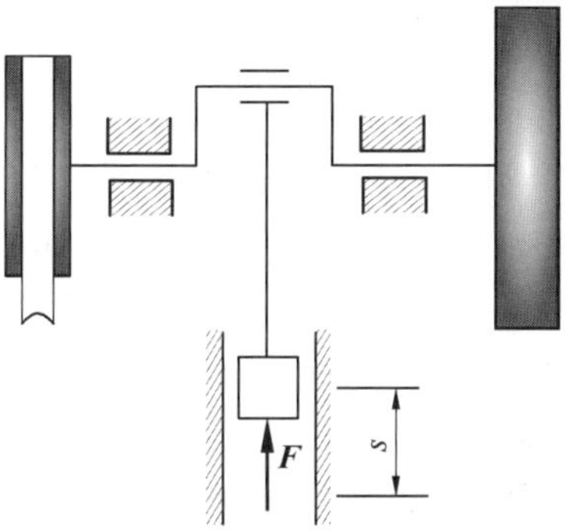

Fig. 10.74 Diagram of Exercise 11
(图 10.74 习题 11 图)

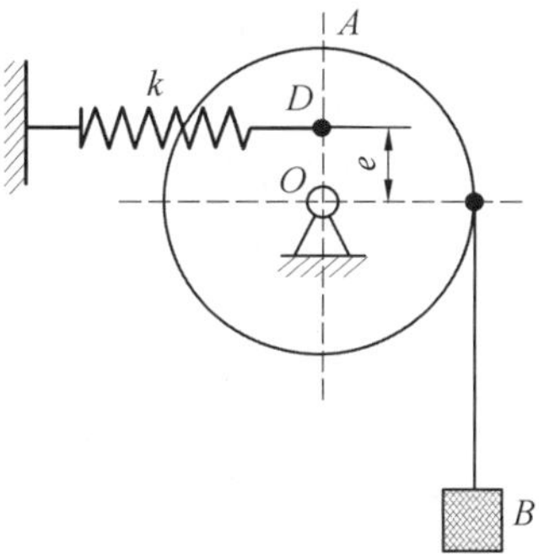

Fig. 10.75 Diagram of Exercise 12
(图 10.75 习题 12 图)

13. As shown in Fig. 10.76, a planetary gear mechanism with external meshing is located in the horizontal plane, and the wheel Ⅱ is fixed. The constant torque M_0 is applied on the crank OA to drive the gear Ⅰ to roll along the fixed gear Ⅱ. The masses of wheels Ⅰ and Ⅱ is m_1 and m_2, respectively, and can be regarded as homogeneous discs with radius r_1 and r_2. The crank has a mass of m and can be regarded as a homogeneous thin rod. Assuming that the mechanism starts to move from rest. Try to find the relationship between the angular velocity ω of the crank and its angle φ, regardless of friction. (如图 10.76 所示, 一个外啮合的行星齿轮机构位于水平面内, 轮Ⅱ固定. 在曲柄 OA 上作用不变转矩 M_0 来驱动齿轮Ⅰ沿定齿轮Ⅱ纯滚动. 已知轮Ⅰ和Ⅱ的质量分别是 m_1 和 m_2,并可看成半径是 r_1 和 r_2 的均质圆盘. 曲柄质量是 m,并可看成均质细杆. 假设机构由静止开始运动,试求曲柄的角速度 ω 与其转角 φ 之间的关系,摩擦不计.)

14. As shown in Fig. 10.77, there are two homogeneous thin rods hinged at point B in the vertical plane, both the rods have length b and mass m. A constant couple moment M is now applied to the AB rod and released from the shown position at rest, regardless of friction. Find the velocity of end A when A touches support O. (如图 10.77 所示,铅直平面内有两均质细杆在 B 点铰接,两杆的长度均为 b,质量均为 m. 今在 AB 杆上作用一不变力偶矩 M,并从图示位置静止释放,不计摩擦. 求当 A 碰到支座 O 时,A 端的速度.)

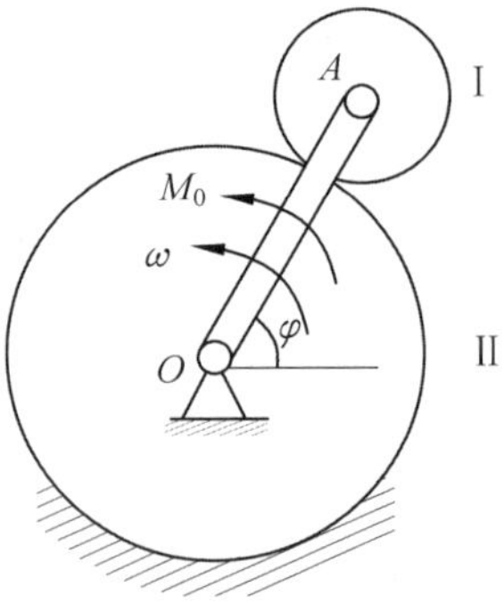

Fig. 10.76 Diagram of Exercise 13
(图 10.76 习题 13 图)

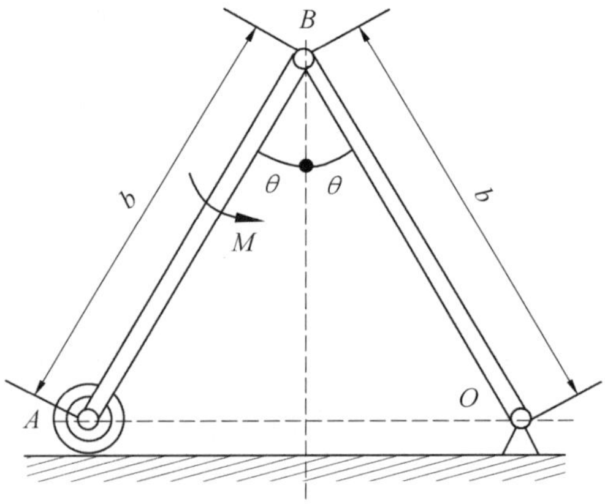

Fig. 10.77 Diagram of Exercise 14
(图 10.77 习题 14 图)

15. The pumping water of a pump is $Q = 0.06\ \mathrm{m^3/s}$, the head is $h = 20$ m. If the total efficiency of the pump $\eta = 0.6$, what power motor should be selected? If a 20 kW motor is selected to match this pump, how much time will it take to complete the pumping task of 5 000 $\mathrm{m^3}$? (一水泵抽水量 $Q=0.06\ \mathrm{m^3/s}$,扬程 $h=20$ m.假如抽水机的总效率 $\eta=0.6$,需选用多大功率的电动机? 假如选用一台 20 kW 的电动机与此水泵配套,需要多少时间才能完成抽水 5 000 $\mathrm{m^3}$的任务?)

Answers of Exercises(习题答案)

第 1 章习题答案

1.

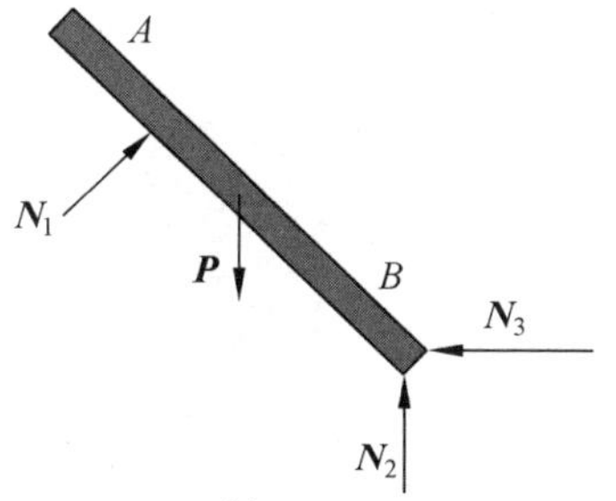

(a)

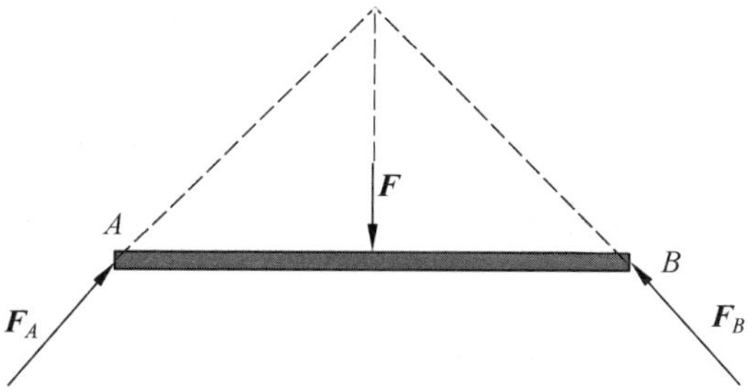

(b)

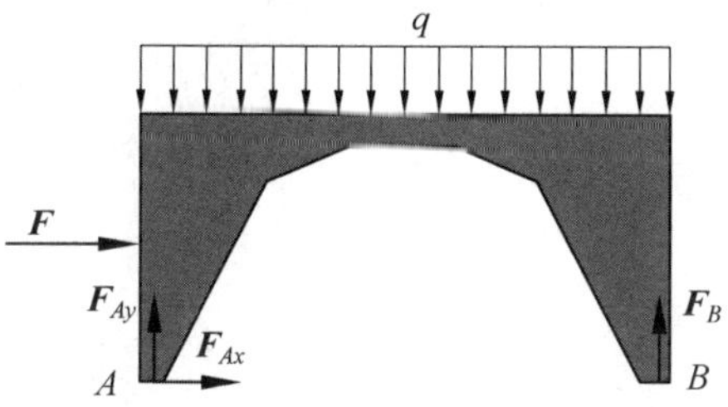

(c)

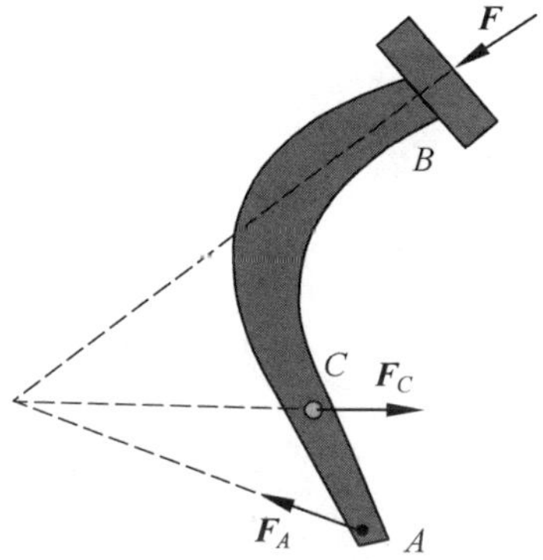

(d)

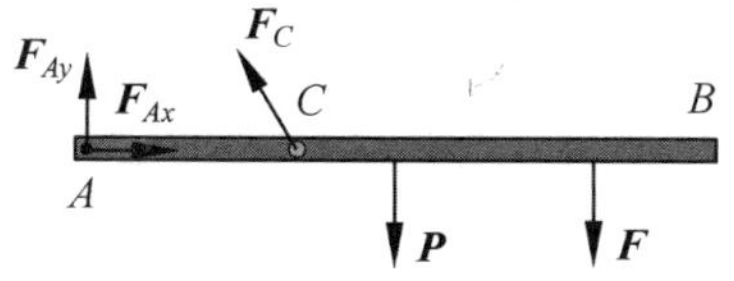

(e)

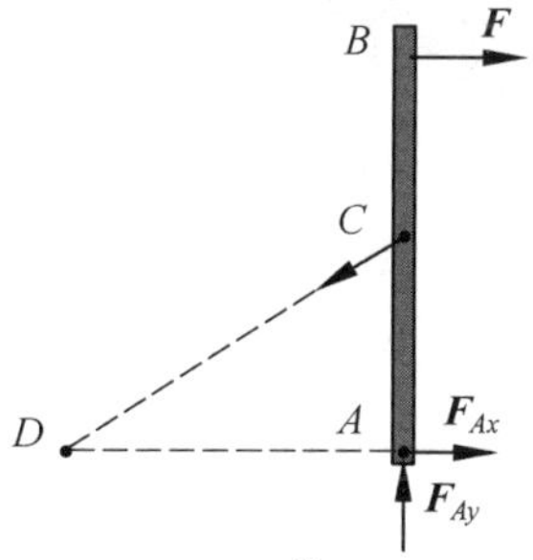

(f)

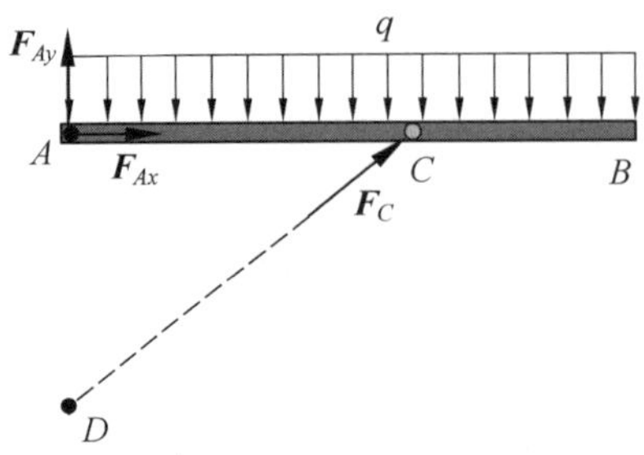
F_{Ay}
q
A
F_{Ax}
C
B
F_C
D

(g)

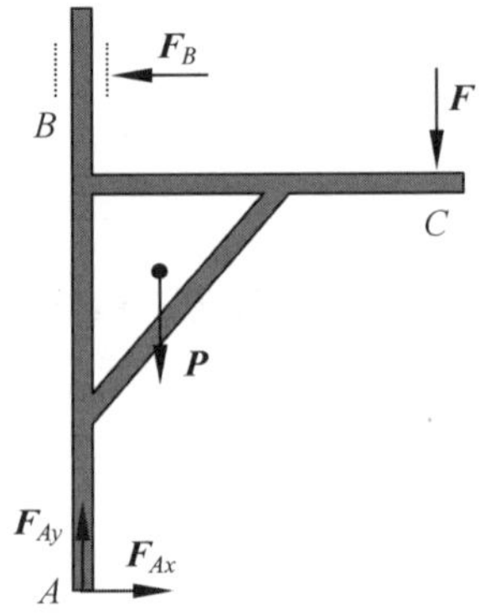
F_B
F
B
C
P
F_{Ay}
F_{Ax}
A

(h)

2.

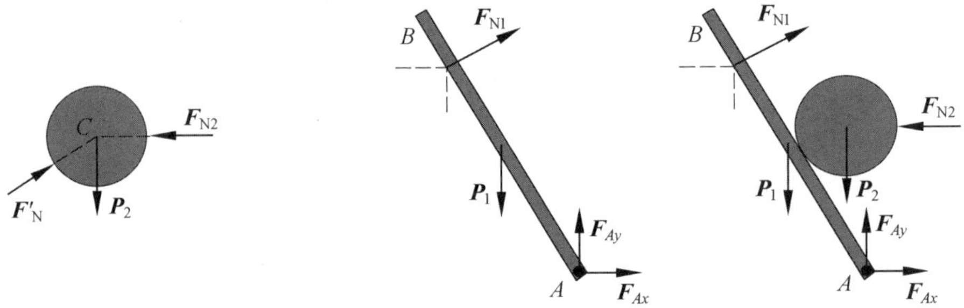
C
F_{N2}
F'_N
P_2
B
F_{N1}
P_1
F_{Ay}
A
F_{Ax}
B
F_{N1}
F_{N2}
P_1
P_2
F_{Ay}
A
F_{Ax}

(a)

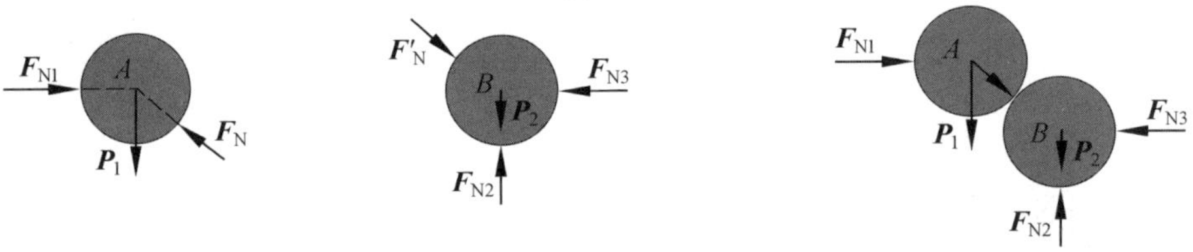
F_{N1}
A
P_1
F_N
F'_N
B
P_2
F_{N3}
F_{N2}
F_{N1}
A
P_1
B
P_2
F_{N3}
F_{N2}

(b)

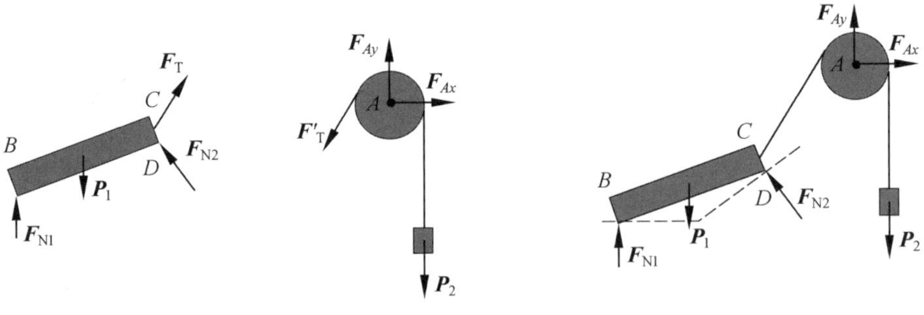
F_T
C
B
D
F_{N2}
P_1
F_{N1}
F_{Ay}
F_{Ax}
A
F'_T
P_2
F_{Ay}
F_{Ax}
A
C
B
D
F_{N2}
P_1
F_{N1}
P_2

(c)

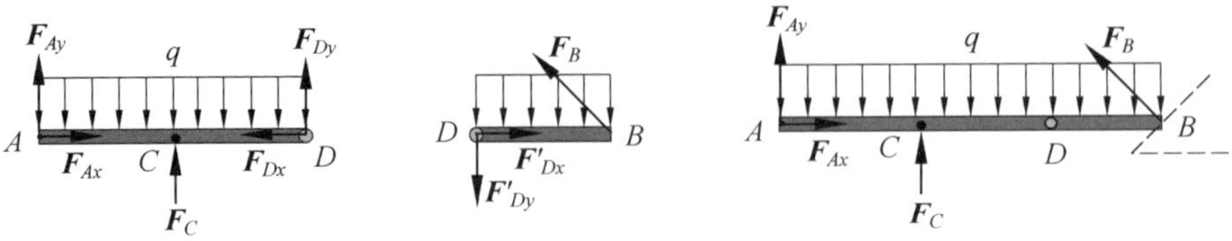
F_{Ay}
q
F_{Dy}
A
F_{Ax}
C
F_{Dx}
D
F_C
F_B
D
B
F'_{Dx}
F'_{Dy}
F_{Ay}
q
F_B
A
F_{Ax}
C
D
B
F_C

(d)

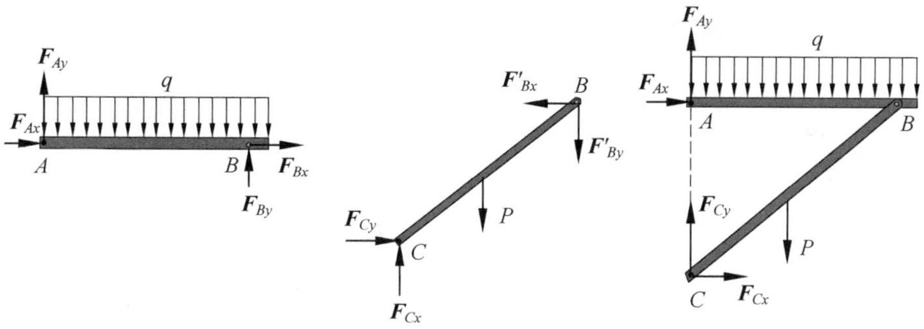
F_{Ay}
q
F_{Ax}
A
B
F_{Bx}
F_{By}
F'_{Bx}
B
F'_{By}
P
F_{Cy}
C
F_{Cx}
F_{Ay}
q
F_{Ax}
A
B
F_{Cy}
P
C
F_{Cx}

(e)

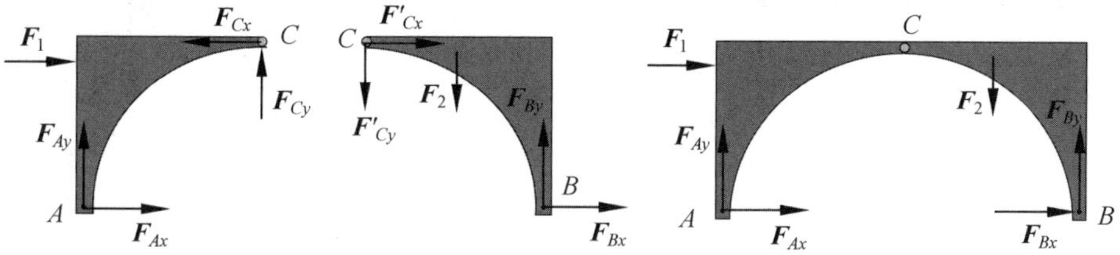
F_1
F_{Cx}
C
F_{Cy}
F_{Ay}
A
F_{Ax}
F'_{Cx}
C
F'_{Cy}
F_2
F_{By}
B
F_{Bx}
F_1
C
F_2
F_{Ay}
F_{By}
A
F_{Ax}
F_{Bx}
B

(f)

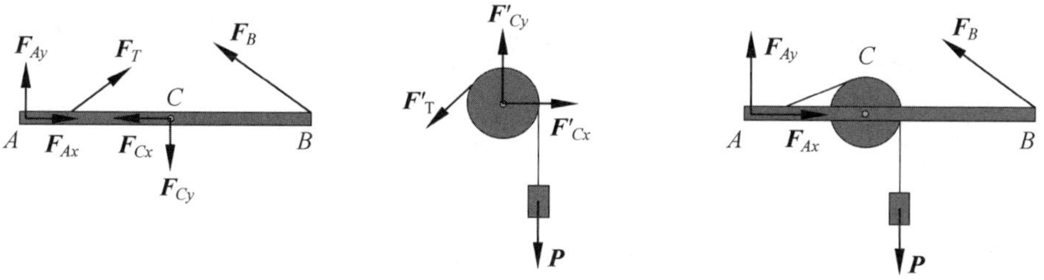
F_{Ay}
F_T
F_B
C
A
F_{Ax}
F_{Cx}
B
F_{Cy}
F'_{Cy}
F'_T
F'_{Cx}
P
F_{Ay}
C
F_B
A
F_{Ax}
B
P

(g)

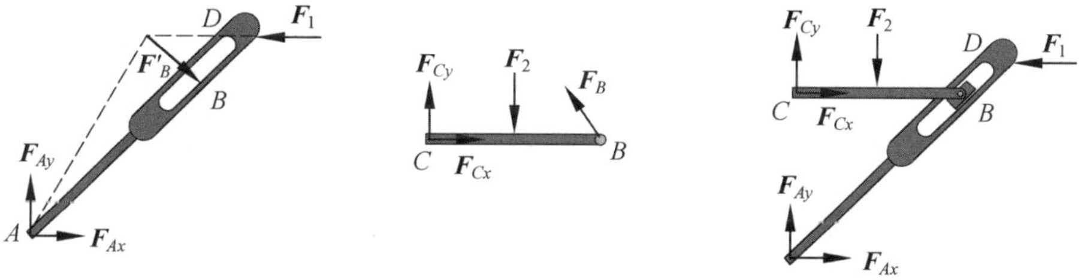
D
F_1
F'_B
B
F_{Ay}
A
F_{Ax}
F_{Cy}
F_2
F_B
C
F_{Cx}
B
F_{Cy}
F_2
D
F_1
C
F_{Cx}
B
F_{Ay}
F_{Ax}

(h)

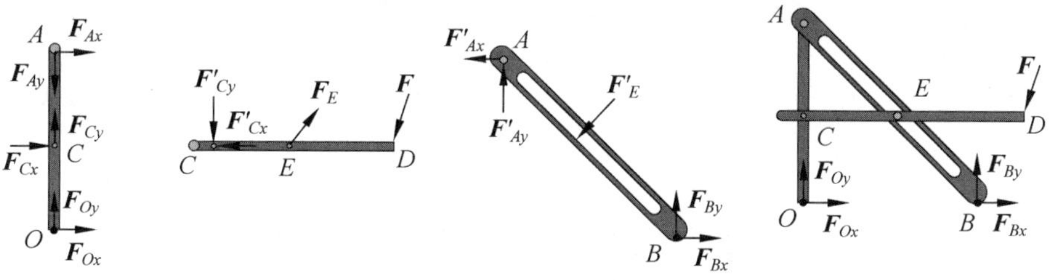
A
F_{Ax}
F_{Ay}
F_{Cy}
F_{Cx}
C
F_{Oy}
O
F_{Ox}
F'_{Cy}
F_E
F
F'_{Cx}
C
E
D
F'_{Ax}
A
F'_E
F'_{Ay}
F_{By}
B
F_{Bx}
A
E
F
D
C
F_{By}
F_{Oy}
O
F_{Ox}
B
F_{Bx}

(i)

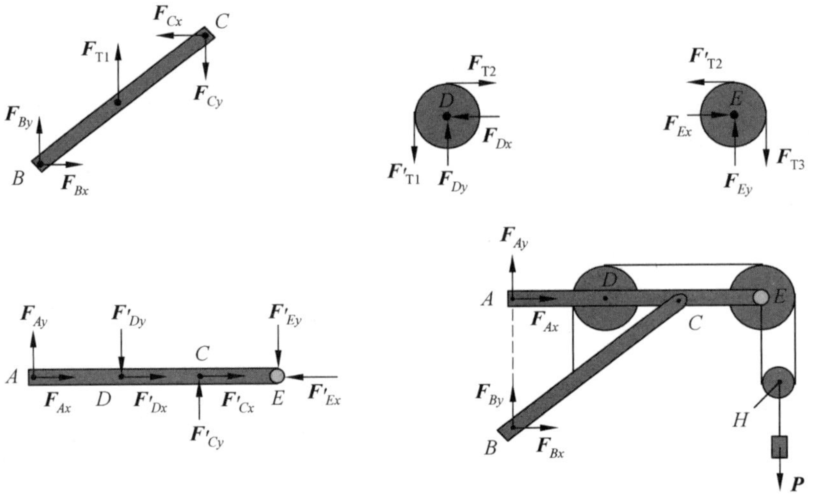
F_{Cx}
C
F_{T1}
F_{Cy}
F_{By}
B
F_{Bx}
F_{T2}
D
F_{Dx}
F'_{T1}
F_{Dy}
F'_{T2}
E
F_{Ex}
F_{T3}
F_{Ey}
F_{Ay}
F'_{Dy}
F'_{Ey}
A
C
F_{Ax}
D
F'_{Dx}
F'_{Cx}
E
F'_{Ex}
F'_{Cy}
F_{Ay}
A
D
E
F_{Ax}
C
F_{By}
H
B
F_{Bx}
P

(j)

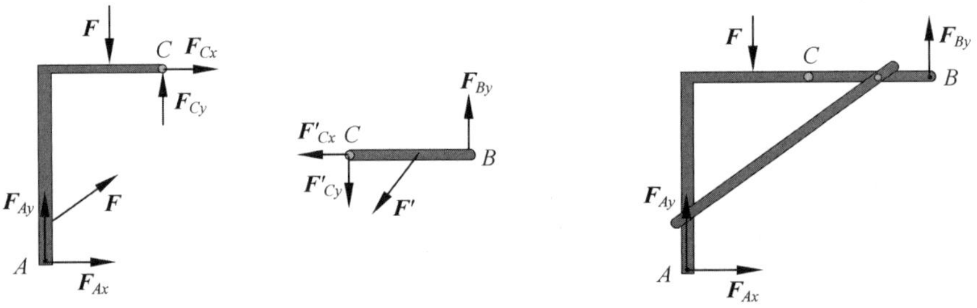
F
C
F_{Cx}
F_{Cy}
F_{Ay}
F
A
F_{Ax}
F_{By}
F'_{Cx} C
B
F'_{Cy}
F'
F
C
F_{By}
B
F_{Ay}
A
F_{Ax}

(k)

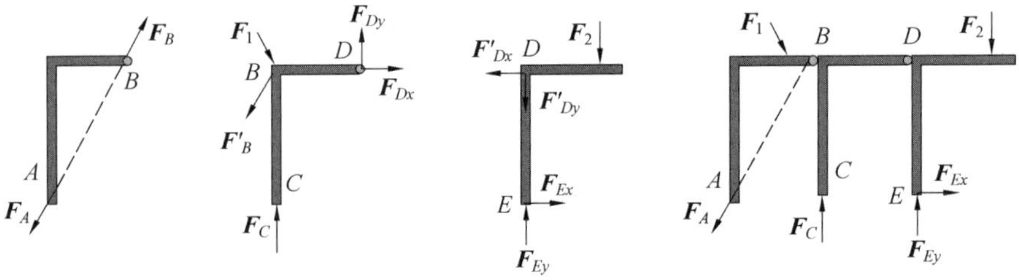
F_B
B
A
F_A
F_{Dy}
F_1
D
B
F_{Dx}
F'_B
C
F_C
F'_{Dx} D
F_2
F'_{Dy}
F_{Ex}
E
F_{Ey}
F_1
B
D
F_2
A
C
E
F_{Ex}
F_A
F_C
F_{Ey}

(l)

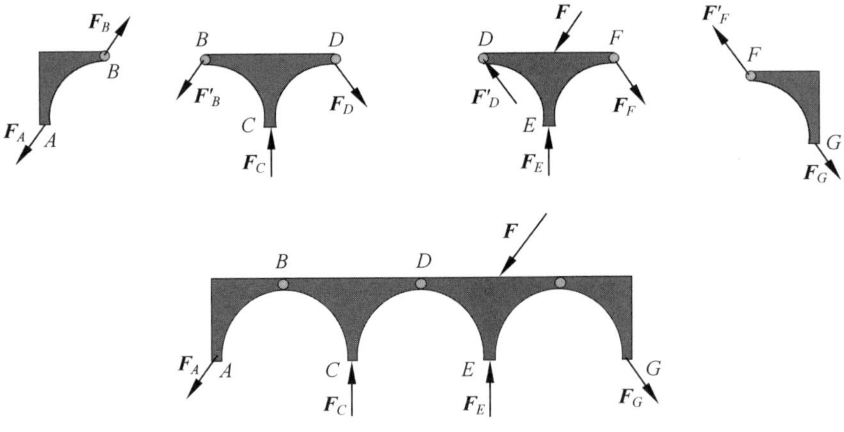
F_B
B
F_A
A
B
D
F'_B
F_D
C
F_C
F
D
F
F'_D
F_F
E
F_E
F'_F
F
G
F_G
F
B
D
F_A
A
C
E
G
F_C
F_E
F_G

(m)

3.

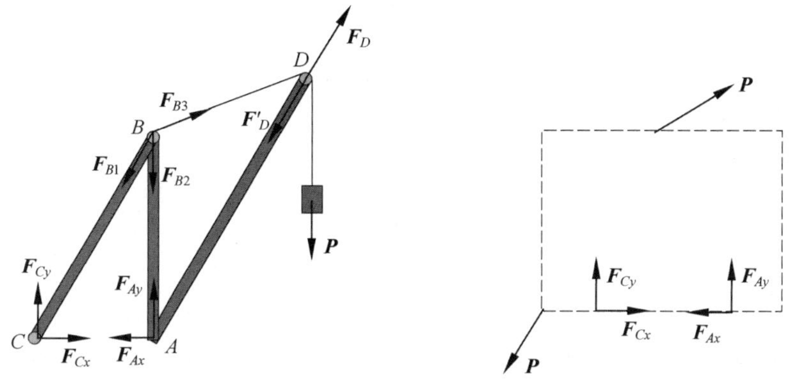
F_D
D
F_{B3}
F'_D
B
F_{B1}
F_{B2}
P
F_{Cy}
F_{Ay}
C
F_{Cx}
F_{Ax}
A
P
F_{Cy}
F_{Ay}
F_{Cx}
F_{Ax}
P

(a)

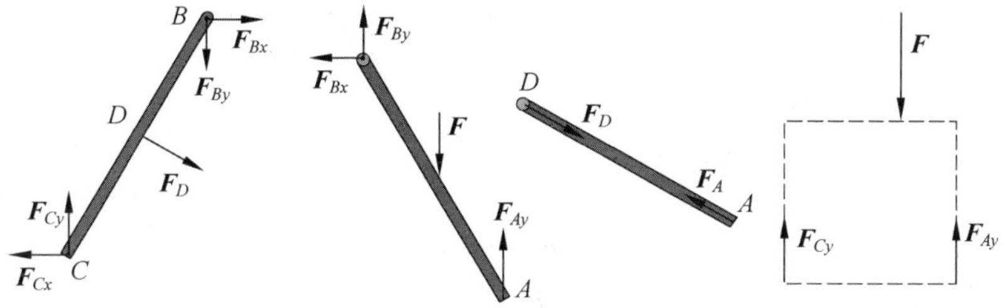
B
F_{Bx}
F_{By}
D
F_D
F_{Cy}
F_{Cx}
C
F_{By}
F_{Bx}
F
F_{Ay}
A
D
F_D
F_A
A
F
F_{Cy}
F_{Ay}

(b)

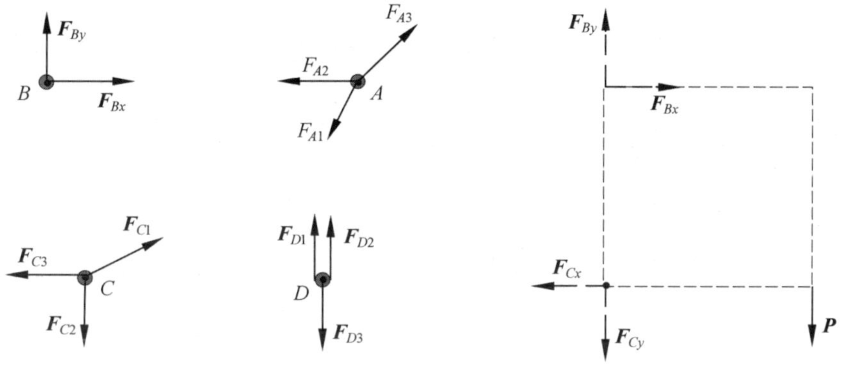
F_{By}
B
F_{Bx}
F_{A3}
F_{A2}
A
F_{A1}
F_{By}
F_{Bx}
F_{C1}
F_{C3}
C
F_{C2}
F_{D1}
F_{D2}
D
F_{D3}
F_{Cx}
F_{Cy}
P

(c)

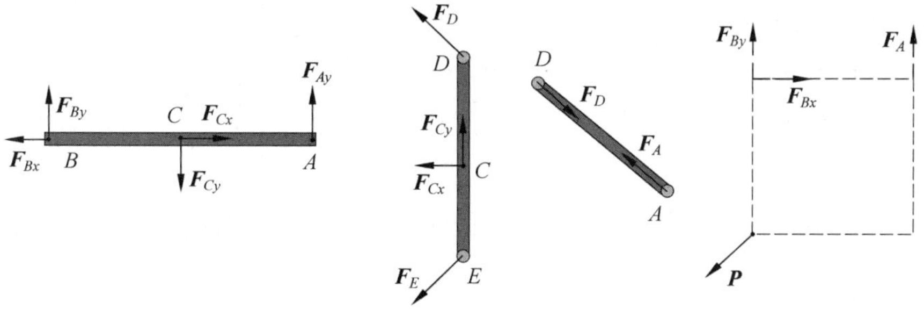
F_{Ay}
F_{By}
C
F_{Cx}
F_{Bx}
B
A
F_{Cy}
F_D
D
F_{Cy}
F_{Cx}
C
F_E
E
D
F_D
F_A
A
F_{By}
F_A
F_{Bx}
P

(d)

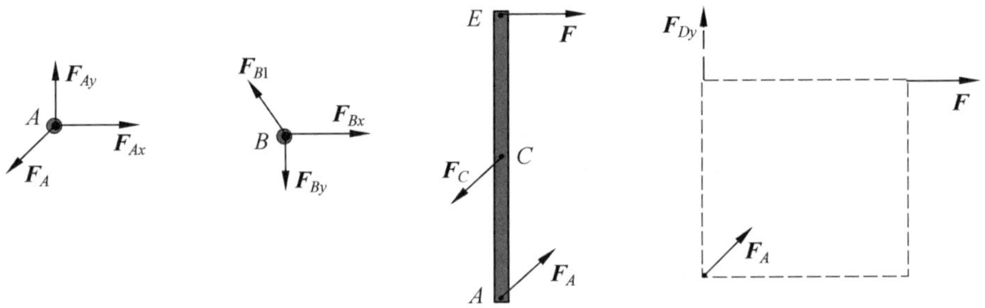

(e)

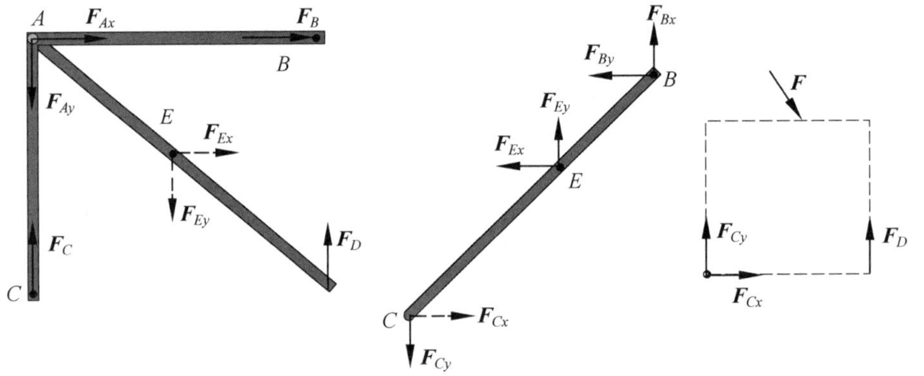

(f)

第 2 章习题答案

1. $\boldsymbol{F}_A=-\dfrac{\sqrt{5}}{2}\boldsymbol{F},\boldsymbol{F}_D=-\dfrac{1}{2}\boldsymbol{F}(\uparrow)$

2. $F_{BC}=5\ \text{kN}$(BC 杆受压),$F_{RA}=5\ \text{kN}$

3. $F_{BC}=-74.64\ \text{kN}$(压),$F_{BA}=54.64\ \text{kN}$(拉)

4. $\theta=30°,P_B=100\ \text{N}$

5. (a)$M_O(\boldsymbol{F})=0$

(b)$M_O(\boldsymbol{F})=Fl$

(c)$M_O(\boldsymbol{F})=-Fb$

(d)$M_O(\boldsymbol{F})=Fl\sin\theta$

6. $M_A=-Fb\cos\theta,M_B(F)=F(a\sin\theta-b\cos\theta)$

7. $F_{RB}=F_{RC}=\dfrac{M}{2\sqrt{2}a}$

8. $M_2=\dfrac{r_2}{r_1}M_1,F_{R1}=F_N=F_{R2}=\dfrac{M_1}{r_1\cos\theta}$,$F_{R1}$指向第四象限,$F_{R2}$指向第二象限

9. $F_{AB}=500\ \text{N}$(拉),$M_2=300\ \text{N}\cdot\text{m}$

10. $F_{Ax}=0,F_{Ay}=6\ \text{kN},M_A=12\ \text{kN}\cdot\text{m}$

11. $F_{BC}=848.5\ \text{kN},F_{Ax}=2\ 400\ \text{N},F_{Ay}=1\ 200\ \text{N}$

12. (a) $F_{Ax}=0, F_{Ay}=-\frac{1}{2}\left(F+\frac{M}{a}\right), F_{NB}=\frac{1}{2}\left(3F+\frac{M}{a}\right)$

(b) $F_{Ax}=0, F_{Ay}=-\frac{1}{2}\left(F+\frac{M}{a}-\frac{5}{2}qa\right), F_{NB}=\frac{1}{2}\left(3F+\frac{M}{a}-\frac{1}{2}qa\right)$

13. (a) $F_{Ax}=\frac{M}{a}\tan\theta, F_{Ay}=-\frac{M}{a}, M_A=-M, F_{NC}=\frac{M}{a\cos\theta}$

(b) $F_{Ax}=\frac{1}{2}qa\tan\theta, F_{Ay}=\frac{1}{2}qa, M_A=\frac{1}{2}qa^2, F_{NC}=\frac{qa}{2\cos\theta}$

14. $M=\frac{rr_1r_3}{r_2r_4}P, F_{3x}=\frac{r}{r_4}P\tan\theta, F_{3y}=\left(1-\frac{r}{r_4}\right)P$

15. $F_{Ax}=0, F_{Ay}=-\frac{M}{2a}, F_{Dx}=0, F_{Dy}=\frac{M}{a}, F_{Bx}=0, F_{By}=-\frac{M}{2a}$

16. $F_{Ax}=1\ 200$ N, $F_{Ay}=150$ N, $F_{NB}=1\ 050$ N, $F_{BC}=-1\ 500$ N(压)

17. $F_{Ax}=0$, $F_{Ay}=15.1$ kN, $M_A=68.4$ kN · m, $F_{Bx}=-22.8$ kN, $F_{By}=-17.85$ kN, $F_{Cx}=22.8$ kN, $F_{Cy}=4.55$ kN

18. $F_{BC}=848.5$ N, $F_{Ax}=2\ 400$ N, $F_{Ay}=1\ 200$ N

第3章习题答案

1. $F_{OB}=F_{OC}=707$ N(拉), $F_{OA}=1\ 414$ N(压)
2. $F_1=F_2=-5$ kN(压), $F_3=707$ kN(压), $F_4=F_5=5$ kN(拉), $F_6=-10$ kN(压)
3. $M_2(F)=-101.4$ N · m
4. $M_{AB}(F)=Fa\sin\beta\sin\theta$
5. $M_x(F)=\frac{F}{4}(h-3R), M_y(F)=\frac{\sqrt{3}}{4}F(h+R), M_z(F)=-\frac{1}{2}RF$
6. $F'_R=-100\mathbf{k}$ kN, $M_C=-(12.5\mathbf{i}+5\mathbf{j})$ kN · m
7. $F'_R=0, M_O=-40(\mathbf{i}+\mathbf{j})$ N · m
8. $F'_R=0, M_A=(-32\mathbf{i}-30\mathbf{j}+24\mathbf{k})$ N · m
9. $x_C=90$ mm, $y_C=0$ mm
10. $x_C=78.26$ mm, $y_C=59.53$ mm

第4章习题答案

1. $F_T=\frac{P\sin(\beta+\varphi_f)}{\cos(\beta-\varphi_f)}, F_{T\min}=P\sin(\beta+\varphi_f)$
2. $f_s=0.223$
3. $f_s=\frac{1}{2\sqrt{3}}$
4. $s=0.456l$

5. $b_{\min}=\frac{1}{3}f_{s}h$,门重与最小宽度无关

6. $P=500\ \text{N}$

7. $b<d\left(1-\sqrt{\frac{1}{1+f_{s}^{2}}}\right)=7.84\ \text{mm}$

8. $\frac{M\sin(\theta-\varphi_{f})}{l\cos\theta\cos(\theta-\varphi_{f})}\leqslant F\leqslant\frac{M\sin(\theta+\varphi_{f})}{l\cos\theta\cos(\theta+\varphi_{f})}$

9. $F_{\min}=240\ \text{N}$

10. $\varphi_{A}=16°6'$,$\varphi_{B}=\varphi_{C}=30°$

11. $F=\frac{P(\delta_{1}+\delta_{2})+2P_{1}\delta_{2}}{2r}$

第5章习题答案

1. 若此人垂直到达对岸,船行方向与水流方向成126.8°夹角;假如他要在最短的时间到达对岸,则船行的方向是与水流成90°夹角

2. 16 s

3. $x=200\cos\frac{\pi}{5}t$(式中x以mm计),$y=200\sin\frac{\pi}{5}t$(式中x以mm计),轨迹$\frac{x^{2}}{40\ 000}+\frac{y^{2}}{10\ 000}=1$

4. $\frac{(x-a)^{2}}{(b+l)^{2}}+\frac{y^{2}}{l^{2}}=1$

5. $y=e\sin\omega t+\sqrt{R^{2}-e^{2}\cos^{2}\omega t}$,$v=e\omega\left(\cos\omega t+\frac{e\sin 2\omega t}{2\sqrt{R^{2}-e^{2}\cos^{2}\omega t}}\right)$

6. $v=ak$,$v_{r}=ak\sin kt$

7. $\rho=\frac{v_{0}}{\omega_{0}}\varphi$

8. $a_{\max}=\sqrt{16\pi^{4}f^{4}z_{0}^{2}+\omega^{4}r^{2}}$

9. $v'_{M}=v_{0}\sec^{2}\frac{v_{0}}{R}t$,$a'_{M}=\frac{2v_{0}^{2}}{R}\frac{\sin\frac{v_{0}}{R}t}{\cos^{3}\frac{v_{0}}{R}t}$

10. 证明过程略

第6章习题答案

1. $v_{M}=157\ \text{mm/s}$,$a_{M}=227.2\ \text{mm/s}^{2}$

2. M点的轨迹为一个半径为R的圆,其轨迹形状与A点、B点的轨迹平行,$v_{M}=\frac{Rn\pi}{30}$,

$a_M=\frac{Rn^2\pi^2}{900}$

3. $\varphi=\frac{\sqrt{3}}{3}\ln\frac{1}{1-\sqrt{3}\omega_0 t},\omega=\omega_0 e^{\sqrt{3}\varphi}$

4. $N=30\ 000$ r

5. $\varphi=\frac{1}{30}t$(式中 φ 以 rad 计)，$x^2+(y+0.8)^2=1.5^2$

6. $\omega=\frac{v}{2l},\alpha=-\frac{v^2}{2l^2}$

7. $a_M=\arctan\frac{\sin\omega_0 t}{\frac{h}{r}-\cos\omega_0 t}$

8. $\omega=2k,\alpha=1.5k,a_C=(-388.9\mathbf{i}+1763.8\mathbf{j})\ \mathrm{mm/s^2}$

9. (1) $\boldsymbol{v}_G=(-400\mathbf{i}-400\mathbf{j}+200\mathbf{k})\ \mathrm{mm/s},v_G=600\ \mathrm{mm/s}$

(2) $\boldsymbol{a}_{Gn}=(2\ 400\mathbf{i}-1\ 200\mathbf{j}+2\ 400\mathbf{k})\ \mathrm{mm/s^2},a_{Gn}=3\ 600\ \mathrm{mm/s^2}$

(3) $\boldsymbol{a}_{G\tau}=(200\mathbf{i}+200\mathbf{j}-100\mathbf{k})\ \mathrm{mm/s^2},a_{G\tau}=300\ \mathrm{mm/s^2}$

(4) $\boldsymbol{a}_G=(2\ 600\mathbf{i}-1\ 000\mathbf{j}+2\ 300\mathbf{k})\ \mathrm{mm/s^2},a_G=3\ 613\ \mathrm{mm/s^2}$

10. $\boldsymbol{v}=(7\mathbf{i}+6\mathbf{j}+9\mathbf{k})\ \mathrm{m/s},a_\tau=5.12\ \mathrm{mm/s^2},a_n=3.71\ \mathrm{mm/s^2}$

11. $v=(19.959\ 6\ e_\theta+1.27e_z)\ (\mathrm{m/s}),a=-795.13e_r(\mathrm{mm/s^2})$

第7章习题答案

1. $v=40$ km/h

2. $l=200$ m, $u=20$ m/min; $v=12$ m/min

3. $x'=v_e t,y'=a\cos(kt+\beta),y'=a\cos\left(\frac{k}{v_e}x'+\beta\right)$

4. $a_{AB}=a_0\tan\varphi$，铅垂向上

5. $v_A=\frac{lav}{x^2+a^2}$

6. $v_r=63.62$ mm/s，$\angle(v_r,v)v_r=80°57'$

7. $r=\frac{v}{\omega}\varphi$

8. $\omega_{AB}=\frac{\sqrt{r^2-e^2}}{l}\omega$

9. $v_r=36.74\ \mathrm{mm/s},a_r=30.62\ \mathrm{mm/s^2},\omega=0.5\ \mathrm{rad/s},a=-0.5\ \mathrm{rad/s^2}$

10. $v=0.1\ \mathrm{m/s},a=0.346\ \mathrm{mm/s^2}$

11. $v_C=0.173\ \mathrm{m/s},a_C=0.05\ \mathrm{m/s^2}$

12. $v_A=(\sqrt{3}-1)v_0,a_A=\sqrt{2}(2-\sqrt{3})\frac{v_0^2}{r},\omega=\frac{1}{2}\sqrt{2}(\sqrt{3}-1)\frac{v_0^2}{r},\alpha=(2-\sqrt{3})\frac{v_0^2}{r^2}$

13. $a_1=r\omega^2-\frac{v^2}{r}-2\omega v, a_2=\sqrt{\left(r\omega^2+\frac{v^2}{r}+2\omega v\right)^2+4r^2\omega^4}$

14. $a_M=255.5\ \mathrm{mm/s^2}$

第8章习题答案

1. $x_C=r\cos\omega_0 t, y_C=r\sin\omega_0 t, \varphi=\omega_0 t$

2. $x_A=0, y_A=\frac{1}{3}gt^2, \varphi=\frac{g}{3r}t^2$

3. $x_A=(R+r)\cos\frac{\alpha t^2}{2}, y_A=(R+r)\sin\frac{\alpha t^2}{2}, \varphi_A=\frac{1}{2r}(R+r)\alpha t^2$

4. $v_{BC}=2.513\ \mathrm{m/s}$

5. $\omega_B=3.62\ \mathrm{rad/s}, \alpha_B=2.2\ \mathrm{rad/s^2}$

6. $\omega_{OB}=3.75\ \mathrm{rad/s}, \omega_{\mathrm{I}}=6\ \mathrm{rad/s}$

7. $a_C=2r\omega_0^2$

8. $a_B^n=8\sqrt{2}\ \mathrm{m/s^2}, a_B^\tau=-4\sqrt{2}\ \mathrm{m/s^2}, \omega_{BC}=4\ \mathrm{rad/s}, \alpha_{BC}=-8\ \mathrm{rad/s^2}$(逆时针)

9. $v_M=0.098\ \mathrm{m/s}, a_M=0.013\ \mathrm{m/s^2}$

10. $v_C=\frac{3}{2}r\omega_0, a_C=\frac{\sqrt{3}}{12}r\omega_0^2$

11. $\omega=-1\ \mathrm{rad/s}, \alpha=2\ \mathrm{rad/s^2}, v_C=0.05\ \mathrm{m/s}(\uparrow), a_C=0.1\ \mathrm{m/s^2}(\downarrow), v_D=0.2\ \mathrm{m/s}(\uparrow), a_D=0.427\ \mathrm{m/s^2}(\searrow), v_E=0.1\ \mathrm{m/s}(\downarrow), a_E=0.25\ \mathrm{m/s^2}(\nwarrow)$

12. $v_O=\frac{R}{R-r}v, a_O=\frac{R}{R-r}a$

13. $v_B=2\ \mathrm{m/s}, v_C=2.828\ \mathrm{m/s}, a_B=8\ \mathrm{m/s^2}, a_C=11.31\ \mathrm{m/s^2}$

14. (1) $v_C=0.4\ \mathrm{m/s}, v_\mathrm{r}=0.2\ \mathrm{m/s}$

(2) $a_C=0.159\ \mathrm{m/s^2}, a_\mathrm{r}=0.139\ \mathrm{m/s^2}$

15. $\omega_1=\frac{\sqrt{3}}{2}\frac{v}{r}$(顺时针),$\omega=\frac{\sqrt{3}}{6}\frac{v}{r}$(逆时针)

16. $v_{\mathrm{r}1}=0.6\ \mathrm{m/s}, v_{\mathrm{r}2}=0.9\ \mathrm{m/s}, v_M=0.459\ \mathrm{m/s}, a_{\mathrm{r}1}=2.816\ \mathrm{m/s^2}, a_{\mathrm{r}2}=4.592\ \mathrm{m/s^2}, a_M=2.5\ \mathrm{m/s^2}$

第9章习题答案

1. $v=\sqrt{gl(\cos\varphi+1-\sqrt{3})}, F_\mathrm{T}=mg(3\cos\varphi+2-2\sqrt{3})$

2. $F_\mathrm{T}=100\ \mathrm{kN}, \varphi=8.19°$

3. $\Delta m=\frac{2ma}{g-a}$

4. $F_{\mathrm{Nmin}}=461.7\ \mathrm{N}, F_{\mathrm{Nmax}}=741.3\ \mathrm{N}$

5. $t=\sqrt{\frac{h}{g}\frac{m_1+m_2}{m_1-m_2}}$

6. $F=-bv$

7. $F_T=mg(3-2\cos\varphi_0)$

8. $F_{\text{Nmin}}=F+m\left(g-\frac{2\pi^2v^2\delta}{l^2}\right)$, $F_{\text{Nmax}}=F+m\left(g+\frac{2\pi^2v^2\delta}{l^2}\right)$

9. (1) $F_{\text{Nmax}}=m(g+e\omega^2)$, $F_{\text{Nmin}}=m(g-e\omega^2)$, (2) $\omega_{\max}=\sqrt{\frac{g}{e}}$

10. $F=0.002v-0.0098$

11. $v=\sqrt{Rg\frac{\tan\theta+f_s}{1-f_s\tan\theta}}=13\ \text{m/s}$

12. $f_{\text{smin}}=\frac{a\cos\theta}{g+a\sin\theta}$

13. $F_T=m\left(g+\frac{l^2v_0^2}{x^3}\right)\frac{\sqrt{l^2+x^2}}{x}$

14. $t=2.02\ \text{s}$, $s=7.069\ \text{m}$

15. $v_0=\sqrt{2gR}=11.2\ \text{km/s}$

16. $v=\sqrt{\frac{k}{m}(\ln R^2-\ln r^2)}$

17. 轨迹为一个椭圆,圆心在(0,0),长短半轴分别为 x_0 和 $v_0\sqrt{\frac{m}{k}}$

18. $y=\frac{x}{v_0\cos\varphi}\left(v_0\sin\varphi+\frac{1}{k}\right)+\frac{1}{k^2g}\ln\left(\frac{v_0\cos\varphi-kgx}{v_0\cos\varphi}\right)$

19. 运动方程为 $\begin{cases}x=\frac{v_0\cos\varphi}{kg}(1-e^{-kgt})\\ y=\frac{1}{kg}\left(v_0\sin\varphi+\frac{1}{k}\right)(1-e^{-\log t})-\frac{t}{k}\end{cases}$,轨迹方程为 $y=\frac{x}{v_0\cos\varphi}(v_0\sin\varphi+\frac{1}{k})+\frac{1}{k^2g}\ln\frac{v_0\cos\varphi-kgx}{v_0\cos\varphi}$

20. $v=\frac{F+mg}{kA}(1-e^{-\frac{kA}{m}t})$, $H=\frac{F+mg}{kA}\left[t+\frac{m}{kA}(e^{-\frac{kA}{m}t}-1)\right]$

第10章习题答案

一、动量定理习题答案

1. 最初 3 s,$I=5\ \text{N}\cdot\text{s}$;最初 4 s,$I=0\ \text{N}\cdot\text{s}$

2. 作用在墙面的冲量 $I_{Bx}=200.2\ \text{N}\cdot\text{s}$,作用在地面的冲量 $I_{Ay}=246.7\ \text{N}\cdot\text{s}$

3. $F=1\ 068\ \text{N}$

4. $F=100\ \text{N}$

5. $F_{max}=4.23\ kN$

6. $t=0.102\ s$

7. $t=2.06\ s$

8. $\Delta v=0.2\ m/s$

9. (1)$v_1=0.14\ m/s$,(2)$x_M=-\frac{10\times5}{100+5}=-0.48\ m$

10. (1)$v_1=3.31\ m/s$,(2)$\Delta s=0.114\ m$

11. (1)$\begin{cases}x_C=\dfrac{m_3 l}{2(m_1+m_2+m_3)}+\dfrac{m_1+2m_2+2m_3}{2(m_1+m_2+m_3)}l\cos\ \omega t\\ y_C=\dfrac{m_1+2m_2}{2(m_1+m_2+m_3)}l\sin\omega t\end{cases}$,(2)$F_{Ox\max}=\frac{1}{2}(m_1+m_2+m_3)l\omega^2$

12. $s=0.138\ m$

13. $\Delta x=-\frac{1}{4}(a-b)$

14. (1)$\ddot{x}_1=\frac{m_2\sin\ \theta\cos\ \theta}{m_2+m_1\ \sin^2\theta}g$,(2)$\ddot{x}_2=-\frac{m_1\sin\ \theta\cos\ \theta}{m_2+m_1\ \sin^2\theta}g$,(3)$F_1=\frac{m_1 m_2\cos\ \theta}{m_2+m_1\ \sin^2\theta}g$,(4)$F_2=\frac{m_2(m_1+m_2)}{m_2+m_1\ \sin^2\theta}g$

15. $F_{x\max}=\frac{P_1}{g}e\omega^2$,$\omega_{\min}=\sqrt{\frac{P+P_1}{P_1 e}g}$

16. $\begin{cases}F_{Ox}=\dfrac{P}{g}(a_C^n\cos\ \varphi+a_C^\tau\sin\varphi)=\dfrac{P}{g}l(\omega^2\cos\ \varphi+\alpha\sin\ \varphi)\\ F_{Oy}=\dfrac{P}{g}l(\omega^2\sin\ \varphi-\alpha\cos\ \varphi)+P\end{cases}$

17. $a=0.1\ m$

二、动量矩定理习题答案

1. $L_O=2abpm\ \cos^3 pt$

2. $n_1=480\ r/min$

3. $\omega=8.4\ rad/s$

4. $\omega=\frac{m_1\omega_0}{m_1+4m_2}$

5. $t=\frac{l}{k}\ln 2$

6. $v=20\ m/s$

7. $L_x=0$,$L_y=-\frac{1}{3}ml^2\omega\sin\ \theta\cos\ \theta$,$L_z=\frac{1}{3}ml^2\omega\ \sin^2\theta$,$|L_O|=\frac{1}{3}ml^2\omega\sin\ \theta$

8. $0=\frac{m_1R^2}{2}\omega+m_2(r\omega-at)r$

9. $\omega=\frac{2m_2art}{R^2m_1+2m_2r^2}$,$\alpha=\dot{\omega}=\frac{2m_2ar}{R^2m_1+2m_2r^2}$

10. $J_x=\frac{1}{3}ma^2, J_y=\frac{1}{3}mb^2, J_{z_{C'}}=\frac{1}{12}m(a^2+b^2)$

11. B 端物体以 v_r 向上运动

12. $\alpha=\frac{m_1r_1-m_2r_2}{m_1r_1^2+m_2r_2^2}g$

13. $J=1\ 060\ \mathrm{kg\cdot m^2}, M_f=6.024\ \mathrm{N\cdot m}$

14. $r=\sqrt{r_0^2+\frac{M_O\sin\omega t}{2m\omega^2}}$

15. $F_T=89.2\ \mathrm{N}$

16. 均质细杆先达到水平位置,原因略

17. $J_O=10.09\ \mathrm{kg\cdot m^2}$

18. $T_1=\frac{2\pi l}{a}\sqrt{\frac{m}{k}}, T_2=\frac{2\pi l}{a}\sqrt{\frac{m}{3k}}$

19. $F_s=\left(1-\frac{Rh}{\rho^2+R^2}\right)F$,当 $h<\frac{\rho^2+R^2}{R}$ 时,方向向左,反之向右

20. (1) $a=\frac{4}{5}g$, (2) $M>2mgr$

三、动能定理习题答案

1. $W=20.3\ \mathrm{J}$

2. $W=40\ \mathrm{J}$

3. $h=0.395\ \mathrm{m}$

4. $v=6.71\ \mathrm{m/s}$

5. 第一种 $x=160\ \mathrm{mm}$,第二种 $x=207\ \mathrm{mm}$

6. $T=\frac{3}{4}m(R-r)^2\dot{\varphi}^2$

7. (1) $T=\frac{1}{3}(33m_1+8m_2)r^2\omega^2$, (2) $W_3=T_{32}-T_{31}=0$

8. $v_A=1.294\ \mathrm{m/s}$

9. $v=1.146\ \mathrm{m/s}$

10. 经过 2.35 r

11. $n=385\ \mathrm{r/min}$

12. $\ddot{\theta}+\frac{2ke^2}{(m_A+2m_B)r^2}\theta=0$

13. $\omega=\frac{2}{R+r}\sqrt{\frac{3M\varphi}{9m_1+2m_2}}$

14. $v_A=\sqrt{\frac{3}{m}[M\theta-mgl(1-\cos\theta)]}$

15. $t=22\ \mathrm{h}41\ \mathrm{min}7\ \mathrm{s}$

References(参考文献)

[1] VEREINA L I, KRASNOV M M. Technical mechanics[M]. Moscow: Academia Publishing Center, 2013.

[2] KREKNIN A I. Theoretical mechanics Part 1:Statics[M]. Tyumen: GOU VPO RIO, 2010.

[3] ZAGORUIKO M G, MARADUDIN, A M, PERETYATKO A V, et al. Theoretical mechanics dynamics[M]. Saratov: FGOU VO "Saratov SAU", 2016.

[4] BEREZKIN E N, The solution of problems in theoretical mechanics. Part 2[M]. Moscow: Moscow State University Press, 1974.

[5] AISENBERG T B, VORONKOV I M, OSETSKY V M. Handbook for solving problems in theoretical mechanics[M].6th ed. Moscow: High School,1968.

[6] SHAMOLIN M V. Methods of analysis of dynamic systems with variable dissipation in solid state dynamics[M]. Moscow: Examin, 2007.

[7] ZHURAVLEV V F. Fundamentals and methodical aspects of mechanics[M]. Moscow: Institute of Mechanics Problems of the Russian Academy of Sciences, 1985.

[8] KOZLOV V. The methods of qualitative analysis in dynamics of solids[M]. 2nd ed. Izhevsk: Research Centre "Regular and Chaotic Dynamics", 2000.

[9] GERTZ G. Principles of mechanics stated in a new relation[M]. Moscow:USSR Academy of Sciences, 1959.

[10] BUCHHOLZ N N. The basic course of theoretical mechanics. Volume 1: Kinematics, statics, dynamics of a material point[M]. 6th ed. Moscow: Science, 1965.

[11] VESELOVSKY I N. Dynamics[M]. Moscow: National Publisher of Technical and Theoretical Literature, 1941.

[12] APPEL P. Theoretical mechanics,dynamics of a system,analytic mechanics[M]. Moscow: Fizmatlit, 1960.

[13] 哈尔滨工业大学理论力学教研室. 理论力学 [M]. 8 版. 北京:高等教育出版社,2016.

[14] 郝桐生,殷祥超,赵玉成,等. 理论力学 [M]. 4 版. 北京:高等教育出版社,2017.

[15] 周培源. 理论力学[M]. 北京:科学出版社,2012.

[16] 贾启芬,刘习军,王春敏. 理论力学[M]. 天津:天津大学出版社,2003.

[17] 朱照宣,周起钊,殷金生. 理论力学(上、下册)[M]. 北京:北京大学出版社,1982.

[18] 刘然慧,闵国林,李翠赟. 理论力学[M]. 成都:西南交通大学出版社,2017.

[19] 哈尔滨工业大学等六校理论力学教研室. 理论力学解题指导及习题集 [M]. 3 版. 北京:高等教育出版社,2005.

[20] 浙江大学理论力学教研室. 理论力学 [M]. 3 版. 北京:高等教育出版社, 1997.

[21] 郭应征,李兆霞. 应用力学基础[M]. 北京:高等教育出版社, 2000.